Sobolev Spaces on Metric Measure Spaces

Analysis on metric spaces emerged in the 1990s as an independent research field providing a unified treatment of first-order analysis in diverse and potentially nonsmooth settings. Based on the fundamental concept of the upper gradient, the notion of a Sobolev function was formulated in the setting of metric measure spaces supporting a Poincaré inequality.

This coherent treatment from first principles is an ideal introduction to the subject for graduate students and a useful reference for experts. It presents the foundations of the theory of such first-order Sobolev spaces, then explores the geometric implications of the critical Poincaré inequality and indicates numerous examples of spaces satisfying this axiom. A distinguishing feature of the book is its focus on vector-valued Sobolev spaces. The final chapters include proofs of several landmark theorems, including Cheeger's stability theorem for Poincaré inequalities under Gromov–Hausdorff convergence and the Keith–Zhong self-improvement theorem for Poincaré inequalities.

JUHA HEINONEN (1960–2007) was Professor of Mathematics at the University of Michigan. His principal areas of research interest included quasiconformal mappings, nonlinear potential theory, and analysis on metric spaces. He was the author of over 60 research articles, including several published posthumously, and two textbooks. A member of the Finnish Academy of Science and Letters, Heinonen received the Excellence in Research Award from the University of Michigan in 1997 and gave an invited lecture at the International Congress of Mathematicians in Beijing in 2002.

PEKKA KOSKELA is Professor of Mathematics at the University of Jyväskylä, Finland. He works in Sobolev mappings and in the associated nonlinear analysis, and he has authored over 140 publications. He gave invited lectures at the European Congress of Mathematics in Barcelona in 2000 and at the International Congress of Mathematicians in Hyderabad in 2010. Koskela is a member of the Finnish Academy of Science and Letters. He received the Väisälä Award in 2001 and the Magnus Ehrnrooth Foundation prize in 2012.

NAGESWARI SHANMUGALINGAM is Professor of Mathematics at the University of Cincinnati. Her research interests include analysis in metric measure spaces, potential theory, functions of bounded variation, and quasiminimal surfaces in metric setting. She developed the foundational part of the structure of Sobolev spaces in metric setting in her Ph.D. thesis in 1999, and she has also contributed to the development of potential theory in metric setting. Her research contributions were recognized by the College of Arts and Sciences at the University of Cincinnati with a McMicken Dean's Award in 2008.

JEREMY T. TYSON is Professor of Mathematics at the University of Illinois at Urbana-Champaign, working in analysis in metric spaces, geometric function theory, and sub-Riemannian geometry. He has authored over 40 research articles and coauthored two other books. Tyson has received awards for teaching from the University of Illinois at both the departmental and college level. He is a Fellow of the American Mathematical Society.

NEW MATHEMATICAL MONOGRAPHS

All the titles listed below can be obtained from good booksellers or from Cambridge University Press. For a complete series listing visit www.cambridge.org/mathematics.

1 M. Cabanes and M. Enguehard *Representation Theory of Finite Reductive Groups*
2 J. B. Garnett and D. E. Marshall *Harmonic Measure*
3 P. Cohn *Free Ideal Rings and Localization in General Rings*
4 E. Bombieri and W. Gubler *Heights in Diophantine Geometry*
5 Y. J. Ionin and M. S. Shrikhande *Combinatorics of Symmetric Designs*
6 S. Berhanu, P. D. Cordaro and J. Hounie *An Introduction to Involutive Structures*
7 A. Shlapentokh *Hilbert's Tenth Problem*
8 G. Michler *Theory of Finite Simple Groups I*
9 A. Baker and G. Wüstholz *Logarithmic Forms and Diophantine Geometry*
10 P. Kronheimer and T. Mrowka *Monopoles and Three-Manifolds*
11 B. Bekka, P. de la Harpe and A. Valette *Kazhdan's Property (T)*
12 J. Neisendorfer *Algebraic Methods in Unstable Homotopy Theory*
13 M. Grandis *Directed Algebraic Topology*
14 G. Michler *Theory of Finite Simple Groups II*
15 R. Schertz *Complex Multiplication*
16 S. Bloch *Lectures on Algebraic Cycles (2nd Edition)*
17 B. Conrad, O. Gabber and G. Prasad *Pseudo-Reductive Groups*
18 T. Downarowicz *Entropy in Dynamical Systems*
19 C. Simpson *Homotopy Theory of Higher Categories*
20 E. Fricain and J. Mashreghi *The Theory of H(b) Spaces I*
21 E. Fricain and J. Mashreghi *The Theory of H(b) Spaces II*
22 J. Goubault-Larrecq *Non-Hausdorff Topology and Domain Theory*
23 J. Śniatycki *Differential Geometry of Singular Spaces and Reduction of Symmetry*
24 E. Riehl *Categorical Homotopy Theory*

Sobolev Spaces on Metric Measure Spaces

An Approach Based on Upper Gradients

JUHA HEINONEN
Late, University of Michigan

PEKKA KOSKELA
University of Jyväskylä, Finland

NAGESWARI SHANMUGALINGAM
University of Cincinnati

JEREMY T. TYSON
University of Illinois at Urbana-Champaign

CAMBRIDGE
UNIVERSITY PRESS

CAMBRIDGE
UNIVERSITY PRESS

University Printing House, Cambridge CB2 8BS, United Kingdom

Cambridge University Press is part of the University of Cambridge.

It furthers the University's mission by disseminating knowledge in the pursuit of education, learning and research at the highest international levels of excellence.

www.cambridge.org
Information on this title: www.cambridge.org/9781107092341

First published 2015

A catalogue record for this publication is available from the British Library

Library of Congress Cataloguing in Publication data
Sobolev spaces on metric measure spaces : an approach based on upper gradients /
Juha Heinonen, Pekka Koskela, Nageswari Shanmugalingam, Jeremy T. Tyson.
pages cm. – (New mathematical monographs)
Includes bibliographical references and indexes.
ISBN 978-1-107-09234-1 (Hardback)
1. Metric spaces. 2. Sobolev spaces. I. Heinonen, Juha. II. Koskela, Pekka.
III. Shanmugalingam, Nageswari. IV. Tyson, Jeremy T., 1972–
QA611.28.S63 2015
515′.7–dc23 2014027794

ISBN 978-1-107-09234-1 Hardback

In memory of Frederick W. Gehring
(1925–2012)

Contents

Preface

The aim of this book is to present a coherent and essentially self-contained treatment of the theory of first-order Sobolev spaces on metric measure spaces, based on the notion of *upper gradients*.

The project of writing this book was initiated by Juha Heinonen in 2000. His premature passing in 2007 significantly delayed progress in its preparation. We wish to thank Karen E. Smith for securing for us valuable private material of Juha Heinonen pertaining to this text.

Over the years of preparation of the manuscript, we have benefitted from discussions with, and advice from, many colleagues. Amongst them, we wish to give special thanks to the following individuals. We thank Luigi Ambrosio, Piotr Hajłasz, Ilkka Holopainen, Riikka Korte, Jan Malý, Anton Petrunin, and Stephen Semmes for valuable contributions to the mathematical content of this book. Bruce Hanson and Pietro Poggi-Corradini provided detailed comments and corrections of various drafts of the manuscript. We also acknowledge Sita Benedict, Anders Björn, Jana Björn, Estibalitz Durand Cartagena, Nicola Gigli, Changyu Guo, Nijjwal Karak, Aapo Kauranen, Panu Lahti, Marcos Lopez, Marie Snipes, and Thomas Zürcher for reading the manuscript and providing useful feedback. The Mathematica code used to create Figures 14.3 and 14.4 was written by Anton Lukyanenko.

The authors are grateful to Karen E. Smith and Kai Rajala for their encouragement in completing this project.

Our contributions to the field of research summarized in this book have been supported by grants from the US National Science Foundation, the Simons Foundation, and the Academy of Finland. We wish to acknowledge these agencies for their support over the years. Also, we would like to thank the Institute for Pure and Applied Mathematics and the University of Jyväskylä for their kind hospitality during some of the intensive writing periods. The final editing of the manuscript was completed during a snowy weekend in January 2014 at the Clifton Gaslight Bed and Breakfast in Cincinnati, Ohio. We wish to thank Scott and Maria Crawford for their hospitality.

J. H. (Ann Arbor, deceased)

P. K. (Jyväskylä)

N. S. (Cincinnati)

J. T. T. (Urbana)

1

Introduction

Analysis in metric spaces, in the sense that we are considering in this book, emerged as an independent research field in the late 1990s. Its origins lie in the search for an abstract context suitable to recover a substantial component of the classical Euclidean geometric function theory associated with quasiconformal and quasisymmetric mappings. Such a context, identified in the paper [125], consists of *doubling metric measure spaces* supporting a Poincaré inequality.

Over the past 15 years the subject of analysis in metric spaces has expanded dramatically. A significant part of that development has been a detailed study of abstract first-order Sobolev spaces and their relation to variational problems and partial differential equations as well as their role as a tool in, e.g., function theory, dynamics, and related fields. The subject advanced to the point where a careful treatment from first principles, in textbook form, appeared to be needed. This book is intended to serve that purpose.

The concept of an *upper gradient* plays a critical role in both the notion of Sobolev spaces considered in this book and the concomitant framework of metric measure spaces supporting a Poincaré inequality. This concept, also proposed originally in [125], provides an effective replacement for the gradient, or more precisely, the *norm of the gradient* of a smooth function. A nonnegative Borel function g (possibly taking the value $+\infty$) on a metric space (X, d) is said to be an upper gradient of a real-valued function u if the inequality

$$|u(x) - u(y)| \leq \int_\gamma g\, ds \tag{1.1}$$

is satisfied for all rectifiable curves γ joining x to y in X. Here the integral of g on the right-hand side of (1.1) is computed with respect to the arc length measure along γ induced by the metric d. We review the theory of path integrals along rectifiable curves in metric spaces in Chapter 5; Chapter 6 is devoted to the basic properties of upper gradients in metric spaces. It is worth emphasizing that no smoothness assumption on u is imposed a priori in the definition. (Indeed, it is not clear what such an assumption would entail in the metric space setting.) However, as we will see in this book, the existence of a well-behaved upper gradient for a function u necessarily implies certain regularity properties for u itself.

With the notion of upper gradient in hand it is natural to inquire about the existence of a theory of Sobolev spaces based on such gradients. The classical Sobolev space $W^{1,p}(\Omega)$, where Ω is a domain in $\mathbb{R}^n$, can be adapted to the setting of a metric measure space (X, d, μ) by introducing the space of p-integrable functions which admit a p-integrable upper gradient. The foundations for such a theory were laid in the thesis [247] and the accompanying paper [248]. In the literature this space is often referred to as the *Newtonian space* and denoted $N^{1,p}(X)$. This terminology highlights the essential role played by the upper gradient inequality (1.1), which in turn serves as an abstract counterpart of the fundamental theorem of calculus.

Chapters 7, 8, and 9 form the heart of this book. In these chapters we introduce and give a detailed study of the Sobolev space $N^{1,p}$. In these chapters, we show among other results that $N^{1,p}$ is a Banach function space, we study the pointwise properties of Sobolev functions (both scalar- and vector-valued), and we discuss the density of Lipschitz functions in the Sobolev space.

In this book we consistently employ the terminology *Sobolev space*, although we retain the notation $N^{1,p}(X)$ both in homage to the origins of the concept and to distinguish this space from other abstract versions of the classical Sobolev space. In Chapter 10 we review several alternate approaches to abstract Sobolev-type spaces on metric measure spaces. Under suitable assumptions, some of or all these spaces coincide, either as sets or (up to a linear isomorphism or even up to isometry) as Banach spaces.

One version of the classical Poincaré inequality on the Euclidean space $\mathbb{R}^n$ states that

$$\frac{1}{|B|}\int_B |u - u_B| \leq Cr\,\frac{1}{|B|}\int_B |\nabla u|. \tag{1.2}$$

Here u denotes a C^∞-function on $\mathbb{R}^n$ and B denotes a ball of radius r. The notation $u_B = |B|^{-1}\int_B u$ denotes the mean value of u on B. The constant C depends only on the dimension n, i.e., it is independent of B and u.

Using the notion of the upper gradient one can reformulate the Poincaré inequality (1.2) in the metric measure space context, by replacing $|\nabla u|$ by any fixed upper gradient g of a given function u. Actually the story is more subtle. It follows trivially from Hölder's inequality that (1.2) implies the corresponding inequality where the integral on the right-hand side is replaced by the L^p-norm of $|\nabla u|$ with respect to the Lebesgue measure on B (normalized by the volume of B as in (1.2)). Moreover, one can replace the ball B by any larger concentric ball λB ($\lambda > 1$), at the cost of possibly changing the constant C. We say that a metric measure space (X, d, μ) *supports a weak p-Poincaré inequality* if there exist constants $C > 0$ and $\lambda \geq 1$ such that the inequality

$$\frac{1}{\mu(B)} \int_B |u - u_B| \, d\mu \leq Cr \left(\frac{1}{\mu(\lambda B)} \int_{\lambda B} g^p \, d\mu \right)^{1/p} \tag{1.3}$$

holds for all balls B in X and all function–upper-gradient pairs (u, g). As before, r denotes the radius of B while λB denotes the ball with the same center as B and with radius λr.

The importance of the abstract Poincaré inequality (1.3) lies in the fact that it imposes an additional relation between functions and their upper gradients, at the level of the volume measure μ rather than at the level of the length measure along curves. The length–volume principle (usually known as the *length–area principle*) lies at the core of classical Euclidean geometric function theory. In our setting the interplay between the upper gradient inequality (1.1) and the Poincaré inequality (1.3) is a principal driving force. When coupled with the *doubling condition* for the measure μ (namely, the assumption that $\mu(2B) \leq C\mu(B)$ for all balls B in X, where the constant C is independent of B), the Poincaré inequality becomes a powerful tool with both analytic and geometric consequences.

The reader may wonder why we complicate the story by distinguishing the Poincaré inequality according to the value of the exponent p as well as by allowing for dilated balls λB in the definition. In Euclidean space, as already observed, the Poincaré inequality holds with $p = 1$ and $\lambda = 1$ (and this is the strongest form of the inequality). In an abstract setting, it is not necessarily the case that a space supporting a Poincaré inequality for some $1 \leq p < \infty$ and with some dilation constant $\lambda \geq 1$ necessarily supports a Poincaré inequality for better choices of this data. Under rather mild conditions, however, the dilation parameter λ can always be chosen to be 1. We discuss this and other self-improvement phenomena relating to Sobolev–Poincaré inequalities in Chapter 9.

It is a much deeper fact of the theory that if the underlying metric space is complete and the measure μ is doubling then the exponent p on the right-hand side of (1.3) can be improved. In other words, if such a space (X, d, μ) supports a p-Poincaré inequality for some $p > 1$ then it supports a q-Poincaré inequality for some $1 \leq q < p$. This fact, due to Keith and Zhong, is a highlight of the modern theory of analysis on metric spaces. Chapter 12 contains a detailed and self-contained proof of the Keith–Zhong theorem as well as a discussion of its numerous implications and corollaries. Examples of doubling spaces supporting a p-Poincaré inequality for some, but not all, values of p in the range $[1, \infty)$ are described in Chapters 13 and 14.

Of comparable importance is the landmark theorem of Cheeger on the almost everywhere differentiability of Lipschitz functions on doubling spaces supporting a Poincaré inequality. This result, an abstract reformulation of the famous Rademacher differentiation theorem for Euclidean Lipschitz functions, demonstrates that doubling metric measure spaces supporting a Poincaré inequality possess a rich infinitesimal "linear" structure not immediately apparent from the indefinition. Indeed, on such spaces it is possible to define not only the norm of the gradient of a Lipschitz function but (in a suitable sense) the *gradient* or differential itself, acting as a linear operator. Chapter 13 contains a proof of Cheeger's differentiation theorem.

One of our aims in preparing this book has been to present self-contained proofs of these two key theorems of Keith and Zhong and Cheeger.

Another major theme of this book is our consistent emphasis on the class of *vector-valued functions*, that is to say, functions taking values in a Banach space V. The integrability theory for vector-valued functions goes back to the work of Bochner and Pettis; we review this theory in Chapter 3. Our standard setting is that of V-valued Bochner integrable functions u defined on a metric measure space (X, d, μ). (Note, however, that the upper gradients of such functions u, as analogs of the norm of the classical gradient, remain real-valued functions.) The theory of first-order Sobolev spaces is, with a few notable expections, no more difficult to develop in the vector-valued case than in its scalar-valued counterpart. Moreover, there are important reasons why one wishes to have a theory in such a context. Every metric space admits an isometric embedding into some Banach space. (See Chapter 4 for a summary of classical embedding and extension theorems.) Taking advantage of such embeddings one can define metric space-valued Sobolev mappings. The space of Sobolev mappings from a metric measure space (X, d, μ) into another metric space (Y, d') plays a key analytic role in the theory of quasisymmetric maps as well as in nonlinear geometric variational problems. While we do not investigate those subjects in this book, we can remark that the analytic definition of quasisymmetric maps in terms of metric space-valued Sobolev mappings, as developed in our paper [129], was a primary impetus for this book. A brief survey of the theory of quasiconformal and quasisymmetric mappings on metric spaces can be found in Section 14.1.

In Chapter 14 we describe various examples of metric measure spaces supporting a Poincaré inequality and, although we do not provide proofs of the relevant inequality for these examples, we do give copious references to the literature in case the reader wishes to pursue such matters further. It is also useful to know that the collection of doubling metric measure spaces supporting a Poincaré inequality, with uniform constants, is closed under a suitable notion of convergence (e.g., convergence in the *Gromov–Hausdorff* sense). We discuss Gromov–Hausdorff convergence and prove the preceding claim in Chapter 11. This observation expands the class of example spaces for our theory by including suitable Gromov–Hausdorff limit spaces.

The following references are recommended to readers who wish to learn more about the subject. The short books [120] and [8] are good introductions to the field of analysis in metric spaces. Hajłasz's survey articles [109] and [112] focus specifically on the theory of Sobolev spaces on metric spaces; these two articles are well suited for readers wishing to learn more about alternative notions of Sobolev spaces as discussed in Chapter 10. For a general historical survey of nonsmooth calculus, see [122]. The book by A. and J. Björn [31] is a comprehensive treatment of nonlinear potential theory, especially the theory of *p-harmonic functions* on metric measure spaces; this book serves as a valuable counterpart to the present volume. Other topics closely related to the subject matter of this book, and that are currently under active study, include abstract notions of curvature (as in the books [8] and [276]) and analysis on fractals (as in the book [155]).

This book is intended as a graduate textbook. We have endeavored to include detailed proofs of virtually all the major results and to present the material in such a way as to minimize the assumed background. However, prior knowledge of abstract measure theory and functional analysis, at the level of a standard introductory graduate course, is highly recommended. We review the basic tools of functional analysis needed for this book in Chapter 2, while in Chapter 3 we review the foundations of Borel and Radon measures, the integration of Banach space-valued functions, and the basic tools of harmonic analysis such as the Hardy–Littlewood maximal function. A prior exposure to Sobolev spaces (e.g., as can be found in a graduate partial differential equation course) could help the reader to place the topics of this book in a broader context.

Throughout this book, we let C denote any positive constant whose particular value is not of interest to us; thus, even within the same line, two occurrences of C may refer to two different values. However, C will always be assumed to be a positive constant.

Our style of exposition has undoubtedly been influenced by the works of our mathematical fathers and grandfathers, including Olli Martio, Olli

Lehto, and Rolf Nevanlinna. Besides this we wish to acknowledge Jussi Väisälä, whose lecture notes on quasiconformal mappings attracted each of us to the subject. Finally, we have benefited tremendously from the inspiring atmosphere generated by Lois and Fred Gehring, and from the mentoring which we have all received from Fred. We dedicate this book with great appreciation to his memory.

2

Review of basic functional analysis

The theory of Sobolev spaces as developed in this book requires only a small amount of elementary functional analysis. In this chapter we present the required background material. For the sake of completeness, we have included proofs for all but the most standard facts. Anyone with a good working knowledge of analysis can safely skip this chapter. Alternatively, one can quickly glance through the chapter to check the notation and return to it later as needed.

We assume that the reader is familiar with the basic measure theory of real-valued functions and Lebesgue integration. The integration theory for Banach space-valued functions will be developed in Chapter 3.

2.1 Normed and seminormed spaces

Let V be a vector space over the real numbers. A *norm* on V is a function

$$|\cdot| : V \to \mathbb{R}$$

that satisfies the following three conditions:

$$|v| > 0 \qquad \text{for all } v \in V \setminus \{0\}, \tag{2.1.1}$$

$$|\lambda v| = |\lambda|\,|v| \qquad \text{for all } v \in V \text{ and } \lambda \in \mathbb{R}, \tag{2.1.2}$$

$$|v + w| \leq |v| + |w| \qquad \text{for all } v, w \in V. \tag{2.1.3}$$

Here and throughout this book, $|\lambda|$ denotes the absolute value of a real number λ. The notational similarity between absolute values and general norms should not cause any confusion.

If $|\cdot|$ is a norm and $v \in V$, it follows from the definition that $|v| \geq 0$ and that $|v| = 0$ if and only if $v = 0$. A function $|\cdot| : V \to \mathbb{R}$ is called a *seminorm* on V if it satisfies (2.1.2), (2.1.3), and in place of (2.1.1) the following weaker version:

$$|v| \geq 0 \qquad \text{for all } v \in V. \tag{2.1.4}$$

If $|\cdot|$ is a norm on V, the pair $(V, |\cdot|)$ is called a *normed space*. Analogously, $(V, |\cdot|)$ is a *seminormed space* if $|\cdot|$ is a seminorm.

The n-dimensional space $\mathbb{R}^n$, $n \geq 1$, is most commonly equipped with the *Euclidean norm*

$$|x| = (x_1^2 + \cdots + x_n^2)^{1/2}, \quad x = (x_1, \ldots, x_n). \tag{2.1.5}$$

We always assume, unless otherwise explicitly stated, that $\mathbb{R}^n$ comes equipped with the norm (2.1.5). There are, however, many other norms in $\mathbb{R}^n$. For $1 \leq p \leq \infty$ we have the p-norms

$$|x|_p := (|x_1|^p + \cdots + |x_n|^p)^{1/p}, \qquad 1 \leq p < \infty, \tag{2.1.6}$$

and

$$|x|_\infty := \max\{|x_1|, \ldots, |x_n|\}. \tag{2.1.7}$$

Thus $|x| = |x|_2$ for $x \in \mathbb{R}^n$.

The norms $|\cdot|_p$ can be defined as extended real-valued functions on the vector space of infinite sequences $\mathbb{R}^\infty := \{(x_1, x_2, \ldots) : x_i \in \mathbb{R}\}$ in an obvious way,

$$|x|_p := \left(\sum_{i=1}^{\infty} |x_i|^p\right)^{1/p}, \qquad |x|_\infty := \sup\{|x_i| : i = 1, 2, \ldots\}. \tag{2.1.8}$$

Then a family of norms can be defined by restricting $|x|_p$ to the vector subspace of $\mathbb{R}^\infty$ consisting of those $x \in \mathbb{R}^\infty$ for which $|x|_p < \infty$. In this way we construct the *l^p-spaces*,

$$l^p = l^p(\mathbb{N}) := \{x \in \mathbb{R}^\infty : |x|_p < \infty\}, \qquad 1 \leq p \leq \infty. \tag{2.1.9}$$

More generally, let (X, μ) be a measure space, where μ is a measure on the set X (see Section 3.1 for a review of the basic terminology) and define, for $1 \leq p < \infty$ and measurable $f : X \to [-\infty, \infty]$,

$$\|f\|_p := \left(\int_X |f|^p \, d\mu\right)^{1/p}. \tag{2.1.10}$$

Then $\|\cdot\|_p$ is a seminorm on the vector space of measurable functions f for which $\|f\|_p < \infty$. It is not always a norm, for the integral in (2.1.10) vanishes whenever f vanishes almost everywhere. If we identify two functions that agree almost everywhere then, for the resulting equivalence classes $[f]$, we can define $\|[f]\|_p$ unambiguously via (2.1.10) by using a representative. In this way we arrive at the *L^p-spaces*,

$$L^p = L^p(X) = L^p(X, \mu), \qquad 1 \leq p < \infty, \tag{2.1.11}$$

consisting of the equivalence classes $[f]$ of measurable functions on X with $\|[f]\|_p < \infty$. It is customary in L^p-theory to speak about *functions* in L^p rather than equivalence classes, and to use the notation f and $\|f\|_p$ rather than $[f]$ and $\|[f]\|_p$. We will follow the same practice. In the theory of Sobolev spaces, the issue of the identification of functions arises in a more subtle way; this will be discussed in detail in Chapters 5 and 6. Functions in $L^p(X)$ are also referred to as *p-integrable functions* on X.

The *sup norm* for a measurable function $f : X \to [-\infty, \infty]$ is given by

$$\|f\|_\infty := \sup\{\lambda \in \mathbb{R} : \mu(\{x \in X : |f(x)| > \lambda\}) \neq 0\}. \tag{2.1.12}$$

Upon following the preceding identification convention for functions, we obtain the normed space

$$L^\infty = L^\infty(X) = L^\infty(X, \mu). \tag{2.1.13}$$

This is the space of *essentially bounded functions*, consisting of those (equivalence classes of) measurable functions for which the expression $\|f\|_\infty$ is finite.

In the case when $X = \mathbb{N}$ and μ is the counting measure, we recover the l^p-spaces as in (2.1.9).

For an arbitrary set A, with no assigned measure, one can define a normed space

$$l^\infty(A) \tag{2.1.14}$$

consisting of all bounded functions $f : A \to \mathbb{R}$ with the norm

$$\|f\|_\infty := \sup_{a \in A} |f(a)|. \tag{2.1.15}$$

We will use the shorthand notation $l^\infty = l^\infty(\mathbb{N})$, $\|x\|_\infty = |x|_\infty$ for $x \in \mathbb{R}^\infty$, which is in accordance with (2.1.8) and (2.1.9).

Remark 2.1.16 The procedure of passing to the equivalence classes of functions in L^p-spaces is an example of a general procedure whereby a seminormed space can be turned into a normed space. To wit, let $(V, |\cdot|_S)$ be a seminormed space. For $v \in V$ we consider the *equivalence class* $[v]$ given by the equivalence relation $\sim$, where $v \sim w$ if and only if $|v - w|_S = 0$. By setting

$$\big|[v]\big| = |v|_S \tag{2.1.17}$$

we obtain a norm in the vector space of equivalence classes $[v]$. Put differently, if V_S denotes the vector subspace of V consisting of those vectors v for which $|v| = 0$ then the map $|\cdot|_S : V \to \mathbb{R}$ factors through the canonical projection $V \to V/V_S$ as a norm $|\cdot| : V/V_S \to \mathbb{R}$.

Lebesgue measure We denote the n-dimensional Lebesgue measure on $\mathbb{R}^n$ by m_n and the corresponding Lebesgue spaces by $L^p(\mathbb{R}^n)$. More generally, if $A \subset \mathbb{R}^n$ is a Lebesgue measurable set then the short notation

$$L^p(A) = L^p(A, m_n)$$

is used, where m_n is restricted to A in a natural manner.

Metric spaces

A *metric space* is a pair (X, d), where X is a set and $d : X \times X \to [0, \infty)$ is a function called a *distance* or *metric* satisfying the following three conditions:

$$d(x, y) = d(y, x) \qquad \text{for all } x, y \in X, \tag{2.1.18}$$

$$d(x, y) = 0 \qquad \text{if and only if } x = y, \tag{2.1.19}$$

$$d(x, y) \leq d(x, z) + d(z, y) \qquad \text{for all } x, y, z \in X. \tag{2.1.20}$$

Both (2.1.3) and (2.1.20) are commonly referred to as the *triangle inequality*. We assume that the reader is familiar with the basic theory of metric spaces, including standard topological notions such as completeness and compactness. A reasonable discussion of this basic theory can be found in [214].

It follows from the definitions that every normed space $(V, |\cdot|)$ is naturally a metric space with distance function $d(v, w) = |v - w|$. Unless otherwise stated, all topological notions on a normed space $V = (V, |\cdot|)$ will be based on this metric. For example, "the sequence (v_i) converges to v in V", or "$v_i \to v$ in V", means $\lim_{i\to\infty} |v_i - v| = 0$. We will, however, consider other modes of convergence in V later (see Section 2.3).

A metric space is *separable* if it possesses a countable dense subset. A normed space is said to be separable if it is separable as a metric space.

The space $L^p(X)$ for $1 \leq p < \infty$ is separable under some mild conditions on the measure space $X = (X, \mu)$. For example, $L^p(\mathbb{R}^n)$ is separable for $1 \leq p < \infty$. (See Proposition 3.3.49 for a statement in the main context of this book.) On the other hand, the space $L^\infty(X)$ is rarely separable, and $l^\infty(A)$ is separable if and only if A is a finite set.

Banach spaces

A normed space $(V, |\cdot|)$ is said to be a *Banach space* if it is complete as a metric space. We also use the self-explanatory term *complete norm* in this case.

The spaces introduced earlier, $(\mathbb{R}^n, |\cdot|_p)$ and $L^p(X)$ for $1 \leq p \leq \infty$ as well as $l^\infty(A)$, are all examples of Banach spaces.

Every normed space $(V, |\cdot|)$ can be completed and this completion $\overline{(V, |\cdot|)}$ is a Banach space. The elements in $\overline{(V, |\cdot|)}$ are equivalence classes of Cauchy sequences (v_i) in $(V, |\cdot|)$, where $(v_i) \sim (w_i)$ if and only if $|v_i - w_i| \to 0$ as $i \to \infty$. The norm $|\cdot|$ is extended to the completion by setting $|(v_i)| = \lim_{i\to\infty} |v_i|$. The limit exists because (2.1.3) shows that $(|v_i|)$ is Cauchy in $\mathbb{R}$. Moreover, the limit value is independent of the representative (v_i). The elements in the completion of a normed space can often be identified more concretely. For example, let (X, μ) be a measure space and let S be the vector space of simple functions s on X of the type

$$s = \sum_{i=1}^{N} a_i \chi_{A_i}, \tag{2.1.21}$$

where $a_i \in \mathbb{R}$, the sets $A_i \subset X$ are pairwise disjoint and measurable with $\mu(A_i) < \infty$, and χ_A denotes the characteristic function of a set $A \subset X$. We equip S with the norm

$$|s| = \sum_{i=1}^{N} |a_i| \mu(A_i);$$

then the completion of $(S, |\cdot|)$ can be identified with the Lebesgue space $L^1(X, \mu)$. (Compare Section 3.2.)

A vector subspace of a normed space is itself a normed space with the induced norm. A subspace S of a Banach space V is said to be *dense* in V if the completion of S equals V. For example, the space S of simple functions as in (2.1.21) is dense in every $L^p(X)$, $1 \leq p < \infty$.

There is a useful characterization of Banach spaces, among all normed spaces, in terms of summable series. Namely, *a normed space $(V, |\cdot|)$ is a Banach space if and only if every absolutely summable series converges in the norm.* Here, a series $\sum_{n=1}^{\infty} v_n$ of elements $v_n \in V$ is said to be *absolutely summable* if $\sum_{n=1}^{\infty} |v_n| < \infty$, and it is said to be *convergent in the norm* if the partial sums $\sum_{n=1}^{N} v_n$ converge to an element $v \in V$ as $N \to \infty$. This characterization is easy to verify from the definitions; see for example [86, p. 144].

Hilbert spaces

Hilbert spaces are important special classes of Banach spaces. To wit, a Hilbert space is a Banach space whose norm is induced by an inner product. An *inner product* on a vector space V is a function

$$\langle \cdot, \cdot \rangle : V \times V \to \mathbb{R} \tag{2.1.22}$$

that is symmetric ($\langle v, w \rangle = \langle w, v \rangle$ for all $v, w \in V$), bilinear ($\langle \alpha v + \beta w, z \rangle = \alpha \langle v, z \rangle + \beta \langle w, z \rangle$ and $\langle z, \alpha v + \beta w \rangle = \alpha \langle z, v \rangle + \beta \langle z, w \rangle$ for $v, w, z \in V$ and $\alpha, \beta \in \mathbb{R}$), and satisfies

$$\langle v, v \rangle > 0 \qquad \text{for all } v \in V \setminus \{0\}. \tag{2.1.23}$$

If $\langle \cdot, \cdot \rangle$ is an inner product on V then the expression

$$|v| := \langle v, v \rangle^{1/2} \tag{2.1.24}$$

defines a norm on V. A vector space equipped with an inner product $\langle \cdot, \cdot \rangle$ is called an *inner product space*. It is a *Hilbert space* if it is, in addition, complete in the induced norm (2.1.24).

An inner product on a vector space provides extra structure not available on general Banach spaces; one can talk about angles and orthogonality. The Euclidean space $\mathbb{R}^n$ has its standard inner product

$$\langle x, y \rangle := x_1 y_1 + \cdots + x_n y_n, \qquad \begin{array}{l} x = (x_1, \ldots, x_n), \\ y = (y_1, \ldots, y_n), \end{array} \tag{2.1.25}$$

which induces the Euclidean norm (2.1.5). More generally, if (X, μ) is a measure space, then $L^2(X)$ is a Hilbert space with inner product

$$\langle f, g \rangle := \int_X f \cdot g \, d\mu. \tag{2.1.26}$$

The spaces $L^p(X)$ for $p \neq 2$ cannot be equipped with an inner product inducing the p-norm (2.1.10).

Norms that arise from inner products are characterized by the *parallelogram law*,

$$|v + w|^2 + |v - w|^2 = 2(|v|^2 + |w|^2), \tag{2.1.27}$$

in the following sense: every norm in a vector space V that arises from an inner product satisfies (2.1.27) for all $v, w \in V$, and if a norm $|\cdot|$ satisfies (2.1.27) for all $v, w \in V$ then we can define an inner product on V by the formula

$$\langle v, w \rangle := \frac{1}{4}\left(|v + w|^2 - |v - w|^2 \right).$$

Finally, a *semi-inner product* on V is a map as in (2.1.22) for which we replace (2.1.23) by $\langle v, v \rangle \geq 0$ for all $v \in V$. In this case, the expression $|v| = \langle v, v \rangle^{1/2}$ defines a seminorm on V.

Hilbert spaces do not play a special role in this book, but they will be mentioned from time to time, mostly in examples and remarks.

2.2 Linear operators and dual spaces

A map $T : V \to W$ between two normed spaces is also called an *operator*. Such an operator is said to be a *bounded operator* if there is a constant $C \geq 0$ such that

$$|Tv| \leq C|v| \tag{2.2.1}$$

for all $v \in V$. The least number C is called the *operator norm* (or norm) of T and is denoted by $|T|$. When T is linear, we call it a *linear operator*. A linear operator is bounded if and only if it is continuous as a map between metric spaces.

The vector space $B(V, W)$ of all bounded operators $T : V \to W$ is itself a normed space with the norm $T \mapsto |T|$ defined above. It is a Banach space if W is a Banach space.

An operator $T : V \to W$ is bounded precisely when it maps bounded sets in V to bounded sets in W.

Open mapping theorem A basic result about bounded operators is the *open mapping theorem*, which asserts that *every surjective bounded linear operator between Banach spaces is an open mapping*. The proof of this theorem is a standard application of the Baire category theorem. Recall that an *open mapping* between topological spaces is a mapping that takes open sets to open sets.

Dual spaces

The Banach space of all bounded linear operators from a normed space V to $\mathbb{R}$ is called the *dual space* of V. We use the standard notation $V^* = B(V, \mathbb{R})$ and call the elements of V^* *bounded linear functionals* on V. Moreover, we usually write

$$\langle v^*, v\rangle \tag{2.2.2}$$

for the numerical value $v^*(v)$, where $v \in V$ and $v^* \in V^*$. The Banach norm in V^* is given by

$$|v^*| = \sup_{|v|\leq 1} |\langle v^*, v\rangle|,$$

in accordance with (2.2.1). This notation does not claim that $\langle v^*, v\rangle$ comes from an inner product. However, should V happen to be a Hilbert space then, by the Riesz representation theorem, V^* can be identified with V via the inner product on V: for $v^* \in V$, the map $v \mapsto \langle v^*, v\rangle$ is a bounded linear operator on V and every bounded operator has such a representation.

It follows from the Hahn–Banach theorem (see below) that every normed space V admits a canonical isometric embedding to its *double dual*:

$$V \hookrightarrow V^{**} := (V^*)^* \tag{2.2.3}$$

where, for $v \in V$, we define a linear functional on V^* by

$$v^* \mapsto \langle v^*, v \rangle, \qquad v^* \in V^*. \tag{2.2.4}$$

A Banach space is called *reflexive* if $V = V^{**}$ in the sense that the above embedding is surjective.

Hilbert spaces are reflexive, as follows from the Riesz representation theorem. (See also Theorem 2.4.9 below.) The dual of $L^p(X)$, $1 < p < \infty$, is $L^q(X)$, where $q = p/(p-1)$. Thus L^p-spaces for $1 < p < \infty$ are reflexive. For σ-finite measure spaces (X, μ) we have $L^\infty(X) = L^1(X)^*$. The dual of $L^\infty(X)$ has a description as a space of finitely additive signed measures on X. In general, the spaces L^∞ and L^1 are not reflexive.

We note the following special cases of the above dualities: $l^q = (l^p)^*$ for $1 \leq p < \infty$ and $q = p/(p-1)$, with the usual understanding that $q = \infty$ if $p = 1$. Although $L^1(\mathbb{R}^n)$ is not the dual of any Banach space, we have that

$$l^1 = c_0^*, \tag{2.2.5}$$

where c_0 is the Banach subspace of l^∞ consisting of all sequences $x = (x_i)$ such that $x_i \to 0$ as $i \to \infty$. Indeed, the duality in (2.2.5) is a special case of the duality

$$M(S) = c_0(S)^*, \tag{2.2.6}$$

where $c_0(S)$ is the space of all continuous functions on a locally compact Hausdorff space S that "vanish at infinity" and $M(S)$ is the space of all finite Borel regular measures on S. This follows from another Riesz representation theorem; see [237, Theorem 2.14 on p. 40]. We will forgo giving the precise definitions as they are not needed in this book.

It is important to recognize reflexive Banach spaces, for these enjoy some strong properties commonly used in analysis. See Theorem 2.4.1, for example.

The Hahn–Banach theorem

A *sublinear map* on a vector space V is a map $p : V \to \mathbb{R}$ that satisfies

$$p(v + w) \leq p(v) + p(w) \qquad \text{and} \qquad p(\lambda v) = \lambda p(v) \tag{2.2.7}$$

for all $v, w \in V$ and $\lambda \geq 0$. In particular, every seminorm on V determines a sublinear map. More generally, if C is a convex open neighborhood of 0 in a normed space V then the formula

$$p_C(v) := \inf\{\lambda > 0 : \lambda^{-1}v \in C\}, \qquad v \in V, \tag{2.2.8}$$

determines a sublinear map called the *Minkowski functional* associated with the convex set C.

Let $p : V \to \mathbb{R}$ be a sublinear map, let W be a vector subspace of V, and let $v^* : W \to \mathbb{R}$ be a linear map such that

$$v^*(w) \leq p(w)$$

for all $w \in W$. The *Hahn–Banach theorem* asserts that *there exists a linear map* $\overline{v}^* : V \to \mathbb{R}$ *such that* $\overline{v}^*|_W = v^*$ *and that*

$$\overline{v}^*(v) \leq p(v)$$

for all $v \in V$.

The Hahn–Banach theorem is most often applied in a situation where one needs to extend a bounded linear functional from a subspace of a normed space. However, the general formulation in terms of sublinear maps is crucial for some basic facts.

We record two immediate corollaries of the Hahn–Banach theorem.

Given a nonzero vector v *in a normed space* V, *there is an element* v^* *in the dual space* V^* *such that* $\langle v^*, v\rangle = |v|$ *and* $|v^*| = 1$. In particular, the dual of a normed space is never trivial. It also follows that the canonical embedding (2.2.3) is isometric. This corollary follows on applying the Hahn–Banach theorem with $v \mapsto |v|$ as the sublinear map and $\lambda v \mapsto \lambda|v|$ as the linear map, defined on the one-dimensional subspace of V spanned by v.

Given a convex open neighborhood C *of* 0 *in a normed space* V *and a vector* $v \notin C$, *there is an element* v^* *in the dual space* V^* *such that*

$$\langle v^*, w\rangle < \langle v^*, v\rangle \tag{2.2.9}$$

for all $w \in C$. This corollary follows on applying the Hahn–Banach theorem with the Minkowski functional p_C and with the linear map $\lambda v \mapsto \lambda p_C(v)$ defined on the one-dimensional subspace of V spanned by v. Indeed, we have that (2.2.9) holds for a linear map v^* and, because C contains a neighborhood of 0, we also have that v^* is bounded as required (compare (2.2.1)).

There is another, less immediate, corollary of the Hahn–Banach theorem called Mazur's lemma, which will play an important role in the development of Sobolev space theory in this book. We will review and prove Mazur's lemma in Section 2.3.

2.3 Convergence theorems

Principle of uniform boundedness

Another basic theorem of functional analysis, which we quote without proof, is the *principle of uniform boundedness*: *if* $\{T_\alpha : \alpha \in A\}$ *is a collection of bounded linear operators from a Banach space* V *into a normed space* W, *and if*

$$\sup_{\alpha \in A} |T_\alpha(v)| < \infty \tag{2.3.1}$$

for each $v \in V$, *then*

$$\sup_{\alpha \in A} |T_\alpha| < \infty.$$

Recall that $|T|$ is the operator norm of T as defined in Section 2.2.

A standard application of the principle of uniform boundedness is the following. *If* (T_i) *is a sequence of bounded linear operators from a Banach space* V *into a normed space* W *such that*

$$\lim_{i \to \infty} T_i(v) \tag{2.3.2}$$

exists in W *for every* $v \in V$ *then the expression (2.3.2) determines a bounded linear operator from* V *to* W.

The principle of uniform boundedness is also known as the *Banach–Steinhaus theorem.* While the proof of the uniform boundedness principle in its complete generality will not be needed here, we will need the following weaker version of it in Chapter 11.

Theorem 2.3.3 *If* V *is a separable Banach space,* $\mathcal{T} = \{T_\alpha : \alpha \in A\}$ *is a collection of bounded linear operators from* V *into* $\mathbb{R}$, *and for all* $v \in V$ *we have*

$$\sup_{\alpha \in A} |T_\alpha(v)| < \infty$$

then there is a sequence (T_{α_j}) *in* $\mathcal{T}$ *such that* $T(v) := \lim_j T_{\alpha_j}(v)$ *exists for each* $v \in V$, *with* T *a bounded linear operator on* V.

Proof The proof relies on the fact that bounded subsets of $\mathbb{R}$ are precompact. A sketch of the proof is as follows. Since V is separable, we can choose a countable dense subset S of V. For each $v_i \in S$ we can find a sequence $(T_{\alpha_{j,i}})_j$ such that $\lim_j T_{\alpha_{j,i}}(v_i)$ exists. A Cantor diagonalization argument gives a sequence (T_{α_j}) such that for each $i \in \mathbb{N}$ the limit $\lim_j T_{\alpha_j}(v_i) =: T(v_i)$ exists. Extend T to the linear span of S.

We now show that $\sup_{j\in\mathbb{N}} |T_{\alpha_j}| < \infty$. To do so, note that for each j, by the continuity of T_{α_j}, the set $T_{\alpha_j}^{-1}([-1, 1])$ is a closed subset of V. Set

$$E := \bigcap_{j\in\mathbb{N}} T_{\alpha_j}^{-1}([-1, 1]).$$

Then E is a closed subset of V. We claim that E has nonempty interior. Suppose that $\mathrm{int}(E)$ is empty. For $r > 0$ we set $rE := \{rv : v \in E\}$. Then for each $n \in \mathbb{N}$ we know that nE has empty interior but, by the assumption on the family of operators, $V = \bigcup_{n\in\mathbb{N}} nE$. Then the open set $V_n := V \setminus nE$ is dense in V because nE has empty interior.

For $v \in V$ and $r > 0$ we let $B(v, r) = \{w \in V : |w - v| < r\}$ denote the ball centered at v with radius r. Since V_1 is dense in V, we know that $B(0, 1) \cap V_1$ is nonempty (but this intersection is also open in V); so we can find $x_1 \in V$ and $0 < r_1 < 2^{-1}$ such that $B(x_1, r_1) \subset V_1 \cap B(0, 1)$. Now, $V_2 \cap B(x_1, r_1)$ is a nonempty (because V_2 is dense in V) open set; thus we can find $x_2 \in V$ and $0 < r_2 < 2^{-2}$ such that $B(x_2, r_2) \subset V_2 \cap B(x_1, r_1)$. Proceeding by induction, we can find $x_n \in V$ and $0 < r_n < 2^{-n}$ such that $B(x_n, r_n) \subset B(x_{n-1}, r_{n-1}) \cap V_n$.

It is easy to see that (x_n) is a Cauchy sequence in V and so has a limit $x_\infty \in V$. Because $x_\infty \in B(x_n, r_n)$ for each $n \in \mathbb{N}$, it follows that $\bigcap_{n\in\mathbb{N}} V_n$ is nonempty; this violates the fact that $V = \bigcup_{n\in\mathbb{N}} nE$. Thus E has nonempty interior; that is, there is some $v \in V$ and $r > 0$ such that $B(v, r) \subset E$. For each $y \in B(v, r)$ we know that $\sup_j |T_{\alpha_j}(y)| \le 1$. It follows that, for all $z \in B(0, r)$ and all $j \in \mathbb{N}$,

$$|T_{\alpha_j}(z)| - |T_{\alpha_j}(v)| \le |T_{\alpha_j}(v - z)| \le 1,$$

that is, $\sup_{j\in\mathbb{N}} |T_{\alpha_j}(z)| \le 2$, which directly verifies that $\sup_{j\in\mathbb{N}} |T_{\alpha_j}| < \infty$. It follows that the constructed functional T on the span of S is also a bounded linear map; an application of the Hahn–Banach theorem, together with the density of S in V, gives a unique extension of T to V. It is now directly verifiable that for each $v \in V$ the limit $\lim_j T_{\alpha_j}(v)$ exists and equals $T(v)$; we leave the details to the interested reader. □

The above proof also gives an indirect proof that a Banach space is necessarily of Baire category 2; see [238].

Weak convergence

A sequence (v_i) in a normed space V is said to *converge weakly to an element* $v \in V$ if

$$\lim_{i\to\infty} \langle v^*, v_i\rangle = \langle v^*, v\rangle \tag{2.3.4}$$

for each $v^* \in V^*$. In this case, the vector v is called the *weak limit* of the sequence (v_i). Note that the weak limit, if exists, is unique; namely, if v and v' are two weak limits of a sequence in V, then

$$\langle v^*, v - v' \rangle = 0$$

for each $v^* \in V^*$, and it follows from the Hahn–Banach theorem together with the isometric nature of the embedding (2.2.3) that $v = v'$.

If $(v_i) \subset V$ is a sequence that converges to $v \in V$ in the norm, then

$$|\langle v^*, v_i \rangle - \langle v^*, v \rangle| \leq |v^*|\,|v_i - v| \to 0$$

so that $v_i \to v$ weakly as well. The converse is not true in general. For example, the sequence $\{\sin(ix) : i = 1, 2, \dots\}$ converges weakly to 0 in $L^p([0, 2\pi])$ for $1 \leq p < \infty$. This follows from the dualities in 2.2 (cf. 2.3) and from the Riemann–Lebesgue lemma [237, p. 109]. The fundamental result about weak convergence is Theorem 2.4.1 below.

The second assertion in the following proposition is often called the *lower semicontinuity of norms.*

Proposition 2.3.5 *Weakly convergent sequences are norm bounded. Moreover, if $v_i \to v$ weakly then*

$$|v| \leq \liminf_{i \to \infty} |v_i|. \tag{2.3.6}$$

Proof Let $v_i \to v$ weakly in a normed space V. Each v_i determines an element in the double dual V^{**} of V, as explained in Section 2.2. Because $(\langle v^*, v_i \rangle)$ is bounded for each $v^* \in V^*$, the first assertion follows from the principle of uniform boundedness. Furthermore,

$$|\langle v^*, v \rangle| = \lim_{i \to \infty} |\langle v^*, v_i \rangle| \leq \liminf_{i \to \infty} |v_i|\,|v^*|,$$

whence (2.3.6) follows upon invoking the first corollary of the Hahn–Banach theorem as in Section 2.2. This proves the proposition. □

Remark 2.3.7 Often, in the literature, a sequence (v_i) in a normed space V is said to be *weakly convergent* if the limit

$$\lim_{i \to \infty} \langle v^*, v_i \rangle$$

exists for each $v^* \in V^*$. In general, a weakly convergent sequence need not converge weakly to a vector in V (see, for example, [286, p. 120] or [197, p. 20]), although it can always be thought of as converging weakly to a vector in the double dual V^{**} by the principle of uniform boundedness. In this book, we will consider only weakly convergent sequences that converge weakly to a vector $v \in V$. Note, however, that the first assertion in Proposition 2.3.5 is true under the weaker meaning of the term "weakly convergent".

The following result will be used repeatedly in this book.

Mazur's lemma *Let (v_i) be a sequence in a normed space V converging weakly to an element $v \in V$. Then v belongs to the norm closure of the convex hull of the sequence (v_i).*

The *convex hull* of a set A in a normed space V is the intersection of all convex sets in V that contain A. Thus, if $v_i \to v$ weakly in V, Mazur's lemma guarantees the existence of a sequence $(\tilde{v}_k)$ of convex combinations

$$\tilde{v}_k = \sum_{i=k}^{m_k} \lambda_{i,k} v_i, \quad \lambda_{i,k} \geq 0, \quad \lambda_{k,k} + \cdots + \lambda_{m_k,k} = 1, \tag{2.3.8}$$

converging to v in the norm.

Given a metric space (X, d), a set $A \subset X$, and a point $x \in X$, we denote the distance from x to A by

$$\operatorname{dist}(x, A) := \inf\{d(x, a) : a \in A\}.$$

Proof of Mazur's lemma Let H be the convex hull of (v_i). By replacing the sequence (v_i) by a sequence $(v_i - h)$ for some $h \in H$, we may assume that $0 \in H$. Assume now that there exists $\epsilon > 0$ such that

$$|v - w| > 2\epsilon$$

for each $w \in H$. Then, in particular, $v \neq 0$. Since $|a - a'| < \epsilon$ and $|b - b'| < \epsilon$ imply $|(ta+(1-t)b)-(ta'+(1-t)b')| < \epsilon$ for $a, a', b, b' \in V$ and $0 \leq t \leq 1$, the ϵ-neighborhood

$$H_\epsilon := \{w \in V : \operatorname{dist}(w, H) < \epsilon\}$$

of H is convex; it is also an open neighborhood of 0 in V and consequently defines a Minkowski functional

$$|w|_\epsilon := \inf\{\lambda > 0 : \lambda^{-1} w \in H_\epsilon\}, \quad w \in V, \tag{2.3.9}$$

as explained in Section 2.2. By the Hahn–Banach theorem (see (2.2.9)), applied to the linear map $tv \mapsto t|v|_\epsilon$ defined on the one-dimensional subspace of V spanned by v and our Minkowski functional, there exists a linear functional $\overline{v}^* : V \to \mathbb{R}$ such that $\langle \overline{v}^*, v\rangle = |v|_\epsilon$ and $\langle \overline{v}^*, w\rangle \leq |w|_\epsilon$ for all $w \in V$. It follows that

$$1 < |v|_\epsilon = \lim_{i\to\infty} \langle \overline{v}^*, v_i \rangle \le \liminf_{i\to\infty} |v_i|_\epsilon \le 1,$$

which is absurd. The lemma follows. □

Remark 2.3.10 We will frequently employ Mazur's lemma in the following formulation: *if a sequence (v_i) in a normed space V converges weakly to an element v then a sequence $(\tilde{v}_j)$ of convex combinations of the vectors v_i converges to v in the norm.*

A pedantic reading of this formulation would allow the situation where the sequence $(\tilde{v}_j)$ consists of a constant sequence $\tilde{v}_j \equiv v$ for every j, in the case where v appears as a member of the sequence (v_i). However, with a slight abuse of terminology, in the preceding formulation of Mazur's lemma the following additional requirement is assumed throughout this book: for every $n \ge 1$, all but finitely many of the members in the sequence $(\tilde{v}_j)$ are convex combinations of the vectors v_i for $i \ge n$.

Weak convergence in L^p

Let $X = (X, \mu)$ be a σ-finite measure space, and let $1 \le p < \infty$. Then a sequence (f_i) in $L^p(X)$ converges weakly to $f \in L^p(X)$ if and only if

$$\lim_{i\to\infty} \int_X g \cdot f_i \, d\mu = \int_X g \cdot f \, d\mu$$

for all $g \in L^q(X)$, where $q = p/(p-1)$ if $1 < p < \infty$ and $q = \infty$ if $p = 1$. This follows from the dualities explained in Section 2.2.

The following result is often useful in recognizing weak limits in L^p-spaces.

Proposition 2.3.11 *Let $X = (X, \mu)$ be a measure space, let $1 \le p \le \infty$, and let (f_i) in $L^p(X)$ be a sequence converging weakly to $f \in L^p(X)$. If*

$$\lim_{i\to\infty} f_i(x) = g(x) \tag{2.3.12}$$

for almost every $x \in X$ then $g = f$ almost everywhere.

Proof By Mazur's lemma (see above), a sequence $(\tilde{f}_k)$ of convex combinations of the f_i converges to f in $L^p(X)$. By passing to a subsequence we may assume that $\tilde{f}_k \to f$ pointwise almost everywhere in X (see Proposition 2.3.13 below). Because also $\tilde{f}_k \to g$ almost everywhere, by (2.3.12), we have that $f = g$ and the proposition follows. □

The following well-known result from Lebesgue theory was used in the preceding proof. A similar argument will appear later in a different context (see Egoroff's theorem in Section 3.1 and Proposition 7.3.1), and to emphasize this relation we provide a proof.

Proposition 2.3.13 *Let (X, μ) be a measure space, let $1 \leq p \leq \infty$, and let $(f_i) \subset L^p(X)$ be a sequence converging to f in $L^p(X)$. Then (f_i) has a subsequence (f_{i_j}) with the following property: for every $\epsilon > 0$ there exists a set $E_\epsilon \subset X$ such that $\mu(E_\epsilon) < \epsilon$ and $f_{i_j} \to f$ uniformly in $X \setminus E_\epsilon$. In particular, (f_{i_j}) converges to f pointwise almost everywhere in X.*

Proof The statement for $p = \infty$ is straightforward, with a stronger conclusion (there is no need to pass to a subsequence and the convergence is uniform outside a set of measure zero). Thus assume that $1 \leq p < \infty$. The proof in this case splits naturally into three parts. First one shows that (f_i) is *Cauchy in measure*, which means the following:

$$\lim_{i,j\to\infty} \mu(\{x \in X : |f_i(x) - f_j(x)| > \epsilon\}) = 0 \tag{2.3.14}$$

for each $\epsilon > 0$. Then it is a fact, independent of L^p-theory, that a subsequence converges pointwise almost everywhere to a function g; moreover, the convergence is uniform outside a set of arbitrarily small measure. Finally, Fatou's lemma implies that $g \in L^p(X)$ and that $f_i \to g$ in L^p.

To prove (2.3.14) we simply observe that

$$\epsilon^p \mu(\{x \in X : |f_i(x) - f_j(x)| > \epsilon\}) \leq \int_X |f_i - f_j|^p \, d\mu \to 0, \quad i, j \to \infty,$$

whenever $\epsilon > 0$. Next, by passing to a subsequence and using (2.3.14), we may assume that $\mu(E_i) < 2^{-i}$, where

$$E_i = \{x \in X : |f_i(x) - f_{i+1}(x)| > 2^{-i}\}.$$

Thus

$$\mu(F_j) < 2^{-j+1}, \tag{2.3.15}$$

where

$$F_j = \bigcup_{i=j}^{\infty} E_i,$$

while for $x \in X \setminus F_j$ we have $|f_i(x) - f_k(x)| \leq 2^{-i+1}$ for all $j \leq i \leq k$. This implies that (f_i) converges uniformly in $X \setminus F_j$ to a function g, and because of (2.3.15) we have pointwise convergence almost everywhere.

It remains to show that $f_i \to g$ in L^p (which in particular implies that $g = f$). This follows because (f_i) is a Cauchy sequence in L^p and because, by Fatou's lemma,

$$\int_X |g - f_i|^p \, d\mu \leq \liminf_{j\to\infty} \int_X |f_j - f_i|^p \, d\mu$$

for each i. The proof of the proposition is complete. □

Remarks 2.3.16 (a) If, in Proposition 2.3.13, the set X is a topological space and the sequence (f_i) consists of continuous functions then the sets F_j defined in the proof are open. Therefore, if continuous functions are dense in $L^p(X)$ then every function in $L^p(X)$ has a representative with the following property: for every $\epsilon > 0$ there is an open set $O \subset X$ such that $\mu(O) < \epsilon$ and the restriction of the function to the complement of O is continuous. See Corollary 3.3.51 for a result of this kind in the main context of this book.

For Sobolev functions, in many cases a similar statement where the underlying measure is replaced with a different (outer) measure called the capacity, is true. See Theorems 7.4.2 and 8.2.1.

(b) Recall that Egoroff's theorem (see e.g. [83, 2.3.7] or [81, Theorem 3, p. 16]) asserts that if $\mu(X) < \infty$ and if (f_i) is a sequence of real-valued measurable functions on X converging pointwise almost everywhere to a real-valued function f then the sequence converges uniformly to f outside a set of arbitrarily small prescribed measure. Proposition 2.3.13 shows that, upon passing to a subsequence, the hypothesis that the measure of X is finite can be omitted in the presence of L^p-convergence.

Later, in Section 3.1, we state and prove a vector-valued version of Egoroff's theorem.

(c) Another useful fact about integrable functions is the following *absolute continuity of integrals*: if A is a measurable subset of a Euclidean space and $f \in L^1(A)$ then for every $\epsilon > 0$ we can find a $\delta > 0$ such that whenever $E \subset A$ with the Lebesgue measure $|E| < \delta$ we have $\int_E |f| < \epsilon$. This fact holds also in the context of complete measure spaces.

Weak* convergence

Let V^* be the dual space of a normed space V. A sequence (v_i^*) in V^* is said to *converge weakly** to an element $v^* \in V^*$ if

$$\lim_{i\to\infty} \langle v_i^*, v \rangle = \langle v^*, v \rangle \tag{2.3.17}$$

for each $v \in V$. In this case, the vector v^* is called the *weak* limit* of the sequence (v_i^*). Note that weak* convergence is nothing other than pointwise convergence for sequences in V^*. Obviously, the weak* limit, if exists, is unique.

As in the case of weak convergence, if $v_i^* \to v^*$ in the norm of V^* then $v_i^* \to v^*$ weakly* as well. The converse is not true in general. If V is a reflexive Banach space then the notions of weak* convergence and weak convergence in V^* agree. In an arbitrary dual space V^*, weak convergence implies weak* convergence; this is simply because for weak* convergence one only tests pointwise convergence for elements in V, which is naturally a subspace of V^{**}. In general, weak* convergence does not imply weak convergence. For an example, consider the duality $l^1 = c_0^*$ given in (2.2.5). The sequence (e_i), where

$$e_i = (0, \ldots, 0, \underbrace{1}_{\substack{i\text{th} \\ \text{spot}}}, 0, \ldots), \qquad i = 1, 2, \ldots,$$

converges weakly* to 0 in l^1, because

$$\langle e_i, (x_j) \rangle = x_i \to 0$$

for each sequence (x_j) in c_0. However, we have that

$$\langle e_i, (1, 1, 1 \ldots) \rangle = 1$$

for all i, where $(1, 1, 1, \ldots) \in l^\infty = (l^1)^*$.

In fact, one can show that a sequence in l^1 converges weakly if and only if it converges strongly. This result is known as Schur's lemma [286, p. 122]. We will not need it here.

Remark 2.3.18 A sequence (v_i^*) in V^* is said to be *weakly* convergent* if the limit

$$\lim_{i \to \infty} \langle v_i^*, v \rangle \tag{2.3.19}$$

exists for each $v \in V$. (Compare Remark 2.3.7.) We will not use or need this concept in this book. In any event, if V is a Banach space, every sequence that is weakly* convergent in the above sense converges weakly* to some element of V^*. This follows from the principle of uniform boundedness; the expression (2.3.19) defines a bounded linear operator on V.

Both weak and weak* convergences are, in essence, pointwise convergences tested on each element of the spaces V^* or V, respectively. It helps to know that in both cases one needs only test the convergence on norm dense subsets, provided that the sequences are *a priori* known to be bounded.

Proposition 2.3.20 *Let V be a normed space. A sequence (v_i) in V converges weakly to an element $v \in V$ if and only if (v_i) is a bounded sequence in V such that*

$$\lim_{i \to \infty} \langle v^*, v_i \rangle = \langle v^*, v \rangle$$

for each v^* in some norm dense subset D^* of V^*, that is, $\overline{D^*} = V^*$.

Furthermore, a sequence (v_i^*) *in* V^* *converges weakly* to an element* $v^* \in V^*$ *if and only if* (v_i^*) *is a bounded sequence in* V^* *such that*

$$\lim_{i\to\infty} \langle v_i^*, v\rangle = \langle v^*, v\rangle$$

for each v *in some norm dense subset* D *of* V.

Proof The necessity part is trivial for both statements. The proofs for the sufficiency part are in turn similar, and so we demonstrate only the proof of the first assertion. Thus, let $v_0^* \in V^*$, let $\epsilon > 0$, and pick $v_\epsilon^* \in D^*$ such that $|v_0^* - v_\epsilon^*| < \epsilon$. Then

$$\begin{aligned}
|\langle v_0^*, v_i\rangle - \langle v_0^*, v\rangle| &\le |\langle v_0^*, v_i\rangle - \langle v_\epsilon^*, v_i\rangle| + |\langle v_\epsilon^*, v_i\rangle - \langle v_\epsilon^*, v\rangle| \\
&\quad + |\langle v_\epsilon^*, v\rangle - \langle v_0^*, v\rangle| \\
&\le \epsilon\, |v_i| + |\langle v_\epsilon^*, v_i\rangle - \langle v_\epsilon^*, v\rangle| + \epsilon\, |v| \\
&\le \epsilon\, \sup_i |v_i| + |\langle v_\epsilon^*, v_i\rangle - \langle v_\epsilon^*, v\rangle| + \epsilon\, |v|.
\end{aligned}$$

Because the factor $\sup_i |v_i|$ in the last line is finite by assumption, and because the second term converges to zero as $i \to \infty$, we obtain that $\langle v_0^*, v_i\rangle \to \langle v_0^*, v\rangle$ as required. The proposition follows. □

2.4 Reflexive spaces

Recall from Section 2.2 that a Banach space V is called reflexive if the canonical embedding of V into V^{**} is a surjection, i.e., it is onto. In this section we first prove the following important property of reflexive spaces.

Theorem 2.4.1 *Every bounded sequence in a reflexive Banach space has a weakly convergent subsequence.*

The proof of Theorem 2.4.1 is rather straightforward for separable Banach spaces. The nonseparable case can be reduced to the separable case by invoking some auxiliary results.

Lemma 2.4.2 *Let* W *be a subspace of a normed space* V, *and let* $v \in V \setminus W$ *be such that* $\operatorname{dist}(v, W) > 0$. *Then there exists an element* $v^* \in V^*$ *such that* $\langle v^*, v\rangle = 1$ *and* $\langle v^*, w\rangle = 0$ *for each* $w \in W$.

Proof Let Z be the linear subspace of V spanned by W and $\{v\}$. Then each $z \in Z$ can be written uniquely as $z = w + \lambda v$, where $w \in W$ and $\lambda \in \mathbb{R}$. The linear map $w + \lambda v \mapsto \lambda$ is bounded on Z, because

$$|w + \lambda v| = |\lambda| \left|\frac{w}{\lambda} + v\right| \ge |\lambda| \operatorname{dist}(v, W)$$

and because $\mathrm{dist}(v, W) > 0$. The Hahn–Banach theorem with the sublinear map p on V given by $p(v) = C\,|v|$, where C is the norm of our linear map on Z, now provides a map v^* as desired and the lemma follows. □

Lemma 2.4.3 *A normed space is separable if its dual is separable. In particular, the dual of a reflexive and separable Banach space is separable.*

Proof Let (v_i^*) be a countable dense subset of the dual space V^* of a normed space V. Pick a sequence $(v_i) \subset V$ such that $|v_i| \leq 1$ and $|v_i^*| \leq 2\langle v_i^*, v_i\rangle$ for each i. The linear subspace of V spanned by the sequence (v_i) is clearly separable. If it is not dense in V then there exists a nonzero element $v^* \in V^*$ such that $\langle v^*, v_i\rangle = 0$ for each i (Lemma 2.4.2). Assuming that $v_{i_j}^* \to v^*$ in V^*, we find that

$$|v_{i_j}^*| \leq 2\langle v_{i_j}^*, v_{i_j}\rangle = 2\langle v_{i_j}^* - v^*, v_{i_j}\rangle \leq 2|v_{i_j}^* - v^*| \to 0.$$

This gives $v^* = 0$, which is a contradiction, and the lemma follows. □

Proposition 2.4.4 *Every closed subspace of a reflexive Banach space is reflexive.*

Proof Let $V = V^{**}$ be a reflexive Banach space, and let $W \subset V$ be a closed subspace. We have the natural bounded linear maps, obtained via the restriction of bounded linear maps from V to W

$$\alpha : V^* \to W^*, \qquad \alpha(v^*)(w) = \langle v^*, w\rangle,$$

and

$$\beta : W^{**} \to V = V^{**}, \qquad \beta(w^{**})(v^*) = \langle w^{**}, \alpha(v^*)\rangle.$$

If there exists $v_0 \in \beta(W^{**}) \setminus W$ then, because W is closed, by Lemma 2.4.2 there exists $v_0^* \in V^*$ such that $\langle v_0^*, v_0\rangle \neq 0$ and $\langle v_0^*, w\rangle = 0$ for each $w \in W$. Then $\alpha(v_0^*) = 0$. However, $v_0 = \beta(w_0^{**})$ for some $w_0^{**} \in W^{**}$, so that

$$0 = \langle w_0^{**}, \alpha(v_0^*)\rangle = \beta(w_0^{**})(v_0^*) = \langle v_0^*, v_0\rangle \neq 0,$$

which is absurd. It follows that $\beta(W^{**}) \subset W$.

Next, pick an arbitrary element $w^{**} \in W^{**}$. By the previous paragraph, $\beta(w^{**}) = w \in W$. Let $w_1^* \in W^*$ and let v_1^* denote an extension of w_1^* to V^* (which exists by the Hahn–Banach theorem). Then

$$\langle w^{**}, w_1^*\rangle = \langle w^{**}, \alpha(v_1^*)\rangle = \beta(w^{**})(v_1^*) = \langle v_1^*, w\rangle = \langle w_1^*, w\rangle,$$

which implies that $w^{**} = w$ as required. The proposition follows. □

Proof of Theorem 2.4.1 Let V be a reflexive Banach space and let (v_i) be a bounded sequence in V. Denote by V' the completion of the linear span of the sequence (v_i) in V. Then V' is separable by construction and reflexive by Proposition 2.4.4. Consequently, by Lemma 2.4.3 we have that the dual of V' is separable.

We now pick a countable norm dense subset (v_j^*) of $(V')^*$ and proceed with a diagonalization argument. The sequence $(\langle v_1^*, v_i\rangle)$ is bounded and hence contains a subsequence

$$\{\langle v_1^*, v_{i_1^1}\rangle, \langle v_1^*, v_{i_2^1}\rangle, \langle v_1^*, v_{i_3^1}\rangle, \ldots\}$$

such that

$$\lim_{k\to\infty} \langle v_1^*, v_{i_k^1}\rangle$$

exists. Similarly, $(\langle v_2^*, v_{i_k^1}\rangle)$ contains a convergent subsequence $(\langle v_2^*, v_{i_k^2}\rangle)$. Continuing in this manner, we find that for the diagonal sequence $(v_{i_k^k}) =: (v_{i_k})$ the limit

$$\lim_{k\to\infty} \langle v_j^*, v_{i_k}\rangle$$

exists for all j. We claim that

$$\lim_{k\to\infty} \langle v^*, v_{i_k}\rangle \tag{2.4.5}$$

exists for each $v^* \in (V')^*$. This is shown analogously to the method in the proof of Proposition 2.3.20. Fix $v^* \in (V')^*$ and let $\epsilon > 0$. Then choose v_j^* such that $|v^* - v_j^*| < \epsilon$. We find that

$$\begin{aligned}
|\langle v^*, v_{i_k}\rangle - \langle v^*, v_{i_l}\rangle| &\le |\langle v^* - v_j^*, v_{i_k}\rangle| + |\langle v_j^*, v_{i_k}\rangle - \langle v_j^*, v_{i_l}\rangle| \\
&\quad + |\langle v_j^*, v_{i_l}\rangle - \langle v^*, v_{i_l}\rangle| \\
&\le 2\epsilon \sup_i |v_i| + |\langle v_j^*, v_{i_k}\rangle - \langle v_j^*, v_{i_l}\rangle|.
\end{aligned}$$

This shows that $(\langle v^*, v_{i_k}\rangle)$ is a Cauchy sequence and hence that the limit in (2.4.5) exists.

By the principle of uniform boundedness, (2.3.2), the expression (2.4.5) determines a bounded linear functional on $(V')^*$; this is given by an element v of V', by the reflexivity of V'. Thus, $v_{i_k} \to v$ weakly in V'.

To finish the proof, we observe that (v_{i_k}) converges weakly to v in V also. Indeed, if $v^* \in V^*$ then obviously v^* restricts to an element in $(V')^*$. Thus the limit (2.4.5) exists and equals $\langle v^*, v\rangle$, as required. This completes the proof of Theorem 2.4.1. □

Remark 2.4.6 The *Banach–Alaoglu theorem* asserts that the closed unit ball in the dual space V^* of a normed space V is compact in the weak* topology [238, p. 66]. We will not require the general form of the Banach–Alaoglu theorem, but rather its corollary for reflexive spaces, Theorem 2.4.1. We have also omitted the definitions for weak and weak* topologies, as they are not needed in this book.

One should note, however, that the proof of Theorem 2.4.1 can be used essentially verbatim to obtain the following form of the Banach–Alaoglu theorem: *Every bounded sequence in the dual space V^* of a separable Banach space V contains a weakly* convergent subsequence.*

Uniformly convex Banach spaces

We will prove later in this book that certain Sobolev spaces are reflexive. Towards this end, we next discuss a useful reflexivity criterion.

A Banach space V is said to be *uniformly convex* if, for every $\epsilon > 0$, there is a $\delta > 0$ such that

$$|v| = |w| = 1 \qquad \text{and} \qquad |v - w| > \epsilon \tag{2.4.7}$$

implies that

$$|\tfrac{1}{2}(v + w)| < 1 - \delta \tag{2.4.8}$$

for every pair of vectors v and w in V.

Hilbert spaces are uniformly convex, as follows readily from the parallelogram law (2.1.27). We will show later, in Proposition 2.4.19, that L^p-spaces are uniformly convex for $1 < p < \infty$.

Theorem 2.4.9 *Uniformly convex Banach spaces are reflexive.*

First we require a lemma.

Lemma 2.4.10 *Let $v_1^*, \ldots, v_n^*$ be elements in the dual space V^* of a Banach space V, and let $t_1, \ldots, t_n$ be real numbers such that*

$$\left|\sum_{i=1}^{n} \lambda_i t_i\right| \le \left|\sum_{i=1}^{n} \lambda_i v_i^*\right| \tag{2.4.11}$$

whenever $\lambda_1, \ldots, \lambda_n$ are real numbers, where the expression on the right-hand side denotes the operator norm of $\lambda_1 v_1^* + \cdots + \lambda_n v_n^* \in V^*$. Then for every $\epsilon > 0$ there exists a vector $v_\epsilon \in V$ such that $|v_\epsilon| < 1 + \epsilon$ and $\langle v_i^*, v_\epsilon \rangle = t_i$ for each $i = 1, \ldots, n$.

Proof It is easy to see that no loss of generality is entailed in assuming that the v_i^* are linearly independent elements of V^*. Indeed, if

$$v_k^* = \sum_{i \neq k} \mu_i v_i^*$$

then (2.4.11), applied with $\lambda_i = \mu_i$ for $i \neq k$ and $\lambda_k = -1$, implies that

$$\sum_{i \neq k} \mu_i t_i = t_k.$$

In particular, if v_ϵ has been found for the vectors v_i^*, $i \neq k$, then

$$\langle v_k^*, v_\epsilon \rangle = \sum_{i \neq k} \mu_i \langle v_i^*, v_\epsilon \rangle = \sum_{i \neq k} \mu_i t_i = t_k ,$$

as required.

Recall that $v^* \in V^*$ is a bounded linear functional on V; we write $v^*(w)$ for $\langle v^*, w \rangle$ to simplify the notation in the definition of the operator T below. We proceed under the assumption that $v_1^*, \ldots, v_n^*$ are linearly independent. Then the mapping $T : V \to \mathbb{R}^n$ defined by $T(v) = (v_1^*(v), \ldots, v_n^*(v))$ is linear, bounded, and surjective. By the open mapping theorem in Section 2.2, the image under T of every ball

$$B_\epsilon := \{v \in V : |v| < 1 + \epsilon\}, \quad \epsilon > 0,$$

is an open neighborhood of 0 in $\mathbb{R}^n$. Moreover, as B_ϵ is convex, $T(B_\epsilon)$ is convex as well.

Assume now, towards a contradiction, that $t = (t_1, \ldots, t_n)$ does not lie in $T(B_\epsilon)$ for some $\epsilon > 0$. It follows from the second corollary to the Hahn–Banach theorem (see (2.2.9)) that there is a vector $\lambda = (\lambda_1, \ldots, \lambda_n) \in \mathbb{R}^n = \mathbb{R}^{n*}$ such that

$$\begin{aligned} \sum_{i=1}^n \lambda_i t_i = \langle \lambda, t \rangle &\geq \sup_{v \in B_\epsilon} \langle \lambda, T(v) \rangle \\ &= \sup_{v \in B_\epsilon} \sum_{i=1}^n \lambda_i v_i^*(v) = (1+\epsilon) \left| \sum_{i=1}^n \lambda_i v_i^* \right|, \end{aligned}$$

where $\langle \lambda, t \rangle$ denotes the standard Euclidean inner product. In the last equality above we used the first corollary to the Hahn–Banach theorem and the fact that the closed unit ball in V^* is contained in B_ϵ. The above lower bound contradicts the hypotheses, and the lemma follows. □

Proof of Theorem 2.4.9 Let V be a uniformly convex Banach space and let $w^{**} \in V^{**}$. We need to show that there exists $v \in V$ such that

$$\langle w^{**}, v^* \rangle = \langle v^*, v \rangle \tag{2.4.12}$$

for every $v^* \in V^*$. To achieve this, we may assume that $|w^{**}| = 1$. Pick a sequence (v_i^*) from V^* such that $|v_i^*| = 1$ and $\langle w^{**}, v_i^* \rangle \geq 1 - i^{-1}$ for $i = 1, 2, \ldots$ Fix a positive integer n. Then

$$\left| \sum_{i=1}^{n} \lambda_i \langle w^{**}, v_i^* \rangle \right| = \left| \langle w^{**}, \sum_{i=1}^{n} \lambda_i v_i^* \rangle \right| \leq \left| \sum_{i=1}^{n} \lambda_i v_i^* \right|$$

whenever $\lambda_1, \ldots, \lambda_n$ are real numbers. Lemma 2.4.10 now implies that there is, for each $n \geq 1$, a vector $v_n \in V$ such that $|v_n| < 1 + n^{-1}$ and that

$$\langle w^{**}, v_i^* \rangle = \langle v_i^*, v_n \rangle \tag{2.4.13}$$

for each $i = 1, \ldots, n$. It follows that

$$1 - n^{-1} \leq \langle w^{**}, v_n^* \rangle = \langle v_n^*, v_n \rangle \leq |v_n| \leq 1 + n^{-1}$$

and hence that $\lim_{n\to\infty} |v_n| = 1$. We claim that the sequence (v_n) converges in the norm.

Suppose, towards a contradiction, that there exists $\epsilon > 0$ together with arbitrarily large indices n, m such that $|v_n - v_m| > \epsilon$. Recalling that $\lim_{n\to\infty} |v_n| = 1$, we conclude with $\| v_n|^{-1} v_n - |v_m|^{-1} v_m| > \epsilon$ for suitable sufficiently large n, m. Consequently, the uniform convexity conditions (2.4.7) and (2.4.8) give

$$|v_n + v_m| < 2(1 - \delta) \tag{2.4.14}$$

for some $\delta > 0$, for arbitrarily large indices n, m. But we also have that

$$\langle w^{**}, v_i^* \rangle = \langle v_i^*, v_n \rangle \to 1$$

as $i, n \to \infty$, $i \leq n$, which in combination with

$$\langle v_i^*, v_n + v_m \rangle \leq |v_n + v_m|$$

contradicts (2.4.14). It follows that the sequence (v_n) is Cauchy, and, since V is Banach, it converges in the norm to an element $v \in V$. We claim that this vector v satisfies (2.4.12).

To establish the claim, observe first that (2.4.13) gives

$$\langle w^{**}, v_i^* \rangle = \langle v_i^*, v \rangle$$

for each i. Then pick an arbitrary element $v_0^* \in V^*$ with $|v_0^*| = 1$, and apply the preceding argument to the sequence $v_0^*, v_1^*, v_2^*, \ldots$ We similarly pick an element $v' \in V$ such that $|v'| = 1$ and

$$\langle w^{**}, v_i^* \rangle = \langle v_i^*, v' \rangle \tag{2.4.15}$$

for each $i = 0, 1, 2, \ldots$ If $v' \neq v$ then the uniform convexity guarantees that there exists a $\delta > 0$ such that

$$|v + v'| < 2(1 - \delta).$$

However, we have for $i \geq 1$ that

$$2 - 2i^{-1} \leq \langle w^{**}, v_i^* + v_i^* \rangle = \langle v_i^*, v \rangle + \langle v_i^*, v' \rangle \leq |v + v'|,$$

which is a contradiction as $i \to \infty$. Thus $v = v'$. Because v_0^* was an arbitrary element of V^* with unit norm, we conclude from this and from (2.4.15) that (2.4.12) holds. The proof of the theorem is complete. □

We next record the following corollary to the above discussion.

Corollary 2.4.16 *Every closed and convex set in a reflexive Banach space contains an element of smallest norm; if the Banach space is uniformly convex then there is only one such element.*

Proof It follows from Theorem 2.4.1 and Mazur's lemma that every closed and convex subset of a reflexive space has an element of smallest norm. By uniform convexity, such an element is obviously unique. □

The proof of Theorem 2.4.9 essentially contained an argument for the following proposition. We now state and prove this important proposition although it is not used later in the book.

Proposition 2.4.17 *Let (v_i) be a sequence in a uniformly convex Banach space converging weakly to an element $v \in V$. If also $\lim_{i\to\infty} |v_i| = |v|$ then $v_i \to v$ in the norm.*

Proof We may clearly assume that $v \neq 0$ and hence that $v_i \neq 0$ for each i. Write $w_i = v_i/|v_i|$ and $w = v/|v|$. Then $|w_i| = |w| = 1$ and $w_i \to w$ weakly. To prove the lemma, it suffices to show that $|w_i - w| \to 0$.

Assuming the opposite, and using uniform convexity, we find that there exist a $\delta > 0$ and infinitely many indices i such that

$$|w_i + w| < 2(1 - \delta). \tag{2.4.18}$$

By the Hahn–Banach theorem, we can pick an element $v^* \in V^*$, with unit norm, such that $\langle v^*, w\rangle = 1$. Then

$$2 = 2\langle v^*, w\rangle = \lim_{i\to\infty} \langle v^*, w_i + w\rangle \le \liminf_{i\to\infty} |w_i + w|,$$

which contradicts (2.4.18). The proposition follows. □

We know that L^p-spaces for $1 < p < \infty$ are reflexive. The fact that they are also uniformly convex comes in handy sometimes. It is easy to see that the spaces L^1 and L^∞ are not uniformly convex (except under some trivial circumstances).

Proposition 2.4.19 *Let (X, μ) be a measure space and let $1 < p < \infty$. Then $L^p(X)$ is uniformly convex.*

Proof Let $\epsilon > 0$. It suffices to show that there exists some $\delta > 0$ such that

$$\|\tfrac{1}{2}(f + g)\|_p^p > 1 - \delta \tag{2.4.20}$$

implies that

$$\|\tfrac{1}{2}(f - g)\|_p^p \le 2\epsilon^p$$

whenever $f, g \in L^p(X)$ with $\|f\|_p = \|g\|_p = 1$.

To this end, we first record the inequality

$$\int_E |\tfrac{1}{2}(f - g)|^p \, d\mu \le \epsilon^p \int_X |\tfrac{1}{2}(f + g)|^p \, d\mu \le \epsilon^p,$$

where $E = \{|f - g| \le \epsilon |f + g|\}$. The proof will be completed by exhibiting $\delta > 0$ such that

$$\int_{X\setminus E} |\tfrac{1}{2}(f - g)|^p \, d\mu \le \epsilon^p \tag{2.4.21}$$

holds in the presence of (2.4.20).

Indeed, it follows from the *strict* convexity of the function $s \mapsto |s|^p$ that

$$\lambda \mapsto \tfrac{1}{2}(|\lambda + 1|^p + |\lambda - 1|^p) - |\lambda|^p$$

is continuous and positive on $\mathbb{R}$. Hence there exists a t, depending on ϵ and p, such that

$$\tfrac{1}{2}(|\lambda + 1|^p + |\lambda - 1|^p) - |\lambda|^p \ge t \tag{2.4.22}$$

whenever $\lambda \in [-1/\epsilon, 1/\epsilon]$. Applying inequality (2.4.22) for the choice

$$\lambda = \frac{f(x) + g(x)}{f(x) - g(x)},$$

where $x \notin E$, we conclude that

$$\tfrac{1}{2}\Big(|f|^p + |g|^p\Big) \geq t|\tfrac{1}{2}(f-g)|^p + |\tfrac{1}{2}(f+g)|^p$$

holds on $X \setminus E$. By integrating the preceding inequality over $X \setminus E$ and the inequality

$$\tfrac{1}{2}\Big(|f|^p + |g|^p\Big) \geq |\tfrac{1}{2}(f+g)|^p$$

over E we obtain

$$1 = \int_X \tfrac{1}{2}\Big(|f|^p + |g|^p\Big)\, d\mu \geq \int_X |\tfrac{1}{2}(f+g)|^p \, d\mu + \int_{X\setminus E} t|\tfrac{1}{2}(f-g)|^p \, d\mu.$$

Therefore, by choosing $\delta = t\epsilon^p$ we have that (2.4.21) holds, and the proposition follows. □

Two norms $|\cdot|$ and $|\cdot|'$ on a vector space V are said to be *equivalent* if there exists a constant $C \geq 1$ such that

$$C^{-1}|v| \leq |v|' \leq C|v| \tag{2.4.23}$$

for each vector $v \in V$.

The following proposition follows straightforwardly from the definitions.

Proposition 2.4.24 *Let $|\cdot|$ and $|\cdot|'$ be equivalent complete norms on a vector space V. If $(V, |\cdot|)$ is reflexive then so is $(V, |\cdot|')$.*

Proposition 2.4.24 implies that reflexive spaces need not be uniformly convex. Indeed, $\mathbb{R}^n$ equipped with either of the norms $|\cdot|_1$ or $|\cdot|_\infty$ is reflexive but obviously not uniformly convex. It is much harder to give examples of reflexive spaces that do not possess equivalent uniformly convex norms at all; see the notes at the end of this chapter.

We next show that in finite-dimensional spaces every norm is close to an inner product norm. This fact will be used later in connection with Cheeger's differentiation theorem for Lipschitz functions on metric measure spaces. See Theorem 13.5.7.

Theorem 2.4.25 *Let $(V, \|\cdot\|)$ be an n-dimensional normed vector space. Then there exists an inner product $\langle\cdot,\cdot\rangle$ on V such that*

$$\|v\| \leq |v| \leq \sqrt{n}\|v\| \tag{2.4.26}$$

for all $v \in V$, where $|v| = \langle v, v\rangle^{1/2}$ denotes the norm induced by the inner product.

Proof As a vector space, we identify V with the Euclidean space $\mathbb{R}^n$ via the inverse of the linear map obtained by setting $T(e_i) = e_i^V$, where e_i^V, $i = 1, \ldots, n$ are basis vectors for V and the e_i are the canonical basis vectors of $\mathbb{R}^n$. Then clearly T is continuous and hence $B_0 = T^{-1}(\{v \in V : \|v\| \leq 1\})$ is a closed and convex set, invariant under the symmetry $v \mapsto -v$. Let E be a Euclidean ellipsoid centered at the origin of maximal volume (Lebesgue measure) inscribed in B_0. We see that such an ellipsoid exists by maximizing the determinant function among all linear transformations that map the Euclidean unit ball into B_0. (One can show further that such an ellipsoid is unique [23, Theorem V.2.2, p. 207], but this fact is not trivial and, moreover, is unnecessary for the proof here.) Every ellipsoid in $\mathbb{R}^n$ induces an inner product if the ellipsoid is declared to be the closed unit ball in the associated norm with principal axes orthogonal. Consequently, to establish (2.4.26) it suffices to show that $B_0 \subset \sqrt{n}\, E$. Moreover, it involves no loss of generality to assume that $E = \{v \in \mathbb{R}^n : |v| \leq 1\}$ is the Euclidean unit ball corresponding to the Euclidean norm $|\cdot|$, (2.1.5).

The preceding being understood, we argue by contradiction and assume that B_0 contains a point v with $|v| > \sqrt{n}$. Since B_0 is convex, it contains the convex hull C of $E \cup \{\pm v\}$. We claim that C contains an ellipsoid whose volume exceeds that of E, which is the desired contradiction. Now the volume of the ellipsoid

$$E_{a,b} = \left\{(x_1, \ldots, x_n) : x_1^2/a^2 + \sum_{i=2}^{n} x_i^2/b^2 \leq 1\right\}, \qquad a, b > 0,$$

is ab^{n-1} times the volume of E, and so it suffices to find a and b such that $E_{a,b}$ is contained in C and that $ab^{n-1} > 1$. To this end, we can assume without loss of generality that $v = (r, 0, \ldots, 0)$ for some $r > \sqrt{n}$ and claim that

$$a = \frac{r}{\sqrt{n}}, \qquad b = \sqrt{\frac{1 - 1/n}{1 - 1/r^2}}$$

will do.

Indeed, a direct computation reveals that $ab^{n-1} > 1$. Next, to show that $E_{a,b} \subset C$ we can work in two dimensions, as both $E_{a,b}$ and C are radially symmetric with respect to the variables $x_2, \ldots, x_n$. It is appropriate to switch to complex notation, so that

$$E_{a,b} = \{z = x + iy \in \mathbb{C} : x^2/a^2 + y^2/b^2 \leq 1\}$$

and C is the convex hull of $\mathbb{D} \cup \{\pm r\}$, where $\mathbb{D} = \{z \in \mathbb{C} : |z| \leq 1\}$ is the closed unit disk. Note that $C = \mathbb{D} \cup T \cup (-T)$, where T is the triangle with vertices r, $e^{i\theta}$, and $e^{-i\theta}$ with $\theta = \arccos(1/r)$. Next, let

$$\Sigma = \{z = x + iy \in \mathbb{C} : |y| \le \sin\theta = \sqrt{1 - 1/r^2}\}$$

be a horizontal strip and let

$$S^{\pm} = \{z \in \mathbb{C} : \arg(r - e^{i\theta}) < \arg(r \mp z) \le \arg(r - e^{-i\theta})\}$$

be sectors based at the vertices $\pm r$. Then

$$C \cap \Sigma = S^+ \cap S^- \cap \Sigma \qquad \text{and} \qquad C \setminus \Sigma = \mathbb{D} \setminus \Sigma.$$

The proof now reduces to the following two claims: (i) $E_{a,b} \subset S^+$ and $E_{a,b} \subset S^-$ and (ii) $E_{a,b} \setminus \Sigma \subset \mathbb{D}$; (i) shows that $E_{a,b} \cap \Sigma \subset C \cap \Sigma$ and (ii) shows that $E_{a,b} \setminus \Sigma \subset C \setminus \Sigma$.

To prove (i), we note that $z = x + iy \in S^+$ if and only if

$$|y| \le \frac{r - x}{\sqrt{r^2 - 1}}. \tag{2.4.27}$$

If $z \in E_{a,b}$ then the Cauchy–Schwarz inequality and the definitions of a and b give

$$\frac{x}{r} + \frac{|y|\sqrt{r^2 - 1}}{r} \le \left(\frac{x^2}{a^2} + \frac{y^2}{b^2}\right)^{1/2} \left(\frac{a^2}{r^2} + \frac{b^2(r^2 - 1)}{r^2}\right)^{1/2} \le 1. \tag{2.4.28}$$

Solving for $|y|$ in (2.4.28) gives (2.4.27). The proof that $E_{a,b} \subset S^-$ is similar.

Finally, we prove (ii). If $z = x + iy \in E_{a,b} \setminus \Sigma$ then $x^2/a^2 + y^2/b^2 \le 1$ and $|y| > \sqrt{1 - 1/r^2}$. Hence the choices of a, b, and r, yield

$$x^2 + y^2 = \left(x^2 + \frac{a^2}{b^2}y^2\right) - \left(\frac{a^2}{b^2} - 1\right)y^2 \le a^2 - \left(\frac{a^2}{b^2} - 1\right)\left(1 - \frac{1}{r^2}\right) \le 1.$$

This completes the proof of Theorem 2.4.25. □

2.5 Notes to Chapter 2

The material in Chapter 2 is standard and can be found in most textbooks of real and functional analysis. In particular, the open mapping theorem, the Hahn–Banach theorem, and the principle of uniform boundedness can be found in each of the following sources: [77], [86], [135], [206], [237], [238], [242], [286]. A good source for more advanced facts is [285]. These references also contain ample historical comments.

A good discussion about the dual space of $L^\infty(X, \mu)$, as well as the failure of $L^1(X, \mu)^* = L^\infty(X, \mu)$ in general, can be found in [135, V.20]; see also

[86, p. 183]. For the duality (2.2.6), see, e.g., [86, p. 216] or [237, p. 138]. For the fact that the parallelogram law (2.1.27) characterizes inner product spaces, see [286, p. 39].

As mentioned in the text, Theorem 2.4.1 is customarily derived from the general Banach–Alaoglu theorem. The approach that we have taken here can be found in e.g., [77] or [242]. For a readable, elementary discussion on reflexive spaces, see [242, Chapter 8]. The uniform convexity of L^p-spaces for $1 < p < \infty$ is usually proved by using the *Clarkson inequalities* [61], [135, pp. 225ff]. We learned the short proof of Proposition 2.4.19 from Jan Malý. A similar proof appears in [206] and is based on the proof by McShane in [205]. See [77, p. 473] and [206, Chapter 5] for more remarks about uniform convexity. The first example of a reflexive space that does not admit a comparable uniformly convex norm was found by Day [70].

Theorem 2.4.25 is due to John [142]; the proof given here is from [208]. See [23, Chapter V] and [26, p. 299, Appendices A and G] for more information about convexity and norms.

3

Lebesgue theory of Banach space-valued functions

In this chapter we review some topics in the classical Lebesgue theory for functions valued in a Banach space. We will study the basic properties of measurable vector-valued functions as defined by Bochner and Pettis. The L^p-spaces of Banach space-valued functions are introduced and studied. Along the way, we recall many fundamental notions of measure theory. It is assumed that the reader is familiar with basic Lebesgue theory for real-valued functions. (Knowledgeable readers who are only interested in the real-valued theory may proceed directly to Section 3.3.) We also define what is meant by a metric measure space in this book and discuss at some length the relationship between Borel regular measures and Radon measures in the context of metric measure spaces. Finally, we discuss covering theorems, Lebesgue differentiation theory, and maximal functions.

3.1 Measurability for Banach space-valued functions

In the first two sections of this chapter we assume that (X, μ) is a complete and σ-finite measure space and that V is a Banach space.

At this juncture, by a *measure* on a set X we mean a countably additive set function μ that is defined in some σ-algebra $\mathcal{M}$ of measurable subsets of X such that $\emptyset \in \mathcal{M}$ and that takes values in $[0, \infty]$ with $\mu(\emptyset) = 0$; later, in Section 3.3, we give this term a wider meaning. A measure on X is *σ-finite* if X admits a partition into countably many measurable sets of finite measure, and it is *complete* if every subset of a set of measure zero is measurable. Every measure can be completed by enlarging, if necessary, the σ-algebra of measurable sets. A function $f : X \to [-\infty, \infty]$ is *measurable* if $f^{-1}([-\infty, a))$ is a measurable set for every $a \in \mathbb{R}$.

A function $f : X \to V$ is called *simple* if it has finite range and if the preimage of every point is a measurable set. Thus f is simple if and only if there exist vectors $v_1, \ldots, v_n$ in V and a partition of X into measurable sets $E_1, \ldots, E_n$ such that

$$f = \sum_{i=1}^{n} v_i \chi_{E_i}.$$

A function $f : X \to V$ is defined to be *measurable* if it is the pointwise almost-everywhere limit of a sequence of simple functions. It is clear that the set of measurable V-valued functions on X forms a vector space.

Remark 3.1.1 It is a standard fact in measure theory that the two notions of measurability coincide when $V = \mathbb{R}$; see also the Pettis measurability theorem below. Furthermore, given a simple function f, the function $|f| : X \to \mathbb{R}$ defined by setting $|f|(x) := |f(x)|$ is clearly measurable. Consequently, this is also the case whenever $f : X \to V$ is measurable. We also record here the following elementary fact: *for every nonnegative measurable function $f : X \to [0, \infty]$ there exists an increasing sequence (f_i) of nonnegative simple functions such that $f(x) = \lim_{i\to\infty} f_i(x)$ for every $x \in X$.* It is a routine matter to verify this assertion directly from the definitions.

Next, we say that $f : X \to V$ is *weakly measurable* if

$$\langle v^*, f \rangle : X \to \mathbb{R},$$

given by the map $x \mapsto \langle v^*, f(x) \rangle$, is measurable for each v^* in the dual space V^*. In the literature, weakly measurable functions are sometimes called *scalarly measurable* (see, e.g. [74]). The ensuing *Pettis measurability theorem*, a basic result in the subject, asserts that weakly measurable and *essentially separably valued* functions are measurable. A function $f : X \to V$ is said to be essentially separably valued if there exists a set $N \subset X$ of measure zero such that $f(X \setminus N)$ is a separable subset of V. It is immediate from the definition of measurable functions that a measurable function is essentially separably valued.

Pettis measurability theorem *The following are equivalent for a function $f : X \to V$:*

(i) *f is measurable;*
(ii) *f is essentially separably valued and $f^{-1}(U)$ is measurable for each open set U in V;*
(iii) *f is essentially separably valued and weakly measurable.*

For separable target spaces we have the following neat corollary.

Corollary 3.1.2 *Let V be separable. The following are equivalent for a function $f : X \to V$:*

(i) *f is measurable;*
(ii) *$f^{-1}(U)$ is measurable for each open set U in V;*
(iii) *f is weakly measurable.*

Remarks 3.1.3 (a) In practice, the equivalence between (i) and (ii) in Corollary 3.1.2 can be assumed to hold always. Namely, one can show that it holds provided that V has a dense subset whose cardinality is an *Ulam number*. Every *accessible cardinal* is an Ulam number, and the statement that all cardinal numbers are accessible is independent of the usual axioms of set theory. We refer to [83, 2.1.6 and 2.3.6] for the definitions of both Ulam numbers and accessible cardinals and for the equivalence between (i) and (ii) under the aforesaid condition.

(b) The function $f : [0, 1] \to L^\infty([0, 1])$, given by $f(t) = \chi_{[0,t]}$, is weakly measurable but not essentially separably valued. This follows easily from the characterization of the dual of $L^\infty([0, 1])$ as the space of finitely additive signed measures that are absolutely continuous with respect to Lebesgue measure [135, V.20], [77, IV 8.16]. Hence, by the Pettis measurability theorem, f is not measurable. Therefore one cannot drop the assumption that V is separable in the equivalence between (i) and (iii) in Corollary 3.1.2.

For the proof of the Pettis measurability theorem, we require *Egoroff's theorem* for Banach space-valued functions; cf. Remarks 2.3.16(b).

Egoroff's theorem *Assume that $\mu(X) < \infty$. Let $f, f_1, f_2, \ldots$ be measurable functions from X to V such that*

$$\lim_{i\to\infty} f_i(x) = f(x) \tag{3.1.4}$$

for almost every $x \in X$. Then for every $\epsilon > 0$ there exists a measurable set $A \subset X$ such that $\mu(A) < \epsilon$ and $f_i \to f$ uniformly in $X \setminus A$.

Proof Let $E \subset X$ be the collection of all $x \in X$ for which (3.1.4) fails. Fix $\epsilon > 0$. Define

$$E_{jk} := \bigcup_{i=j}^{\infty} \{x \in X : |f_i(x) - f(x)| > 2^{-k}\},$$

where $j, k = 1, 2, \ldots$ The sets E_{jk} are all measurable by Remark 3.1.1. For each fixed $k \geq 1$, we have that $E_{1k} \supset E_{2k} \supset \cdots$ and that

$$\bigcap_{j=1}^{\infty} E_{jk} \setminus E = \emptyset .$$

In particular, because $\mu(X) < \infty$ there exists an integer j_k such that $\mu(E_{j_k k}) < \epsilon\, 2^{-k}$. By setting $A := \bigcup_{k=1}^{\infty} E_{j_k k}$, we have that $\mu(A) < \epsilon$. However, if $k \geq 1$ and $x \in X \setminus A$, we have that $|f_i(x) - f(x)| \leq 2^{-k}$ whenever $i \geq j_k$. The proposition follows. □

In fact, Egoroff's theorem can be upgraded in such a way that we do not need to assume a priori that f is measurable.

Corollary 3.1.5 *If f_i, $i = 1, 2, \ldots$, is a sequence of measurable functions from X to V such that $f_j \to f : X \to V$ almost everywhere in X then f is measurable on X.*

Proof Recall that μ is σ-finite; therefore there is a countable collection of measurable sets X_k with $\mu(X_k) < \infty$ and $X = \bigcup_k X_k$. Without loss of generality we may assume that these sets are pairwise disjoint. Fix one of the sets X_k.

By the definition of measurability, for each i we have a sequence $(h_{i,j})$ of simple functions on X_k that converges pointwise almost everywhere on X_k to f_i. An application of Egoroff's theorem (with f_i playing the role of f and $h_{i,j}$ playing the role of f_j in that theorem) gives a measurable set $D_{k,i}$ with $\mu(D_{k,i}) < 2^{-i-1-k}$ such that $(h_{i,j})$ converges to f_i uniformly in $X_k \setminus D_{k,i}$. Now, for each i we can find a positive integer j_i such that $|f_i - h_{i,j_i}| < 2^{-i}$ on $X_k \setminus D_{k,i}$. For $m = 1, 2, \ldots$ and points $x \in X_k \setminus \bigcup_{i=m}^{\infty} D_{k,i}$, we have

$$\begin{aligned} |f(x) - h_{i,j_i}(x)| &\leq |f(x) - f_i(x)| + |f_i(x) - h_{i,j_i}(x)| \\ &< 2^{-i} + |f(x) - f_i(x)| \end{aligned}$$

whenever $i > m$. Since $\lim_i |f(x) - f_i(x)| = 0$ when $x \in X_k \setminus \bigcup_{i=m}^{\infty} D_{k,i}$, we can conclude that the subsequence (h_{i,j_i}) of simple functions converges to f on $X_k \setminus \bigcup_{i=m}^{\infty} D_{k,i}$. Because the measure of $\bigcup_{i=m}^{\infty} D_{k,i}$ is at most 2^{1-m}, it follows that (h_{i,j_i}) converges pointwise almost everywhere in X_k to f. To simplify the notation, call this subsequence $(h_{k,j})$. Clearly, we may assume that $h_{k,j}(x) = 0$ for all $x \in X \setminus X_k$.

Because f is the pointwise almost-everywhere limit of the sequence $(h_{k,j})$ of simple functions on X_k, for each k, it follows that this also holds for f, which is approximated pointwise almost everywhere by the sequence of simple functions $g_m := \sum_{k=1}^{m} h_{k,m}$ on X. □

Remark 3.1.6 We can deduce the following from the proof of the preceding corollary. Given measurable pairwise disjoint sets X_k of finite measure, with $X = \bigcup_k X_k$, a function $f : X \to V$ is measurable if and only if f is measurable on each X_k.

Proof of the Pettis measurability theorem First we prove the implication (i) $\Rightarrow$ (ii). Let

$$f_i = \sum_{k=1}^{N(i)} v_{ik} \chi_{E_{ik}}$$

be a sequence of simple functions such that $f_i(x) \to f(x)$ for all x in the complement of a set $N \subset X$ of measure zero. We assume that the sets E_{ik} form a partition of X for each fixed i. Then $\{v_{ik}\}$ is a countable set whose closure contains $f(X \setminus N)$. This implies that f is essentially separably valued. Next, let $U \subset V$ be open. We observe that

$$N \cup f^{-1}(U) = N \cup \bigcup_{n=1}^{\infty} \bigcup_{j=1}^{\infty} \bigcap_{i=j}^{\infty} f_i^{-1}(U_n),$$

where $U_n := \{v \in U : \operatorname{dist}(v, V \setminus U) > 1/n\}$. Because $f_i^{-1}(U_n) \subset X$ is measurable, we have that $N \cup f^{-1}(U)$ is measurable. Therefore, by the completeness of μ,

$$f^{-1}(U) = (N \cup f^{-1}(U)) \setminus (N \cap (X \setminus f^{-1}(U)))$$

is measurable.

The implication (ii) $\Rightarrow$ (iii) is clear because the elements in V^* are continuous.

To prove the implication (iii) $\Rightarrow$ (i), assume that f is weakly measurable and that $f(X \setminus N)$ is separable for some set $N \subset X$ of measure zero. Let $D = \{v_1, v_2, \dots\}$ be a countable dense set of distinct points in $f(X \setminus N)$. Then the difference set

$$D - D := \{v_i - v_j : v_i, v_j \in D\}$$

is a countable dense set in the difference set

$$f(X \setminus N) - f(X \setminus N) := \{f(x) - f(y) : x, y \in X \setminus N\}.$$

By the Hahn–Banach theorem, we have elements $v_{ij}^* \in V^*$ such that $|v_{ij}^*| = 1$ and $\langle v_{ij}^*, v_i - v_j \rangle = |v_i - v_j|$ for each pair of distinct indices i, j. It follows that, for each $v_j \in D$, the function

$$x \mapsto |f(x) - v_j| = \sup_i |\langle v_{ij}^*, f(x) - v_j \rangle|$$

is a measurable (real-valued) function on X by the hypotheses. Fix $\epsilon > 0$, and put

$$A_1 := \{x \in X : |f(x) - v_1| < \epsilon\}$$

and

$$A_j := \left\{x \in X : |f(x) - v_k| \geq \epsilon \text{ for all } k = 1, \dots, j-1 \text{ and } |f(x) - v_j| < \epsilon\right\},$$

that is,

$$A_j = \{x \in X \ : \ |f(x) - v_j| < \epsilon\} \setminus \bigcup_{i=1}^{j-1} A_i$$

for $j \geq 2$. Then (A_j) is a pairwise disjoint collection of measurable sets in X, and

$$X \setminus N \subset \bigcup_{j=1}^{\infty} A_j .$$

In particular, the countably valued measurable function

$$g := \sum_{j=1}^{\infty} v_j \chi_{A_j} \tag{3.1.7}$$

satisfies $|f(x) - g(x)| < \epsilon$ for all $x \in X \setminus N$. It follows that $f|_{X\setminus N}$ can be approximated uniformly by (countably valued) measurable functions. An application of Corollary 3.1.5 completes the proof. □

The proof of the Pettis measurability theorem can be used to obtain additional useful consequences. We record these in a separate proposition as follows.

Proposition 3.1.8 *A function $f : X \to V$ is measurable if and only if there is a set $N \subset X$ of measure zero such that the restriction $f|_{X\setminus N}$ can be approximated uniformly by countably valued measurable functions $X \to V$. Moreover, $\varphi \circ f : X \to W$ is measurable whenever $f : X \to V$ is measurable, W is a Banach space, and $\varphi : V \to W$ is continuous.*

Proof The first assertion is explicitly contained in the proof of the implication (iii) $\Rightarrow$ (i) in the Pettis measurability theorem.

The second assertion is clear by the same theorem, once we observe that $\varphi \circ f : X \to W$ is essentially separably valued under the given hypotheses and satisfies the condition (ii) of the theorem.

The proposition follows. □

We return to the approximation of measurable functions in the context of metric measure spaces later in this chapter. In that context, something more can be said. See in particular Section 3.3.

3.2 Integrable functions and the spaces $L^p(X : V)$

In this section we continue to assume that $X = (X, \mu)$ is a complete, σ-finite measure space and that V is a Banach space.

Bochner integrability

Suppose that

$$f : X \to V, \qquad f = \sum_{i=1}^{n} v_i \chi_{E_i},$$

is a simple function, with the sets E_i measurable and pairwise disjoint, and suppose that $v_i = 0$ for each i with $\mu(E_i) = \infty$. We define the *integral of f over X* as follows:

$$\int_X f \, d\mu := \sum_{i=1}^{n} \mu(E_i) v_i.$$

Then $\int_X f \, d\mu$ is well defined as an element of V, $|f|$ is measurable, and we observe that

$$\left| \int_X f \, d\mu \right| \le \int_X |f| \, d\mu = \sum_{i=1}^{n} \mu(E_i)|v_i| < \infty. \tag{3.2.1}$$

We call such a simple function f *integrable.*

Integrals for general measurable functions can now be defined in analogy with the scalar-valued case. A measurable function $f : X \to V$ is said to be *(Bochner) integrable* if there exists a sequence (f_j) of integrable simple functions such that

$$\lim_{j \to \infty} \int_X |f - f_j| \, d\mu = 0;$$

notice that the real-valued function $x \mapsto |f(x) - f_j(x)|$ is measurable by Remark 3.1.1. If f is integrable then the *(Bochner) integral* of f over X is defined as follows:

$$\int_X f \, d\mu = \lim_{j \to \infty} \int_X f_j \, d\mu.$$

It is straightforward to verify using (3.2.1) that this limit exists as an element of V, independently of the choice of the sequence (f_j).

If $E \subset X$ is measurable and $f : E \to V$ is a function, we say that f is *integrable over* E if the function $f\chi_E : X \to V$ is integrable, where (with a slight abuse of notation) the function $f\chi_E$ is defined by $f\chi_E(x) := f(x)$ for $x \in E$ and $f\chi_E(x) := 0$ for $x \notin E$. Then we set

$$\int_E f\, d\mu := \int_X f\chi_E\, d\mu\,.$$

We could also consider E, together with the restriction of the measure of X to E, to be a measure space in its own right; the corresponding Bochner integral $\int_E f\, d\mu$ is consistent with the above construction of $\int_E f\, d\mu$. By using the preceding definitions and (3.2.1) we find, by a simple limit argument, that

$$\left|\int_E f\, d\mu\right| \leq \int_E |f|\, d\mu \tag{3.2.2}$$

for each measurable set $E \subset X$ and for each function f that is integrable over E. In particular, if f is integrable then

$$\lim_{\mu(E)\to 0} \int_E f\, d\mu = 0 \tag{3.2.3}$$

by standard Lebesgue theory.

The following observation can be used in applications where we need to reduce to the case when X has finite measure.

Proposition 3.2.4 *Assume that $X = \bigcup_{k=1}^{\infty} X_k$, where X_k, $k = 1, 2, \ldots,$ are pairwise disjoint measurable sets, that $f : X \to V$ is integrable over each X_k and that*

$$\sum_{k=1}^{\infty} \int_{X_k} |f|\, d\mu < \infty. \tag{3.2.5}$$

Then f is integrable over X, and

$$\int_X f\, d\mu = \sum_{k=1}^{\infty} \int_{X_k} f\, d\mu. \tag{3.2.6}$$

Proof By Remark 3.1.6 we see that f is measurable on X. Let $\epsilon > 0$ and choose n such that

$$\sum_{k=n+1}^{\infty} \int_{X_k} |f|\, d\mu < \epsilon.$$

For each $k = 1, \ldots, n$, let g_k be a simple function on X_k with

$$\int_{X_k} |f - g_k|\, d\mu < 2^{-k}\epsilon.$$

Then $g = \sum_{k=1}^{n} g_k \chi_{X_k}$ is a simple function and

$$\begin{aligned}\int_X |f - g| \, d\mu &= \sum_{k=1}^{\infty} \int_{X_k} |f - g| \, d\mu \\ &= \sum_{k=1}^{n} \int_{X_k} |f - g_k| \, d\mu + \sum_{k=n+1}^{\infty} \int_{X_k} |f| \, d\mu \\ &< \sum_{k=1}^{n} 2^{-k}\epsilon + \epsilon < 2\epsilon.\end{aligned}$$

Thus f is integrable over X. To see that (3.2.6) holds, we can easily deduce from the definitions, from (3.2.2), and from (3.2.5) that

$$\lim_{n\to\infty} \left| \int_X f \, d\mu - \sum_{k=1}^{n} \int_{X_k} f \, d\mu \right| \leq \lim_{n\to\infty} \sum_{k=n+1}^{\infty} \int_{X_k} |f| \, d\mu = 0.$$

The proposition follows. □

Proposition 3.2.7 *Bochner-integrable functions are precisely those measurable functions f whose norm $|f|$ is integrable.*

Proof If f is Bochner integrable and (f_j) is an approximating sequence of simple functions then

$$\int_X |f| \, d\mu \leq \int_X |f - f_j| \, d\mu + \int_X |f_j| \, d\mu < \infty$$

for all j.

To prove the converse, suppose that f is measurable with $\int_X |f| \, d\mu < \infty$. By Proposition 3.2.4 there is no loss of generality in assuming that $\mu(X) < \infty$. Fix $\epsilon > 0$. By Proposition 3.1.8 we may choose a countably valued measurable function $g : X \to V$ such that $|f(x) - g(x)| < \epsilon$ for every $x \in X \setminus N$, where $N \subset X$ has measure zero. Then

$$\int_X |g| \, d\mu \leq \int_X |f| \, d\mu + \epsilon \, \mu(X) < \infty.$$

Hence we can choose a $\delta > 0$ such that $\int_F |g| \, d\mu < \epsilon$ whenever $F \subset X$ is measurable and satisfies $\mu(F) < \delta$. Next, partition X measurably, i.e. set $X = E \cup F$, in such a way that $g_1 := g\chi_E$ has finite range and $\mu(F) < \delta$. Then

$$\begin{aligned}\int_X |f - g_1| \, d\mu &\leq \int_X |f - g| \, d\mu + \int_X |g - g_1| \, d\mu \\ &\leq \int_X |f - g| \, d\mu + \int_F |g| \, d\mu \leq \epsilon \, \mu(X) + \epsilon \, .\end{aligned}$$

This shows that f is integrable, and the proof of the proposition is complete. □

Remark 3.2.8 If $T : V \to W$ is a bounded linear operator between Banach spaces and if $f : X \to V$ is integrable then $T \circ f : X \to W$ is also integrable and

$$T\left(\int_X f \, d\mu\right) = \int_X T \circ f \, d\mu. \tag{3.2.9}$$

This follows directly from the definitions for simple functions and via approximations for more general functions. In particular, we have that

$$\left\langle v^*, \int_X f \, d\mu \right\rangle = \int_X \langle v^*, f \rangle \, d\mu \tag{3.2.10}$$

for each $v^* \in V^*$ and integrable f.

Mean value Let $E \subset X$ be measurable with finite and positive measure, and let $f : E \to V$ be integrable over E. The *mean value of f over E* is the vector

$$f_E := \fint_E f \, d\mu := \frac{1}{\mu(E)} \int_E f \, d\mu. \tag{3.2.11}$$

It is easy to see that the mean value f_E is always in the closure of the convex hull of $f(E)$ in V. Indeed, the closure of the convex hull of a set A in V can be characterized as the set of those vectors $w \in V$ such that

$$\inf_{v \in A} \langle v^*, v \rangle \leq \langle v^*, w \rangle \leq \sup_{v \in A} \langle v^*, v \rangle \tag{3.2.12}$$

for all $v^* \in V^*$. The preceding claim follows from this characterization and from (3.2.10).

By an argument similar to that for (3.2.9), one concludes that if $T : V \to W$ is a bounded linear map then

$$T\left(\fint_E f \, d\mu\right) = \fint_E T \circ f \, d\mu.$$

L^p-spaces of vector-valued functions

The classes of V-valued p-integrable functions are defined in the usual manner. For $1 \leq p < \infty$, we denote by

$$L^p(X : V) = L^p(X, \mu : V)$$

the vector space of all equivalence classes of measurable functions $f : X \to V$ for which $\int_X |f|^p \, d\mu < \infty$. Two functions are declared equivalent if they agree almost everywhere. As in the case of real-valued functions, we speak about functions, rather than equivalence classes, in L^p and make no notational distinction. This is done with the understanding that such functions are well defined only up to sets of measure zero. Typically, members of the various L^p-spaces are called *p-integrable functions* in this book.

Endowed with the norm

$$\|f\|_p := \|f\|_{L^p(X:V)} := \left(\int_X |f|^p \, d\mu\right)^{1/p}$$

we have that $L^p(X : V)$ is a Banach space. Using the characterization of complete norms in Section 2.1, the proof of this assertion is identical with the proof for the real-valued case [86, p. 175].

By Proposition 3.2.7, $L^1(X : V)$ coincides with the class of Bochner-integrable functions. We have a similar characterization in terms of simple functions for p-integrable V-valued functions. The following proposition is proved in just the same way as Proposition 3.2.7, with some obvious modifications.

Proposition 3.2.13 *Let $1 \leq p < \infty$. A measurable function $f : X \to V$ belongs to $L^p(X : V)$ if and only if there exists a sequence (f_k) of p-integrable simple functions, $f_k : X \to V$, such that*

$$\lim_{k\to\infty} \int_X |f - f_k|^p \, d\mu = 0. \tag{3.2.14}$$

For $p = \infty$ we denote by $L^\infty(X : V)$ the vector space of (equivalence classes of) essentially bounded measurable functions $f : X \to V$, endowed with the norm

$$\begin{aligned}\|f\|_\infty = \|f\|_{L^\infty(X:V)} &:= \operatorname{ess\,sup}_{x\in X} |f(x)| \\ &= \sup\big\{\lambda \in \mathbb{R} : \mu(\{x \in X : |f(x)| > \lambda\}) \neq 0\big\}.\end{aligned}$$

Then $L^\infty(X : V)$ is a Banach space as well.

If $V = \mathbb{R}$, we simply write $L^p(X) = L^p(X : \mathbb{R})$ in accordance with (2.1.11).

We say that a measurable function $f : X \to V$ is *locally (Bochner) integrable* if every point in X has a neighborhood on which f is integrable. Also, f is *locally p-integrable* if $|f|^p$ is locally integrable as a real-valued function. The self-explanatory notation $L^p_{\mathrm{loc}}(X : V)$ and $L^p_{\mathrm{loc}}(X)$ is used.

The following analog of Proposition 2.3.13 holds for Banach space-valued functions; the proof of Proposition 2.3.13 applies verbatim.

Proposition 3.2.15 *Let $1 \leq p \leq \infty$, and let $(f_i) \subset L^p(X : V)$ be a sequence converging to f in $L^p(X : V)$. Then (f_i) has a subsequence (f_{i_j}) with the following property: for every $\epsilon > 0$ there exists a set $E_\epsilon \subset X$ such that $\mu(E_\epsilon) < \epsilon$ and $f_{i_j} \to f$ uniformly in $X \setminus E_\epsilon$. In particular, (f_{i_j}) converges to f pointwise almost everywhere in X.*

Remark 3.2.16 Analogously to Remark 2.3.16(a), we observe the following by examining the argument of the proof of Proposition 2.3.13 (applied in the case of Proposition 3.2.15): if X is a topological space and the sequence (f_i) consists of continuous functions then the sets F_j defined in the proof are open. In particular, if continuous functions are dense in $L^p(X : V)$ then every function in $L^p(X : V)$ has a representative with the following property: for every $\epsilon > 0$ there is an open set $O \subset X$ such that $\mu(O) < \epsilon$ and the restriction of the function to the complement of O is continuous. See Proposition 3.3.49 and Corollary 3.3.51 below for a statement in the context of metric measure spaces.

Duality for the spaces $\mathbf{L^p(X : V)}$

It is not difficult to see that $L^q(X : V^*)$ embeds isometrically in $L^p(X : V)^*$, where $1 \leq p < \infty$ and $q = p/(p-1)$. Indeed, if $g \in L^q(X : V^*)$, then the linear operator

$$f \mapsto \int_X \langle g(x), f(x) \rangle \, d\mu(x)$$

defines a bounded operator on $L^p(X : V)$ with norm equal to $\|g\|_q$. Moreover, if $p = 2$ and H is a Hilbert space, then

$$L^2(X : H)^* = L^2(X : H^*) = L^2(X : H).$$

However, when $p \neq 2$ the Banach spaces $L^q(X : V^*)$ and $L^p(X : V)^*$ need not be isometrically equivalent, although this is true for a large class of Banach spaces; it suffices to have V reflexive, for example. For further details and examples, we refer the reader to [75, Chapter 4, Section 1].

We will mostly not need the dual spaces $L^p(X : V)^*$ in this book outside the familiar case $V = \mathbb{R}$.

Pettis integral

In this book, we will work exclusively with measurable functions and the Bochner integral. Before moving on, however, we should point out that there exist alternate integration theories where it is possible to assign values to the integrals of weakly measurable but not necessarily measurable functions. One example of this is the *Pettis integral*, which we will describe briefly. The results of this subsection will not be needed or used elsewhere in this book.

We call a function $f : X \to V$ *weakly integrable* if it is weakly measurable and $\langle v^*, f \rangle \in L^1(X)$ for each $v^* \in V^*$. Given a weakly integrable function f and a measurable set $E \subset X$, consider the linear operator

$$v^* \mapsto \langle v^*, \chi_E f \rangle = \chi_E \langle v^*, f \rangle \tag{3.2.17}$$

from V^* to $L^1(X)$. It is easy to see (using Proposition 2.3.13) that if $v_i^* \to v^*$ in V^* and if $\langle v_i^*, \chi_E f\rangle \to g$ in $L^1(X)$ then $g = \langle v^*, \chi_E f\rangle$. The closed graph theorem, [286, p. 79] or [238, p. 51, Theorem 2.15], applied to the linear map $\Lambda : V^* \to L^1(X)$ given by (3.2.17), now implies that the operator in (3.2.17) is bounded, that is,

$$\left|\int_E \langle v^*, f\rangle \, d\mu\right| \le C|v^*|$$

for some constant $C < \infty$ independent of v^*, and hence the map

$$v^* \mapsto \int_E \langle v^*, f\rangle \, d\mu$$

defines an element of V^{**}. Recalling the canonical embedding of V in V^{**}, we say that f is *Pettis integrable* over E if this element, denoted by

$$(P) - \int_E f \, d\mu,$$

actually lies in V. Because the Pettis integral of f over a set E, if it exists, is characterized by the validity of the identity

$$\left\langle v^*, (P) - \int_E f \, d\mu\right\rangle = \int_E \langle v^*, f\rangle \, d\mu$$

for all $v^* \in V^*$, the uniqueness of the Pettis integral follows from the Hahn–Banach theorem.

Proposition 3.2.18 *Bochner-integrable functions are Pettis integrable, and the two integrals agree for a function that is Bochner integrable.*

Proof Let $f : X \to V$ be a Bochner-integrable function; then f is weakly integrable, by (3.2.10). Moreover, by (3.2.10) we have that

$$\left\langle v^*, \int_E f \, d\mu\right\rangle = \int_E \langle v^*, f\rangle \, d\mu$$

for all $v^* \in V^*$, where $\int_E f \, d\mu$ denotes the Bochner integral. Thus the proposition follows from the discussion immediately before it. □

It is easy to see that the function $f : [0, 1] \to L^\infty([0, 1])$ in Remark 3.1.3(b) is Pettis integrable.

Here is a standard example of a weakly integrable function that is not Pettis integrable. Define $f : (0, 1) \to c_0$ by

$$f(t) = (\chi_{(0,1)}(t), 2\chi_{(0,1/2)}(t), \ldots, n\chi_{(0,1/n)}(t), \ldots),$$

and let $v^* = (\alpha_1, \alpha_2, \ldots, \alpha_n, \ldots) \in c_0^* = l^1$. Then

$$\int_0^1 \langle v^*, f(t)\rangle\, dt = \int_0^1 \sum_{n=1}^{\infty} \alpha_n n \chi_{(0,1/n)}(t)\, dt$$
$$= \langle v^*, (1, 1, \ldots, 1, \ldots)\rangle,$$

where $(1, 1, \ldots, 1, \ldots) \in l^\infty \setminus c_0 = c_0^{**} \setminus c_0$. This example is typical in the following sense: if V contains no isometric copy of c_0 then every weakly integrable function $f : X \to V$ is Pettis integrable [75, Chapter 2, Theorem 7].

3.3 Metric measure spaces

In this section, we define and discuss metric measure spaces. The concept of a metric measure space plays a central role in this book, and it is worthwhile to spend some time on the basics. In particular, we will discuss in considerable detail the theory of Borel measures in topological and metric spaces.

Measures and outer measures

A collection $\mathcal{M}$ of subsets of a set X is said to be a *σ-algebra* if X belongs to $\mathcal{M}$, $X \setminus A \in \mathcal{M}$ whenever $A \in \mathcal{M}$, and $\mathcal{M}$ is closed under countable unions. In books on probability theory, σ-algebras are often called σ-fields. A *measure* on X is a nonnegative function μ defined on a σ-algebra $\mathcal{M}$ such that $\mu(\emptyset) = 0$ and μ is *countably additive*, that is, $\mu(\bigcup_i A_i) = \sum_i \mu(A_i)$ whenever A_i, $i \in \mathcal{F} \subset \mathbb{N}$, is a countable pairwise disjoint subcollection of the elements of $\mathcal{M}$. An *outer measure* on a set X is a function μ that is defined on all subsets of X, takes values in $[0, \infty]$, satisfies $\mu(\emptyset) = 0$, and is *countably subadditive*:

$$\mu(A) \leq \sum_{E \in \mathcal{F}} \mu(E)$$

whenever $\mathcal{F}$ is a countable collection of subsets of X whose union contains A. For $A \subset X$, the number $\mu(A) \in [0, \infty]$ is called the *measure of* A, or the *μ-measure of* A if μ needs to be mentioned. We also use self-explanatory phrases such as "A has finite measure" or "A is of measure zero". We say that μ is *nontrivial* if $\mu(X) > 0$.

Every outer measure has its σ-algebra of *μ-measurable sets*; these are the sets $A \subset X$ that satisfy

$$\mu(T) = \mu(T \cap A) + \mu(T \setminus A) \tag{3.3.1}$$

for each $T \subset X$. If the measure μ is clear from the context, we speak of *measurable sets* for simplicity. Note that, by subadditivity, a set $A \subset X$ is measurable if and only if

$$\mu(T) \geq \mu(T \cap A) + \mu(T \setminus A) \tag{3.3.2}$$

for each $T \subset X$. Also note that sets of measure zero are always measurable and that we can integrate with respect to μ only over measurable sets.

Every (countably additive) measure μ defined on a σ-algebra $\mathcal{M}$ of measurable subsets of a set X can be extended to an outer measure in a canonical way, by setting

$$\mu(E) := \inf\{\mu(A) : E \subset A \in \mathcal{M}\}. \tag{3.3.3}$$

It is readily checked that the σ-algebra of measurable sets for this extension always contains $\mathcal{M}$.

It is common practice in modern geometric analysis not to make a distinction between an outer measure and a measure. In this book, we will follow the same practice. Thus, abusing the preceding terminology, from now on:

By a measure we mean an outer measure.

In particular, the term "outer" will not be used hereafter in this context.

For example, it is understood that Lebesgue measure in $\mathbb{R}^n$ is defined on all subsets of $\mathbb{R}^n$ via formula (3.3.3). More generally, every measure μ that is perhaps defined only on some σ-algebra $\mathcal{M}$ is automatically understood as defined on all sets by the formula (3.3.3).

If μ is a measure on a set X and $f : X \to [-\infty, \infty]$ is a function, we say that f is *μ-measurable*, or just *measurable*, if f is measurable with respect to the σ-algebra of μ-measurable sets; that is, we require that $f^{-1}([-\infty, a))$ is μ-measurable for every $a \in \mathbb{R}$. Similarly, if X can be written as a countable union of μ-measurable sets each of finite measure then the measurability of a Banach space-valued function on X is understood with respect to the σ-algebra of μ-measurable sets as in Section 3.1.

A measure on a set X is said to be *σ-finite* if X can be expressed as a countable union of measurable sets each of which has finite measure.

A measure on a topological space is said to be *locally finite* if every point in the space has a neighborhood of finite measure.

Borel sets

Every topological space has its natural σ-algebra of *Borel sets*; this is the σ-algebra generated by open sets. That is, the σ-algebra of Borel sets is the smallest σ-algebra containing all the open subsets. A *Borel partition* of a topological space is a decomposition of the space into pairwise disjoint Borel sets.

We record the following observation about Borel sets on subspaces.

Lemma 3.3.4 *If Y is a subspace of a topological space Z then the Borel sets of Y are precisely of the form $B \cap Y$, where $B \subset Z$ is Borel.*

Proof It is readily checked that the collection

$$\{B \cap Y : B \subset Z \text{ is a Borel set}\}$$

is a σ-algebra of subsets of Y, containing every open subset of Y. On the other hand, it is also readily checked that the collection

$$\{B \subset Z : B \cap Y \text{ is a Borel set in } Y\}$$

is a σ-algebra of subsets of Z, containing every open subset of Z. The lemma follows from these two remarks. □

Borel measures

A measure on a topological space is called a *Borel measure* if the Borel sets are measurable; this is tantamount to saying that the open sets are measurable. A Borel measure is further called *Borel regular* if every set is contained in a Borel set of equal measure.

Essentially all Borel measures that arise in geometry and analysis are Borel regular measures. For example, Lebesgue measure in $\mathbb{R}^n$ is Borel regular and so are all Hausdorff measures in a metric space. (See Section 4.3.) Also, it should be noted that if μ is a measure initially defined on the σ-algebra of Borel sets of a topological space then its (automatic) extension, given in (3.3.3), is Borel regular. Some Borel measures that are not Borel regular are given in Examples 3.3.17(a) below.

The *Carathéodory criterion* can be used to establish whether a measure is Borel in the context of metric spaces.

Lemma 3.3.5 *A measure μ on a metric space (X, d) is a Borel measure if and only if*

$$\mu(E_1 \cup E_2) = \mu(E_1) + \mu(E_2) \tag{3.3.6}$$

for every pair of sets $E_1, E_2 \subset X$ such that $\operatorname{dist}(E_1, E_2) > 0$.

Proof Let μ be a measure in a metric space (X, d). Assume first that μ is a Borel measure. Then pick two sets $E_1, E_2 \subset X$ with $\operatorname{dist}(E_1, E_2) > 0$. Let O

be an open set containing E_1 such that $O \cap E_2 = \emptyset$. Because O is measurable, we have

$$\mu(E_1 \cup E_2) = \mu((E_1 \cup E_2) \cap O) + \mu((E_1 \cup E_2) \setminus O) = \mu(E_1) + \mu(E_2),$$

which gives (3.3.6).

Assume next that (3.3.6) holds. To show that μ is a Borel measure, it suffices to show that closed sets are measurable. Thus, pick a closed set $C \subset X$ closed, and let $T \subset X$ be arbitrary; we will prove that

$$\mu(T) \geq \mu(T \cap C) + \mu(T \setminus C), \tag{3.3.7}$$

which suffices by (3.3.2). To this end, define

$$T_i := \{x \in T \setminus C : \text{dist}(x, C) \geq 1/i\}$$

for $i = 1, 2, \ldots$ Then $T_i \subset T_{i+1} \subset T \setminus C$ for every $i \geq 1$ and $T \setminus C = \bigcup_{i=1}^{\infty} T_i$ because C is closed. Because $\text{dist}(T \cap C, T_i) > 0$, we have by the assumption (3.3.6) that

$$\mu(T) \geq \mu((T \cap C) \cup T_i) = \mu(T \cap C) + \mu(T_i) \tag{3.3.8}$$

for each $i \geq 1$. We now claim that

$$\lim_{i\to\infty} \mu(T_i) \geq \mu(T \setminus C). \tag{3.3.9}$$

By (3.3.8), inequality (3.3.7) follows from (3.3.9). To prove (3.3.9), write $A_i := T_{i+1} \setminus T_i$ and observe that $T \setminus C = T_{2k} \cup A_{2k} \cup A_{2k+1} \cup \cdots$ for every $k \geq 1$. Inequality (3.3.9) obviously holds if $\mu(T_i) = \infty$ for some $i \geq 1$; we may thus assume that $\mu(A_i) < \infty$ for every $i \geq 1$. Next, we have that

$$\mu(T \setminus C) \leq \mu(T_{2k}) + \sum_{i=k}^{\infty} \mu(A_{2i}) + \sum_{i=k}^{\infty} \mu(A_{2i+1}). \tag{3.3.10}$$

If both sums appearing in (3.3.10) converge then (3.3.9) follows. Assume then that

$$\sum_{i=1}^{\infty} \mu(A_{2i}) = \infty.$$

Therefore, by using the fact that $\text{dist}(A_{2i}, A_{2j}) > 0$ whenever $i \neq j$, we have by the assumption (3.3.6) that

$$\infty = \lim_{k\to\infty} \sum_{i=1}^{k} \mu(A_{2i}) = \lim_{k\to\infty} \mu(A_2 \cup \cdots \cup A_{2k}) \leq \lim_{k\to\infty} \mu(T_{2k+1}).$$

Thus (3.3.9) again holds. Finally, if the second sum in (3.3.10) diverges, we argue similarly and conclude that (3.3.9) holds in all cases. This completes the proof of (3.3.7), and the lemma follows. □

Restriction and extension of Borel measures

Every measure μ on a set Z determines a measure on each subset Y of Z simply by restricting μ to the subsets of Y. We will call such a measure on Y the *restriction of* μ *to* Y and when necessary use the notation μ_Y.

Lemma 3.3.11 *Let μ be a Borel measure on a topological space Z and let $Y \subset Z$. Then the restriction μ_Y is a Borel measure on Y. If moreover μ is Borel regular then so is μ_Y.*

Proof To prove the first assertion, let $U \subset Y$ be an open set and let $T \subset Y$. Then $U = O \cap Y$ for some open set $O \subset Z$, and we have that

$$\begin{aligned}\mu_Y(T) &= \mu(T) = \mu(T \cap O) + \mu(T \setminus O)\\ &= \mu(T \cap U) + \mu(T \setminus U) = \mu_Y(T \cap U) + \mu_Y(T \setminus U).\end{aligned}$$

This proves that U is measurable. Next, suppose that μ is Borel regular; let $E \subset Y$ and let $B \subset Z$ be a Borel set containing E such that $\mu(E) = \mu(B)$. Then $B \cap Y$ is a Borel set in Y (Lemma 3.3.4), $E \subset B \cap Y$, and

$$\mu_Y(E) \leq \mu_Y(B \cap Y) \leq \mu(B) = \mu(E) = \mu_Y(E).$$

The lemma follows from these remarks. □

A different type of restriction can be defined as follows. Let μ be a measure on a set Z and let $Y \subset Z$. Then we can define a measure $\mu \lfloor Y$ on Z as follows:

$$\mu \lfloor Y(E) := \mu(E \cap Y) \tag{3.3.12}$$

for $E \subset Z$.

Lemma 3.3.13 *Let μ be a Borel measure on a topological space Z and let $Y \subset Z$. Then the measure $\mu \lfloor Y$ is a Borel measure on Z. If μ is also Borel regular then $\mu \lfloor Y$ is Borel regular if and only if Y admits a Borel partition $Y = B_0 \cup N$ such that B_0 is a Borel set in Z and $\mu(N) = 0$.*

Proof The proof of the first assertion is similar to that of Lemma 3.3.11 and is thus left to the reader. Assume next that μ is Borel regular. If $\mu \lfloor Y$ is Borel regular then there is a Borel set B containing $Z \setminus Y$ such that $\mu \lfloor Y(B) = \mu \lfloor Y(Z \setminus Y) = 0$. Hence $B_0 := Z \setminus B$ and $N := Y \setminus B_0$ provide the desired partition. Conversely, assume that such a partition, $Y = B_0 \cup N$, exists and let $E \subset Z$. Because μ is Borel regular, we can choose Borel sets $B \supset E \cap Y$

and $B' \supset N$ such that $\mu(B) = \mu(E \cap Y)$ and $\mu(B') = \mu(N) = 0$. Thus $B_1 := B' \cup B \cup (Z \setminus B_0)$ is a Borel set in Z containing E, and

$$\mu\lfloor Y(E) \leq \mu\lfloor Y(B_1) \leq \mu\lfloor Y(B') + \mu\lfloor Y(B) + \mu\lfloor Y(Z \setminus B_0)$$
$$\leq \mu(B') + \mu(B) + \mu(N) = \mu(E \cap Y) = \mu\lfloor Y(E).$$

This completes the proof. □

Next we consider extensions of measures. Every measure μ on a subset Y of a set Z can be extended to a measure $\overline{\mu}$ on Z by the formula

$$\overline{\mu}(E) := \mu(E \cap Y) \tag{3.3.14}$$

for $E \subset Z$. Observe that if μ is a measure on Z and $Y \subset Z$ then

$$\mu\lfloor Y = \overline{\mu_Y}, \tag{3.3.15}$$

so that $\mu\lfloor Y$ equals "the extension of the restriction".

The next lemma is a special case of Proposition 3.3.21 below. We also note that in turn Lemma 3.3.13 is a special case of Lemma 3.3.16, by (3.3.15); in fact, the proofs are similar. We refer to Proposition 3.3.21 for the proof of Lemma 3.3.16.

Lemma 3.3.16 *Let μ be a Borel measure on a subset Y of a topological space Z. Then the extension $\overline{\mu}$ as given in* (3.3.14) *determines a Borel measure on Z. If μ is also Borel regular then the extension $\overline{\mu}$ is Borel regular if and only if Y admits a Borel partition $Y = B_0 \cup N$ such that B_0 is a Borel set in Z and $\mu(N) = 0$.*

Examples 3.3.17 (a) Let Y be a non-Lebesgue-measurable subset of $Z = [0, 1]$ and let μ be the restriction of the Lebesgue measure m_1 to Y. Then μ is Borel regular by Lemma 3.3.11. Because Y does not admit a decomposition into a Borel set and a Lebesgue null set, as in Lemma 3.3.16, it follows that the extension $\overline{\mu}$ of μ is not Borel regular. It also follows that $m_1\lfloor Y = \overline{\mu}$ is not Borel regular on $[0, 1]$.

(b) In general, we cannot choose the set N in Lemma 3.3.16 to be a Borel subset of Z. For example, let $Y \subset [0, 1]$ be a Lebesgue-measurable set that is not Borel, and let μ be the restriction of the Lebesgue measure m_1 to Y. Then μ is Borel regular on Y by Lemma 3.3.11. By choosing a Borel set $B \subset [0, 1]$ such that $m_1(B) = m_1([0, 1] \setminus Y)$ and $[0, 1] \setminus Y \subset B$, we find that $\mu(B \cap Y) = 0$, which gives the decomposition $Y = ([0, 1] \setminus B) \cup (B \cap Y)$ into a Borel subset of $[0, 1]$ and a set of μ-measure zero. Lemma 3.3.16 therefore implies that the extension $\overline{\mu}$ of μ to $[0, 1]$ is Borel regular; however, we can see that Y, and hence $Y \cap B$, is not a Borel subset of $[0, 1]$.

Given the above discussion of extensions and restrictions of measures, it is natural to ask about extensions of measurable functions.

Lemma 3.3.18 *Suppose that U is a measurable subset of X and that $f : U \to [-\infty, \infty]$ is a measurable function. Then the zero-extension of f to X, given by $F : X \to [-\infty, \infty]$ with $F(x) = f(x)$ if $x \in U$ and $F(x) = 0$ if $x \in X \setminus U$, is also measurable.*

Proof To show that F is measurable, it suffices to show that for each $t \in \mathbb{R}$ the super level set $\{x \in X : F(x) > t\}$ is a measurable subset of X. Observe that if $t \geq 0$ then this super level set is merely $\{x \in U : f(x) > t\}$, and if $t < 0$ then it is $\{x \in U : f(x) > t\} \cup (X \setminus U)$. Hence, to prove that F is measurable it suffices to show that if $E \subset U$ is a μ_U-measurable subset of U then it is a μ-measurable subset of X (recall the definition of the restricted measure μ_U from Lemma 3.3.11). To this end, let $A \subset X$. Then, because E is μ_U-measurable, by (3.3.1) we know that

$$\mu(A \cap U \cap E) + \mu(A \cap U \setminus E) = \mu(A \cap U).$$

Therefore, because $A \cap E = (A \cap U) \cap E$ and $A \setminus E = (A \setminus U) \cup (A \cap U \setminus E)$, we have

$$\begin{aligned}
\mu(A) &\leq \mu(A \cap E) + \mu(A \setminus E) \\
&= \mu(A \cap U \cap E) + \mu((A \setminus U) \cup (A \cap U \setminus E)) \\
&\leq \mu(A \cap U \cap E) + \mu(A \setminus U) + \mu(A \cap U \setminus E) \\
&= \mu(A \cap U) + \mu(A \setminus U) \\
&= \mu(A),
\end{aligned}$$

where the last equality follows from the fact that U is a μ-measurable subset of X. Thus we see that for each $A \subset X$ the equality $\mu(A \cap E) + \mu(A \setminus E) = \mu(E)$ holds; therefore E is a μ-measurable subset of X. This completes the proof of the lemma. □

For future use, we record the following lemma and its proof here.

Lemma 3.3.19 *Fix $1 < p < \infty$, let Ω_n, $n = 1, \dots,$ be a sequence of measurable subsets of X with $\Omega_n \subset \Omega_{n+1}$ for each positive integer n, and suppose that we have a corresponding sequence of measurable functions $g_n : \Omega_n \to [-\infty, \infty]$ such that $\int_{\Omega_n} |g_n|^p \, d\mu \leq 1$ for each positive integer n. Then, for $\Omega = \bigcup_{n \in \mathbb{N}} \Omega_n$, there is a subsequence (g_{n_k}) and a function $g_\infty \in L^p(\Omega)$ such that for each $k_0 \in \mathbb{N}$ the sequence $(g_{n_{k+k_0}})$ converges weakly to g_∞ in $L^p(\Omega_{k_0})$. Furthermore, $\int_\Omega g_\infty^p \, d\mu \leq 1$.*

Proof Since each Ω_n is a measurable set and g_n is a measurable function on Ω_n, by Lemma 3.3.18 it follows that the zero-extension of g_n to X is measurable; therefore the restriction of this extension to Ω is a measurable function on Ω. We denote this function by G_n. Note that $\int_\Omega |G_n|^p \, d\mu = \int_{\Omega_n} |g_n|^p \, d\mu \le 1$; that is, the sequence (G_n) is a bounded sequence in $L^p(\Omega)$. Because $1 < p < \infty$, we may apply Proposition 2.4.19 together with Theorem 2.4.1 to this sequence to obtain a subsequence (G_{n_k}) and a function $G_\infty \in L^p(\Omega)$ such that (G_{n_k}) converges weakly to G_∞ in $L^p(\Omega)$. The proof of the lemma is now complete upon noting that $(G_{n_{k+k_0}})$ also weakly converges to G_∞ in Ω and hence also in Ω_{k_0}. □

Borel functions

A function from one topological space into another is said to be a *Borel function* if the preimage of every open set is a Borel set. Under a Borel function the preimage of every Borel set is a Borel set. To see this, we generalize the argument given earlier: if $f : Y \to Z$ is a Borel function then the collection of all sets B in Z for which $f^{-1}(B)$ is a Borel set in Y is a σ-algebra; because this collection contains all open sets by definition, it contains all Borel sets.

We have, in particular, that continuous functions are Borel functions and that the composition of two Borel functions is a Borel function.

A function f is said to be a *Borel bijection* if it is a bijection between topological spaces such that both f and f^{-1} are Borel functions. For example, homeomorphisms are always Borel bijections.

When we consider extended real-valued functions, i.e., functions with values in $[-\infty, \infty]$, this target space is understood to have the natural topology extending the topology from $\mathbb{R}$. Thus, a basis for this topology consists of sets, or intervals, of the form (a, b), $[-\infty, a)$, and $(b, \infty]$, where $a, b \in \mathbb{R}$. It follows that an extended real-valued function f on a topological space is a Borel function if and only if the preimage under f of any of the three types of basis interval is a Borel set.

Push-forward measures

With every function f from a set Y to a set Z, and with every measure μ on Y, we can associate a *push-forward measure* $f_\# \mu$ on Z:

$$f_\# \mu(E) := \mu(f^{-1}(E)) \tag{3.3.20}$$

for $E \subset Z$. For example, the extension $\overline{\mu}$ given in (3.3.14) is precisely the push-forward of μ under the inclusion $Y \to Z$.

The next proposition explains some properties of Borel measures and their push-forwards; compare Lemma 3.3.16 and Examples 3.3.17.

Proposition 3.3.21 *Let Y and Z be topological spaces, let μ be a Borel measure on Y, and let $f : Y \to Z$ be a Borel function. Then $f_{\#}\mu$ is a Borel measure on Z. If $f_{\#}\mu$ is also Borel regular then Y admits a Borel partition $Y = B_0 \cup N$ such that $f(B_0)$ is a Borel set in Z and $\mu(N) = 0$.*

Finally, assume that μ is Borel regular, that f determines a Borel bijection between Y and $f(Y)$, and that Y admits a Borel partition $Y = B_0 \cup N$ such that $f(B_0)$ is a Borel set in Z and $\mu(N) = 0$. Then $f_{\#}\mu$ is Borel regular.

Proof It is clear that $f_{\#}\mu$ is a measure on Z. To check that open sets are measurable, let $T \subset Z$ and let $U \subset Z$ be open; then

$$\begin{aligned} f_{\#}\mu(T) &= \mu(f^{-1}(T)) = \mu(f^{-1}(T) \cap f^{-1}(U)) + \mu(f^{-1}(T) \setminus f^{-1}(U)) \\ &= \mu(f^{-1}(T \cap U)) + \mu(f^{-1}(T \setminus U)) = f_{\#}\mu(T \cap U) + f_{\#}\mu(T \setminus U) \end{aligned}$$

as desired. Thus $f_{\#}\mu$ is always a Borel measure.

Assume next that $f_{\#}\mu$ is Borel regular. Then there exists a Borel set B in Z containing $Z \setminus f(Y)$ such that

$$\mu(f^{-1}(B)) = f_{\#}\mu(B) = f_{\#}\mu(Z \setminus f(Y)) = 0\,.$$

Now we can set $B_0 := f^{-1}(Z \setminus B)$ and $N := f^{-1}(B)$.

Finally, assume that μ is Borel regular, that $f : Y \to f(Y)$ is a Borel bijection, and that a partition $Y = B_0 \cup N$ is given as in the hypotheses. Let $E \subset Z$. Because μ is Borel regular, there is a Borel set $B' \subset Y$ containing $f^{-1}(E)$ such that $\mu(B') = \mu(f^{-1}(E))$. Because f is a Borel bijection, we deduce from Lemma 3.3.4 that $f(B') = B'' \cap f(Y)$ for some Borel set B'' in Z. Set $B := (Z \setminus f(B_0)) \cup B''$. Then B is a Borel set in Z, $E \subset B$, and

$$\begin{aligned} f_{\#}\mu(E) &\leq f_{\#}\mu(B) = \mu(f^{-1}(B)) \\ &\leq \mu(N) + \mu(B') = \mu(f^{-1}(E)) = f_{\#}\mu(E). \end{aligned}$$

This proves that $f_{\#}\mu$ is Borel regular as required.

The proposition follows. □

Embeddings

In this book we typically use push-forward measures obtained via embeddings of metric spaces. Recall that an *embedding* of a topological space Y into a topological space Z is a map $f : Y \to Z$ that determines a homeomorphism between Y and $f(Y)$. In particular, an embedding $f : Y \to Z$ always determines a Borel bijection between Y and $f(Y)$.

For example, every metric space X canonically embeds into its *metric completion* $\overline{X}$. The embedding $X \to \overline{X}$ is moreover isometric (see Section 4.1).

With a slight abuse of notation, throughout this book we view X as a subset of its completion and thus write $X \subset \overline{X}$.

We will discuss the embeddings of metric spaces in more detail in Chapter 4.

Borel and μ-representatives

Let μ be a Borel measure on a topological space X. Given a function f from X to another topological space, Z, a function $g: X \to Z$ is said to be a *Borel representative* of f, if g is a Borel function that equals f almost everywhere on X. Similarly, if μ is σ-finite and if f is a function from X to some Banach space V, we say that a function $g: X \to V$ is a *μ-representative*, or just a *representative*, of f, if g is measurable (see Section 3.1) and equals f almost everywhere on X. Here, as always, measurability refers to the σ-algebra of μ-measurable sets. Note that if f has a μ-representative then f is itself measurable.

We will also use the self-explanatory term *Lebesgue representative* for extended real-valued functions, or for V-valued functions, that are defined on subsets of $\mathbb{R}^n$ equipped with the Lebesgue n-measure.

Proposition 3.3.22 *Let μ be a Borel-regular σ-finite measure on a topological space X and let $f: X \to [-\infty, \infty]$ be a function. If f is a Borel function then f is the pointwise limit of a sequence of simple (or arbitrary) Borel functions outside a Borel set of measure zero. Conversely, if f admits such an approximation then f can be modified in this Borel set of measure zero in such a way as to become a Borel function in X. Moreover, if f is Borel and real-valued then there is a Borel set $N \subset X$ of measure zero such that $f|_{X \setminus N}$ can be approximated uniformly by countably valued Borel functions $X \to \mathbb{R}$.*

Proof If f is a Borel function, it is in particular measurable and hence the pointwise almost everywhere limit of a sequence of measurable simple functions; these simple functions can be assumed to be Borel by the definition of Borel regular measures, and we can also enclose the set of points of non-convergence in a Borel set of measure zero.

Suppose that f is the pointwise limit of a sequence of simple Borel functions outside a Borel set N of measure zero. We may assume that these simple functions take the constant value zero on N. Then this sequence converges everywhere to a function g which coincides with f outside N. Now the proof of the implication (i) $\Rightarrow$ (ii) in the Pettis measurability theorem in Section 3.1 shows that g is a Borel function in X.

Next, the last assertion follows from the proof of the implication (iii) $\Rightarrow$ (i) in in the Pettis measurability theorem; the functions g_k can be assumed to be

Borel functions and the set N can be assumed to be a Borel set by the Borel regularity.

Finally, assume that f is the pointwise limit of a sequence of arbitrary Borel functions outside a Borel set of measure zero. It is easy to reduce the proof to the case where f is real-valued. Then the assertion is proved in the same way as the corresponding assertion for general measurable functions in Proposition 3.1.8 or Corollary 3.1.5. Note that Egoroff's theorem, see Section 3.1, admits an obvious Borel version, needed in the aforementioned proof.

The proposition follows. □

Proposition 3.3.23 *Let μ be a Borel-regular σ-finite measure on a topological space X. Then every measurable extended real-valued function on X can be modified in a set of measure zero in such a way as to become Borel measurable.*

Proof Let f be measurable; then f is the pointwise limit of a sequence (f_k) of simple functions outside a set of measure zero. Since μ is Borel regular, we may modify these simple functions on sets of measure zero in such a way as to become simple Borel functions. This new sequence will still converge to f outside a set of measure zero. Using the Borel regularity of μ one more time, we may assume further that the convergence occurs outside a Borel set of measure zero. The claim follows from Proposition 3.3.22. □

There are analogous results for Banach space-valued functions, with analogous proofs.

Proposition 3.3.24 *Let μ be a Borel-regular σ-finite measure on a topological space X, let V be a Banach space, and let $f : X \to V$ be a function. Then f has an essentially separably valued Borel representative if and only if f is the pointwise limit of a sequence of simple (or arbitrary) Borel functions outside a Borel set of measure zero. Moreover, if f is an essentially separably valued Borel function then there is a Borel set $N \subset X$ of measure zero such that $f|_{X \setminus N}$ can be approximated uniformly by countably valued Borel functions $X \to V$.*

Balls in metric spaces

We denote open balls in a metric space $X = (X, d)$ by $B(x, r)$. Thus

$$B(x, r) = \{y \in X : d(x, y) < r\},$$

where $x \in X$ is the *center* and $0 < r < \infty$ is the *radius* of $B(x, r)$. We emphasize that a ball as a set does not, in general, uniquely determine center and radius. Therefore, "center" and "radius" actually refer to the notation $B(x, r)$.

Closed balls are denoted by $\overline{B}(x, r)$. Thus

$$\overline{B}(x, r) = \{y \in X : d(x, y) \leq r\},$$

where we assume $r > 0$ so that a closed ball is always a neighborhood of its center. The closed ball $\overline{B}(x, r)$ may be a larger set than the (topological) *closure* $\overline{B(x, r)}$ of $B(x, r)$. If $B = B(x, r)$ is a ball, with center and radius understood, and $\lambda > 0$, we write

$$\lambda B = B(x, \lambda r). \tag{3.3.25}$$

With a small abuse of notation we write $\mathrm{rad}(B)$ for the radius of a ball B; this notation entails that $B = B(x, \mathrm{rad}(B))$ for some $x \in X$, and it is used when the center x is unimportant for the context at hand. We always have

$$\mathrm{diam}(B) \leq 2\,\mathrm{rad}(B),$$

and the inequality is strict in general. Here, and in the rest of the book, the *diameter* of a nonempty set $A \subset X$ is given by

$$\mathrm{diam}(A) := \sup\{d(y, z) : y, z \in A\}.$$

When we use the generic term *ball*, it is understood that this may be open or closed.

5B-covering lemma

We next state and prove a fundamental covering lemma. This lemma will be used on several occasions in this book.

5B-covering lemma *Every family $\mathcal{F}$ of balls of uniformly bounded diameter in a metric space X contains a pairwise disjoint subfamily $\mathcal{G}$ such that for every $B \in \mathcal{F}$ there exists $B' \in \mathcal{G}$ with $B \cap B' \neq \emptyset$ and* $\mathrm{diam}(B) < 2\,\mathrm{diam}(B')$. *In particular, we have that*

$$\bigcup_{B \in \mathcal{F}} B \subset \bigcup_{B \in \mathcal{G}} 5B. \tag{3.3.26}$$

If X is separable then the family $\mathcal{G}$ is necessarily countable.

Proof The last assertion is obvious; there cannot be uncountably many pairwise disjoint balls in a separable metric space.

To prove the first assertion, consider pairwise disjoint subfamilies $\mathcal{B}$ of balls from $\mathcal{F}$ with the following property: if $B \in \mathcal{F}$ then either $B \cap B' = \emptyset$ for all $B' \in \mathcal{B}$ or there is a ball $B' \in \mathcal{B}$ such that $B \cap B' \neq \emptyset$ and

$\operatorname{diam}(B) < 2\operatorname{diam}(B')$. There is at least one such family, namely the one-ball family $\{B'\}$, where $B' \in \mathcal{F}$ is chosen such that

$$\sup_{B \in \mathcal{F}} \operatorname{diam}(B) < 2\operatorname{diam}(B') .$$

Moreover, such families form a partially ordered set by inclusion and every chain of such families has an upper bound, namely the union of all the families in the chain. By the Hausdorff maximality principle, or Zorn's lemma, there is a family $\mathcal{G}$ as required. (See [237, p. 87] or [214, p. 69] for a discussion of the Hausdorff maximality principle, known also as the maximum principle.) Indeed, if there is a ball $B \in \mathcal{F}$ such that no ball in $\mathcal{G}$ intersects B, we can then choose a $B_0 \in \mathcal{F}$ that does not intersect any ball from the collection $\mathcal{G}$ and at the same time satisfies $\operatorname{diam}(B) \leq 2\operatorname{diam}(B_0)$ whenever $B \in \mathcal{F}$ does not intersect any ball from the collection $\mathcal{G}$. Now the larger collection, $\mathcal{G} \cup \{B_0\}$, would violate the maximality of $\mathcal{G}$. □

Note that the $5B$-covering lemma is a purely metric result; no measures are involved.

We also require the following well-known point set topology result (also a kind of covering theorem). Recall that a topological space is said to have the *Lindelöf property* if from every open cover of the space we can extract a countable subcover.

Lemma 3.3.27 *Every separable metric space has the Lindelöf property.*

Proof We will give a quick proof using the $5B$-covering lemma. Let $X = (X, d)$ be a separable metric space and let $\mathcal{U} = \{U\}$ be an open cover of X. Fix a positive integer k. For every point $x \in X$ choose an open set $U \in \mathcal{U}$ and a ball $B_x = B(x, r_x)$ such that $r_x \leq k$ and $5B_x \subset U$. Then, from the collection $\mathcal{F} := \{B_x : x \in X\}$ we can pick a countable subcollection $\mathcal{G} = \{B_i\}$ such that

$$X = \bigcup_{B_x \in \mathcal{F}} B_x \subset \bigcup_{B_i \in \mathcal{G}} 5B_i .$$

The required countable subcover of $\mathcal{U}$ is obtained by choosing, for each i, a set $U_i \in \mathcal{U}$ such that $5B_i \subset U_i$. The lemma follows by taking the countable union (one per positive integer k) of these countable subcovers. □

From now on, we will use without further mention the elementary fact that *every subset of a separable metric space is separable.* In particular, subspaces of separable metric spaces have the Lindelöf property; that is, separable metric spaces are hereditarily Lindelöf.

Metric measure spaces

A *metric measure space* is defined to be a triple (X, d, μ), where (X, d) is a separable metric space and μ is a nontrivial locally finite Borel-regular measure on X.

Recall that locally finite means in this context (see the text after (3.3.3)) that for every point $x \in X$ there is an $r > 0$ such that $\mu(B(x, r)) < \infty$.

We often write simply X in place of the triple (X, d, μ) or in place of either of the pairs (X, d) and (X, μ) if the distance and measure are either understood from the context or do not need specific labeling.

The Lindelöf property of separable metric spaces (Lemma 3.3.27) immediately yields the following.

Lemma 3.3.28 *Every metric measure space can be written as a countable union of balls each of which has finite measure. In particular, every metric measure space is σ-finite.*

It follows from Lemma 3.3.28, and from the fact that sets of measure zero are always measurable, that the conventions adopted in Sections 3.1 and 3.2 are valid in the context of metric measure spaces.

It is not true that every ball in a metric measure space has finite measure. For example, consider the open unit interval $X = (0, 1)$ equipped with the standard metric and with measure $d\mu(x) = x^{-1}\, dm_1(x)$, where m_1 is the standard Lebesgue measure restricted to $(0, 1)$.

A metric measure space *is not assumed to be complete or even locally complete as a metric space*. Indeed, assuming completeness would exclude many natural metric spaces from our discussion, e.g. Euclidean domains. Also note that while every Borel-regular measure μ on a metric space X admits a natural extension $\overline{\mu}$ to a Borel measure on the metric completion $\overline{X}$ of X by formula (3.3.14), i.e.,

$$\overline{\mu}(E) = \mu(E \cap X), \qquad E \subset \overline{X}, \tag{3.3.29}$$

this extension may fail to be Borel regular. However, if X is a Borel subset of $\overline{X}$, for example if X is locally compact, then $\overline{\mu}$ is Borel regular by Lemma 3.3.13. We elaborate further on extensions below.

Restrictions and extensions

If (X, d, μ) is a metric measure space and if $Y \subset X$ then restriction of the measure μ and the metric d to Y always determines a metric measure space (Y, d, μ) (Lemma 3.3.11). In other words, subsets of metric measure spaces can be regarded naturally as metric measure spaces on their own right. It also follows from this observation that we can express every metric measure space as a countable union of pairwise disjoint measurable subsets each of which constitutes a metric measure space in its own right (the restriction of the ambient measure to each subset being Borel regular by Lemma 3.3.11.) Moreover, the partition can be made in such a way that each piece in the partition has finite measure (Lemma 3.3.28).

However, if (X, d, μ) is a metric measure space and f is an embedding of X in a separable metric space Z then Z equipped with the push-forward measure $f_{\#}\mu$ is not always a metric measure space even if $f_{\#}\mu$ is Borel regular. For example, consider $f : [0, \infty) \to [0, 1], f(x) = (2/\pi)\arctan(x)$. Then the push-forward $f_{\#}m_1$ of Lebesgue measure is not locally finite in $[0, 1]$. On the other hand, if $f(X)$ is a closed subset of Z then Z equipped with the push-forward measure $f_{\#}\mu$ is a metric measure space; see Proposition 3.3.21. Indeed, by this proposition we see that $f_{\#}\mu$ is a Borel regular measure on Z and, because $f(X)$ is a closed subset of Z, every point in $Z \setminus f(X)$ has a neighborhood on which $f_{\#}\mu$ is zero. Furthermore, since μ is locally finite in X and f is an embedding, $f_{\#}\mu$ is locally finite on $f(X)$. Similarly, if (X, d, μ) is a metric measure space, where X is a subset of a separable metric space Z, then Z equipped with the extension $\overline{\mu}$ given in (3.3.14) is not necessarily a metric measure space; see Lemma 3.3.16 and, in particular, Example 3.3.17(a)). Thus some caution is required in extending measures. In practice, however, situations like the one in Example 3.3.17(a) rarely arise.

The assumption that metric measure spaces be separable is a mild one; it is satisfied in most geometrically and analytically interesting cases. Moreover, we have the following fact. (See also Remark 3.3.35 below.)

Lemma 3.3.30 *If μ is a Borel measure on a metric space X, where X can be written as a countable union of open sets with finite measure and where every open ball has positive measure, then X is separable as a metric space.*

Proof Fix, for every positive integer $n = 1, 2, \ldots$, a maximal $1/n$-separated set $N_n = \{x_{ni}\}$. That is, $d(x_{ni}, x_{nj}) \geq 1/n$ for $x_{ni} \neq x_{nj}$ and $X = \bigcup_i B(x_{ni}, 1/n)$. The existence of such a set N_n is easily established by the aid of the Hausdorff maximality principle; see [237, p. 87]. Let $X = \bigcup_k X_k$ be a countable union of open sets with $\mu(X_k) < \infty$. Then $N_n \cap X_k$ must be countable because

$$\infty > \mu(X_k) \geq \sum_{x_{ni} \in N_n \cap X_k} \mu(B(x_{ni}, 1/2n) \cap X_k)$$

with $\mu(B(x_{ni}, 1/2n) \cap X_k) > 0$ whenever x_{ni} are points in $N_n \cap X_k$. It follows that N_n must be countable, and from the construction that the set $\cup_{n=1}^{\infty} N_n$ is dense in X. The lemma follows. □

We also have that every set of locally zero measure in a metric measure space has zero measure.

Lemma 3.3.31 *Let E be a subset of a metric measure space $X = (X, d, \mu)$. If every point $x \in E$ has a neighborhood U_x such that $\mu(E \cap U_x) = 0$ then $\mu(E) = 0$.*

Proof Use the Lindelöf property of separable metric spaces (Lemma 3.3.27) together with countable subadditivity. □

Remark 3.3.32 The statement in Lemma 3.3.31 is not true if (X, d) is allowed to be nonseparable. Let X be an uncountable set equipped with the discrete metric; that is, $d(x, y) = 1$ if $x \neq y$. Then the Borel regular measure μ defined by $\mu(E) = 0$ if E is countable and $\mu(E) = \infty$ if E is uncountable is locally zero, but $\mu(X) = \infty$.

Support of a measure

If (X, d, μ) is a metric measure space, we define the *support* of μ by

$$\operatorname{spt}(\mu) := X \setminus \cup\{O : \; O \subset X \text{ open and } \mu(O) = 0\}. \tag{3.3.33}$$

By Lemma 3.3.31 we have that

$$\mu(X \setminus \operatorname{spt}(\mu)) = 0. \tag{3.3.34}$$

Then $(\operatorname{spt}(\mu), d, \mu)$ is a metric measure space with the property that every ball in it has positive measure. Moreover, if we consider $(\operatorname{spt}(\mu), d, \mu)$ as a metric measure space in its own right then the extension $\overline{\mu}$ of μ from $\operatorname{spt}(\mu)$ to X coincides with the original measure μ on X. Noting that $\operatorname{spt}(\mu)$ is closed in X, this assertion follows from (3.3.34) and Lemma 3.3.16.

Remark 3.3.35 Let μ be a Borel measure on a metric space X such that X can be written as a countable union of open sets with finite measure. Then the support $\operatorname{spt}(\mu)$ of μ is separable by Lemma 3.3.30. However, it is not obvious that $\mu(X \setminus \operatorname{spt}(\mu)) = 0$ in general. We have that $\mu(X \setminus \operatorname{spt}(\mu)) = 0$ if for example X has a dense subset whose cardinality is an Ulam number. For this fact and comments on Ulam numbers, see [83, 2.2.16 and 2.1.6]. Observe that a metric space equipped with a Borel measure need not in general be a metric measure space in the sense of Section 3.3.

Examples of metric measure spaces

The concept of a metric measure space is very general. It embraces most naturally occurring geometric measure spaces in analysis and geometry. Here is a short list of specific examples.

If $X \subset \mathbb{R}^n$ is any subset of positive Lebesgue n-measure then X equipped with the Euclidean distance and with Lebesgue measure m_n is a metric measure space (Lemma 3.3.11). More generally, if X is any subset of a Riemannian manifold of positive Riemannian measure then X equipped with the Riemannian distance and measure is a metric measure space.

If (X, d) is a separable metric space equipped with locally finite and nontrivial α-Hausdorff measure for some $\alpha > 0$, then $(X, d, \mathcal{H}_\alpha)$ is a metric measure space. See Section 4.3 below and [83, Section 2.10.1].

Every locally compact and separable group equipped with a left-invariant metric and left-invariant Haar measure is a metric measure space. See [83, Chapter 2.7] or [86, Section 10.1].

A Banach space equipped with a Gaussian measure constitutes a metric measure space. See [26, Section 6.2].

If (X, d, μ) is any metric measure space and f is a locally integrable non-negative function on X such that $\mu(\{x \in X : f(x) > 0\}) > 0$ then we have a metric measure space (X, d, μ_f), where

$$\mu_f(A) := \int_A f \, d\mu \tag{3.3.36}$$

for measurable $A \subset X$. In this case the function f is called a *density*, or *weight*, and the metric measure space (X, d, μ_f) is said to be *weighted* by f. We also use the notation $f \, d\mu$ for the measure μ_f and write $(X, d, f \, d\mu)$ instead of (X, d, μ_f).

In the preceding, as always when needed, the measures can be extended to all subsets of the space in question by formula (3.3.3).

Approximation of Borel-regular measures

The following proposition records some fundamental approximation properties of Borel-regular measures in metric measure spaces.

Proposition 3.3.37 *Let (X, d, μ) be a metric measure space. Then*

$$\mu(A) = \sup\{\mu(C) : C \subset A, \ C \subset X \text{ closed}\} \tag{3.3.38}$$

for every measurable set $A \subset X$, and

$$\mu(E) = \inf\{\mu(O) : E \subset O, \ O \subset X \text{ open}\} \tag{3.3.39}$$

for every set $E \subset X$. In particular, every measurable set $A \subset X$ contains an F_σ-set F such that $\mu(A) = \mu(F)$, and every set $E \subset X$ is contained in a G_δ-set G such that $\mu(E) = \mu(G)$.

Proof We begin by observing that for every measurable set A of finite measure in X there are Borel sets B and B' in X such that $B' \subset A \subset B$

and $\mu(B') = \mu(A) = \mu(B)$. Indeed B exists, by the definition of Borel regularity; similarly, there is a Borel set $B'' \subset X$ containing $B \setminus A$ such that $\mu(B'') = \mu(B \setminus A) = 0$, and we can set $B' := B \setminus B''$. Moreover, to prove the proposition we may assume that $\mu(X) < \infty$, for the general case can easily be reduced to this case by using Lemma 3.3.28.

The preceding understood, to prove (3.3.38) by replacing A with B' if necessary we may assume that A is a Borel set in X and that $\mu(X) < \infty$. Let us consider the family $\mathcal{F}$ consisting of all subsets of X for which (3.3.38) holds. Obviously this family contains all closed sets. It also contains all open sets, because these can be written as countable unions of closed sets (closed balls in fact) by separability; using $\mu(X) < \infty$, the measure of a countable union of closed sets can be approximated by the measures of finite unions of these closed sets. Next, it is easy to see, using the assumption $\mu(X) < \infty$, that $\mathcal{F}$ is closed under countable unions and intersections. It follows that the family

$$\mathcal{G} := \{A \in \mathcal{F} : X \setminus A \in \mathcal{F}\}$$

is a σ-algebra containing all closed sets. Therefore, $\mathcal{G}$ must contain all Borel sets, and (3.3.38) follows.

To prove (3.3.39), we find by Borel regularity a Borel set E_0 with $E \subset E_0$ and $\mu(E) = \mu(E_0)$. As earlier, we may assume that $\mu(X) < \infty$. Fix $\epsilon > 0$. By (3.3.38) we may choose a closed set $C \subset X \setminus E_0$ such that $\mu(C) > \mu(X \setminus E_0) - \epsilon$. Then $O = X \setminus C$ is open, contains E, and satisfies $\mu(O) < \mu(E_0) + \epsilon = \mu(E) + \epsilon$. This proves (3.3.39).

The proof of the proposition is complete. □

Radon measures

A Borel-regular measure μ on a metric space X is called a *Radon measure* if $\mu(K) < \infty$ for every compact $K \subset X$, if (3.3.39) holds, and if

$$\mu(O) = \sup\{\mu(K) : K \subset O\,,\ K \subset X \text{ compact}\} \tag{3.3.40}$$

for every open set $O \subset X$.

Notice that each Borel measure that satisfies (3.3.40) is actually Borel regular. The counting measure on a topological space is Borel regular, but is a Radon measure only if compact sets are finite sets. We next clarify the relation between Radon measures and general Borel-regular measures in the context of metric measure spaces.

Our first result is the following improvement of (3.3.40) and (3.3.38).

Proposition 3.3.41 *Let $X = (X, d, \mu)$ be a metric measure space with μ a Radon measure. Then*

$$\mu(A) = \sup\{\mu(K) : K \subset A \text{ compact}\} \tag{3.3.42}$$

for every measurable set $A \subset X$.

Proof Again using Lemma 3.3.28, we may assume that $\mu(X) < \infty$. Let $A \subset X$ be measurable and let $\epsilon > 0$. Because μ is Borel regular, it follows from Proposition 3.3.37 that there is a closed set C contained in A and an open set O containing A such that $\mu(O \setminus C) < \epsilon$. By the definition of Radon measures, we can further find a compact set $K \subset O$ such that $\mu(K) > \mu(O) - \epsilon$. Then the compact set $K \cap C \subset A$ satisfies

$$\mu(A) \geq \mu(K \cap C) = \mu(K) - \mu(K \setminus C) > \mu(O) - 2\epsilon \geq \mu(A) - 2\epsilon.$$

The proposition follows. □

Later, in Proposition 3.3.46, we will characterize the metric measure spaces (X, d, μ), where μ is indeed a Radon measure. Notice that in this context the only difference between the two concepts is the difference between (3.3.38) and (3.3.40), since in a metric measure space compact sets always have finite measure because of the local finiteness of μ. Moreover, by (3.3.38) one only needs to verify (3.3.42) for closed sets A.

Remark 3.3.43 If (X, d, μ) is a metric measure space with μ a Radon measure, and if $f : X \to [0, \infty]$ is a locally integrable function, then the measure μ_f given by (3.3.36) is a Radon measure on X. This follows readily from the definitions and from the basic properties of integrals. In the literature on measure theory, μ_f is also denoted $f\, d\mu$.

Proposition 3.3.44 *Let $X = (X, d, \mu)$ be a metric measure space with (X, d) complete. Then μ is a Radon measure. In particular, X can be expressed as a countable union of compact sets plus a set of measure zero.*

Proof First observe that, because μ is locally finite (see Section 3.3), a compact set can be covered by finitely many balls each of which has finite measure; it follows that every compact subset of X has finite measure.

By the comment made just before Remark 3.3.43, we need only show that (3.3.42) holds for closed sets A in X. Moreover, using Lemma 3.3.28, we may assume that $\mu(A) < \infty$.

The preceding understood, let $A \subset X$ be closed and such that $\mu(A) < \infty$ and let $\epsilon > 0$. Because X is separable, we can find, for each $n = 1, 2, \ldots$, a countable collection of closed balls $\overline{B}_{n1}, \overline{B}_{n2}, \ldots$ with centers in A and each with radius n^{-1}, such that

$$A \subset \bigcup_{i=1}^{\infty} \overline{B}_{ni} .$$

Now we choose i_n such that $\mu(A \cap C_n) > \mu(A) - \epsilon\, 2^{-n}$, where $C_n := \overline{B}_{n1} \cup \cdots \cup \overline{B}_{ni_n}$. Next, set

$$K := \bigcap_{n=1}^{\infty} C_n .$$

We claim both that K is compact and that

$$\mu(A \cap K) \geq \mu(A) - \epsilon . \tag{3.3.45}$$

Because A is closed, the proposition follows from these two claims.

To prove the latter claim, observe first that, by monotone convergence,

$$\mu(A \cap K) = \lim_{m \to \infty} \mu(A \cap C_1 \cap \cdots \cap C_m) ,$$

and then that

$$\begin{aligned} \mu(A) &\leq \mu(A \cap C_1 \cap \cdots \cap C_m) + \sum_{n=1}^{m} \mu(A \setminus C_n) \\ &\leq \mu(A \cap C_1 \cap \cdots \cap C_m) + \epsilon . \end{aligned}$$

Hence (3.3.45) follows. Next, to prove that K is compact it suffices to prove that, whenever $\delta > 0$ and $T \subset K$ satisfies $d(x, y) \geq \delta$ for all distinct $x, y \in T$, T is finite. Indeed, this means that K is *totally bounded* and, because K is also closed and hence complete, K must be compact (see [214, Theorem 3.1, p. 275].) To this end, let $\delta > 0$ and T be as above. Now $T \subset C_n$ for each n. In particular, if n is such that $2/n < \delta$ then no two distinct points from T can belong to just one of the finitely many balls $\overline{B}_{ni}$ whose union is C_n. This implies that T is finite as desired.

To complete the proof of the proposition, we note that, since X is separable and μ is locally finite, we can cover X by countably many balls (B_i), each of finite measure. Because μ is Radon, for each positive integer j we can find a compact set $K_{i,j} \subset B_i$ such that $\mu(K_{i,j}) \geq \mu(B_i) - 1/j$. Since both B_i and $K_{i,j}$ are measurable, $\mu(B_i \setminus K_{i,j}) \leq 1/k$. Therefore $\mu(B_i \setminus \bigcup_j K_{i,j}) = 0$, and so we have a countable decomposition $(K_{i,j})$ of X by compact sets such that $\mu(X \setminus \bigcup_i \bigcup_j K_{i,j}) = 0$.

The proof of the proposition is now complete. □

For the next proposition, recall that $\overline{X}$ denotes the metric completion of a metric space X.

Proposition 3.3.46 *Let $X = (X, d, \mu)$ be a metric measure space. Then the following are equivalent:*

(i) *μ is a Radon measure on X;*
(ii) *$\overline{\mu}$ is a Borel regular measure on $\overline{X}$, where $\overline{\mu}$ is given in (3.3.29);*
(iii) *X admits a Borel partition $X = B_0 \cup N$ such that B_0 is a Borel set in $\overline{X}$ and $\mu(N) = 0$;*
(iv) *there exists an embedding f of X in a complete and separable metric space Z such that X admits a Borel partition $X = B_0 \cup N$ where $f(B_0)$ is a Borel set in Z and $\mu(N) = 0$.*

Note in particular that condition (iii) in Proposition 3.3.46 holds if X is a Borel subset of $\overline{X}$. Similarly, condition (iv) holds if $f(X)$ is a Borel subset of Z. See Example 3.3.17(b) for the role of the set N.

Recall that a metric space is *locally complete* if every point in the space has a neighborhood that is complete in the induced topology. For example, every locally compact metric space is locally complete. We have the following noteworthy corollary to Proposition 3.3.46.

Corollary 3.3.47 *Let $X = (X, d, \mu)$ be a metric measure space with (X, d) locally complete. Then μ is a Radon measure.*

Proof of Proposition 3.3.46 Assume first that μ is a Radon measure. Then, because X is separable and μ is locally finite by our standing assumption in Section 3.3, X can be expressed as a countable union of compact sets plus a set of measure zero, proving the implication (i) $\Rightarrow$ (iii). The equivalence of (ii) and (iii) follows from Lemma 3.3.16, and condition (iii) trivially implies (iv).

Assume next that (iv) holds. By Proposition 3.3.21 we have that $f_{\#}\mu$ is a Borel-regular measure on Z. Because also $f_{\#}\mu(Z \setminus f(X)) = 0$ we have that $f(X) \subset Z$ is $f_{\#}\mu$-measurable. Now if $f_{\#}\mu$ were locally finite on Z, it would follow from Proposition 3.3.44 that $f_{\#}\mu$ is a Radon measure on Z and hence that $f(X)$ can be written as a countable union of compact sets plus a set N' such that $\mu(f^{-1}(N')) = 0$. Since f is an embedding, the preimage of a compact set is a compact subset of X; it would be easy to deduce from this that μ is Radon. However, as was pointed out earlier, push-forward measures need not be locally finite. This problem is avoided by the following technical reduction. Fix an arbitrary closed ball $B \subset X$ such that $\mu(B) < \infty$. Because X can be expressed as a countable union of such balls, it follows that μ is a Radon measure if we can show that

$$\mu(C) = \sup\{\mu(K) : K \subset C \text{ compact}\} \tag{3.3.48}$$

for every closed $C \subset B$. (Compare the discussion just before Remark 3.3.43.)

The preceding discussion understood, we can replace X by the closed, and hence complete, metric space B and consider an embedding $f : B \to Z$. The hypothesis on the Borel partition remains intact, for $B = (B_0 \cap B) \cup (N \cap B)$, where $f(B_0 \cap B) = f(B_0) \cap f(B)$ is a Borel set in Z and $\mu(N \cap B) = 0$. Thus, without loss of generality we may assume that the push-forward measure $f_{\#}\mu$ is finite and hence Radon on Z. As explained in the previous paragraph, it follows from this fact that $f(B)$ can be expressed as a countable union of compact sets plus a set whose preimage under f has zero μ-measure. In particular, B can be expressed as a countable union of compact sets plus a set of μ-measure zero. This in turn yields (3.3.48) as required.

We have now completed the proof for the last remaining implication (iv) $\Rightarrow$ (i), and the proposition follows. □

Examples There are metric measure spaces (X, d, μ) such that μ is not a Radon measure, although it follows from Proposition 3.3.46 that such examples can be considered pathological. To be specific, let $X \subset [0, 1]$ be a dense non-Lebesgue-measurable set. Then X equipped with the usual distance and Lebesgue measure m_1 is a metric measure space. However, m_1 is not a Radon measure on X; see Example 3.3.17(a).

Density of continuous functions in L^p-spaces and separability

We next apply the preceding approximation results to prove some simple structural properties of the spaces $L^p(X : V)$.

Proposition 3.3.49 *Let $X = (X, d, \mu)$ be a metric measure space and let $1 \le p < \infty$. Then for every $f \in L^p(X : V)$ and for every $\epsilon > 0$ there exists a continuous function $g : X \to V$ such that $\|f - g\|_p < \epsilon$. Assume next, in addition, that μ is a Radon measure. Then for every function $f \in L^p(X : V)$ and for every $\epsilon > 0$ there exists a function $g : X \to V$ with compact support such that $\|f - g\|_p < \epsilon$.*

Recall that a function h from a topological space Z to a vector space W is said to have *compact support* if the closure of the set $\{z \in Z : h(z) \neq 0\}$ is compact in Z. We also say that h is *compactly supported* in this case. The closure of the preceding set $\{h \neq 0\}$ is called the *support of* h and denoted by $\operatorname{spt}(h)$.

Proof Because simple Borel functions are dense in $L^p(X : V)$, and because the measure is Borel regular, in the first assertion we may assume that f is of the form $f = v \cdot \chi_U$ for some vector $v \in V$ and for some open set $U \subset X$ of finite measure. (See Propositions 3.2.13, 3.3.24, and 3.3.37.) Then, for $\epsilon > 0$, the functions

$$f_\epsilon(x) := v \cdot \min\left\{\frac{1}{\epsilon}\operatorname{dist}(x, X \setminus U), 1\right\} \tag{3.3.50}$$

are continuous and satisfy $f_\epsilon \to f$ in $L^p(X : V)$ as $\epsilon \to 0$. If μ is a Radon measure, there is a sequence (K_i) of compact sets $K_i \subset U$ such that $\lim_{i\to\infty} \mu(K_i) = \mu(U)$. In particular, the functions $v \cdot \chi_{K_i}$ converge to f in $L^p(X : V)$. The proposition follows. □

As discussed in Remark 3.2.16, the preceding proposition together with Proposition 3.2.15 yield the following corollary by the fact that uniform limits of sequences of continuous functions are continuous.

Corollary 3.3.51 *Let $X = (X, d, \mu)$ be a metric measure space and let $1 \le p < \infty$. Then for every $f \in L^p(X : V)$ and for every $\epsilon > 0$ there exists an open set $O \subset X$ such that $\mu(O) < \epsilon$ and $f|_{X\setminus O}$ is continuous.*

It is clear that we cannot, in general, assume in the second assertion of Proposition 3.3.49 that g is continuous and of compact support. Indeed, the support of a continuous function that is not identically zero has nonempty interior and X need not be locally compact. However, we have the following proposition.

Proposition 3.3.52 *Let $X = (X, d, \mu)$ be a metric measure space such that (X, d) is locally compact, and let $1 \le p < \infty$. Then for every $f \in L^p(X : V)$ and for every $\epsilon > 0$ there exists a continuous compactly supported function $g : X \to V$ such that $\|f - g\|_p < \epsilon$.*

Proof As in the proof of Proposition 3.3.49 we may assume that f is of the form $v \cdot \chi_U$ for some $v \in V$ and for some open set $U \subset X$ of finite measure. Fix $\epsilon > 0$. Because X is locally compact, μ is a Radon measure by Corollary 3.3.47 and we can pick a compact set $K \subset U$ such that $\mu(U \setminus K) < \epsilon$. Cover K by finitely many open balls $B_1, \ldots, B_N$ such that $U' := B_1 \cup \cdots \cup B_N \subset U$ and U' has compact closure in X. Set $\delta := \operatorname{dist}(K, X \setminus U') > 0$. Then the continuous function

$$g(x) := v \cdot \min\left\{\frac{1}{\delta}\operatorname{dist}(x, X \setminus U'), 1\right\} \tag{3.3.53}$$

has compact support and satisfies

$$\|f - g\|_p \le |v|\, \mu(U \setminus K)^{1/p} < |v|\, \epsilon^{1/p}.$$

This completes the proof. □

Remark 3.3.54 As a consequence of the above density result, we can obtain the continuity of the translation operators on $L^p(\mathbb{R}^n : V)$. Indeed, for each

$h \in \mathbb{R}^n$ we define the translation operator $\tau_h : \mathbb{R}^n \to \mathbb{R}^n$ by $\tau_h(x) = x + h$. For uniformly continuous functions $g \in L^p(\mathbb{R}^n : V)$ we easily see that

$$\lim_{|h|\to 0} \|g \circ \tau_h - g\|_{L^p(\mathbb{R}^n : V)} = 0.$$

The density of uniformly continuous functions in $L^p(X : V)$ together with the Minkowski inequality shows that, for each $f \in L^p(\mathbb{R}^n : V)$,

$$\lim_{|h|\to 0} \|f \circ \tau_h - f\|_{L^p(\mathbb{R}^n : V)} = 0.$$

Proposition 3.3.55 *Let $X = (X, d, \mu)$ be a metric measure space and let $1 \leq p < \infty$. Then $L^p(X : V)$ as a Banach space is separable if and only if V is separable.*

Proof Assume first that V is separable. Fix a countable dense set $D := \{v_1, v_2, \dots\}$ in V. Similarly, fix a countable dense set $G := \{x_1, x_2, \dots\}$ in X. Denote by $\mathcal{S}$ the countable collection of subsets of X that are finite unions of the form

$$\overline{B}(x_{i_1}, q_{i_1}) \cup \cdots \cup \overline{B}(x_{i_k}, q_{i_k}),$$

where each $B(x_{i_j}, q_{i_j})$ is a ball with $x_{i_j} \in G$ and $q_{i_j} \in \mathbb{Q}$. We claim that the collection of simple functions that are finite sums of the functions

$$v\chi_S, \quad v \in D,\ S \in \mathcal{S},$$

is dense in $L^p(X : V)$. Indeed, as in the proof of Proposition 3.3.49 we need only consider functions that are of the form $v\chi_U$ for some vector $v \in V$ and for some open set $U \subset X$ of finite measure. Because every such U can be expressed as a countable union of sets in $\mathcal{S}$, the claim is easily verified by noting that the measure of U can be approximated by the measures of sets from $\mathcal{S}$ contained in U. This proves that $L^p(X : V)$ is separable if V is separable.

If V is not separable, there is an $\epsilon > 0$ and an uncountable set $G \subset V$ such that $B(v, \epsilon) \cap B(w, \epsilon) = \emptyset$ whenever $v, w \in G$ are distinct points. (This assertion is an easy consequence of Zorn's lemma; see [237, p. 87] or [214, p. 69].) Now fix a ball $B \subset X$ of positive and finite measure, and observe that the collection $\{v \cdot \chi_B : v \in G\}$ of p-integrable functions is uncountable, with

$$\|v\chi_B - w\chi_B\|_{L^p(X:V)} \geq 2\epsilon\mu(B)^{1/p} > 0$$

whenever $v, w \in G$ are distinct points. This shows that $L^p(X : V)$ cannot be separable.

The proposition is proved. □

3.4 Differentiation

Classical differentiation theorems involving integrals of functions extend rather directly to Bochner-integrable vector-valued functions. In this section we treat such extensions, including the differentiation of measures on metric spaces. Covering theorems are pivotal to Lebesgue-type differentiation, and we begin the discussion by introducing classes of metric measure spaces where such covering theorems hold. Both the covering theorems and the Lebesgue differentiation theorem, especially in the context of a doubling metric measure space, will feature prominently in later sections of this book.

Vitali measures

The *Vitali covering theorem* for Lebesgue measure asserts that from every fine covering $\mathcal{B}$ of a set $A \subset \mathbb{R}^n$ by closed balls we can extract a pairwise disjoint subcollection $\mathcal{C} \subset \mathcal{B}$ such that the measure of $A \setminus \bigcup_{B\in\mathcal{C}} B$ is zero. A covering $\mathcal{B}$ of a set A by closed balls is called *fine* if

$$\inf\{r : r > 0 \text{ and } \overline{B}(x, r) \in \mathcal{B}\} = 0, \tag{3.4.1}$$

for each $x \in A$.

Turning Vitali's theorem into a definition, we call a metric measure space (X, d, μ) a *Vitali metric measure space*, and the measure μ a *Vitali measure*, if the corresponding result holds for fine coverings of subsets of X. Thus, X is a Vitali metric measure space if and only if for every subset A of X and for every covering $\mathcal{B}$ of A by closed balls satisfying (3.4.1) for each $x \in A$ there exists a pairwise disjoint subcollection $\mathcal{C} \subset \mathcal{B}$ such that

$$\mu\left(A \setminus \bigcup_{B\in\mathcal{C}} B\right) = 0. \tag{3.4.2}$$

The following theorem provides a large class of Vitali measures.

Theorem 3.4.3 *Let (X, d, μ) be a metric measure space such that*

$$\limsup_{r\to 0} \frac{\mu(B(x, 2r))}{\mu(B(x, r))} < \infty \tag{3.4.4}$$

for almost every $x \in X$. Then X is a Vitali metric measure space.

Proof Let $A \subset X$ and let $\mathcal{B}$ be a fine covering of A by closed balls. We need to find a pairwise disjoint subcollection $\mathcal{C}$ of $\mathcal{B}$ such that (3.4.2) holds. To this end, we may assume that (3.4.4) holds for every $x \in A$. We may also assume that the balls in the covering $\mathcal{B}$ have uniformly bounded radii; this allows us to use repeatedly the $5B$-covering lemma, see the text before (3.3.26), for $\mathcal{B}$ and for its various subfamilies.

For $x \in X$, denote the numerical value of the limit superior in (3.4.4) by $D(x)$. Next, use Lemma 3.3.28 and write

$$X = \bigcup_{k=1}^{\infty} D_k ,$$

where each D_k is open and of finite measure such that $D_k \subset D_{k+1}$. Then put

$$A_k := \{x \in A : D(x) < 2^k\} \cap D_k .$$

Clearly, $A_k \subset A_{k+1}$ and $A = \bigcup_{k=1}^{\infty} A_k$.

We will inductively choose finite pairwise disjoint subfamilies $\mathcal{C}_l \subset \mathcal{B}$ such that $\mathcal{C}_l \subset \mathcal{C}_{l+1}$ and

$$\mu\left(A_k \setminus \bigcup_{B \in \mathcal{C}_l} B\right) \le 2^{-l}\mu(A_k) \tag{3.4.5}$$

whenever $1 \le k \le l$. It is then clear that the pairwise disjoint family

$$\mathcal{C} := \bigcup_{l=1}^{\infty} \mathcal{C}_l = \{B : B \in \mathcal{C}_l \text{ for some } l\}$$

satisfies (3.4.2).

To this end, we let $\mathcal{B}_1$ consist of all balls $\overline{B}(x, r)$ from $\mathcal{B}$ with $r \le 1$ such that $x \in A_1$, $\overline{B}(x, r) \subset D_1$, and

$$\mu(\overline{B}(x, 5r)) \le 2^3 \mu(\overline{B}(x, r)) . \tag{3.4.6}$$

The family $\mathcal{B}_1$ is a fine covering of A_1 by closed balls of uniformly bounded radii. We use the $5B$-covering lemma and choose a pairwise disjoint subfamily $\mathcal{C}'_1$ of balls from $\mathcal{B}_1$ with the following property: if $B \in \mathcal{B}_1$ then there exists a ball $B' \in \mathcal{C}'_1$ such that $B \cap B' \neq \emptyset$ and $\operatorname{diam}(B) < 2\operatorname{diam}(B')$. Enumerate $\mathcal{C}'_1 = \{B^1_1, B^1_2, \dots\}$. Because the balls B^1_i are closed, and because the covering $\mathcal{B}_1$ is fine, there is a positive integer N such that we can find for every $x \in A_1 \setminus (B^1_1 \cup \cdots \cup B^1_N)$ a ball $B \in \mathcal{B}_1$, centered at x, that does not meet the union $B^1_1 \cup \cdots \cup B^1_N$. Such a ball, therefore, meets a ball $B^1_i \in \mathcal{C}'_1$ for some $i > N$ such that $\operatorname{diam}(B) < 2\operatorname{diam}(B^1_i)$. In particular, we find that

$$A_1 \setminus (B^1_1 \cup \cdots \cup B^1_N) \subset \bigcup_{i \ge N+1} 5B^1_i .$$

It follows from this and from (3.4.6) that

$$\mu(A_1 \setminus (B^1_1 \cup \cdots \cup B^1_N)) \le \sum_{i \ge N+1} \mu(5B^1_i) \le 2^3 \sum_{i \ge N+1} \mu(B^1_i) .$$

The second right-hand side in the preceding inequality tends to zero as $N \to \infty$, because the family $\mathcal{C}_1'$ is pairwise disjoint and consists of balls lying in a fixed subset of X, namely D_1, of finite measure. We choose N_1 such that

$$\mu(A_1 \setminus (B_1^1 \cup \cdots \cup B_{N_1}^1)) \leq \tfrac{1}{2}\mu(A_1)$$

and set

$$\mathcal{C}_1 := \{B_1^1, \ldots, B_{N_1}^1\}.$$

Next, assume that finite pairwise disjoint families of balls $\mathcal{C}_1 \subset \cdots \subset \mathcal{C}_l$ from $\mathcal{B}$ have been selected in such a way that (3.4.5) holds for $1 \leq k \leq l$. Let $\mathcal{B}_{l+1}$ consist of all balls $\overline{B}(x,r)$ from $\mathcal{B}$ such that $x \in A_{l+1}$, $\overline{B}(x,r) \subset D_{l+1} \setminus \bigcup_{B \in \mathcal{C}_l} B$, and

$$\mu(\overline{B}(x, 5r)) \leq 2^{3(l+1)}\mu(\overline{B}(x,r)).$$

Then the family $\mathcal{B}_{l+1}$ is a fine covering of $A_{l+1} \setminus \bigcup_{B \in \mathcal{C}_l} B$ by closed balls of uniformly bounded radii and, by the $5B$-covering lemma, we can choose a pairwise disjoint subfamily $\mathcal{C}_{l+1}'$ of balls from $\mathcal{B}_{l+1}$ with the following property: if $B \in \mathcal{B}_{l+1}$ then there exists a ball $B' \in \mathcal{C}_{l+1}'$ such that $B \cap B' \neq \emptyset$ and $\operatorname{diam}(B) < 2\operatorname{diam}(B')$. Enumerating $\mathcal{C}_{l+1}' = \{B_1^{l+1}, B_2^{l+1}, \ldots\}$ and arguing as earlier, we find that

$$A_{l+1} \Bigg\backslash \left[\bigcup_{B \in \mathcal{C}_l} B \cup (B_1^{l+1} \cup B_2^{l+1} \cup \cdots \cup B_N^{l+1}) \right] \subset \bigcup_{i \geq N+1} 5B_i^{l+1},$$

and thus that

$$\mu\left(A_{l+1} \Bigg\backslash \left[\bigcup_{B \in \mathcal{C}_l} B \cup (B_1^{l+1} \cup B_2^{l+1} \cup \cdots \cup B_N^{l+1}) \right] \right) \\ \leq 2^{3(l+1)} \sum_{i \geq N+1} \mu(B_i^{l+1}). \tag{3.4.7}$$

The right-hand side of the preceding inequality tends to zero as $N \to \infty$, because the family $\mathcal{C}_{l+1}'$ is pairwise disjoint and consists of balls lying in D_{l+1}. Now pick an integer N_{l+1} such that when $N = N_{\ell+1}$ the expression on the right in (3.4.7) for is less than $2^{-l-1}\mu(A_k)$ for every $1 \leq k \leq l+1$ for which $\mu(A_k) > 0$, that is,

$$\sum_{i \geq N_{l+1}+1} \mu(B_i^{l+1}) \leq 2^{-4(l+1)} \min\{\mu(A_k) : 1 \leq k \leq l+1, \mu(A_k) > 0\}.$$

Then set

$$\mathcal{C}_{l+1} := \mathcal{C}_l \cup \{B_1^{l+1}, \ldots, B_{N_{l+1}}^{l+1}\}.$$

It follows from the construction that $\mathcal{C}_{l+1}$ is a collection as required; this completes the induction step.

The proof of the theorem is now complete. □

Doubling measures

A Borel regular measure μ on a metric space (X, d) is called a *doubling measure* if every ball in X has positive and finite measure and there exists a constant $C \geq 1$ such that

$$\mu(B(x, 2r)) \leq C\mu(B(x, r)) \tag{3.4.8}$$

for each $x \in X$ and $r > 0$. In particular, X is separable as a metric space (Lemma 3.3.30) and (X, d, μ) is a metric measure space. We call the triple (X, d, μ) a *doubling metric measure space* if μ is a doubling measure on X.

The *doubling constant* of a doubling measure μ is the smallest constant $C \geq 1$ for which inequality (3.4.8) holds. We denote this constant by C_μ. By iterating (3.4.8), we deduce the growth estimate

$$\mu(B(x, \lambda r)) \leq C_\mu \circ \lambda^{\log_2 C_\mu} \mu(B(x, r)) \tag{3.4.9}$$

for each $x \in X$, $\lambda \geq 1$, and $r > 0$. The number $\log_2 C_\mu$ sometimes takes the role of a "dimension" for a doubling metric measure space X. Note that for $\mu = m_n$, the Lebesgue n-measure, we have $\log_2 C_\mu = n$. Moreover, $\log_2 C_\mu > 0$ unless X is a one-point space.

We reiterate that in this book it is not assumed in general that metric measure spaces are complete as metric spaces. In particular, doubling metric measure spaces need not be complete. For example, every open ball in $\mathbb{R}^n$ equipped with the Lebesgue n-measure is a doubling metric measure space.

It is clear that doubling measures satisfy (3.4.4), so doubling measures are Vitali measures.

Doubling metric measure spaces will play a central role in the later chapters of this book. Additional examples of such spaces are given in Chapter 14.

Examples The hypothesis (3.4.4) in Theorem 3.4.3 is considerably weaker than the assumption that X is a doubling metric measure space. For example, it follows from the theorem, and from basic Riemannian geometry, that every Riemannian manifold with the Riemannian distance and measure is a Vitali metric measure space. Moreover, if (X, d, μ) satisfies (3.4.4) and f is a locally integrable nonnegative function on X then (X, d, μ_f) is a Vitali metric measure space, where $d\mu_f = f\, d\mu$. This follows from a combination of Theorem 3.4.3 and the Lebesgue differentiation theorem below. A complete Riemannian manifold, with Ricci curvature nonnegative outside a compact subset of the manifold, satisfies the doubling condition, as demonstrated in [184, Lemma 1.3].

If μ is a Borel-regular measure on a Riemannian manifold (X, d) with support S (as defined in (3.3.33)) then (S, d, μ) is a Vitali metric measure space. This follows from [83, Sections 2.8.9 and 2.8.18]. See also [197, Theorem 2.8] for the case $X = \mathbb{R}^n$.

Vitali-type covering theorems were originally devised to treat various differentiation theorems. The *Lebesgue differentiation theorem* is a central result of analysis; it is crucial to the theory developed in this book.

Lebesgue differentiation theorem *Let (X, d, μ) be a Vitali metric measure space, let V be a Banach space, and let $f : X \to V$ be a locally integrable function. Then almost every point $x \in X$ is a Lebesgue point of f, that is,*

$$\lim_{r\to 0} \fint_{\overline{B}(x,r)} |f(y) - f(x)|\, d\mu(y) = 0 \tag{3.4.10}$$

for almost every $x \in X$. In particular,

$$\lim_{r\to 0} \fint_{\overline{B}(x,r)} f(y)\, d\mu(y) = f(x) \tag{3.4.11}$$

for almost every $x \in X$.

We note that $\mu(\overline{B}(x, r)) > 0$ for every $r > 0$ whenever $x \in \operatorname{spt}(\mu)$ (see section 3.3), so that the expressions in (3.4.10) and in (3.4.11) make sense at μ-almost every point.

Proof Note that the second statement follows from the first by (3.2.2) and the fact that if $0 < \mu(\overline{B}(x, r)) < \infty$ then $\fint_{\overline{B}(x,r)} f(x)\, d\mu(y) = f(x)$. However, we begin by establishing (3.4.11) in the scalar-valued case $V = \mathbb{R}$. To do this, we may assume that f is nonnegative (by considering the positive and the negative parts separately) as well as integrable over X (by using Lemma 3.3.28 and replacing f with $f\chi_B$ for an appropriate open ball $B \subset X$). Likewise, we may assume that $\mu(X) < \infty$ and that $\mu(B(x, r)) > 0$ for every $x \in X$ and $r > 0$.

The preceding being understood, we will first show that

$$\limsup_{r\to 0} \fint_{\overline{B}(x,r)} f\, d\mu < \infty \tag{3.4.12}$$

for almost every $x \in X$. For $c > 0$ define

$$F_c := \left\{ x \in X : \limsup_{r \to 0} \fint_{\overline{B}(x,r)} f \, d\mu > c \right\} .$$

Fix $c > 0$ and let O be any open set containing F_c. The family

$$\mathcal{B} := \{ \overline{B}(x,r) \subset O : \; x \in F_c \text{ and} \fint_{\overline{B}(x,r)} f \, d\mu > c \}$$

is a fine covering of F_c. We can therefore pick a countable pairwise disjoint subcollection $\mathcal{C}$ of balls from $\mathcal{B}$ that covers almost all F_c. It follows that

$$c\, \mu(F_c) \le c \sum_{B \in \mathcal{C}} \mu(B) \le \sum_{B \in \mathcal{C}} \int_B f \, d\mu \le \int_O f \, d\mu \le \int_X f \, d\mu < \infty .$$

In particular, (3.4.12) holds.

Next, for $c > 0$ define

$$E_c := \left\{ x \in X : \liminf_{r \to 0} \fint_{\overline{B}(x,r)} f \, d\mu < c \right\} .$$

The set of all those points $x \in X$ for which the limit on the left-hand side of (3.4.11) fails to exist is contained in the countable union of sets of the form $G_{s,t} := E_s \cap F_t$, where $s < t$ are rational numbers. We claim that for every such set we have

$$t\mu(G_{s,t}) \le s\, \mu(G_{s,t}) . \tag{3.4.13}$$

Because $\mu(G_{s,t}) \le \mu(X) < \infty$, this gives $\mu(G_{s,t}) = 0$ as required. To establish (3.4.13), fix $G_{s,t}$ and fix a Borel set A containing $G_{s,t}$ such that $\mu(G_{s,t}) = \mu(A)$. For any open set O containing A, the covering argument as earlier yields

$$t\mu(G_{s,t}) \le \int_O f \, d\mu \le \int_A f \, d\mu + \int_{O \setminus A} f \, d\mu ,$$

and, by taking the infimum over all such open sets O and by using the Borel regularity formula (3.3.39) together with the absolute continuity of the integral, we obtain

$$t\mu(G_{s,t}) \le \int_A f \, d\mu . \tag{3.4.14}$$

Now fix $\epsilon > 0$ and let $0 < \delta < \epsilon$ be such that $\int_H f \, d\mu < \epsilon$ whenever $H \subset X$ is measurable and such that $\mu(H) < \delta$. By the Borel regularity formula (3.3.39), we can find an open set O containing A such that $\mu(O) \le \mu(G_{s,t}) + \delta$. We apply again a covering argument and the Vitali property of μ to find a countable pairwise disjoint collection $\mathcal{C}$ of closed balls contained in O such that

$$\int_B f \, d\mu < s\,\mu(B),$$

for every $B \in \mathcal{C}$, and $\mu(G_{s,t} \setminus \cup_{B\in\mathcal{C}} B) = 0$. In particular, since

$$\mu(G_{s,t}) \le \mu\left(G_{s,t} \setminus \bigcup_{B\in\mathcal{C}} B\right) + \mu\left(\bigcup_{B\in\mathcal{C}} B\right) = \mu\left(\bigcup_{B\in\mathcal{C}} B\right)$$

we have

$$\begin{aligned}\mu\left(A \setminus \bigcup_{B\in\mathcal{C}} B\right) &\le \mu(O) - \mu\left(\bigcup_{B\in\mathcal{C}} B\right) \\ &\le \mu(G_{s,t}) + \delta - \mu\left(\bigcup_{B\in\mathcal{C}} B\right) \le \delta .\end{aligned}$$

Hence

$$\begin{aligned}\int_A f \, d\mu &\le \int_{A \setminus \cup_{B\in\mathcal{C}} B} f \, d\mu + \sum_{B\in\mathcal{C}} \int_B f \, d\mu \\ &\le \epsilon + s\mu(O) \le \epsilon + s\mu(G_{s,t}) + s\delta ,\end{aligned}$$

which gives, by letting $\epsilon \to 0$, that

$$\int_A f \, d\mu \le s\mu(G_{s,t}) . \tag{3.4.15}$$

Now (3.4.13) follows, on combining (3.4.14) and (3.4.15).

We have thus proved that the limit on the left-hand side in (3.4.11) exists and is finite for almost every $x \in X$ whenever f is a locally integrable nonnegative function on X. At this juncture, the preceding assumptions understood, we state and prove the following lemma.

Lemma 3.4.16 *Denote by g the almost everywhere defined limit on the left-hand side in (3.4.11). Then g is measurable.*

As before, we notice that it suffices to establish the claim on each open ball B over which f is integrable.

Proof Because g is the pointwise almost everywhere limit of the functions

$$g_n(x) := ⨍_{\overline{B}(x,1/n)} f \, d\mu$$

as $n \to \infty$, it suffices to show that for a fixed $\delta > 0$ the functions

$$u(x) := \mu(\overline{B}(x,\delta)) \quad \text{and} \quad v(x) := \int_{\overline{B}(x,\delta)} f \, d\mu$$

are measurable on $U_\delta = \{x : B(x, 2\delta) \subset B\}$. Towards this, fix $x \in U_\delta$ and let (x_i) be a sequence in X converging to x. Fix an open neighborhood O of $\overline{B}(x,\delta)$. The balls $\overline{B}(x_i,\delta)$ lie in O for all large i, which gives that

$$\limsup_{i\to\infty} u(x_i) \le \mu(O)$$

and also that

$$\limsup_{i\to\infty} v(x_i) \le \int_O f \, d\mu \, .$$

By taking the infimum over all such O, we obtain $\limsup_{i\to\infty} u(x_i) \le u(x)$ and also $\limsup_{i\to\infty} v(x_i) \le v(x)$. This shows that both u and v are upper semicontinuous functions (cf. Section 4.2) and hence measurable. Thus g is measurable as asserted. □

We next verify that g equals f almost everywhere, where g is as in Lemma 3.4.16. To this end, we will show that their integrals over every measurable set in X agree. Thus, let A be a measurable set in X. Fix $t > 1$ and note that up to a set of measure zero A can be expressed as the disjoint union of the measurable sets

$$\begin{aligned} A_n &:= A \cap \{x \in X : t^n \le g(x) < t^{n+1}\}\,, \\ A_{-n-1} &:= A \cap \{x \in X : t^{-n-1} \le g(x) < t^{-n}\}\,, \\ A_\infty &:= A \cap \{x \in X : g(x) = 0\}\,, \end{aligned} \tag{3.4.17}$$

where $n = 0, 1, 2, \ldots$ Note that

$$\int_{A_\infty} f \, d\mu = 0 = \int_{A_\infty} g \, d\mu, \tag{3.4.18}$$

by (3.4.15). The covering argument used to prove (3.4.14) and (3.4.15) applies to every measurable subset A of the set F_c or of the set E_c, giving in particular that

$$t^n \mu(A_n) \le \int_{A_n} f \, d\mu \tag{3.4.19}$$

and that

$$\int_{A_n} f \, d\mu \le t^{n+1} \mu(A_n)\,. \tag{3.4.20}$$

(Strictly speaking, when proving (3.4.19) we need to observe that $A_n \subset \{g > s^n\}$ for every $s < t$, and then let $s \to t$.) Thus

$$t^{-1}\int_{A_n} f\,d\mu \;\le\; t^n\mu(A_n) \;\le\; \int_{A_n} g\,d\mu$$
$$\le\; t^{n+1}\mu(A_n) \;\le\; t\int_{A_n} f\,d\mu\,.$$

A similar argument shows that

$$t^{-1}\int_{A_{-n-1}} f\,d\mu \;\le\; \int_{A_{-n-1}} g\,d\mu \;\le\; t\int_{A_{-n-1}} f\,d\mu\,.$$

By summing over n, and by observing (3.4.18), we deduce from the preceding inequalities that

$$t^{-1}\int_A f\,d\mu \;\le\; \int_A g\,d\mu \;\le\; t\int_A f\,d\mu\,.$$

Finally, by letting $t \to 1$, we find that

$$\int_A f\,d\mu \;=\; \int_A g\,d\mu\,.$$

Because A was arbitrary it follows that f and g agree almost everywhere in X, as required.

We have thus proved (3.4.11) for real-valued locally integrable functions f.

Now we turn to the general case and prove (3.4.10) for a locally integrable function $f : X \to V$. Let $Z \subset X$ be a set of measure zero such that $f(X \setminus Z)$ is a separable subset of V (by Pettis' theorem; see Section 3.1). Pick a dense subset $D = \{v_1, v_2, \ldots\}$ of $f(X \setminus Z)$ and consider the real-valued functions $f_j(x) := |f(x) - v_j|$ for $j = 1, 2, \ldots$ Since, by Proposition 3.2.7, the functions f_j are locally integrable it follows from what was proved earlier that

$$|f(x) - v_j| = \lim_{r\to 0} ⨍_{B(x,r)} |f(y) - v_j|\,d\mu(y)$$

for every $x \in X \setminus Z_j$, where $Z_j \subset X$ has measure zero. Thus

$$Z' = Z \cup \bigcup_{j=1}^{\infty} Z_j$$

has measure zero, while for each $x \in X \setminus Z'$ and for each j we have that

$$\limsup_{r\to 0} ⨍_{B(x,r)} |f(y) - f(x)|\,d\mu(y)$$
$$\le \limsup_{r\to 0} ⨍_{B(x,r)} |f(y) - v_j|\,d\mu(y) + |f(x) - v_j| = 2|f(x) - v_j|.$$

Since D is dense in $f(X \setminus Z)$, we find that (3.4.10) holds.

As was remarked in the beginning of the proof, (3.4.10) implies (3.4.11). The proof of the theorem is thereby complete. □

Differentiation of measures

The preceding proof of the Lebesgue differentiation theorem applies to the general differentiation of measures. The results of the above discussion are used in the study of rectifiable curves in the metric setting; see for instance the proof of Theorem 4.4.8 below. However, these results are also used in other contexts such as the study of functions of bounded variation in metric measure spaces, as explained in for example [6]. Let (X, d, μ) be a metric measure space, and let ν be a Borel regular locally finite measure on X. The *derivative of* ν *with respect to* μ *at a point* $x \in X$ is the limit

$$\lim_{r\to 0} \frac{\nu(\overline{B}(x,r))}{\mu(\overline{B}(x,r))} =: \frac{d\nu}{d\mu}(x), \tag{3.4.21}$$

provided that the limit exists and is finite.

Because $\mu(\overline{B}(x,r)) > 0$ for every $x \in \operatorname{spt}(\mu)$ and for every $r > 0$ (see Section 3.3), the existence of a limit as in (3.4.21) can be investigated μ almost-everywhere.

Lebesgue–Radon–Nikodym theorem *Let (X, d, μ) be a Vitali metric measure space and let ν be a locally finite Borel-regular measure on X. There exist unique locally finite Borel-regular measures ν^s and ν^a on X with the following two properties:*

$$\nu(A) = \nu^s(A) + \nu^a(A) \tag{3.4.22}$$

for every Borel set $A \subset X$; there exists a Borel set $D \subset X$ such that $\nu^s(D) = 0$, $\mu(X \setminus D) = 0$, and $\nu^a = \nu\lfloor D$. Moreover, the derivatives of both ν and ν^a with respect to μ exist at μ-almost every (μ-a.e.) point in X and they are μ-measurable and locally integrable in X, with

$$\nu^a(A) = \int_A \frac{d\nu}{d\mu}(x)\, d\mu(x) = \int_A \frac{d\nu^a}{d\mu}(x)\, d\mu(x) \tag{3.4.23}$$

for every Borel set $A \subset X$. In particular,

$$\frac{d\nu}{d\mu} = \frac{d\nu^a}{d\mu} \tag{3.4.24}$$

μ-almost everywhere.

In (3.4.22) we have the *Lebesgue decomposition* of the measure ν into its *singular part* ν^s and *absolutely continuous part* ν^a, with respect to μ. Recall the definition of the measure $\nu\lfloor B$ from (3.3.12).

Proof We first describe a decomposition as in (3.4.22). For $E \subset X$, set

$$\nu^a(E) := \inf \nu(B)\,,$$

where the infimum is taken over all Borel sets $B \subset X$ such that $\mu(E \setminus B) = 0$. It is easy to see that ν^a is a measure; we leave this to the reader. Note that $\nu^a(E) \leq \nu(E)$ for every set E by Borel regularity, and that $\nu^a(N) = 0$ for every set such that $\mu(N) = 0$. In particular, ν^a is locally finite. To check that ν^a is a Borel measure, we use the Carathéodory criterion (3.3.6). Thus, let E_1 and E_2 be two sets in X such that $\operatorname{dist}(E_1, E_2) > 0$. Let $B \subset X$ be a Borel set such that $\mu((E_1 \cup E_2) \setminus B) = 0$. Choose open sets O_1 and O_2, containing E_1 and E_2 respectively, such that $\operatorname{dist}(O_1, O_2) > 0$, and set $B_1 := B \cap O_1$ and $B_2 := B \cap O_2$. Then $\mu(E_1 \setminus B_1) = \mu(E_2 \setminus B_2) = 0$, and therefore

$$\nu(B) \geq \nu(B_1 \cup B_2) = \nu(B_1) + \nu(B_2) \geq \nu^a(E_1) + \nu^a(E_2) .$$

Taking the infimum over all Borel sets B as above, we conclude that the Carathéodory criterion holds and hence that ν^a is a Borel measure.

To verify that ν^a is also Borel regular, let $E \subset X$. We may assume that $\nu^a(E) < \infty$. Pick a decreasing sequence of Borel sets $B_1 \supset B_2 \supset \cdots$ such that $\mu(E \setminus B_j) = 0$ for all $j \geq 1$ and $\lim_{j\to\infty} \nu(B_j) = \nu^a(E)$. Such a sequence can be found, because $\mu(E \setminus B') = \mu(E \setminus B'') = 0$ implies that $\mu(E \setminus (B' \cap B'')) = 0$. There is a Borel set B_0 containing $\bigcup_{j=1}^\infty (E \setminus B_j)$ such that $\mu(B_0) = 0$. Set $B := (\bigcap_{j=1}^\infty B_j) \cup B_0$. Then B is a Borel set containing E, and

$$\nu^a(E) \leq \nu^a(B) \leq \nu^a\left(\bigcap_{j=1}^{\infty} B_j\right) \leq \nu^a(B_k) \leq \nu(B_k)$$

for every $k = 1, 2, \ldots$ This gives $\nu^a(E) = \nu^a(B)$, establishing Borel regularity. Note also that, with $D = \bigcap_{j=1}^\infty B_j$, we have that D is Borel and $\mu(E \setminus D) = 0$. It follows that $\nu^a(E) \leq \nu(D) \leq \lim_j \nu(B_j) = \lim_j \nu^a(B_j) = \nu^a(E)$, that is, $\nu^a(E) = \nu(D)$.

Now, if E is a Borel set then write $D' = D \cap E$ where D is as in the above paragraph. Then D' also is a Borel set, with $\mu(E \setminus D') = \mu(E \setminus D) = 0$, and so $\nu^a(E) \leq \nu(D') \leq \nu(D) = \nu^a(E)$, that is, $\nu^a(E) = \nu(D')$. We will now prove that

$$\nu^a(B) = \nu(B \cap D') \tag{3.4.25}$$

for every Borel set $B \subset E$. Indeed, if B is a Borel set in D then on the one hand (cf. Lemma 3.3.4)

$$\nu^a(B) + \nu^a(E \setminus B) = \nu^a(D) = \nu(D') = \nu(B \cap D') + \nu(D' \setminus B) ;$$

on the other hand, because $\mu(B \setminus D') = 0$ and $\mu((E \setminus B) \setminus (D' \setminus B)) \leq \mu(E \setminus D') = 0$, we have that $\nu^a(B) \leq \nu(B \cap D')$ and that $\nu^a(D \setminus B) \leq \nu(D' \setminus B)$, whence (3.4.25) follows.

We now apply the preceding construction to a decomposition of X into countably many pairwise disjoint Borel sets (D_i), such that both $\mu(D_i)$ and $\nu(D_i)$ are finite for each i (Lemma 3.3.28), to obtain Borel sets $D_i' \subset D_i$ such that $\mu(D_i') = \mu(D_i)$ and $\nu^a(B) = \nu(B \cap D_i')$ whenever $B \subset D_i$ is a Borel set. We set $D := \bigcup_i D_i'$. Then $\mu(X \setminus D) = 0$, and we claim that $\nu^a(E) = \nu(E \cap D)$ for every $E \subset X$. To prove the claim, fix $E \subset X$ and fix a Borel set B containing E such that $\nu^a(E) = \nu^a(B)$. Then

$$\nu^a(E) = \sum_i \nu^a(B \cap D_i) = \sum_i \nu(B \cap D_i') = \nu(B \cap D) \geq \nu(E \cap D),$$

while $\mu(B \setminus D) = 0$ and so $\nu^a(E \setminus D) \leq \nu^a(B \setminus D) = 0$, from which it follows that

$$\nu^a(E) \leq \nu^a(E \cap D) + \nu^a(E \setminus D) = \nu^a(E \cap D) \leq \nu(E \cap D),$$

as desired. Thus, $\nu^a = \nu \lfloor D$.

Finally, we set $\nu^s(E) := \nu(E) - \nu^a(E)$. The preceding discussion shows that $\nu^s(D) = 0$. It is routine to verify that ν^s is a measure; it is also Borel regular because ν and ν^a are Borel regular. We leave this to the reader. The uniqueness of the Lebesgue decomposition (3.4.22) is straightforward to check directly, but it also follows from (3.4.23).

We will next show that the derivative of ν with respect to μ exists at μ-almost-every point in X. The proof of this assertion is analogous to that of the Lebesgue differentiation theorem; see above. In fact the argument is slightly more straightforward, as the Borel regularity can be used to finesse measurability issues that arose with integration. First, by restricting both measures μ and ν appropriately, we may assume that both $\mu(X)$ and $\nu(X)$ are finite and that $\mu(\overline{B}(x,r)) > 0$ for every $x \in X$ and $r > 0$. Next, for $c > 0$ define

$$E_c := \left\{ x \in X : \liminf_{r \to 0} \frac{\nu(\overline{B}(x,r))}{\mu(\overline{B}(x,r))} < c \right\}$$

and

$$F_c := \left\{ x \in X : \limsup_{r \to 0} \frac{\nu(\overline{B}(x,r))}{\mu(\overline{B}(x,r))} > c \right\}.$$

Fix $c > 0$, and let E_c' be any subset of E_c. Then fix $\epsilon > 0$ and choose an open set O containing E_c' such that $\mu(O) \leq \mu(E_c') + \epsilon$. Using a fine covering of E_c' consisting of all closed balls $\overline{B}(x,r) \subset O$, $x \in E_c'$, such that $\nu(\overline{B}(x,r)) < c\mu(\overline{B}(x,r))$ and the hypothesis that μ is a Vitali measure, we conclude that there exists a countable pairwise disjoint collection of closed balls $\mathcal{C} = \{B\}$

in O such that $\nu(B) < c\mu(B)$ for every $B \in \mathcal{C}$ and $\mu(E'_c \setminus \bigcup_{B\in\mathcal{C}} B) = 0$. Hence

$$\nu^a(E'_c) \leq \nu\left(\bigcup_{B\in\mathcal{C}} B\right) \leq c \sum_{B\in\mathcal{C}} \mu(B) \leq c\mu(O) \leq c\mu(E'_c) + c\epsilon\,.$$

By letting $\epsilon \to 0$, we conclude that

$$\nu^a(E'_c) \leq c\mu(E'_c)\,. \tag{3.4.26}$$

Similarly, we claim that

$$c\mu(F'_c) \leq \nu^a(F'_c) \tag{3.4.27}$$

for every $c > 0$ and every subset $F'_c \subset F_c$. To see this, we use the fact that $\nu^a = \nu\lfloor D$ for some $D \subset X$ such that $\mu(X \setminus D) = 0$. Fix $F'_c \subset F_c$, and fix $\epsilon > 0$. Choose an open set O containing $F'_c \cap D$ such that $\nu(O) \leq \nu(F'_c \cap D) + \epsilon$. As earlier, a covering argument and the Vitali property yields that

$$c\mu(F'_c) = c\mu(F'_c \cap D) \leq \nu(O) \leq \nu(F'_c \cap D) + \epsilon = \nu^a(F'_c) + \epsilon\,,$$

whence (3.4.27) follows by letting $\epsilon \to 0$. In particular,

$$c\mu(F_c) \leq \nu^a(F_c) \leq \nu(X) < \infty$$

for every $c > 0$, and hence we find that

$$\limsup_{r\to 0} \frac{\nu(\overline{B}(x,r))}{\mu(\overline{B}(x,r))} < \infty$$

at μ-almost every (μ-a.e.) point $x \in X$. As in the proof of the Lebesgue differentiation theorem, we infer that the set of points $x \in X$ such that the limit in (3.4.21) fails to hold is contained in a countable union of sets of the form $G_{s,t} := E_s \cap F_t$, where $s < t$ are rational numbers. It follows from (3.4.26) and (3.4.27), and from the assumption $\mu(G_{s,t}) \leq \mu(X) < \infty$, that $\mu(G_{s,t}) = 0$.

We conclude that the function $g(x) := d\mu(x)/d\nu(x)$ exists and is finite at μ-a.e. $x \in X$. The measurability of g is established analogously to Lemma 3.4.16. We will now show that

$$\nu^a(A) = \int_A g(x)\,d\mu(x) \tag{3.4.28}$$

for every Borel set $A \subset X$. Again, this is similar to the argument in the proof of the Lebesgue differentiation theorem. Let $A \subset X$ be a Borel set. Fix $t > 1$ and define sets A_n as in (3.4.17). Then, up to a set of μ-measure zero,

A can be expressed as a disjoint union of the sets A_n. We use the inequalities (3.4.26) and (3.4.27) in place of (3.4.15), (3.4.19), and (3.4.20) and conclude that $\nu^a(A_\infty) = 0$ and that

$$t^{-1}\nu^a(A) \le \int_A g \, d\mu \le t\nu^a(A) .$$

Therefore (3.4.28) follows by letting $t \to 1$.

The second equality in (3.4.23) follows from the first equality and from the Lebesgue differentiation theorem. Finally, (3.4.24) follows from (3.4.23).

The Lebesgue–Radon–Nikodym theorem is thus proved. □

Remark 3.4.29 The Lebesgue differentiation theorem was formulated for closed balls, because a Vitali-type covering argument was needed in the proof. Similarly, closed balls were used in the measure differentiation (3.4.21). For technical reasons it is sometimes better to use open balls, however.

If (X, d, μ) is a metric measure space and if $f : X \to V$ is locally integrable then

$$⨍_{B(x,r)} |f(y) - f(x)| \, d\mu(y) \le \frac{\mu(B(x, 2r))}{\mu(B(x, r))} ⨍_{\overline{B}(x,3r/2)} |f(y)-f(x)| \, d\mu(y)$$

for almost every $x \in X$ and for every small enough $r > 0$ depending on x (see Section 3.3). In particular, if μ satisfies the asymptotic doubling property (3.4.4) then we have that

$$\lim_{r\to 0} ⨍_{B(x,r)} |f(y) - f(x)| \, d\mu(y) = 0 \tag{3.4.30}$$

for almost every $x \in X$. Consequently, if μ satisfies (3.4.4) then we also have

$$\lim_{r\to 0} ⨍_{B(x,r)} f(y) \, d\mu(y) = f(x) \tag{3.4.31}$$

for almost every $x \in X$.

By using (3.4.24) and (3.4.31), we infer that

$$\frac{d\nu}{d\mu}(x) = \lim_{r\to 0} \frac{\nu(B(x, r))}{\mu(B(x, r))} \tag{3.4.32}$$

for μ almost every $x \in X$ if μ satisfies (3.4.4) and if ν is a locally finite Borel-regular measure on X.

In particular, (3.4.30), (3.4.31), and (3.4.32) hold on doubling metric measure spaces.

Suprema of measures

Let X be a topological space, and let μ and ν be two Borel measures on X. Define

$$(\mu \vee \nu)(B) := \sup\{\mu(B_1) + \nu(B_2)\} \tag{3.4.33}$$

for every Borel set $B \subset X$, where the supremum is taken over all Borel partitions $B = B_1 \cup B_2$ of B.

We extend $(\mu \vee \nu)$ to all subsets E of X as in (3.3.3):

$$(\mu \vee \nu)(E) := \inf(\mu \vee \nu)(B)\,, \tag{3.4.34}$$

where the infimum is taken over all Borel sets B in X containing E. Note that if E happens to be a Borel set then the two numbers (3.4.33) and (3.4.34) are equal.

Furthermore, notice that $(\mu \vee \nu)(E) \geq \mu(E)$ and $(\mu \vee \nu)(E) \geq \nu(E)$, that is, $(\mu \vee \nu)$ majorizes both μ and ν. It is also straightforward to see that $(\mu \vee \nu) \leq \mu + \nu$, and so $(\mu \vee \nu)$ is locally finite if both μ and ν are locally finite, and that $(\mu \vee \mu) = \mu$.

Lemma 3.4.35 *The set function $\mu \vee \nu$ is a Borel-regular measure on X if both μ and ν are Borel regular.*

Proof From (3.4.34), we can see that $\mu \vee \nu$ is monotonic. If $B_1, B_2, \ldots \subset X$ are Borel sets and if $\bigcup_i B_i = B \cup B'$ is a Borel partition of $\bigcup_i B_i$ then $B_i = (B \cap B_i) \cup (B' \cap B_i)$ is a Borel partition of B_i for each i, whence

$$\mu(B) + \nu(B') \leq \sum_i \mu(B \cap B_i) + \nu(B' \cap B_i) \leq \sum_i (\mu \vee \nu)(B_i)\,.$$

It follows that

$$(\mu \vee \nu)\left(\bigcup_i B_i\right) \leq \sum_i (\mu \vee \nu)(B_i)\,.$$

Next, if $E_1, E_2, \ldots \subset X$ are arbitrary, choose Borel sets $B_i \supset E_i$ such that $\mu(B_i) = \mu(E_i)$ and $\nu(B_i) = \nu(E_i)$. Then $B := \bigcup_i B_i$ is a Borel set containing $\bigcup_i E_i$ and, by (3.4.34),

$$(\mu \vee \nu)\left(\bigcup_i E_i\right) \leq (\mu \vee \nu)\left(\bigcup_i B_i\right) \leq \sum_i (\mu \vee \nu)(B_i)\,.$$

Taking infima over all such B_i, by (3.4.34) we see that

$$(\mu \vee \nu)\left(\bigcup_i E_i\right) \le \sum_i (\mu \vee \nu)(E_i).$$

That is, $\mu \vee \nu$ is subadditive; it follows that $(\mu \vee \nu)$ is a measure on X.

To check that Borel sets are measurable, let $B \subset X$ be a Borel set, and let $T \subset X$ be arbitrary. Fix $\epsilon > 0$ and let B' be a Borel set containing T such that $(\mu \vee \nu)(B') \le (\mu \vee \nu)(T) + \epsilon$. Let $B' \cap B = B_1 \cup B_2$ and $B' \setminus B = B_1' \cup B_2'$ be Borel partitions; then $B' = B_1 \cup B_1' \cup B_2 \cup B_2'$ is a Borel partition and hence

$$\begin{aligned}\mu(B_1) + \nu(B_2) + \mu(B_1') + \nu(B_2') &= \mu(B_1 \cup B_1') + \nu(B_2 \cup B_2') \\ &\le (\mu \vee \nu)(B') \le (\mu \vee \nu)(T) + \epsilon\,.\end{aligned}$$

This gives

$$\begin{aligned}(\mu \vee \nu)(T \cap B) + (\mu \vee \nu)(T \setminus B) &\le (\mu \vee \nu)(B' \cap B) + (\mu \vee \nu)(B' \setminus B) \\ &\le (\mu \vee \nu)(T) + \epsilon\,,\end{aligned}$$

and it follows that $(\mu \vee \nu)$ is a Borel measure. Finally, $(\mu \vee \nu)$ is Borel regular by definition (3.4.34). The lemma follows. □

If $\mu_1, \mu_2, \ldots$ is a sequence of Borel measures on X, we inductively define $\mu_1 \vee \cdots \vee \mu_k = (\mu_1 \vee \cdots \vee \mu_{k-1}) \vee \mu_k$ and then, for the entire sequence of measures, we define

$$\left(\bigvee_{i=1}^{\infty} \mu_i\right)(E) := \lim_{k\to\infty} (\mu_1 \vee \cdots \vee \mu_k)(E)\,, \tag{3.4.36}$$

where we have simplified the notation by writing $((\mu \vee \nu) \vee \zeta) =: (\mu \vee \nu \vee \zeta)$. Note that the limit in (3.4.36) exists, because $\mu(E) \le (\mu \vee \nu)(E)$. It is also clear that $\mu \vee \nu = \nu \vee \mu$ and that

$$\left(\bigvee_{i=1}^{\infty} \mu_i\right)(E) = \sup(\mu_{i_1} \vee \cdots \vee \mu_{i_k})(E)\,;$$

the supremum is taken over all finite collections of indices $\{i_1, \ldots, i_k\}$.

In particular, it follows that if $\{\mu_i : i \in I\}$ is any countable collection of Borel measures on X then we can define

$$\left(\bigvee_{i\in I} \mu_i\right)(E) := \sup(\mu_{i_1} \vee \cdots \vee \mu_{i_k})(E) \tag{3.4.37}$$

unambiguously.

Lemma 3.4.38 *The set function $(\bigvee_{i\in I}\mu_i)$ is a Borel-regular measure on X if all measures μ_i, $i \in I$, are Borel regular, where I is a countable set.*

In view of Lemma 3.4.35, the proof of Lemma 3.4.38 is straightforward and will be left to the reader.

One can define $(\bigvee_{i\in I}\mu_i)$ for any indexing set I as the supremum of the measures $(\bigvee_{j\in J}\mu_j)$, where the supremum is taken over all countable subsets J of I. In the present book we will not need this concept.

Lemma 3.4.39 *Let (X, d, μ) be a Vitali metric measure space and let $\{\mu_i : i \in I\}$ be a countable collection of locally finite Borel regular measures on X. If $(\bigvee_{i\in I}\mu_i)$ is locally finite then*

$$\frac{d(\bigvee_{i\in I}\mu_i)}{d\mu}(x) = \sup_{i\in I}\frac{d\mu_i^a}{d\mu}(x) \tag{3.4.40}$$

for μ-almost-every $x \in X$, where μ_i^a denotes the absolutely continuous part of the measure μ_i with respect to μ.

Proof Put $\nu := (\bigvee_{i\in I}\mu_i)$. We first claim that the absolutely continuous part of ν with respect to μ satisfies

$$\nu^a = \left(\bigvee_{i\in I}\mu_i^a\right).$$

To prove the claim, pick Borel sets $D \subset X$ and $D_i \subset X$ such that $\mu(X \setminus (D \cap (\bigcap_i D_i))) = 0$, $\nu^a = \nu\lfloor D$, and $\mu_i^a = \mu_i\lfloor D_i$ for each $i \in I$ (the Lebesgue–Radon–Nikodym theorem; see (3.4.22)–(3.4.24)). Because $\mu(D \setminus \bigcap_i D_i) = 0$ we have from the properties of the absolutely continuous part of a measure that, for Borel sets E (and hence for all sets E),

$$\begin{aligned}\nu^a(E) &= \nu\left(E \cap D \cap \left(\bigcap_i D_i\right)\right)\\ &= \sup(\mu_{i_1} \vee \cdots \vee \mu_{i_k})\left(E \cap D \cap \left(\bigcap_i D_i\right)\right)\\ &= \sup(\mu_{i_1}^a \vee \cdots \vee \mu_{i_k}^a)(E),\end{aligned}$$

as required. Because $d\nu/d\mu = d\nu^a/d\mu$ almost everywhere, by (3.4.24), we may assume by the preceding claim that all measures under consideration are absolutely continuous with respect to μ.

The preceding being understood, put $\zeta := \sup_{i\in I} d\mu_i/d\mu$. Then ζ is a μ-measurable function on X. We next observe that, for any closed ball

$\overline{B}(x, r)$, for any collection $\{\mu_{i_1}, \ldots, \mu_{i_k}\}$, and for any Borel partition $\overline{B}(x, r) = B_1 \cup \cdots \cup B_k$, we have that

$$\mu_{i_1}(B_1) + \cdots + \mu_{i_k}(B_k) = \sum_{j=1}^{k} \int_{B_j} \frac{d\mu_{i_j}}{d\mu}\, d\mu \le \int_{\overline{B}(x,r)} \zeta \, d\mu \,,$$

which gives

$$\nu(\overline{B}(x, r)) \le \int_{\overline{B}(x,r)} \zeta \, d\mu \,,$$

and hence that

$$\frac{d\nu}{d\mu}(x) \le \zeta(x)$$

for μ-a.e. x by the Lebesgue differentiation theorem, provided that ζ is locally integrable on X, because μ is a Vitali measure. However, if $B \subset X$ is any Borel set then

$$\int_B \frac{d\mu_i}{d\mu}\, d\mu = \mu_i(B) \le \nu(B)\,,$$

which gives

$$\frac{d\mu_i}{d\mu}(x) \le \frac{d\nu}{d\mu}(x)$$

for μ-a.e. x, and hence

$$\zeta(x) \le \frac{d\nu}{d\mu}(x)$$

for μ-almost-every x. This also shows that ζ is locally integrable on X. We conclude that (3.4.40) holds, and the lemma is thereby proved. □

3.5 Maximal functions

We conclude this chapter with a discussion of the Hardy–Littlewood maximal function. In the remainder of the book we will need only the maximal function of a real-valued function but, given the theory developed earlier in this chapter, the proofs are no more involved in the general vector-valued case, which we thus treat for the sake of completeness.

Let (X, d, μ) be a metric measure space and let V be a Banach space. The *Hardy–Littlewood maximal function* Mf of a locally integrable function

$f : X \to V$ is the real-valued function defined by

$$Mf(x) := \sup_{r>0} \fint_{B(x,r)} |f(y)| \, d\mu(y) \, . \tag{3.5.1}$$

The sublinear operator $f \mapsto Mf$ is also called the *Hardy–Littlewood maximal operator*. For brevity, we speak about the *maximal function* of f and the *maximal operator* for f.

It is understood that in (3.5.1), the supremum is taken over only those values of $r > 0$ for which the measure of $B(x, r)$ is finite and positive. The special assumptions in the following lemma are sufficient for the purposes of this book.

If we are working with a fixed metric measure space (X, d, μ), and $f : A \to V$ is an integrable function on a measurable subset $A \subset X$, then it is understood that Mf is the maximal function of f in the metric measure space (A, d, μ) (see Section 3.3).

Lemma 3.5.2 *Assume that every ball in X has finite and positive measure. Then the maximal function Mf of every locally integrable function $f : X \to V$ is measurable.*

Proof It is easy to see that the supremum in (3.5.1) is obtained over positive rational radii r. That is, we have that

$$Mf(x) := \sup_{n} g_n(x) \, ,$$

where

$$g_n(x) := \fint_{B(x,r_n)} |f(y)| \, d\mu(y)$$

for a fixed enumeration $\{r_1, r_2, \dots\}$ of the positive rational numbers. In analogy to Lemma 3.4.16, one can show that for every fixed $r > 0$ the functions $u(x) := \mu(B(x, r))$ and $v(x) := \int_{B(x,r)} |f| \, d\mu$ are lower semicontinuous (see Section 4.2) and hence measurable. Thus each g_n is measurable, and the lemma follows. □

Before we state important theorems on the boundedness of the Hardy–Littlewood maximal operator on L^p-spaces, we recall the definition for weak L^p-spaces.

Let $1 \leq p < \infty$. A measurable function $g : X \to \mathbb{R}$ is said to be in the *weak L^p-space*

$$L^{p,\infty} = L^{p,\infty}(X) = L^{p,\infty}(X, \mu) \tag{3.5.3}$$

if there exists a constant $C > 0$ such that

$$\mu(\{x \in X : |g(x)| > t\}) \leq \frac{C}{t^p} \tag{3.5.4}$$

for all $t > 0$. The least constant C for which the above inequality holds for all $t > 0$ is called the $L^{p,\infty}$-norm of g and is denoted $\|g\|_{L^{p,\infty}(X)}$.

We also require *Cavalieri's principle*:

$$\int_X |g|^p \, d\mu = p \int_0^\infty t^{p-1} \mu(\{x \in X : |g(x)| > t\}) \, dt\,, \tag{3.5.5}$$

which is valid in every measure space (X, μ), for every $0 < p < \infty$, and for every measurable function $g : X \to V$. This principle follows from an application of Tonelli's theorem; see [86, Proposition 6.24, p. 191].

Theorem 3.5.6 *Let X be a doubling metric measure space. The maximal operator maps $L^1(X : V)$ to $L^{1,\infty}(X)$ and $L^p(X : V)$ to $L^p(X)$ for all $1 < p \leq \infty$. More precisely, there exist constants C_p, depending only on p and on the doubling constant of μ, such that*

$$\mu(\{x \in X : Mf(x) > t\}) \leq \frac{C_1}{t} \|f\|_{L^1(X:V)}, \tag{3.5.7}$$

for all $t > 0$, and

$$\|Mf\|_{L^p(X)} \leq C_p \|f\|_{L^p(X:V)} \tag{3.5.8}$$

for all $1 < p \leq \infty$ and for all measurable functions $f : X \to V$.

Proof We begin with the proof of (3.5.7) by applying the $5B$-covering lemma; see Section 3.3. Because this lemma requires the balls to have uniformly bounded diameter, the required estimate is first proved for the *restricted maximal function*

$$M_R f(x) := \sup_{0<r<R} ⨍_{B(x,r)} |f(y)| \, d\mu(y)\,, \tag{3.5.9}$$

for $R > 0$ fixed, with a constant that is independent of R. Then we pass to the limit as $R \to \infty$.

For each x satisfying $M_R f(x) > t$, choose a ball $B(x, r)$, with $0 < r < R$, such that

$$t\mu(B(x, r)) < \int_{B(x,r)} |f| \, d\mu\,.$$

Using the $5B$-covering lemma, we extract from the collection of all such balls a countable pairwise disjoint subcollection $\mathcal{G}$ such that (3.3.26) is satisfied. Then

$$\mu(\{x \in X : M_R f(x) > t\}) \le \sum_{B \in \mathcal{G}} \mu(5B) \le C \sum_{B \in \mathcal{G}} \mu(B)$$
$$\le \frac{C}{t} \sum_{B \in \mathcal{G}} \int_B |f| \, d\mu \le \frac{C}{t} \int_X |f| \, d\mu \,,$$

where $C \ge 1$ depends only on the doubling constant of the measure μ. As mentioned earlier, inequality (3.5.7) follows from this since $M_R f \to Mf$ as $R \to \infty$.

To prove (3.5.8), notice first that the case $p = \infty$ is trivial; we have $C_\infty = 1$. Assume next that $1 < p < \infty$. Let $f \in L^p(X : V)$, fix $t > 0$, and write $f = g + h$, where $g := f \chi_{\{|f| \le t/2\}}$ and $h := f \chi_{\{|f| > t/2\}}$. Then

$$Mf(x) \le Mg(x) + Mh(x) \le \frac{t}{2} + Mh(x) \,,$$

so that

$$\{x \in X : Mf(x) > t\} \subset \left\{x \in X : Mh(x) > \frac{t}{2}\right\}.$$

We note that Mf is measurable, by Lemma 3.5.2, and use Cavalieri's principle (3.5.5) to obtain

$$\int_X |Mf|^p \, d\mu = p \int_0^\infty t^{p-1} \mu(\{x \in X : Mf(x) > t\}) \, dt$$
$$\le p \int_0^\infty t^{p-1} \mu\left(\left\{x \in X : Mh(x) > \frac{t}{2}\right\}\right) dt.$$

This gives, by the weak-type estimate (3.5.7) together with Cavalieri's principle (applied thrice) and Fubini's theorem,

$$\begin{aligned}
&\int_X |Mf|^p \, d\mu \\
&\quad \le C \int_0^\infty t^{p-1} t^{-1} \int_X |h| \, d\mu \, dt \\
&\quad = C \int_0^\infty t^{p-2} \Bigg(\frac{t}{2} \mu\left(\left\{x \in X : |f(x)| > \frac{t}{2}\right\}\right) \\
&\qquad\qquad + \int_{t/2}^\infty \mu(\{x \in X : |f(x)| > s\}) \, ds \Bigg) \, dt \\
&\quad \le C \int_X |f|^p \, d\mu + C \int_0^\infty \int_0^{2s} \mu(\{x \in X : |f(x)| > s\}) t^{p-2} \, dt \, ds \\
&\quad \le C \int_X |f|^p \, d\mu \,.
\end{aligned}$$

Thus (3.5.8) follows and the proof of Theorem 3.5.6 is complete. □

The weak estimate (3.5.7) can be used to obtain the following integral bound for small powers of the (restricted) maximal function. We will need the next lemma later on.

Lemma 3.5.10 *Let X be a doubling metric measure space and let $0 < q < 1$. Then*

$$\left(\fint_B (M_{2\,\mathrm{rad}(B)} f)^q \, d\mu\right)^{1/q} \leq C \fint_{3B} |f| \, d\mu \tag{3.5.11}$$

for every open ball B in X and for every measurable function $f : 3B \to V$, where $C > 0$ depends only on q and the doubling constant of μ.

Proof We have from the weak estimate (3.5.7) that

$$\mu(\{x \in B : M_{2\,\mathrm{rad}(B)} f(x) > t\}) \leq \frac{C_1}{t} \int_{3B} |f| \, d\mu \, .$$

We then compute by the aid of Cavalieri's principle (3.5.5) that

$$\begin{aligned}\int_B (M_{2\,\mathrm{rad}(B)} f)^q \, d\mu &\leq q \int_0^A t^{q-1} \mu(B) \, dt \\ &\quad + q \, C_1 \int_{3B} |f| \, d\mu \int_A^\infty t^{q-2} \, dt \\ &\leq \mu(B) \, A^q + \frac{q \, C_1}{1-q} A^{q-1} \int_{3B} |f| \, d\mu \, .\end{aligned}$$

Putting $A = C_1 \int_{3B} |f| \, d\mu / \mu(B)$ in the preceding expression gives

$$\int_B (M_{2\,\mathrm{rad}(B)} f)^q \, d\mu \leq \frac{C_1^q}{1-q} \mu(B)^{1-q} \left(\int_{3B} |f| \, d\mu\right)^q ,$$

which gives (3.5.11). The lemma follows. □

To close this chapter, we introduce a minor variant of the Hardy–Littlewood maximal operator defined in (3.5.1).

The *noncentered maximal function* of a locally integrable function $f : X \to V$ is the function defined by

$$M_* f(x) := \sup_{x \in B} \fint_B |f(y)| \, d\mu(y) \, . \tag{3.5.12}$$

In other words, the supremum is taken over all balls B in X that contain the given point x, and not just those that are centered at x. Moreover, we observe the same convention as in (3.5.1), namely that the supremum is taken only over those balls that have finite and positive measure. Obviously,

$$Mf(x) \leq M_* f(x) \tag{3.5.13}$$

for $x \in X$. Moreover, if X is a doubling metric measure space then

$$Mf(x) \leq M_* f(x) \leq C\, Mf(x) \tag{3.5.14}$$

for $x \in X$, where $C \geq 1$ depends only on the doubling constant of the measure; this is easy to see.

One technical advantage offered by the noncentered maximal function is that sets of the form $\{x \in X : M_* f(x) > t\}$, $t > 0$, are open in X. (In particular, $M_* f$ is always measurable.) By using this, we can prove the following strengthening of equation (3.5.7).

Proposition 3.5.15 *Let X be a doubling metric measure space. If $f \in L^1(X : V)$ then*

$$\lim_{t\to\infty} t\mu(\{x \in X : Mf(x) > t\}) = \lim_{t\to\infty} t\mu(\{x \in X : M_* f(x) > t\}) = 0. \tag{3.5.16}$$

Furthermore,

$$\mu(\{x \in X : M_* f(x) > t\}) \leq \frac{C}{t} \int_{\{x\in X : M_* f(x) > t\}} |f|\, d\mu.$$

Proof By (3.5.13), it suffices to consider the noncentered maximal function. The argument is essentially contained in the proof of (3.5.7), but for completeness we will repeat the main steps.

We consider first a restricted version of the noncentered maximal function,

$$M_{*,R} f(x) := \sup \fint_B |f(y)|\, d\mu(y),$$

where $R > 0$ and the supremum is taken over all balls B in X containing x with diameter not exceeding R. As in the proof of (3.5.7), we use the $5B$-covering lemma and find that, for fixed $R > 0$ and $t > 0$, the open set $\{x \in X : M_{*,R} f(x) > t\}$ can be covered by a countable collection of balls $\{5B_i\}$ such that

$$t\mu(B_i) < \int_{B_i} |f|\, d\mu$$

and such that the collection $\{B_i\}$ is pairwise disjoint; note that $B_i \subset \{x \in X : M_{*,R} f(x) > t\}$ for each i. Thus

$$\begin{aligned} \mu(\{x \in X : M_{*,R} f(x) > t\}) &\leq \sum_i \mu(5B_i) \leq C \sum_i \mu(B_i) \\ &\leq \frac{C}{t} \int_{\{x\in X : M_{*,R} f(x) > t\}} |f|\, d\mu, \end{aligned}$$

where $C \geq 1$ depends only on the doubling constant of μ. Letting $R \to \infty$ in the preceding inequality yields

$$\mu(\{x \in X : M_*f(x) > t\}) \le \frac{C}{t} \int_{\{x \in X : M_*f(x) > t\}} |f| \, d\mu \, .$$

The assertion (3.5.16) follows upon observing that $\mu(\{x \in X : M_*f(x) > t\}) \to 0$ as $t \to \infty$ by (3.5.14) and (3.5.7). The proposition is proved. □

3.6 Notes to Chapter 3

Vector-valued integration theories sprang up in the 1930s from attempts to understand differentiation theorems for Banach-space-valued functions. A vivid history of the subject can be found in the introduction of the monograph [75] by Diestel and Uhl. See also [83], [26], and [136]. Vector-valued integration has become a central tool in infinite-dimensional stochastic processes and in the geometric theory of Banach spaces, [181], [38]. The introductory material presented here is standard.

The monographs [83], [86], [197], and [237] are recommended sources for the basic measure theory required in this book. Most results in Section 3.3 can be found in these sources and especially in [83]. Standard texts with an emphasis on analysis typically treat Borel and Radon measures on spaces that are locally compact ([138] being a notable exception). More general discussions, as in Section 3.3, are commonly found in texts on probability theory. See for example [243], [251], [29], and [76]. The fact that Borel-regular measures on complete and separable metric spaces are Radon (Proposition 3.3.44) is credited to Oxtoby and Ulam [221] in [243, p. 122]. (In fact, the proof given in this book appears in a footnote in [221, p. 561] and is credited to Ulam there.) We have not found Proposition 3.3.46 explicitly stated in the literature, although its content should be well known; it can be extracted from [243, Part I, Chapter II] for example.

The density and separability results in Section 3.3 are standard, albeit difficult to locate in the literature in the generality given here.

Federer's monograph [83, Chapters 2.8 and 2.9] contains an extensive discussion of various covering and differentiation theorems. The material in Section 3.4 can be found there.

The general notion of a metric measure space has recently gained prominence in many new areas in analysis and geometry. Gromov in particular has emphasized the metric features of Riemannian spaces and the interplay between distance and volume. See [107].

Doubling measures came into vogue in the 1970s through the work of Coifman and Weiss, [63], [64]; see also [62]. It was discovered that large parts of basic real and harmonic analysis go over to *spaces of homogeneous type*, which are (quasi-)metric measure spaces equipped with a doubling measure. The Hardy–Littlewood maximal theorem 3.5.6 in spaces with a doubling measure was proved in 1956 by Smith [252] and by Rauch [229]. For more recent developments, see [257], [69], [107, Appendix B], [7], [120], and the references in these works. However, it has recently been discovered that in fact much of the standard harmonic analysis and Calderón–Zygmund theory previously established in a doubling setting persists even without that assumption on the measure. Notable work in this direction includes that of Nazarov, Treil, and Volberg [217], [218] and Tolsa [267], [268]. Verdera's survey article [275] is highly recommended for a summary of these works and related developments.

4

Lipschitz functions and embeddings

In this chapter, we gather further miscellaneous basic results that will be used in later chapters. We discuss Lipschitz functions, extension and embedding theorems, Hausdorff measures, lower semicontinuous functions, and functions of bounded variation with values in an arbitrary metric space.

4.1 Lipschitz functions, extensions, and embeddings

A function $f : X \to Y$ from a metric space $X = (X, d_X)$ to a metric space $Y = (Y, d_Y)$ is said to be *L-Lipschitz* if there exists a constant $L \geq 0$ such that

$$d_Y(f(a), f(b)) \leq L d_X(a, b) \tag{4.1.1}$$

for each pair of points $a, b \in X$. We also say that a function is *Lipschitz* if it is L-Lipschitz for some L. The smallest L such that (4.1.1) holds for each pair of points $a, b \in X$ is called the *Lipschitz constant* of f.

If $f : X \to Y$ is a Lipschitz bijection whose inverse is also Lipschitz, we say that f is a *bilipschitz map* between X and Y and that X and Y are *bilipschitz equivalent*. The term *L-bilipschitz* is self-explanatory. A 1-bilipschitz map is an *isometry*. Two metric spaces are *isometric* if there is an isometry between them. We also say that X *admits a bilipschitz embedding into* Y if there is a bilipschitz embedding of X into Y. Recall that an embedding is a map that is a homeomorphism onto its image; this concept was used earlier, in Section 3.3. A 1-bilipschitz embedding is called an *isometric embedding*. If X admits an isometric embedding into Y, we often suppress the embedding from the notation and write $X \subset Y$.

We say that a function $f : X \to Y$ is *locally Lipschitz* if every point in X has a neighborhood such that the restriction of f to this neighborhood is Lipschitz. The term *locally L-Lipschitz* means that these restrictions are L-Lipschitz.

Lipschitz functions play a central role in the theory of Sobolev spaces as developed in this book. In particular, in general spaces Lipschitz functions constitute an important substitute for smooth functions. We will prove, for example, that in many interesting cases locally Lipschitz functions are dense in a Sobolev space. To this end, in the present chapter we will study the density of Lipschitz functions in other situations.

We begin with the following simple but important *Lipschitz extension lemma*, also known as the *McShane–Whitney extension lemma*.

McShane–Whitney extension lemma *Let $X = (X, d)$ be a metric space, let $A \subset X$, and let $f : A \to \mathbb{R}$ be an L-Lipschitz function. Then there exists an L-Lipschitz function $F : X \to \mathbb{R}$ such that $F|_A = f$.*

Proof Without loss of generality we can assume that A is nonempty. We define F by the formula

$$F(x) = \inf\{f(a) + Ld(a, x) : a \in A\} \tag{4.1.2}$$

for $x \in X$. For a fixed point $a_0 \in A$, we have

$$\begin{aligned} f(a) + Ld(a, x) &\geq f(a) + Ld(a, a_0) - Ld(a_0, x) \\ &\geq f(a_0) - Ld(a_0, x), \end{aligned} \tag{4.1.3}$$

so that $F(x) > -\infty$ for each $x \in X$. Because the function $x \mapsto f(a) + Ld(a, x)$ is L-Lipschitz for each given $a \in A$, we find that F is everywhere the finite pointwise infimum of L-Lipschitz functions and hence itself L-Lipschitz. More specifically, given $x, y \in X$, for $\epsilon > 0$ we can find $a_\epsilon^y \in A$ such that $F(y) \geq f(a_\epsilon^y) + Ld(a_\epsilon^y, y) - \epsilon$. Noticing that $F(x) \leq f(a_\epsilon^y) + Ld(a_\epsilon^y, y)$ by definition, we conclude that

$$F(x) - F(y) \leq Ld(a_\epsilon^y, y) - Ld(a_\epsilon^y, x) + \epsilon \leq Ld(x, y) + \epsilon.$$

Letting $\epsilon \to 0$ gives $F(x) - F(y) \leq Ld(x, y)$. By symmetry, we also have $F(y) - F(x) \leq Ld(x, y)$, that is, F is L-Lipschitz. Finally, from (4.1.3) it follows that $F(a) = f(a)$ for $a \in A$. The lemma follows. □

Remark 4.1.4 Formula (4.1.2) provides the largest L-Lipschitz extension of f in the sense that if $G : X \to \mathbb{R}$ is L-Lipschitz such that $G|A = f$ then $G \leq F$. Similarly, the formula

$$F(x) = \sup\{f(a) - Ld(a, x) : a \in A\} \tag{4.1.5}$$

defines the smallest L-Lipschitz extension of an L-Lipschitz map $f : A \to \mathbb{R}$.

Kirszbraun's extension theorem asserts that the conclusion of the McShane–Whitney extension lemma remains true if $X = \mathbb{R}^m$ and f maps to $\mathbb{R}^n$ for $n, m \geq 1$. More generally, if H and H' are Hilbert spaces, $A \subset H$, and $f : A \to H'$ is L-Lipschitz then there exists an L-Lipschitz function $F : H \to H'$ such that $F|A = f$. These assertions are harder to prove, and will not be used in this book; see [83, 2.10.43] and [26, Section 1.2]. The following weaker result follows by applying the above lemma to the coordinate functions of an $\mathbb{R}^n$-valued function.

Corollary 4.1.6 *Let $X = (X, d)$ be a metric space, let $A \subset X$, and let $f : A \to \mathbb{R}^n$ be an L-Lipschitz function. Then there exists an $L\sqrt{n}$-Lipschitz function $F : X \to \mathbb{R}^n$ such that $F|A = f$.*

If we employ the l^∞-norm $|\cdot|_\infty$ in $\mathbb{R}^n$ then the factor $\sqrt{n}$ in Corollary 4.1.6 is unnecessary. More generally, we have the following result.

Corollary 4.1.7 *Let $X = (X, d)$ be a metric space, let $A \subset X$, and let $f : A \to L^\infty(Y)$ be an L-Lipschitz function, where Y is any set. Then there exists an L-Lipschitz function $F : X \to L^\infty(Y)$ such that $F|A = f$.*

Proof To each $z \in X$ we associate a function $F(z) \in L^\infty(Y)$. This is done arguing as in the proof of Lemma 4.1. We define

$$F(z)(y) = \inf\{f(a)(y) + Ld(z, a) : a \in A\}$$

when $y \in Y$. Fixing $a_0 \in A$, as in the proof of the lemma, we infer that $F(z) \geq -\|f(a_0)\|_{L^\infty(Y)} - Ld(a_0, z)$, and the definition of $F(z)$ immediately yields that $F(z)(y) \leq f(a_0)(y) + Ld(z, a_0)$ for every $y \in Y$. Thus $\|F(z)\|_{L^\infty(Y)} \leq \|f(a_0)\|_{L^\infty(Y)} + Ld(a_0, z) < \infty$. It is now easy to check that F is L-Lipschitz on X with $F(z) = f(z)$ when $z \in A$. □

The importance of Corollary 4.1.7 lies in the following fact.

Kuratowski embedding theorem *Every metric space Y embeds isometrically into the Banach space $L^\infty(Y)$.*

Proof Fix a point $y_0 \in Y = (Y, d)$. For each $y' \in Y$ define $f_{y'} : Y \to \mathbb{R}$ by

$$f_{y'}(y) = d(y, y') - d(y, y_0). \tag{4.1.8}$$

The triangle inequality implies that $f_{y'}$ is bounded and that on the one hand

$$|f_{y'}(y) - f_{y''}(y)| = |d(y, y') - d(y, y'')| \leq d(y', y'')$$

for every $y \in Y$. On the other hand, upon choosing $y = y'$ we obtain that

$$\|f_{y'} - f_{y''}\|_\infty = d(y', y''),$$

and the theorem follows. □

The target space for the embedding in the Kuratowski theorem depends on the space itself. For separable metric spaces, we can use a universal target.

Fréchet embedding theorem *Every separable metric space embeds isometrically into the Banach space l^∞.*

Proof Choose a countable dense set $\{y_i : i = 0, 1, 2, \dots\}$ in a metric space $Y = (Y, d)$. It is straightforward to check that the map

$$y \mapsto (d(y, y_1) - d(y_1, y_0), d(y, y_2) - d(y_2, y_0), \dots) \tag{4.1.9}$$

determines an isometric embedding of Y in l^∞. □

Remark 4.1.10 There is no canonical isometric embedding of a metric space Y in $L^\infty(Y)$, or in l^∞ in the case where Y is separable. The embedding $y' \mapsto f_{y'}$ in (4.1.8) depends on the chosen base point y_0, and the embedding in (4.1.9) depends on the chosen dense set $\{y_i\}$. For many applications the particular choices are immaterial, but some caution is necessary; see Section 7.6.

While the Fréchet embedding theorem asserts that every separable metric space is isometrically embeddable in l^∞, this is somewhat unsatisfactory as l^∞ is itself not separable. By a theorem of Banach, [21, Théorème 9, p. 185], [121, Theorem 3.6], every separable metric space admits an isometric embedding in the separable Banach space $C([0, 1])$. In this book, we will frequently embed separable metric spaces in Banach spaces, and Banach's theorem could be used to ensure a universal separable target. However, the above embedding theorems of Kuratowski and Fréchet are sufficient for our purposes here. Isometric embeddings of metric spaces in Banach spaces mostly provide a convenient framework analysis for the required in this book, and the specific structure of the receiving space is not so important.

Simple examples show that on the one hand there need not be a Lipschitz (or even continuous) extension of a Lipschitz map $f : A \to Y$, if $A \subset X$ and X and Y are arbitrary metric spaces. This happens, for instance, when $A = Y$ is the circle $S(0, 1)$ in the standard plane X and f is the identity mapping. On the other hand, it follows from the preceding discussion that a Lipschitz extension of f always exists if we think of f mapping into $L^\infty(Y)$, which contains an isometric copy of Y. This point of view is sometimes useful.

Doubling spaces

An *ϵ-separated set*, $\epsilon > 0$, in a metric space is a set such that every two distinct points in the set have distance at least ϵ. A metric space X is called *doubling with constant* N, where $N \geq 1$ is an integer, if, for each ball $B(x, r)$, every $r/2$-separated subset of $B(x, r)$ has at most N points. We also say that X is *doubling* if it is doubling with some constant that need not be mentioned. It is clear that every subset of a doubling space is doubling with the same constant.

It is immediate that if $X = (X, d, \mu)$ is a doubling metric measure space then X is doubling. Indeed, if an $r/2$-separated subset of $B(x, r)$ contains k points $x_1, \ldots, x_k$ then we have, by the doubling property of the measure and by the pairwise disjointedness of the balls $B(x_i, r/4)$, $i = 1, \ldots, k$, that

$$\begin{aligned}\frac{k}{C_\mu^4}\,\mu(B(x, 2r)) &\leq \sum_i C_\mu^{-1}\mu\left(B\left(x_i, \frac{r}{2}\right)\right)\\ &\leq \sum_i \mu\left(B\left(x_i, \frac{r}{4}\right)\right) \leq \mu(B(x, 2r)),\end{aligned}$$

and thus we conclude that $k \leq C_\mu^4$. Here we have used the fact that balls have positive and finite measure, which follows from the fact that μ is doubling, locally finite, and nontrivial. The above argument also shows that if a metric space X is equipped with a nontrivial locally finite doubling measure then X is separable and so gives a metric measure space in the sense of Section 3.3.

However, there are doubling metric spaces, even open subsets of the real line, that do not admit doubling measures. See Section 4.5.

We now record some elementary results about doubling spaces. The first is an alternate characterization of doubling spaces.

Lemma 4.1.11 *If X is a doubling metric space with constant N then every open ball of radius $r > 0$ in X can be covered by N open balls of radius $r/2$. Conversely, if X is a metric space such that every open ball of radius $r > 0$ in X can be covered by M open balls of radius $r/2$ then X is doubling with constant M^2.*

The doubling condition can be applied at small scales as follows.

Lemma 4.1.12 *Let X be a doubling metric space with constant N and let $k \geq 1$ be an integer. Then every $2^{-k}r$-separated set in every ball $B(x, r)$ in X has at most N^k points.*

Lemma 4.1.13 *Every doubling metric space is separable.*

A metric space is said to be *proper* if every closed ball in it is compact.

Lemma 4.1.14 *The metric completion of a doubling metric space is doubling with the same constant. Moreover, a complete doubling metric space is proper.*

Proofs for the preceding four lemmas are left to the reader.

Whitney decomposition

Open subsets of doubling spaces can be covered by balls that constitute a covering akin to the classical *Whitney decomposition* of open subsets of $\mathbb{R}^n$. We next discuss such coverings.

Proposition 4.1.15 *Let $X = (X, d)$ be a doubling metric space with constant N and let Ω be an open subset of X such that $X \setminus \Omega \neq \emptyset$. There exists a countable collection $\mathcal{W}_\Omega = \{B(x_i, r_i)\}$ of balls in Ω such that*

$$\Omega = \bigcup_i B(x_i, r_i) \tag{4.1.16}$$

and

$$\sum_i \chi_{B(x_i, 2r_i)} \leq 2N^5 , \tag{4.1.17}$$

where

$$r_i = \frac{1}{8} \operatorname{dist}(x_i, X \setminus \Omega) . \tag{4.1.18}$$

Above, it is important to consider $\operatorname{dist}(x, X \setminus \Omega)$ rather than $\operatorname{dist}(x, \partial\Omega)$ because it could very well happen in our setting that $B(x, \operatorname{dist}(x, \partial\Omega))$ intersects $X \setminus \Omega$.

Proof For $x \in \Omega$, denote $d(x) := \operatorname{dist}(x, X \setminus \Omega)$. Then for $k \in \mathbb{Z}$ let

$$\mathcal{F}_k := \left\{ B\left(x, \frac{1}{40} d(x)\right) : x \in \Omega \text{ with } 2^{k-1} < d(x) \leq 2^k \right\} .$$

By the $5B$-covering lemma, in Section 3.3, we can pick a countable pairwise disjoint subfamily $\mathcal{G}_k \subset \mathcal{F}_k$ such that

$$\bigcup_{B \in \mathcal{F}_k} B \subset \bigcup_{B \in \mathcal{G}_k} 5B .$$

We claim that

$$\mathcal{W}_\Omega := \bigcup_{k=1}^{\infty} \{5B : B \in \mathcal{G}_k\}$$

satisfies (4.1.16)–(4.1.18).

It is immediate from the construction that (4.1.16) and (4.1.18) hold. To prove (4.1.17), suppose that there is a point in Ω that belongs to M balls of the form $2B$, $B \in \mathcal{W}_\Omega$. We label these balls conveniently as $B(x_1, \frac{1}{4}d(x_1))$, $\ldots$, $B(x_M, \frac{1}{4}d(x_M))$, with $d(x_1) \geq d(x_i)$ for each $i = 1, \ldots, M$. We readily find, by using the triangle inequality, that

$$d(x_i) \geq \frac{3}{5} d(x_1) \tag{4.1.19}$$

and that

$$B\left(x_i, \frac{1}{4} d(x_i)\right) \subset B\left(x_1, \frac{3}{4} d(x_1)\right)$$

for each $i = 1, \ldots, M$. If x_i and x_j are centers of balls belonging to the same family $\mathcal{F}_k$ then

$$d(x_i, x_j) \geq \frac{1}{20} \min\{d(x_i), d(x_j)\} \geq \frac{1}{40} d(x_1)$$

whenever $i \neq j$. In other words, there is a ball of radius $\frac{3}{4}d(x_1)$ containing a $\frac{1}{40}d(x_1)$-separated set of M elements. Lemma 4.1.12 then gives that at most N^5 of these balls can have their centers in $\mathcal{F}_k$ for a fixed k.

Suppose $x_1 \in \mathcal{F}_{k_1}$. From above, we have that $d(x_1) \geq d(x_i) \geq \frac{3}{5}d(x_1)$ for all $i = 2, \ldots, M$ and thus we conclude that all centers are contained in $\mathcal{F}_{k_1 - 1} \cup \mathcal{F}_{k_1}$. Thus (4.1.17) follows.

The proof of the proposition is complete. □

Lipschitz partition of unity

Let X be a doubling metric space with constant N, let Ω be an open subset of X such that $X \setminus \Omega \neq \emptyset$, and let $\mathcal{W}_\Omega$ be a collection of balls in Ω as in Proposition 4.1.15 satisfying (4.1.16)–(4.1.18). Given a ball $B(x_i, r_i) \in \mathcal{W}_\Omega$, define

$$\psi_i(x) := \min\left\{\frac{1}{r_i} \operatorname{dist}(x, X \setminus B(x_i, 2r_i)), 1\right\}.$$

Then ψ_i is $1/r_i$-Lipschitz. Moreover, (4.1.16) and (4.1.17) give that

$$1 \leq \sum_i \psi_i(x) \leq 2N^5.$$

Set

$$\varphi_i(x) := \frac{\psi_i(x)}{\sum_k \psi_k(x)}. \tag{4.1.20}$$

Then the functions φ_i satisfy the following properties for some constant $C \geq 1$ that depends only on the doubling constant of X:

(i) $\varphi_i(x) = 0$ for $x \notin B(x_i, 2r_i)$, and for every $x \in \Omega$ we have that $\varphi_i(x) \neq 0$ for at most C indices i;

(ii) $0 \le \varphi_i \le 1$ and $\varphi_i | B(x_i, r_i) \ge C^{-1}$;
(iii) φ_i is C/r_i-Lipschitz;
(iv) $\sum_i \varphi_i(x) = 1$ for every $x \in \Omega$.

Indeed, it is obvious from (4.1.17) and from the definitions that (i), (ii), and (iv) hold with $C = 2N^5$. A routine argument shows that (iii) holds with $C = 5N^5$.

A collection $\{\varphi_i\}$ as above is called a *Lipschitz partition of unity* of the open set Ω.

Next we formulate and prove the following Lipschitz extension theorem.

Theorem 4.1.21 *Let $X = (X, d)$ be a doubling metric space, let $A \subset X$, and let $f : A \to V$ be an L-Lipschitz function from A into a Banach space V. Then there exists a CL-Lipschitz function $F : X \to V$ such that $F|_A = f$, where $C \ge 1$ is a constant that depends only on the doubling constant of X.*

Proof By considering the metric completion of X we may assume that X is complete, and hence proper by Lemma 4.1.14. We may further assume that A is closed, because an L-Lipschitz function f as in the hypotheses can uniquely be extended to an L-Lipschitz function on the closure of A. Finally, we may assume that neither A nor $\Omega := X \setminus A$ is empty. Let $\mathcal{W}_\Omega = \{B(x_i, r_i)\}$ be a collection of balls in Ω as Proposition 4.1.15, and let $\{\varphi_i\}$ be a Lipschitz partition of unity satisfying the list (i)–(iv) given before the theorem. Select, for each i, a point $y_i \in A$ such that

$$8r_i = \operatorname{dist}(x_i, A) = d(x_i, y_i) .$$

Because X is proper and A is closed, such points y_i can be found.

Recall the convention that we let $C \ge 1$ denote any constant that may depend on the doubling constant of X but not on other parameters.

We set

$$F(x) := \sum_i \varphi_i(x) f(y_i)$$

when $x \in \Omega$ and $F(x) := f(x)$ when $x \in A$, and claim that F is an extension of f as desired. First we observe that F is a well-defined function from X to V by (i) above. Next, fix $y \in A$ and let $x \in \Omega$. Then, by (iv),

$$\begin{aligned} |F(x) - f(y)| &= \left| \sum_i (f(y_i) - f(y)) \varphi_i(x) \right| \\ &\le C \max_{i:2B_i \ni x} |f(y_i) - f(y)| \end{aligned}$$

$$
\begin{aligned}
&\le CL \max_{i:2B_i \ni x} d(y_i, y) \\
&\le CLd(x, y)\,, \qquad (4.1.22)
\end{aligned}
$$

where we use the abbreviation $B_i = B(x_i, r_i)$ and where the last inequality follows from the estimate

$$
\begin{aligned}
d(y, y_i) &\le d(y, x) + d(x, x_i) + d(x_i, y_i) \\
&\le d(x, y) + \frac{5}{4} d(x_i, y_i) \le \frac{8}{3} d(x, y)\,. \qquad (4.1.23)
\end{aligned}
$$

We obtain in particular that $F(x) \to f(y)$ as $\Omega \ni x \to y \in A$. It remains to show that F is CL-Lipschitz. To this end fix $a, b \in X$. By what was proved in the preceding paragraph, we may assume that a and b both lie in Ω. Suppose first that $a, b \in 2B_j$ for some j, where the notation is as before. Then, again by (iv),

$$
\begin{aligned}
|F(a) - F(b)| &= \left| \sum_i f(y_i)(\varphi_i(a) - \varphi_i(b)) \right| \\
&= \left| \sum_i (f(y_i) - f(y_j))(\varphi_i(a) - \varphi_i(b)) \right| \\
&\le CLr_j \sum_i |\varphi_i(a) - \varphi_i(b)| \le CL|a - b|\,, \qquad (4.1.24)
\end{aligned}
$$

where the penultimate inequality follows from an argument analogous to that in (4.1.23), and where the last inequality follows from (iii) and the fact that if $2B_i \cap 2B_j$ is nonempty then the radii r_i, r_j satisfy $r_i \ge \frac{3}{5} r_j$ by (4.1.19).

It remains to consider the case where $a \in B_j$ for some j while $b \notin 2B_k$ for any k with $a \in B_k$. Let $y_a \in A$ be a point such that $\operatorname{dist}(a, A) = d(a, y_a)$, and similarly for b. Using (4.1.22), we obtain the following estimate:

$$
\begin{aligned}
|F(a) - F(b)| &\le |F(a) - F(y_a)| + |F(y_a) - F(y_b)| + |F(y_b) - F(b)| \\
&\le CLd(a, y_a) + Ld(y_a, y_b) + CLd(y_b, b) \\
&\le CLd(a, b) \qquad (4.1.25)
\end{aligned}
$$

where again the last inequality follows from the conditions on a, b that give $d(a, y_a) = d(a) < 4d(a, b)$, $d(b, y_b) < 4d(a, b)$, and $d(y_a, y_b) \le d(a, y_a) + d(a, b) + d(b, y_b)$. The theorem now follows on combining (4.1.22), (4.1.24), and (4.1.25). □

Lipschitz extension pairs A pair of metric spaces (X, Y) is said to have the *Lipschitz extension property* if there is an increasing function

$\Psi : (0, \infty) \to (0, \infty)$ such that the following holds: for every subset $A \subset X$ and for every L-Lipschitz map $f : A \to Y$ there exists a $\Psi(L)$-Lipschitz map $F : X \to Y$ such that $F|_A = f$. If one can choose $\Psi(t) = Ct$ for some constant $C \geq 1$, we also say that the pair (X, Y) has the *Lipschitz extension property with constant* C. We often omit the function Ψ from the terminology and speak of a *Lipschitz extension property* for brevity.

It follows from the results earlier in this chapter that the pairs $(X, \mathbb{R}^n)$ and $(X, l^\infty(Y))$ have the Lipschitz extension property for arbitrary X and Y, as does the pair (X, V) for every Banach space V provided that X is doubling. However, it is not true that (X, l^2) has the Lipschitz extension property for every metric space X [26, Corollary 1.29].

4.2 Lower semicontinuous functions

Let $X = (X, d)$ be a metric space. A function $f : X \to (-\infty, \infty]$ is said to be *lower semicontinuous* if the set $\{x \in X : f(x) > a\}$ is open for each $a \in \mathbb{R}$.

The noncentered maximal function considered in (3.5.12) is a lower semicontinuous function. A typical lower semicontinuous function that is not, in general, continuous is the characteristic function of an open set. It is easy to see that $f : X \to (-\infty, \infty]$ is lower semicontinuous if and only if

$$\liminf_{y \to x} f(y) \geq f(x) \tag{4.2.1}$$

for every $x \in X$. Thus, if f is lower semicontinuous and $f(x) = \infty$ for a point $x \in X$ then f is continuous (in the extended sense) at x.

A function f is said to be *upper semicontinuous* if $-f$ is lower semicontinuous. Thus, a function is continuous if and only if it is both upper and lower semicontinuous.

Lower semicontinuous functions on a given metric space form a positive cone, closed under the pointwise minimum operation; that is, if f and g are lower semicontinuous and if $c \geq 0$ then both $cf + g$ and $\min\{f, g\}$ are lower semicontinuous. Moreover, the pointwise supremum of an arbitrary family of lower semicontinuous functions is lower semicontinuous. These facts are easily verified.

Proposition 4.2.2 *Let (X, d) be a metric space, let $c \in \mathbb{R}$, and let $f : X \to [c, \infty]$ be lower semicontinuous. Then there exists a sequence (f_i) of Lipschitz functions on X such that $c \leq f_i \leq f_{i+1} \leq f$ and $\lim_{i \to \infty} f_i(x) = f(x)$ for each $x \in X$.*

Proof Define, for each $i = 1, 2, \ldots$, a function f_i on X by

$$f_i(x) = \inf\{f(y) + id(x, y) : y \in X\}.$$

Then, following the argument found in the proof of the McShane–Whitney extension lemma (see Section 4.1), we see that each f_i is i-Lipschitz with $c \leq f_i(x) \leq f_{i+1}(x) \leq f(x)$ for each $x \in X$. Fix $x \in X$. Assume first that $f(x) = \infty$. Let $M > 0$, and choose $\epsilon > 0$ such that $f > M$ on the ball $B(x, \epsilon)$. Therefore $f_i(x)$ is at least the minimum of the numbers M and $c + i\epsilon$. For every i sufficiently large that $c + i\epsilon > M$ we have that $f_i(x) \geq M$, which implies that $\lim_{i\to\infty} f_i(x) = \infty = f(x)$.

Next, assume that $f(x) < \infty$. Let $M < f(x)$, and choose $\epsilon > 0$ such that $f > M$ on the ball $B(x, \epsilon)$. As above, we find that $f_i(x) \geq M$ for all large i and hence that $\lim_{i\to\infty} f_i(x) = f(x)$ in this case as well. The proposition follows. □

Proposition 4.2.2 together with the dominated convergence theorem gives the following corollary.

Corollary 4.2.3 *Let $X = (X, d, \mu)$ be a metric measure space, let $1 \leq p < \infty$, and let $f : X \to [0, \infty]$ be a p-integrable lower semicontinuous function. Then there exists a sequence (f_i) of Lipschitz functions on X such that $0 \leq f_i \leq f_{i+1} \leq f$ and $f_i \to f$ both pointwise and in $L^p(X)$ as $i \to \infty$.*

In every metric measure space nonnegative p-integrable functions can be approximated in L^p by a pointwise decreasing sequence of lower semicontinuous functions. This, the so-called *Vitali–Carathéodory theorem*, has turned out to be handy in the geometric theory of Sobolev spaces.

Vitali–Carathéodory theorem *Let $X = (X, d, \mu)$ be a metric measure space and let $1 \leq p < \infty$. For every p-integrable function $f : X \to [0, \infty]$ there exists a pointwise decreasing sequence (g_i) of lower semicontinuous functions on X such that $f \leq g_{i+1} \leq g_i$ and $g_i \to f$ in $L^p(X)$.*

Proof Let $f : X \to [0, \infty]$ be a p-integrable function on X. Pick an increasing sequence (φ_i) of nonnegative simple functions converging pointwise to f (Remark 3.1.1). By using the representation

$$f = \varphi_1 + \sum_{i=2}^{\infty} (\varphi_i - \varphi_{i-1}),$$

we find that f admits an expression as follows:

$$f = \sum_{j=0}^{\infty} a_j \chi_{E_j},$$

where $a_0 = \infty$, $a_j \in (0, \infty)$ for $j \geq 1$, and $E_j \subset X$ is a measurable set for all $j = 0, 1, \ldots$ Note that $\mu(E_0) = 0$.

Next, fix $\epsilon > 0$. By (3.3.39) we can choose for each $j \geq 1$ an open set $U_j \supset E_j$ such that

$$\mu(U_j) \leq \mu(E_j) + \epsilon^p \, 2^{-jp} \, a_j^{-p} \, .$$

Moreover, we can choose a sequence of open sets $V_j \supset E_0$ such that

$$\mu(V_j) \leq \epsilon^p \, 2^{-jp}$$

for $j = 1, 2, \ldots$ Then, for the lower semicontinuous function

$$g = \sum_{j=1}^{\infty} a_j \chi_{U_j} + \sum_{j=1}^{\infty} \chi_{V_j},$$

we have that both $f \leq g$ on X and

$$\|g - f\|_p \leq \sum_{j=1}^{\infty} a_j \mu(U_j \setminus E_j)^{1/p} + \sum_{j=1}^{\infty} \mu(V_j)^{1/p} \leq 2\epsilon.$$

Because $\epsilon > 0$ is arbitrary, and because the minimum of two lower semicontinuous functions (taken to ensure a decreasing sequence) is lower semicontinuous, the theorem follows from what was proved above. □

Theorem 4.2.4 *Let $X = (X, d, \mu)$ be a metric measure space, let $1 \leq p < \infty$, and let V be a Banach space. Then Lipschitz functions are dense in $L^p(X : V)$. If in addition (X, d) is locally compact then Lipschitz functions with compact support are dense in $L^p(X : V)$.*

Proof The first assertion follows from the proof of Proposition 3.3.49, since the functions f_ϵ in (3.3.50) are Lipschitz. A different proof can be given for real-valued functions, by combining the Vitali–Carathéodory theorem with Corollary 4.2.3. The second assertion follows from Proposition 3.3.52, since the function g in (3.3.53) is Lipschitz. □

4.3 Hausdorff measures

We now review the basic theory of Hausdorff measures. Let $X = (X, d)$ be a metric space. Fix a positive real number α. For each $\delta > 0$ and $E \subset X$, set

$$\mathcal{H}_{\alpha,\delta}(E) = \inf \upsilon(\alpha) \sum_i (\mathrm{diam}(E_i))^\alpha, \tag{4.3.1}$$

where $\upsilon(\alpha)$ is a normalizing constant given below in (4.3.4), and where the infimum is taken over all countable covers of E by sets $E_i \subset X$ with diameter less than δ. Recall that $\mathrm{diam}(A)$ is the *diameter* of a set $A \subset X$, defined as the supremum of the numbers $d(a, a')$ where $a, a' \in A$.

When δ decreases, the value of $\mathcal{H}_{\alpha,\delta}(E)$ for a fixed set E increases, and the *α-Hausdorff measure* of E is the number

$$\mathcal{H}_\alpha(E) := \lim_{\delta \to 0} \mathcal{H}_{\alpha,\delta}(E). \tag{4.3.2}$$

The set function $E \mapsto \mathcal{H}_\alpha(E)$ determines a Borel-regular measure on X [83, Section 2.10.2]. It is rarely a Radon measure, however, because it is easy for the limit in (4.3.2) to become infinite on compact sets. For example, $\mathcal{H}_\alpha([0, 1]) = \infty$ for each $0 < \alpha < 1$.

It is important to notice that each Hausdorff measure $\mathcal{H}_\alpha$ depends on the underlying metric space and that this dependence is not visible in our notation.

Remark 4.3.3 The constant $\upsilon(\alpha)$ in (4.3.1) is defined as

$$\upsilon(\alpha) := \frac{2^{-\alpha}\pi^{\alpha/2}}{\Gamma(\alpha/2 + 1)}, \qquad \Gamma(s) = \int_0^\infty e^{-x} x^{s-1}\, dx\,. \tag{4.3.4}$$

The purpose of this constant is to ensure that if $X = \mathbb{R}^n$ then $\mathcal{H}_n$ agrees with the Lebesgue n-measure m_n. In this case we can use the two symbols $\mathcal{H}_n$ and m_n interchangeably. The normalizing constant $\upsilon(\alpha)$ will not play any explicit role in the chapters to come.

The *Hausdorff dimension* of a set E in a metric space is the infimum of the numbers $\alpha > 0$ such that $\mathcal{H}_\alpha(E) = 0$. If no such numbers α exist, the Hausdorff dimension of E is infinite.

If $f : X \to Y$ is L-Lipschitz then it is easy to see from the definitions that

$$\mathcal{H}_\alpha(f(E)) \le L^\alpha\, \mathcal{H}_\alpha(E) \tag{4.3.5}$$

for each $E \subset X$ and $\alpha > 0$. In particular, the Hausdorff dimension of a set is invariant under bilipschitz transformations.

The construction of the Hausdorff measures $\mathcal{H}_\alpha$, $\alpha > 0$, is a special case of *Carathéodory's construction* in measure theory. For the details and proofs for the preceding facts see, for example, [235, pp. 27 and 50], [197, p. 54], or [83, p. 169].

4.4 Functions with bounded variation

We now consider continuous maps $\gamma : [a, b] \to X$, where $[a, b]$ is a compact interval in $\mathbb{R}$ and $X = (X, d)$ is a metric space. Such maps are called *compact curves* in this book. The purpose of this section is to provide background for some technical facts used in Section 5.1 and later. The main result is Theorem 4.4.8.

We will need the ensuing theory of functions of bounded variation only for continuous functions, and hence will adhere to the above simplifying assumption. Functions of bounded variation are typically considered on open intervals, but because every continuous function of bounded variation on a bounded open interval extends continuously to the end points (with values in the metric completion of X), there is little loss of generality in considering compact domains only.

A continuous map $\gamma : [a, b] \to X$ is said to be of *bounded variation* if the supremum of the numbers

$$\sum_{i=1}^{k} d(\gamma(t_i), \gamma(t_{i-1})) \tag{4.4.1}$$

is finite, where the numbers t_i run over all finite sequences of points of the form

$$a = t_0 < t_1 < \cdots < t_k = b\,.$$

If γ is of bounded variation and $O \subset [a, b]$ is open in $\mathbb{R}$, we define $V(\gamma, O)$, the *variation of* γ *on* O, as follows. If $O = (c, d)$ is a single subinterval of $[a, b]$ then $V(\gamma, O)$ is the supremum of the sums in (4.4.1), with a and b replaced by c and d, respectively. In the general case, O is a disjoint union of open intervals I contained in $[a, b]$ and $V(\gamma, O)$ is defined to be the sum

$$\sum V(\gamma, I)$$

over all the subintervals. We also set

$$V(\gamma, [c, d]) = V(\gamma, [c, d)) = V(\gamma, (c, d]) = V(\gamma, (c, d)) \tag{4.4.2}$$

for $[c, d] \subset [a, b]$. Note that the definitions in (4.4.2) are reasonable by the continuity of γ. Finally, the *total variation* of γ is $V(\gamma, [a, b])$.

In Chapter 5, as in most of this book, maps of bounded variation as defined above are called *rectifiable curves*, and the total variation is the *length of the curve*.

We assume familiarity with the classical theory of real-valued functions of bounded variation. However, here is a brief review of this theory.

Let $f : [a, b] \to \mathbb{R}$ be a function of bounded variation such that $f(a) = 0$. Then f can be written as a difference of two increasing functions, $f = f_1 - f_2$ with $f_1(a) = f_2(a) = 0$. Associated with the functions f_1 and f_2 there are two unique Radon measures μ_1 and μ_2 on $[a, b]$ such that $\mu_1([a, x]) = f_1(x)$ and $\mu_2([a, x]) = f_2(x)$ for every $a \le x \le b$, and there is a Borel set $E \subset [a, b]$ such that $\mu_1(E) = 0$ and $\mu_2([a, b] \setminus E) = 0$. In particular, the signed Radon measure $\mu(f) := \mu_1 - \mu_2$ satisfies

$$\mu(f)([c, d]) = f(d) - f(c) \tag{4.4.3}$$

whenever $[c, d] \subset [a, b]$. Moreover, f is differentiable m_1-almost everywhere on $[a, b]$, with

$$f'(t) = \frac{d\mu(f)}{dm_1}(t) = \frac{d\mu(f)^a}{dm_1}(t) \tag{4.4.4}$$

at m_1-almost every $t \in [a, b]$, where $\mu(f)^a$ denotes the absolutely continuous part of $\mu(f)$ with respect to m_1. (See the Lebesgue–Radon–Nikodym theorem in Section 3.4; here the derivative of $\mu(f)$ is understood to be the difference of the derivatives of μ_1 and μ_2, and similarly for the absolutely continuous part $\mu(f)^a$.) If we write $|\mu(f)| := \mu_1 + \mu_2$ for the *total variation* of the signed measure $\mu(f)$, we also have that

$$|f'(t)| = \frac{d|\mu(f)|}{dm_1}(t) = \frac{d|\mu(f)^a|}{dm_1}(t) \tag{4.4.5}$$

at m_1-a.e. $t \in [a, b]$, where $|\mu(f)^a| = \mu_1^a + \mu_2^a = |\mu(f)|^a$, that

$$\int_a^b |f'(t)|\, dt \le |\mu(f)|([a, b]), \tag{4.4.6}$$

and that

$$|\mu(f)|(O) = V(f, O) \tag{4.4.7}$$

for every open set $O \subset [a, b]$. Indeed, $|\mu(f)|$ is the unique Radon measure on $[a, b]$ satisfying (4.4.7) (see Section 3.3).

For the preceding facts, see for example [237, Chapter 8] or [83, Sections 2.5.17 and 2.9.19]. The reader can also easily derive these facts from the results in Section 3.4.

The main goal of this section is to prove the following theorem. The point of the theorem is that the limit on the right-hand side in (4.4.9) exists at almost every point in the interval of definition. Recall that in this book functions of bounded variation are assumed to be continuous.

Theorem 4.4.8 *With each map $\gamma : [a, b] \to X$ of bounded variation we can associate a unique Radon measure ν_γ on $[a, b]$ such that $\nu_\gamma(O) = V(\gamma, O)$ for each open $O \subset [a, b]$ and*

$$\frac{d\nu_\gamma}{dm_1}(t) = \lim_{u \to t, u \neq t} \frac{d(\gamma(t), \gamma(u))}{|t - u|} =: |\gamma'(t)| \tag{4.4.9}$$

for m_1-a.e. $t \in [a, b]$.

We will call the m_1-almost-everywhere-defined function $|\gamma'(t)|$ in (4.4.9) the *metric differential* of a continuous map of bounded variation $\gamma : [a, b] \to X$.

Remark 4.4.10 We emphasize that, the notation $|\gamma'(t)|$ notwithstanding, a map $\gamma : [a, b] \to X$ of bounded variation may not be differentiable in the usual sense anywhere, even if X is a normed space. A standard example is the map $\gamma(t) := \chi_{[0,t]}$ for $0 \leq t \leq 1$, $\gamma : [0, 1] \to L^1([0, 1])$. Note that γ is an isometry, $\|\gamma(t) - \gamma(s)\|_{L^1} = |t - s|$, so that $|\gamma'(t)| \equiv 1$.

Proof of Theorem 4.4.8 Let $\gamma : [a, b] \to X$ be of bounded variation. The uniqueness of ν_γ is clear from the facts that ν_γ is a Radon measure and that $\nu_\gamma(O) = V(\gamma, O)$ for each open O (see (3.3.39)). To show that such a measure exists, we use Carathéodory's construction as in Section 4.3. For $\delta > 0$ and $E \subset [a, b]$, set

$$\nu_{\gamma,\delta}(E) = \inf \sum_i V(\gamma, I_i),$$

where the infimum is taken over all countable collections $\{I_i\}$, of (relatively) open subintervals of $[a, b]$ of diameter at most δ, such that $E \subset \cup_i I_i$.

The limit

$$\nu_\gamma(A) := \lim_{\delta \to 0} \nu_{\gamma,\delta}(A)$$

is a Radon measure on $[a, b]$ (see [197, Chapter 4, pp. 54–55]). Since the notion of Sobolev spaces considered in this book is based on rectifiable curves (that is, mappings of bounded variation as considered in this section), for self-containment we provide the proof of this fact now. First note that $V(\gamma, I_2) \leq V(\gamma, I_1)$ if $I_2 \subset I_1$ are subintervals of $[a, b]$. Therefore if $E_2 \subset E_1$ then $\nu_{\gamma,\delta}(E_2) \leq \nu_{\gamma,\delta}(E_1)$, and so the monotonicity properties of measures hold here. Furthermore, because $V(\gamma, [a, b])$ is finite, for every $\epsilon > 0$ we can find t_ϵ, with $a < t_\epsilon < b$, such that $V(\gamma, [a, t_\epsilon]) < \epsilon$; in particular, $\nu_{\gamma,\delta}(\emptyset) = 0$. The countable subadditivity follows from the fact that if E_j, $j \in J \subset \mathbb{N}$, are subsets of $[a, b]$ then for every $\epsilon > 0$ we can find a cover of each E_j by intervals $I_{j,i} \subset [a, b]$ with $\nu_{\gamma,\delta}(E_j) \geq \sum_i V(\gamma, I_{j,i}) - 2^{-j}\epsilon$. Now the fact that ν_γ is a monotone increasing limit of $\nu_{\gamma,\delta}$ as $\delta \to 0$ immediately yields that ν_γ also has the monotonicity and countable subadditivity properties and that $\nu_\gamma(\emptyset) = 0$; that is, ν_γ and each $\nu_{\gamma,\delta}$ are outer measures on $[a, b]$.

In general $\nu_{\gamma,\delta}$ need not be a Borel measure, but by the Carathéodory criterion 3.3.5 we can see that ν_γ is a Borel measure. Indeed, if $0 < \delta < \mathrm{dist}(E_1, E_2)$ and I_i, $i \in F \subset \mathbb{N}$, is a cover of $E_1 \cup E_2$ by subintervals of $[a, b]$ with diameter no larger than δ then, for each $i \in F$, either I_i does not intersect E_1 or I_i does not intersect E_2. We thus partition this cover into two subcovers, one covering E_1 but not intersecting E_2 and the other covering E_2 but not intersecting E_1. From this, we can directly see that, for sufficiently small δ, $\nu_{\gamma,\delta}(E_1 \cup E_2) = \nu_{\gamma,\delta}(E_1) + \nu_{\gamma,\delta}(E_2)$. Thus in the limit we have that $\nu_\gamma(E_1 \cup E_2) = \nu_\gamma(E_1) + \nu_\gamma(E_2)$, satisfying the criterion.

Moreover, it easily follows from the definitions that $\nu_\gamma(O) = V(\gamma, O)$ for each open $O \subset [a, b]$. In particular, $\nu_\gamma([a, b])$ equals the total variation of γ, which is finite.

Finally, to see that ν_γ is Borel regular we can use the construction above directly. For $E \subset [a, b]$, and for each $j \in \mathbb{N}$ we can find a countable collection $I_{j,i}$ of relatively open intervals from $[a, b]$ covering E, each $I_{j,i}$ of diameter no more than $1/j$, such that $\nu_{\gamma,1/j}(E) \geq \sum_i V\left(\gamma, I_{j_i}\right) - 1/j$. Since each $I_{j,i}$ is an interval and hence a Borel set, it follows that $\bigcup_i I_{j,i}$ is a Borel set containing E with

$$\nu_\gamma(E) \leq \nu_\gamma\left(\bigcup_i I_{j,i}\right) \leq \sum_i \nu_\gamma(I_{j,i}) = \sum_i V(\gamma, I_{j,i}) \leq \nu_{\gamma,1/i}(E) + 1/j.$$

The Borel set $A = \bigcap_j \bigcup_i I_{j,i}$ again contains E, and from above we see that $\nu_\gamma(E) = \nu_\gamma(A)$; that is, ν_γ is Borel regular. Since $\nu_\gamma([a, b])$ is finite and closed subsets of $[a, b]$ are compact, by (3.3.38) we know that ν_γ is a Radon measure on $[a, b]$.

It remains to show that (4.4.9) holds. To this end, we note that the assumptions and the conclusions of the proposition are invariant under isometries and so no generality is lost in assuming that $X = l^\infty$. Indeed, we need to work with only the separable metric subspace of X that is the image of $[a, b]$ under γ, which in turn embeds isometrically into the Banach space l^∞ by the Kuratowski embedding theorem (see Section 4.1). We may further assume that $\gamma(a) = 0 \in l^\infty$.

For every $\varphi \in (l^\infty)^*$ with dual norm $|\varphi| \le 1$ the function $\gamma_\varphi : [a, b] \to \mathbb{R}$,

$$\gamma_\varphi(t) := \langle \varphi, \gamma(t) \rangle, \tag{4.4.11}$$

is a real-valued continuous function of bounded variation with $\gamma_\varphi(a) = 0$. As can be seen from (4.4.5) and (4.4.7), there exists a unique Radon measure ν_φ on $[a, b]$ satisfying $\nu_\varphi(O) = V(\gamma_\varphi, O)$ and, from the theory of real-valued functions of bounded variation (see for example [236, pp. 100–103]),

$$\frac{d\nu_\varphi}{dm_1}(t) = \lim_{u \to t, u \neq t} \frac{|\langle \varphi, \gamma(t) \rangle - \langle \varphi, \gamma(u) \rangle|}{|t - u|} = |\gamma_\varphi'(t)| \tag{4.4.12}$$

for m_1-a.e. $t \in [a, b]$. We also note that

$$\nu_\varphi(E) \le \nu_\gamma(E) \tag{4.4.13}$$

for each Borel set $E \subset [a, b]$. Indeed, to prove (4.4.13) it suffices to assume that E is open; in this case the inequality comes down to a similar inequality between variations, which is obvious because $|\varphi| \le 1$.

Next, consider the set

$$D := \{\gamma(q) - \gamma(r) : q, r \in \mathbb{Q} \cap [a, b]\} \subset l^\infty.$$

Then D is a countable dense set in the difference set $\{\gamma(s) - \gamma(t) : s, t \in [a, b]\}$. For each $v \in D$ choose an element $\varphi_v \in (l^\infty)^*$ such that $|\varphi_v| \le 1$ and $\langle \varphi_v, v \rangle = |v|$ (the choice is possible by the Hahn–Banach theorem; see Section 2.2), and put $\Phi := \{\varphi_v : v \in D\}$. Set

$$\nu_\Phi := \left(\bigvee_{\varphi \in \Phi} \nu_\varphi \right)$$

and recall (3.4.37). It follows from (4.4.13), from Lemmas 3.4.38 and 3.4.39 and from (4.4.12) and (3.4.24) that ν_Φ is a finite Borel-regular measure on $[a, b]$ such that

$$\frac{d\nu_\Phi}{dm_1}(t) = \sup_{\varphi \in \Phi} |\gamma_\varphi'(t)| =: \tau(t) \tag{4.4.14}$$

for m_1-a.e. $t \in [a, b]$. In particular, the function τ is measurable and satisfies

$$\tau(t) \le \frac{d\nu_\gamma}{dm_1}(t) \tag{4.4.15}$$

for m_1-a.e. $t \in [a, b]$ by (4.4.13) (the existence of the derivative on the right is guaranteed by the Lebesgue–Radon–Nikodym theorem; see Section 3.4).

We will show that the limit on the right-hand side of (4.4.9) equals $\tau(t)$ for m_1-a.e. $t \in [a, b]$ and also that

$$\tau(t) = \frac{d\nu_\gamma}{dm_1}(t) \tag{4.4.16}$$

for m_1-a.e. $t \in [a, b]$.

To achieve this, observe first that

$$\liminf_{u\to t, u\neq t} \frac{|\gamma(t) - \gamma(u)|}{|t-u|} \geq \liminf_{u\to t, u\neq t} \frac{|\langle\varphi, \gamma(t)\rangle - \langle\varphi, \gamma(u)\rangle|}{|t-u|} = |\gamma'_\varphi(t)|$$

for m_1-a.e. $t \in [a, b]$, and hence that

$$\liminf_{u\to t, u\neq t} \frac{|\gamma(t) - \gamma(u)|}{|t-u|} \geq \tau(t) \tag{4.4.17}$$

for m_1-a.e. $t \in [a, b]$. Writing $I = (t, u) \subset [a, b]$, we find that

$$|\gamma(t) - \gamma(u)| = \sup_{\varphi\in\Phi} |\langle\varphi, \gamma(t)\rangle - \langle\varphi, \gamma(u)\rangle| \leq \sup_{\varphi\in\Phi} V(\varphi\circ\gamma, I) \leq \nu_\Phi(I)\,,$$

which gives

$$\limsup_{u\to t, u\neq t} \frac{|\gamma(t) - \gamma(u)|}{|t-u|} \leq \frac{d\nu_\Phi}{dm_1}(t) = \tau(t)\,. \tag{4.4.18}$$

It remains to prove (4.4.16). For this we will show that in fact $\nu_\Phi = \nu_\gamma$. The inequality $\nu_\Phi \leq \nu_\gamma$ follows from (4.4.13). To prove the opposite inequality, it suffices to show that

$$\nu_\gamma(O) \leq \nu_\Phi(O) \tag{4.4.19}$$

whenever $O = (c, d) \subset [a, b]$ is an open interval; see (3.3.39). To this end, let $c = t_0 < t_1 < \cdots < t_k = d$ be such that

$$\nu_\gamma(O) - \epsilon = V(\gamma, O) - \epsilon < \sum_{i=1}^{k} |\gamma(t_i) - \gamma(t_{i-1})|$$

and $\gamma(t_i) - \gamma(t_{i-1}) \in D$ for each $i = 1, \ldots, k$. (Note that we use the norm distance in l^∞ here.) For each $i = 1, \ldots, k$ we can pick an element $\varphi_i \in \Phi$ such that

$$|\gamma_{\varphi_i}(t_i) - \gamma_{\varphi_i}(t_{i-1})| = |\langle\varphi_i, \gamma(t_i) - \gamma(t_{i-1})\rangle| = |\gamma(t_i) - \gamma(t_{i-1})|\,.$$

Therefore

$$\nu_\gamma(O) - \epsilon < \sum_{i=1}^{k} |\gamma(t_i) - \gamma(t_{i-1})| = \sum_{i=1}^{k} |\gamma_{\varphi_i}(t_i) - \gamma_{\varphi_i}(t_{i-1})|$$
$$\leq \sum_{i=1}^{k} \nu_{\varphi_i}((t_{i-1}, t_i)) \leq \nu_\Phi(O).$$

This proves (4.4.19) and hence (4.4.16) follows.

The proof of Theorem 4.4.8 is complete. □

The following corollary will be invoked in Section 5.1 in connection with the arc length parametrization of rectifiable curves. (See Proposition 5.1.8.)

Corollary 4.4.20 *Let $\gamma : [a, b] \to X$ be a map of bounded variation such that $V(\gamma, [t, u]) = u - t$ whenever $a \leq t \leq u \leq b$. Then*

$$\lim_{u \to t, u \neq t} \frac{d(\gamma(t), \gamma(u))}{|t - u|} = 1 \tag{4.4.21}$$

for m_1-a.e. $t \in [a, b]$. Moreover, for every set $E \subset [a, b]$ we have that $\mathcal{H}_1(E) > 0$ if $\mathcal{H}_1(\gamma(E)) > 0$.

Proof The first claim is an immediate corollary to Theorem 4.4.8. The second claim follows from the fact that every γ satisfying the hypotheses is 1-Lipschitz. □

Remark 4.4.22 The embedding of the image of γ in a Banach space was not necessary in the proof of Theorem 4.4.8. We could have used the 1-Lipschitz functions $y \mapsto d(x, y)$, $x \in X$, in place of the dual elements φ. However, similar arguments with more substantial use of a linear structure will be employed repeatedly later on.

The converse to the second part of Corollary 4.4.20 does hold true: if $E \subset [a, b]$ satisfies $\mathcal{H}_1(E) > 0$ then $\mathcal{H}_1(\gamma(E)) > 0$. As we will not need this fact, we do not provide a detailed proof but only a sketch of the argument. First, one may reduce the issue to the case $X = \mathbb{R}$. Indeed, we can choose a countable dense subset (z_i) of $\gamma([a, b])$, and then as in (4.4.16) we see that the limit on the left-hand side of (4.4.21) is also m_1-almost everywhere equal to $\sup_i |(d/dt) d(\gamma(t), z_i)|$. For each i, let I_i be the set where $|(d/dt) d(\gamma(t), z_i)| > 3/4$; then $[a, b] \subset \cup_i I_i$ up to a set of m_1-measure zero. Since $\varphi_i = d(\gamma(\cdot), z_i)$ is also a 1-Lipschitz map, to show that $\mathcal{H}_1(\gamma(E \cap I_i)) > 0$ for some i it suffices to show that $\mathcal{H}_1(\varphi_i(E \cap I_i)) > 0$ for a choice of i for which

$\mathcal{H}_1(E \cap I_i) > 0$. This follows from the usual area formula on the real line [81, p. 96].

Absolute continuity

A continuous map $\gamma : [a, b] \to X$ of bounded variation is said to be *absolutely continuous* if for each $\epsilon > 0$ we can find a $\delta > 0$ such that

$$\sum_{i=1}^{k} d(\gamma(b_i), \gamma(a_i)) < \epsilon \tag{4.4.23}$$

whenever (a_i, b_i) are nonoverlapping subintervals of $[a, b]$ with

$$\sum_{i=1}^{k} |b_i - a_i| < \delta\,. \tag{4.4.24}$$

We also say that γ is an *absolutely continuous curve* in this case. The preceding definition is a straightforward extension of the definition for absolutely continuous functions in classical real analysis. For example, locally Lipschitz maps $[a, b] \to X$ are absolutely continuous.

We assume that the reader is familiar with the classical theory of absolutely continuous functions of a single real variable. For example, a function $f : [a, b] \to \mathbb{R}$ of bounded variation is absolutely continuous if and only if the associated signed measure $\mu(f)$ as in (4.4.3) is absolutely continuous with respect to the Lebesgue measure m_1.

The following proposition is a direct consequence of the definitions, basic measure theory, and Theorem 4.4.8.

Proposition 4.4.25 *A continuous map $\gamma : [a, b] \to X$ of bounded variation is absolutely continuous if and only if the associated Radon measure ν_γ is absolutely continuous with respect to the Lebesgue measure m_1. In particular, if $\gamma : [a, b] \to X$ is absolutely continuous, we have that*

$$d(\gamma(a), \gamma(b)) \le \int_a^b |\gamma'(t)|\, dt\,. \tag{4.4.26}$$

A map $\gamma : \mathbb{R} \to X$ is said to be absolutely continuous if its restriction to every compact subinterval is absolutely continuous.

We will continue to discuss absolutely continuous curves in Chapter 5, where the absolute continuity of a curve is linked to the absolute continuity of the length function; see Proposition 5.1.5. We will also show in Chapter 5 that every rectifiable curve admits an arc-length parametrization, which necessarily must be absolutely continuous.

4.5 Notes to Chapter 4

The McShane–Whitney extension lemma is normally credited to McShane [203], but the same argument was found earlier by Whitney [283] as acknowledged in [203, p. 837].

Proposition 4.1.15 was proved by Kuratowski [175] in 1935, while the Fréchet embedding theorem was proved by Fréchet [88] in 1909. Another interesting fact is that every separable metric space admits a 4-bilipschitz embedding in c_0 [26, Theorem 7.11, p. 176]. Note that c_0 is a "smaller" space than l^∞, which is the double dual of c_0; in particular, c_0 is separable. However, the other hand, no reflexive Banach space can have the property that every separable metric space admits a bilipschitz embedding, by [26, Corollary 7.10, p. 176]. See also [26, Notes to Chapter 7, p. 184].

Pairs of metric spaces with the Lipschitz extension property have been studied extensively in connection with the geometric theory of Banach spaces. See [26, Chapter 1] for further discussion and references. Interesting nonlinear examples were considered by Lang, Pavlović, and Schroeder [180] and Lang and Schlichenmaier [179]. As pointed out in this chapter, the Lipschitz extension property in Theorem 4.1.21 may fail for nondoubling metric spaces in a metric-space–Banach-space pair. However, Lee and Naor [182] showed that if the subset of the metric space on which a Lipschitz function f is defined is itself doubling then f can be extended as a Lipschitz function to the ambient metric space. Theorem 4.1.21, in the special case when the metric space source is a finite-dimensional Banach space, was proved by Johnson, Lindenstrauss, and Schechtman [143].

An embedding theorem due to Assouad [18] asserts that upon replacing the metric d of a doubling metric space by a root of d the resulting "snowflaked" space is bilipschitz embeddable into a finite-dimensional Euclidean space. Assouad's embedding theorem is an important tool in the field of analysis on metric spaces, as it often permits Euclidean machinery to be used in the study of abstract doubling spaces. However, in our setting this embedding theorem is of limited use since the snowflaking operation destroys the rectifiability of nonconstant curves. See Section 7.1 for further information and discussion.

Examples of open subsets of $\mathbb{R}^n$ and open subsets of general doubling metric spaces without isolated points, which that carry no doubling measures, were given by Saksman [239]. However, every complete doubling metric space does carry a doubling measure. For compact doubling metric spaces this fact was proved by Vol'berg and Konyagin in [277]. For complete doubling metric spaces this result was extended by Luukkainen and Saksman in [190]. In light of Lemma 4.1.14 one can then conclude that the completion of a doubling metric space supports a doubling measure.

Proposition 4.2.2 appears in [204, pp. 43–44] for functions defined in subsets of Euclidean space, but the proof is the same in metric spaces. In [204, Chapter 2] an integration theory is developed that is based on the notion of semicontinuous functions. This approach is related to the Vitali–Carathéodory theorem, which is difficult to find in modern texts. Sometimes it appears under the additional hypothesis that X is a locally compact space. See [237, p. 54].

For the Lipschitz approximation of Hölder-continuous functions, or more generally functions with a given modulus of continuity, see [244, p. 167] or [107, Appendix B].

Functions of bounded variation go back to the early days of real analysis. In the context of rectifiable curves in metric spaces, they will be discussed in more detail in Section 5.1. Theorem 4.4.8 is due to Ambrosio [6].

5

Path integrals and modulus

The upper gradient approach to Sobolev functions, which is the main theme of this book, relies crucially on the concept of the modulus of a curve family. Although in this context the term "modulus" has its roots in the notion of capacity in electromagnetism, as a mathematical tool it first flourished in function theory. We offer more historical comments in the Notes to this chapter. Here we present the modern theory of modulus, starting from first principles.

5.1 Curves in metric spaces

Let $X = (X, d)$ be a metric space. A *curve* in X is a continuous map $\gamma : I \to X$, where $I \subset \mathbb{R}$ is an interval. We call γ *compact*, *open*, or *half-open*, depending on the type of the parameter interval I. The interval I is allowed to be a single point, in which case γ is a *constant curve*. More generally, every curve γ whose image $\gamma(I)$ is just one point is called a constant curve; otherwise, γ is a *nonconstant curve*. Typically, we abuse the notation by writing $\gamma = \gamma(I)$ for the image set in X. As a warning, observe that in our terminology an open curve can have a compact image. A *subcurve* of γ is the restriction $\gamma|_{I'}$ of γ to a subinterval $I' \subset I$.

Rectifiable curves

Given a compact curve $\gamma : [a, b] \to X$, its *length* is the supremum of the numbers

$$\sum_{i=1}^{k} d(\gamma(t_i), \gamma(t_{i-1})), \tag{5.1.1}$$

where the numbers t_i run over all finite sequences of points of the form

$$a = t_0 < t_1 < \cdots < t_k = b.$$

In the language of Section 4.4, the length is the total variation $V(\gamma, [a, b])$. If γ is not compact, its length is defined to be the supremum of the lengths of the compact subcurves of γ. Thus, every curve has a well-defined length in the extended nonnegative reals, and we denote this quantity by length(γ).

A curve is said to be *rectifiable* if its length is finite and *locally rectifiable* if each of its compact subcurves is rectifiable. For example, curves that are Lipschitz continuous as maps from I to X, *Lipschitz curves* for short, are always (locally) rectifiable. If a curve is not rectifiable, we call it *nonrectifiable*. Thus, a compact curve is rectifiable if and only if it is of bounded variation as defined in the text after (4.4.1). In this book the only rectifiable curves in which we are interested are compact.

If $f : X \to Y$ is an L-Lipschitz map between metric spaces and if $\gamma : I \to X$ is rectifiable then $f \circ \gamma$ is rectifiable and

$$\text{length}(f \circ \gamma) \leq L\,\text{length}(\gamma)\,. \tag{5.1.2}$$

This follows directly from the definitions.

The length of a rectifiable curve γ is bounded from below by the Hausdorff 1-measure $\mathcal{H}_1(\gamma)$ of its image in X. In general the former is greater; however, the length and the Hausdorff measure agree provided that γ as a map is injective (see Proposition 5.1.11 below).

For each rectifiable curve $\gamma : [a, b] \to X$ there is an associated *length function*

$$s_\gamma : [a, b] \to [0, \text{length}(\gamma)]$$

defined by

$$s_\gamma(t) := \text{length}(\gamma|_{[a,t]})\,.$$

We have that

$$d(\gamma(t_2), \gamma(t_1)) \leq \text{length}(\gamma|_{[t_1,t_2]}) = s_\gamma(t_2) - s_\gamma(t_1) \tag{5.1.3}$$

whenever $a \leq t_1 \leq t_2 \leq b$.

Lemma 5.1.4 *The length function s_γ of a rectifiable curve $\gamma : [a, b] \to X$ is increasing and continuous.*

Proof Clearly s_γ is increasing. To prove the continuity, fix $a \leq t_0 \leq b$. Because s_γ is increasing, the one-sided limits $s_\gamma^-(t_0)$ and $s_\gamma^+(t_0)$ exist. Suppose first that $s_\gamma(t_0) - s_\gamma^-(t_0) > \delta > 0$. Then clearly $t_0 > a$. Let $a < t_1 < t_0$. Because

$$\text{length}(\gamma|_{[t_1,t_0]}) = s_\gamma(t_0) - s_\gamma(t_1) > \delta$$

and because $s_\gamma(t_0) - s_\gamma(t_1) = s_{\gamma|[t_1,t_0]}(t_0)$, we have by the continuity of γ that there are numbers $t_1 = a_0 < \cdots < a_k < t_0$ satisfying

$$\sum_{j=1}^{k} d(\gamma(a_j), \gamma(a_{j-1})) > \delta\,.$$

Define $t_2 = a_k$. Then $\text{length}(\gamma|_{[t_1,t_2]}) > \delta$ and

$$\text{length}(\gamma|_{[t_2,t_0]}) = s_\gamma(t_0) - s_\gamma(t_2) > \delta\,.$$

By induction we can find a sequence of values (t_i), $t_1 < t_2 < \cdots < t_i < \cdots < t_0$ such that $\text{length}(\gamma|_{[t_i,t_{i+1}]}) > \delta$. This implies that

$$\text{length}(\gamma|_{[t_1,t_0]}) \geq \text{length}(\gamma|_{[t_1,t_i]}) > (i-1)\delta$$

for every $i = 2, 3, \ldots$, contradicting the rectifiability of γ. We conclude that $s_\gamma^-(t_0) = s_\gamma(t_0)$.

The equality $s_\gamma^+(t_0) = s_\gamma(t_0)$ is proved analogously, and the lemma follows. □

Arc length parametrization

Recall the definition for an absolutely continuous curve from Section 4.4.

The following proposition captures an important and defining property of absolutely continuous curves.

Proposition 5.1.5 *A rectifiable curve $\gamma : [a, b] \to X$ is absolutely continuous if and only if its length function $s_\gamma : [a, b] \to [0, \text{length}(\gamma)]$ is absolutely continuous.*

Proof The absolute continuity of γ follows from the absolute continuity of s_γ by formula (5.1.3).

Next, assume that γ is absolutely continuous. Let $\epsilon > 0$, and let $\delta > 0$ be as in the definition of absolute continuity for this ϵ. Suppose that we are given a family of k nonoverlapping subintervals $[a_i, b_i]$, $i = 1, \ldots, k$, of $[a, b]$, as in (4.4.24) satisfying $\sum_{i=1}^{k} b_i - a_i < \delta$. Recalling that

$$s_\gamma(b_i) - s_\gamma(a_i) = \text{length}(\gamma|_{[a_i,b_i]}) < \infty,$$

we may subdivide each $[a_i, b_i]$ into k_i intervals $[a_i^j, b_i^j]$ such that

$$\sum_{j=1}^{k_i} d(\gamma(b_i^j), \gamma(a_i^j)) > s_\gamma(b_i) - s_\gamma(a_i) - \epsilon/k.$$

Consequently, we have $\sum_{i=1}^{k} \sum_{j=1}^{k_i} b_i^j - a_i^j = \sum_{i=1}^{k} b_i - a_i < \delta$, and so

$$\sum_{i=1}^{k} |s_\gamma(b_i) - s_\gamma(a_i)| \leq 2\epsilon.$$

This proves the absolute continuity of s_γ, and the proposition follows. □

The *arc length parametrization* of a rectifiable curve $\gamma : [a, b] \to X$ is the curve $\gamma_s : [0, \text{length}(\gamma)] \to X$ defined by

$$\gamma_s(t) := \gamma(s_\gamma^{-1}(t)),$$

where, by the continuity of s_γ,

$$s_\gamma^{-1}(t) := \sup\{s : s_\gamma(s) = t\} = \max\{s : s_\gamma(s) = t\}$$

is the one-sided inverse of s_γ. Notice that s_γ^{-1} is an increasing function that is continuous from the right: $\lim_{t \to t_0^+} s_\gamma^{-1}(t) = s_\gamma^{-1}(t_0)$. If $\lim_{t \to t_0^-} s_\gamma^{-1}(t) = s_0 < s_\gamma^{-1}(t_0)$ then γ is constant on $[s_0, s_\gamma^{-1}(t_0)]$. Thus $\gamma_s : [0, \text{length}(\gamma)] \to X$ is the unique curve satisfying

$$\gamma(t) = \gamma_s(s_\gamma(t)) \tag{5.1.6}$$

for each $t \in [a, b]$. It follows from the definitions that

$$\text{length}(\gamma_s|[t, u]) = u - t \tag{5.1.7}$$

for $0 \leq t \leq u \leq \text{length}(\gamma)$.

From (5.1.7) and from Corollary 4.4.20 we obtain the following proposition.

Proposition 5.1.8 *The arc length parametrization γ_s of a compact rectifiable curve γ is 1-Lipschitz continuous, hence absolutely continuous, and satisfies*

$$\lim_{u \to t, u \neq t} \frac{d(\gamma_s(t), \gamma_s(u))}{|t - u|} = 1 \tag{5.1.9}$$

for almost every $t \in [0, \text{length}(\gamma)]$.

The following version of the *Arzelà–Ascoli theorem* for rectifiable curves will be employed later on, in Chapters 7 and 9. As in the case of the usual

Arzelà–Ascoli theorem, the proof consists of choosing a countable dense subset $\{q_1, q_2, \ldots\}$ of $[0, 1]$, then picking a subsequence for which the values converge at q_1, continuing inductively from this subsequence, and finally passing to a diagonal sequence.

Theorem 5.1.10 *Let $\gamma_j : [0, 1] \to X$, $j = 1, 2, \ldots$ be L-Lipschitz maps, where X is a metric space such that bounded and closed sets in X are compact. If $\gamma_j(0) = x_0 \in X$ for each j then there is an L-Lipschitz map $\gamma : [0, 1] \to X$ and a subsequence (γ_j) that converges to γ uniformly on $[0, 1]$.*

Proposition 5.1.11 *Let $\gamma : I \to X$ be a curve. Then*

$$\operatorname{diam}(\gamma) \leq \mathcal{H}_1(\gamma) \leq \operatorname{length}(\gamma)\,. \tag{5.1.12}$$

If γ as a map is injective then $\mathcal{H}_1(\gamma) = \operatorname{length}(\gamma)$.

We recall that by $\mathcal{H}_1(\gamma)$ we mean $\mathcal{H}_1(\gamma(I))$; this is in the spirit of the abuse of notation adopted earlier, where γ also denotes the image set $\gamma(I)$ of γ.

Proof We invoke the real-valued 1-Lipschitz function $x \mapsto d(x, x_1)$ on X to conclude that $d(x_1, x_2) \leq \mathcal{H}_1(\gamma)$ whenever γ contains the points x_1 and x_2, cf. (4.3.5); alternatively, this can be deduced from the definition of $\mathcal{H}_1$ and the triangle inequality. In particular, the first inequality in (5.1.12) follows. To prove the second inequality in (5.1.12) we may assume that γ is rectifiable. Because the arc length parametrization $\gamma_s : [0, \operatorname{length}(\gamma)] \to X$ is 1-Lipschitz, by (4.3.5) we have that

$$\mathcal{H}_1(\gamma) \leq \mathcal{H}_1([0, \operatorname{length}(\gamma)]) = \operatorname{length}(\gamma).$$

Next, suppose that γ is injective. Given a subdivision $a = t_0 \leq t_1 \leq \cdots \leq t_k = b$, we have that

$$\sum_{i=1}^{k} d(\gamma(t_i), \gamma(t_{i-1})) \leq \sum_{i=1}^{k} \mathcal{H}_1(\gamma|_{[t_{i-1}, t_i]}) = \mathcal{H}_1(\gamma)\,,$$

where (5.1.12) was used in the first step and the injectivity of γ was used in the second step. The desired inequality $\operatorname{length}(\gamma) \leq \mathcal{H}_1(\gamma)$ follows on taking the supremum over all subdivisions. The proposition is proved. □

Line integration

Given a rectifiable curve $\gamma : [a, b] \to X$ and a nonnegative Borel function $\rho : X \to [0, \infty]$, the *line integral of ρ over γ* is the expression

$$\int_\gamma \rho \, ds := \int_0^{\operatorname{length}(\gamma)} \rho(\gamma_s(t))\, dt. \tag{5.1.13}$$

Because $\rho \circ \gamma_s$ is a nonnegative Borel function on $[0, \text{length}(\gamma)]$ (see Section 3.3), the integral exists with values in $[0, \infty]$. The line integral

$$\int_\gamma \rho \, ds$$

of a Borel function $\rho : X \to [0, \infty]$ over a locally rectifiable curve γ is defined to be the supremum of the integrals of ρ over all compact subcurves of γ. Another possible definition could be to replace the right-hand side of (5.1.13) for an absolutely continuous γ by

$$\int_a^b \rho(\gamma(t))|\gamma'(t)| \, dt;$$

this would lead to the same value as above. Indeed, $|\gamma'(t)| = s'_\gamma(t)$ by Theorem 4.4.8, s_γ is absolutely continuous by Proposition 5.1.5, and $\gamma = \gamma_s \circ s_\gamma$ by (5.1.6).

Thus, the line integral of a nonnegative Borel function is defined over each locally rectifiable curve, and no line integrals are defined over curves that are not locally rectifiable. Line integrals over constant curves are always zero.

We note the following simple lemma for future reference.

Lemma 5.1.14 *Let $\gamma : [0, L] \to X$ be 1-Lipschitz. Then*

$$\int_\gamma \rho \, ds \le \int_0^L \rho(\gamma(t)) \, dt \tag{5.1.15}$$

for every Borel function $\rho : X \to [0, \infty]$.

Proof Because γ is 1-Lipschitz, it follows that the length function $s_\gamma : [0, L] \to [0, \text{length}(\gamma)]$ is also 1-Lipschitz and in particular absolutely continuous. Therefore

$$\begin{aligned}\int_\gamma \rho \, ds &= \int_0^{\text{length}(\gamma)} \rho(\gamma_s(t)) \, dt = \int_0^L \rho(\gamma_s(s_\gamma(t))) s'_\gamma(t) \, dt \\ &\le \int_0^L \rho(\gamma_s(s_\gamma(t))) \, dt = \int_0^L \rho(\gamma(t)) \, dt,\end{aligned}$$

as required. □

Remark 5.1.16 Simple examples show that the inequality in (5.1.15) can be strict. For example, let $\gamma : [0, 1] \to X$ be constant and let $\rho > 0$.

If γ is a rectifiable curve, its image has finite Hausdorff 1-measure in X by (5.1.12). Consequently, we can integrate Borel functions $\rho : X \to [0, \infty]$

against $\mathcal{H}_1$ on γ. We claim that the inequality

$$\int_\gamma \rho \, d\mathcal{H}_1 \leq \int_\gamma \rho \, ds \tag{5.1.17}$$

holds for every such γ and ρ. To prove (5.1.17) we may assume that γ is a compact curve and that ρ is the characteristic function of an open set $O \subset X$. Next, we note that the restriction of both $\mathcal{H}_1$ and its weighted version $\rho \, d\mathcal{H}_1$ to γ are Radon measures; see (3.3.36), Proposition 3.3.44, and Section 4.3. Now $\gamma_s^{-1}(O)$ is a union of disjoint intervals that are all open except for those that meet the end points of $[0, \text{length}(\gamma)]$. Moreover, for each of these intervals $[a_i, b_i]$,

$$\mathcal{H}_1(\gamma|_{[a_i,b_i]}) \leq \text{length}(\gamma|_{[a_i,b_i]}) = \int_{a_i}^{b_i} \rho(\gamma_s(t)) \, dt$$

for each such interval, by (5.1.12) and (5.1.7). The claim follows by summing over the intervals.

Next, let $E \subset X$ be an arbitrary set. If γ is a rectifiable curve in X, we define the *length of* γ *in* E to be the number

$$\text{length}(\gamma \cap E) := m_1(\gamma_s^{-1}(E)) \,. \tag{5.1.18}$$

Note that if $E = X$ then $\text{length}(\gamma \cap E) = \text{length}(\gamma)$. The definition extends to locally rectifiable curves in a natural way. If E is a Borel set then

$$\text{length}(\gamma \cap E) = \int_\gamma \chi_E \, ds. \tag{5.1.19}$$

Moreover, if E is a Borel set and if γ is a rectifiable curve with length in E equal to zero then

$$\int_\gamma \rho \, ds = \int_\gamma \rho \cdot \chi_{X \setminus E} \, ds \tag{5.1.20}$$

for all Borel functions $\rho : X \to [0, \infty]$.

The inequality

$$\mathcal{H}_1(\gamma \cap E) \leq \text{length}(\gamma \cap E) \tag{5.1.21}$$

follows from (5.1.17) and from the fact that $\mathcal{H}_1$ is a Radon measure on γ. Obviously, strict inequality can take place in (5.1.21).

5.2 Modulus of a curve family

Let $X = (X, d, \mu)$ be a metric measure space as defined in Section 3.3, that is, (X, d) is separable as a metric space and μ is a locally finite Borel regular measure. Let Γ be a family of curves in X and let $p \geq 1$ be a real number. The *p-modulus* of Γ is defined as

$$\operatorname{Mod}_p(\Gamma) := \inf \int_X \rho^p \, d\mu \,, \tag{5.2.1}$$

where the infimum is taken over all Borel functions $\rho : X \to [0, \infty]$ satisfying

$$\int_\gamma \rho \, ds \geq 1 \tag{5.2.2}$$

for every locally rectifiable curve $\gamma \in \Gamma$. Functions ρ satisfying (5.2.2) are called *admissible densities* (or *metrics* or *functions*) for Γ. Note that the modulus has values in $[0, \infty]$. By definition, the modulus of the family of all curves in X that are not locally rectifiable is zero, and the modulus of every family containing a constant curve is infinite.

We gather some basic properties of modulus. These properties will be used repeatedly and usually without special notice in this book. First, we observe that

$$\operatorname{Mod}_p(\emptyset) = 0 \tag{5.2.3}$$

and that

$$\operatorname{Mod}_p(\Gamma_1) \leq \operatorname{Mod}_p(\Gamma_2) \tag{5.2.4}$$

if $\Gamma_1 \subset \Gamma_2$. Equality (5.2.3) follows because the zero function is admissible in this case. Inequality (5.2.4) is equally obvious, because each ρ that is admissible for Γ_2 is also admissible for Γ_1. Next, if Γ_0 and Γ are two curve families in X such that each curve $\gamma \in \Gamma$ has a subcurve $\gamma_0 \in \Gamma_0$ then

$$\operatorname{Mod}_p(\Gamma) \leq \operatorname{Mod}_p(\Gamma_0) \,. \tag{5.2.5}$$

We say in such a situation that Γ *majorizes* Γ_0. To prove inequality (5.2.5), we simply observe that each ρ that is admissible for Γ_0 is also admissible for Γ.

We also have the *subadditivity* of the modulus; that is,

$$\operatorname{Mod}_p\left(\bigcup_{i=1}^{\infty} \Gamma_i\right) \leq \sum_{i=1}^{\infty} \operatorname{Mod}_p(\Gamma_i) \,. \tag{5.2.6}$$

To prove (5.2.6), we may assume without loss of generality that the right-hand side is finite. Fix $\epsilon > 0$, and pick for each i an admissible density ρ_i for Γ_i such that

$$\int_X \rho_i^p \, d\mu \leq \operatorname{Mod}_p(\Gamma_i) + \epsilon \circ 2^{-i}.$$

Then

$$\rho(x) = \left(\sum_{i=1}^{\infty} \rho_i(x)^p\right)^{1/p}$$

is Borel measurable and admissible for each Γ_i since $\rho \geq \rho_i$ for each i. Because

$$\int_X \rho^p \, d\mu \leq \sum_{i=1}^{\infty} \mathrm{Mod}_p(\Gamma_i) + \epsilon,$$

and because $\epsilon > 0$ is arbitrary, we conclude that (5.2.6) holds.

Remark 5.2.7 Expressions (5.2.3), (5.2.4), and (5.2.6) together imply that the set function

$$\Gamma \mapsto \mathrm{Mod}_p(\Gamma)$$

is a measure on the collection of all curve families in X. In general there are no nontrivial measurable curve families for the modulus [134].

Exceptional curve families

A family of curves is called *p-exceptional* if it has p-modulus zero.We say that a property of curves holds for *p-a.e. curve* if the collection of curves for which the property fails to hold is p-exceptional.

We will use without further mention the following simple observation: if Γ' is a p-exceptional subfamily of a curve family Γ then

$$\mathrm{Mod}_p(\Gamma) = \mathrm{Mod}_p(\Gamma \setminus \Gamma').$$

This follows readily from the monotonicity and subadditivity properties of the modulus. (Recall that sets of measure zero are measurable for every measure; see Section 3.3.)

The following lemma provides a handy criterion for p-exceptionality. The property described in the lemma could be taken as the definition of exceptionality.

Lemma 5.2.8 *A family Γ of locally rectifiable curves in X is p-exceptional if and only if there is a p-integrable Borel function $\rho : X \to [0, \infty]$ such that*

$$\int_\gamma \rho \, ds = \infty \tag{5.2.9}$$

for each curve $\gamma \in \Gamma$.

Proof Assume first that $\mathrm{Mod}_p(\Gamma) = 0$. Then for each $i \geq 1$ we find an admissible density ρ_i with

$$\int_X \rho_i^p \, d\mu \leq 2^{-ip}.$$

By setting

$$\rho(x) = \sum_{i=1}^{\infty} \rho_i(x),$$

we obtain a p-integrable nonnegative Borel function ρ satisfying (5.2.9) for each $\gamma \in \Gamma$ as required.

Conversely, assume that $\rho : X \to [0, \infty]$ is a p-integrable Borel function such that (5.2.9) holds for each $\gamma \in \Gamma$. Then $\epsilon\rho$ is admissible for each $\epsilon > 0$, whence $\mathrm{Mod}_p(\Gamma) = 0$. The lemma follows. □

p-exceptional sets

A subset E of X is said to be *p-exceptional* if the family of all nonconstant curves that meet E is p-exceptional. It follows from (5.2.5) that E is p-exceptional if and only if the family of all nonconstant compact curves that meet E is p-exceptional. We will use this latter observation without further mention in this book.

It follows from the remarks made after (5.3.8) below that every singleton, and hence every countable set in $\mathbb{R}^n$, is p-exceptional if $1 \leq p \leq n$ and $n \geq 2$. More generally, under certain circumstances singleton subsets of a metric measure space are p-exceptional; see Corollary 5.3.11. A p-exceptional set need not have zero measure. If a space has no rectifiable curves then every subset is p-exceptional. A less trivial example is obtained by considering $X = \mathbb{R}^n$ with the usual distance but with measure $\mu = m_n + \delta_0$, where δ_0 is the Dirac mass at the origin. Then $\{0\}$ has positive measure but is p-exceptional provided that $1 \leq p \leq n$.

We will also require the following lemma.

Lemma 5.2.10 *A countable union of p-exceptional sets is p-exceptional. In particular, if $E \subset X$ and if every point $x \in X$ has a neighborhood U_x such that $E \cap U_x$ is p-exceptional then E is p-exceptional.*

Proof The first assertion follows from the subadditivity of the modulus, (5.2.6). The second assertion follows from the first and from Lemma 3.3.27. We leave the easy details to the reader. □

In Chapter 2 we proved a standard result from real analysis (Proposition 2.3.13) stating that a convergent sequence in L^p has a pointwise almost-everywhere convergent subsequence. The following lemma of Fuglede shows that there is an analogous result involving p-exceptional families of curves. Fuglede's lemma plays an important role in the development of our Sobolev space theory. Its repeated use allows us to bypass the fact that, in general, Sobolev functions may not be differentiable along a given curve.

Fuglede's lemma *Let (g_i) be a sequence of Borel functions that converges in $L^p(X)$. Then there is a subsequence (g_{i_k}) with the following property: if g is any Borel representative of the L^p-limit of (g_i) then*

$$\lim_{k\to\infty} \int_\gamma |g_{i_k} - g| \, ds = 0$$

for p-a.e. curve γ in X.

Note that such Borel representatives always exist; see Proposition 3.3.23.

Proof Let g be an arbitrary Borel representative of the L^p-limit of (g_i). Choose a subsequence (g_{i_k}) of (g_i) such that

$$\int_X |g_{i_k} - g|^p \, d\mu \le 2^{-(p+1)k} .$$

Note that this subsequence is independent of the particular representative g. Define

$$\rho_k = |g_{i_k} - g|.$$

Let Γ be the family of locally rectifiable curves γ in X for which the statement

$$\lim_{k\to\infty} \int_\gamma \rho_k \, ds = 0$$

fails to hold, and let Γ_k be the family of all locally rectifiable curves γ in X for which

$$\int_\gamma \rho_k \, ds > 2^{-k}.$$

Then on the one hand

$$\Gamma \subset \bigcup_{k=j}^{\infty} \Gamma_k$$

for each $j \ge 1$. On the other hand, $2^k \rho_k$ is admissible for Γ_k for each k, so that

$$\mathrm{Mod}_p(\Gamma_k) \le 2^{pk} \int_X \rho_k^p \, d\mu \le 2^{-k}.$$

Consequently, we have that

$$\mathrm{Mod}_p(\Gamma) \leq \sum_{k=j}^{\infty} \mathrm{Mod}_p(\Gamma_k) \leq 2^{-j+1}$$

for each $j \geq 1$, whence $\mathrm{Mod}_p(\Gamma) = 0$. The lemma is proved. □

Fuglede's lemma is often applied in conjunction with Mazur's lemma (see Section 2.3). We next give an example of this.

Proposition 5.2.11 *Let $p > 1$ and let $\Gamma_1 \subset \Gamma_2 \subset \cdots$ be an increasing sequence of curve families in X. Then*

$$\lim_{i\to\infty} \mathrm{Mod}_p(\Gamma_i) = \mathrm{Mod}_p(\Gamma),$$

where $\Gamma = \bigcup_{i=1}^{\infty} \Gamma_i$.

Proof Notice first that the numbers $\mathrm{Mod}_p(\Gamma_i)$ form an increasing sequence, by (5.2.4), and that $\mathrm{Mod}_p(\Gamma_i) \leq \mathrm{Mod}_p(\Gamma)$. It therefore suffices to show that

$$\mathrm{Mod}_p(\Gamma) \leq \lim_{i\to\infty} \mathrm{Mod}_p(\Gamma_i) =: M,$$

assuming in addition that the limit is finite. To this end, pick for each i an admissible density ρ_i for Γ_i such that

$$\int_X \rho_i^p \, d\mu < M + \frac{1}{i}.$$

Hence the sequence (ρ_i) is bounded in $L^p(X)$ with

$$\lim_{i\to\infty} \|\rho_i\|_p^p = M.$$

Because $p > 1$, $L^p(X)$ is reflexive and so a subsequence of (ρ_i) converges weakly to some ρ in $L^p(X)$ (Theorem 2.4.1). Moreover, by the lower semicontinuity of norms, Proposition 2.3.5, we have that $\|\rho\|_p^p \leq M$. Next, by Mazur's lemma there exists a sequence of convex combinations $(\tilde{\rho}_j)$ of the functions ρ_i such that $\tilde{\rho}_j \to \rho$ in $L^p(X)$ as $j \to \infty$. Moreover, because the sequence of curve families is increasing, and because the defining condition (5.2.2) for admissibility is stable under convex combinations, the sequence $(\tilde{\rho}_j)$ can be chosen in such a way that $\tilde{\rho}_j$ is admissible for Γ_j for each j. It follows that

$$M \leq \lim_{j\to\infty} \|\tilde{\rho}_j\|_p^p = \|\rho\|_p^p \leq M. \tag{5.2.12}$$

Next, by Proposition 3.3.23 and by Fuglede's lemma above, we may assume that ρ is a Borel function and that

$$\int_\gamma \rho \, ds \geq 1 \tag{5.2.13}$$

for p-a.e. $\gamma \in \Gamma$. It thus follows from (5.2.12) and the subadditivity of the modulus that

$$\mathrm{Mod}_p(\Gamma) \le \int_X \rho^p \, d\mu = M,$$

completing the proof of the proposition. □

Example 5.2.14 The assumption $p > 1$ is needed in Proposition 5.2.11, as the following example shows. Let $x \in \mathbb{R}^n$ and $r > 0$. Consider the curve families Γ_i, $i = 1, 2, \ldots$, consisting of all curves in $\mathbb{R}^n$ with one end point in the closed ball $\overline{B}(x, r)$, and the other outside the open ball $B(x, r + 1/i)$. It follows from (5.3.8) below that

$$\mathrm{Mod}_1(\Gamma_i) = \omega_{n-1} \, r^{n-1}$$

for every i. However, a polar coordinate integration shows that there are no admissible 1-integrable densities for the family $\Gamma := \cup_{i=1}^{\infty} \Gamma_i$, since each admissible density must satisfy $\int_L \rho \, ds = \infty$ for any radial half line with an end point in $\partial B(x, r)$.

We next prove two simple but useful lemmas. Recall the terminology in (5.1.19).

Lemma 5.2.15 *Let $E \subset X$ be a set of measure zero. Then, for p-a.e. curve γ in X, the length of γ in E is zero. In particular, for p-a.e. curve γ in X we have that $\mathcal{H}_1(\gamma \cap E) = 0$.*

Proof Let Γ be the family of all locally rectifiable curves γ in X such that the length of γ in E is positive. Let E' be a Borel set of zero measure containing E. The function $\rho := \infty \cdot \chi_{E'}$ is both in $L^p(X)$ and admissible for Γ, and hence the first assertion follows from the fact that $\int_X \rho^p \, d\mu = 0$. The second assertion follows from the first via (5.1.21). □

Lemma 5.2.16 *Let g and h be two nonnegative Borel functions on X such that $g \le h$ almost everywhere. Then*

$$\int_\gamma g \, ds \le \int_\gamma h \, ds$$

for p-a.e. curve γ in X. In particular, if g and h agree almost everywhere then

$$\int_\gamma g \, ds = \int_\gamma h \, ds$$

for p-a.e. curve γ in X.

Proof Let $E = \{x \in X : g(x) > h(x)\}$. Then $\mu(E) = 0$ and so, by Lemma 5.2.15, we have for p-a.e. curve γ, $\mathcal{H}_1(\gamma^{-1}(E)) = 0$. Hence by the definition of line integrals, $\int_\gamma (h - g)\, ds \geq 0$. The linearity of line integrals now yields the desired result. □

5.3 Estimates for modulus

Except in a few cases, it is essentially impossible to give a precise value for the modulus of a curve family. It is in general difficult even to give good estimates. Upper bounds are easier to achieve, for the simple reason that it suffices to exhibit one nontrivial admissible density. We next give some examples; more examples in the Euclidean setting can be found in [273].

Lemma 5.3.1 *Let Γ be a family of curves in a Borel set $A \subset X$ such that each $\gamma \in \Gamma$ satisfies* $\text{length}(\gamma) \geq L > 0$. *Then*

$$\text{Mod}_p(\Gamma) \leq \mu(A)\, L^{-p}.$$

Proof The assertion follows on using the density $\rho(x) = L^{-1}\chi_A$. □

Lemma 5.3.2 *Let Γ be a family of curves in X and let $B_1, B_2, \dots$ be a sequence of Borel sets in X, each of finite measure. If every curve in Γ has a nonrectifiable subcurve in some B_j then* $\text{Mod}_p(\Gamma) = 0$.

Proof For $j \geq 1$, let Γ_j be the collection of curves in Γ that have a nonrectifiable subcurve contained in B_j. By the subadditivity of the modulus, it suffices to show that $\text{Mod}_p(\Gamma_j) = 0$. But this follows from Lemma 5.3.1 and inequality (5.2.5), because each set B_j has finite measure and because locally rectifiable subcurves that are not rectifiable have infinite length. The lemma is proved. □

Lemma 5.3.2 shows that the modulus ignores those nonrectifiable curves whose nonrectifiability is, in some sense, local. It turns out that nonrectifiable curves are unconditionally p-exceptional if the volume growth of X is no worse than polynomial of order p.

Proposition 5.3.3 *Let $p > 1$, and assume that there exists a point $x_0 \in X$ such that*

$$\limsup_{r\to\infty} \frac{\mu(B(x_0, r))}{r^p} < \infty. \tag{5.3.4}$$

Then the p-modulus of the family of all nonrectifiable curves in X is zero.

Proof By the preceding lemma, it suffices to show that $\text{Mod}_p(\Gamma) = 0$, where Γ consists of all unbounded curves in X. For this, define

$$\rho(x) = (d(x, x_0) \log d(x, x_0))^{-1} \chi_{X \setminus B(x_0, 2)}(x)$$

and

$$A_j = \overline{B}(x_0, 2^{j+1}) \setminus B(x_0, 2^j), \quad j \geq 1.$$

For each unbounded curve γ in X there exists an integer j_0 such that γ both meets A_{j_0} and has, for each $j > j_0$, a subcurve γ_j of length at least 2^j contained in A_j. Thus on the one hand

$$\int_\gamma \rho \, ds \geq \sum_{j=j_0}^{\infty} \int_{\gamma_j} \rho \, ds \geq \sum_{j=j_0}^{\infty} 2^j 2^{-j-1} ((j+1) \log 2)^{-1} = \infty.$$

On the other hand, we have a positive integer j_1 such that for $j > j_1$, $\mu(B(x_0, 2^{j+1})) \leq C \, 2^{p(j+1)}$, and so

$$\begin{aligned}
\int_X \rho^p \, d\mu &\leq \sum_{j=1}^{\infty} \int_{A_j} \rho^p \, d\mu \\
&\leq \sum_{j=1}^{\infty} 2^{-pj} (j \log 2)^{-p} \mu(B(x_0, 2^{j+1})) \\
&\leq \sum_{j=1}^{j_1} 2^{-pj} (j \log 2)^{-p} \mu(B(x_0, 2^{j+1})) \\
&\quad + C \sum_{j=j_1}^{\infty} 2^{-pj} (j \log 2)^{-p} 2^{p(j+1)}
\end{aligned}$$

is finite. Hence Lemma 5.2.8 implies that Γ is p-exceptional, and the proof is complete. □

Remark 5.3.5 The assumption $p > 1$ is needed in Proposition 5.3.3; to see this, simply consider $X = \mathbb{R}$ equipped with the usual distance and Lebesgue measure. Likewise, a volume growth condition is needed; to see this, let $X = \mathbb{R}^n$, $n \geq 2$, be equipped with the usual metric and Lebesgue measure. Consider the family Γ consisting of the half lines $\gamma_\omega(t) = t\omega$, $t \geq 1$, indexed by the unit sphere $\mathbb{S}^{n-1} = \{\omega \in \mathbb{R}^n : |\omega| = 1\}$. If ρ is admissible for Γ then

Hölder's inequality gives

$$\begin{aligned} 1 \le \int_{\gamma_\omega} \rho\, ds &= \int_1^\infty \rho(t\omega) t^{(n-1)/p} t^{-(n-1)/p}\, dt \\ &\le \left(\int_1^\infty \rho^p(t\omega) t^{(n-1)}\, dt \right)^{1/p} \left(\int_1^\infty t^{-(n-1)/(p-1)}\, dt \right)^{(p-1)/p}. \end{aligned}$$

If $1 < p < n$ then

$$\int_1^\infty t^{-(n-1)/(p-1)}\, dt = \frac{p-1}{n-p},$$

and we can conclude by integrating in polar coordinates that

$$\int_{\mathbb{R}^n} \rho^p\, d\mu \ge \left(\frac{n-p}{p-1} \right)^{p-1} \omega_{n-1},$$

where ω_{n-1} is the surface measure of $\mathbb{S}^{n-1}$. Thus

$$\mathrm{Mod}_p(\Gamma) \ge \left(\frac{n-p}{p-1} \right)^{p-1} \omega_{n-1}. \tag{5.3.6}$$

In particular, the p-modulus of all unbounded curves in $\mathbb{R}^n$ is positive if $1 < p < n$. In fact, equality holds in (5.3.6) for $1 < p < n$. This can be seen by observing that the function

$$\rho(x) = \left(\frac{n-p}{p-1} \right) |x|^{-(n-1)/(p-1)} \chi_{\{|x|>1\}}(x)$$

is admissible for Γ and by computing the integral of ρ^p.

We will next compute explicitly the moduli of certain curve families using similar arguments.

Given disjoint sets $E, F \subset X$, we write

$$\mathrm{Mod}_p(E, F) = \mathrm{Mod}_p(E, F; X) \tag{5.3.7}$$

for the p-modulus of the collection of all locally rectifiable curves that join E and F in X.

Spherical shells

Consider the modulus

$$\mathrm{Mod}_p(\overline{B}(x, r), \mathbb{R}^n \setminus B(x, R)), \quad 0 < r < R,$$

in $\mathbb{R}^n$, $n \ge 2$. First we assume that $p > 1$. For $\omega \in \mathbb{R}^n$ with $|\omega| = 1$, let γ_ω denote the curve $\gamma_\omega : [r, R] \to \mathbb{R}^n$ given by $\gamma_\omega(t) = t\omega$. As in Remark 5.3.5, when ρ is an admissible function for computing the above modulus, via Hölder's inequality applied to γ_ω we deduce from $\int_{\gamma_\omega} \rho \, ds \geq 1$ that

$$1 \leq C_{r,R} \int_r^R \rho(t\omega)^p \, t^{n-1} \, dt,$$

where

$$\begin{aligned} C_{r,R} &= \left(\int_r^R t^{-(n-1)/(p-1)} \, dt \right)^{p-1} \\ &= \begin{cases} \left(\frac{p-1}{|n-p|} \left| R^{(p-n)/(p-1)} - r^{(p-n)/(p-1)} \right| \right)^{p-1} & \text{if } 1 < p \neq n, \\ \left(\log \left(\frac{R}{r} \right) \right)^{p-1} & \text{if } p = n. \end{cases} \end{aligned}$$

By integrating in polar coordinates over the annulus $B(x, R) \setminus B(x, r)$, we see that

$$\int_{\mathbb{R}^n} \rho(y)^p \, dy \geq \int_{B(x,R) \setminus B(x,r)} \rho(y)^p \, dy \geq \frac{\omega_{n-1}}{C_{r,R}},$$

that is, we obtain the lower bounds for $\mathrm{Mod}_p(\overline{B}(x, r), \mathbb{R}^n \setminus B(x, R))$ stipulated in (5.3.8) below. These lower bounds are also upper bounds, as can be seen by a suitable choice of admissible function ρ, namely

$$\rho(y) = C_{r,R}^{-1/(p-1)} |y - x|^{-(n-1)/(p-1)} \chi_{B(x,R) \setminus B(x,r)}(y).$$

For $p = 1$ the computation is simpler; choose

$$\rho_j(y) = j \chi_{\overline{B}(x, r+1/j) \setminus B(x,r)}(y)$$

and let j tend to infinity. To wit, for $x \in \mathbb{R}^n$ and $0 < r < R$, we have

$$\begin{aligned} &\mathrm{Mod}_p(\overline{B}(x, r), \mathbb{R}^n \setminus B(x, R))) \\ &= \begin{cases} |\frac{n-p}{p-1}|^{p-1} \omega_{n-1} |r^{(p-n)/(p-1)} - R^{(p-n)/(p-1)}|^{1-p}, & p \neq 1, n, \\ \omega_{n-1} \left(\log \frac{R}{r} \right)^{1-n}, & p = n, \\ \omega_{n-1} r^{n-1}, & p = 1. \end{cases} \end{aligned} \tag{5.3.8}$$

Observe that while $\mathrm{Mod}_n(\overline{B}(x, r), \mathbb{R}^n \setminus B(x, 2r))$ is independent of r, the value of $\mathrm{Mod}_p(\overline{B}(x, r), \mathbb{R}^n \setminus B(x, 2r))$ for $p \neq n$ tends to either zero or infinity when r tends to zero or infinity. The behavior is reciprocal depending on whether $p < n$ or $p > n$. These asymptotic properties could also be obtained by using the scaling properties of the Lebesgue measure. The exceptional behavior of the modulus for $p = n$ is accounted for by its *conformal invariance*.

The preceding dichotomy can also be seen in the following fact: the p-modulus of all nonconstant curves in $\mathbb{R}^n$ that go through a fixed point is zero if and only if $1 \leq p \leq n$.

Interestingly, the logarithmic behavior in (5.3.8) for the n-modulus in $\mathbb{R}^n$ has a counterpart in metric spaces whose measure has an appropriate growth condition.

Proposition 5.3.9 *Let $x_0 \in X$. If there exists a constant $C_0 > 0$ such that*

$$\mu(B(x_0, r)) \leq C_0 \, r^p$$

for each $0 < r < R_0$ then

$$\mathrm{Mod}_p(\overline{B}(x_0, r), X \setminus B(x_0, R)) \leq C_1 \left(\log \frac{R}{r}\right)^{1-p} \tag{5.3.10}$$

whenever $0 < 2r < R < R_0$, where C_1 depends only on C_0 and p.

Proof Set

$$\rho(x) = C_2 \left(\log \frac{R}{r}\right)^{-1} d(x, x_0)^{-1} \quad \text{if } x \in B(x_0, R) \setminus B(x_0, r)$$

and $\rho(x) = 0$ otherwise. We now check that, for an appropriate choice of C_2, ρ is an admissible metric for the curve family in question. Let k be the smallest integer satisfying $2^k r \geq R$, and let γ be a rectifiable curve connecting $\overline{B}(x_0, r)$ to $X \setminus B(x_0, R)$. Then for $j = 0, 1, \ldots, k-1$ there is a subcurve γ_j of γ lying in the annulus $B(x_0, 2^j r) \setminus \overline{B}(x_0, 2^{j-1} r)$, with length at least $2^{j-1} r$. It follows that

$$\int_\gamma \rho \, ds \geq \sum_{j=0}^{k-1} \int_{\gamma_j} \rho \, ds \geq C_2 \frac{1}{\log(R/r)} \, k \geq \frac{C_2}{\log 2}.$$

Therefore, by choosing C_2 appropriately, we can deduce that ρ is admissible, whence

$$\mathrm{Mod}_p(\overline{B}(x_0, r), X \setminus B(x_0, R)) \leq \int_X \rho^p \, d\mu \, .$$

To estimate the integral on the right, as above let k be the smallest integer satisfying $2^k r \geq R$. By using the assumption $\mu(B(x_0, r)) \leq C_0 \, r^p$ on the volume growth and the annuli $B(x_0, 2^{j+1}r) \setminus B(x_0, 2^j r)$, $j = 0, \ldots, k-1$, we easily compute

$$\int_X \rho^p \, d\mu \le C_2^p \, C_0 \left(\log \frac{R}{r}\right)^{-p} \sum_{j=0}^{k-1} (2^j r)^{-p} (2^{(j+1)} r)^p \le C_1 \left(\log \frac{R}{r}\right)^{1-p},$$

where C_1 has the asserted form. The proposition follows. □

We now give a metric measure space version of the exceptionality of the family of curves through a fixed point.

Corollary 5.3.11 *Let (X, d, μ) be a metric measure space and let $x_0 \in X$. If there exist $R > 0$, $C_R > 0$, and $Q > 1$ such that*

$$\mu(B(x_0, r)) \le C_R \, r^Q$$

whenever $0 < r < R$ then, for $1 \le p \le Q$, the p-modulus of the collection of all nonconstant rectifiable curves in X that pass through x_0 is zero.

Proof By the subadditivity relation (5.2.6) and the majorization principle for the modulus (5.2.5), it suffices to show that, given $\epsilon > 0$, we can find $0 < r' < r < \epsilon$ such that

$$\mathrm{Mod}_p(\overline{B}(x_0, r'), X \setminus B(x_0, r)) < \epsilon.$$

For $p = Q$ this follows from Proposition 5.3.9 with the choice $r' = r^2$ for $r > 0$ sufficiently small, and for $1 \le p < Q$ one simply takes $r' = r/2$ for sufficiently small r. Indeed, the function $\rho(y) = 2r^{-1}\chi_{B(x_0,r)\setminus B(x_0,r/2)}(y)$ is admissible, and

$$\int_X \rho^p \, d\mu \le 2^p \, C_R r^{Q-p}.$$

The claim follows. □

Cylindrical curve families

Another situation where the modulus can be computed explicitly is the case of a generalized cylinder. Let $I \subset \mathbb{R}$ be a nondegenerate interval and let $Y = X \times I$ be a product metric measure space equipped with the product measure $\nu = \mu \times m_1$. The distance in Y can be anything that restricts to the distances on the factors. A *generalized cylinder* with *base* E and *height* $h > 0$ is a set of the form

$$G = E \times J,$$

where $E \subset X$ is a Borel set and $J \subset I$ is an interval of length h. Let Γ be the family of all rectifiable curves $\gamma_x : J \to G$, $\gamma_x(t) = (x, t)$ for $x \in E$. Then

$$\mathrm{Mod}_p(\Gamma) = \frac{\mu(E)}{h^{p-1}}. \tag{5.3.12}$$

Indeed, Lemma 5.3.1 shows that $\mathrm{Mod}_p(\Gamma)$ is bounded from above by the right-hand side of (5.3.12). To achieve a lower bound, let ρ be an admissible density for Γ. Given a point $x \in E$, consider the corresponding curve γ_x, and compute

$$1 \leq \int_{\gamma_x} \rho \, ds \leq \left(\int_{\gamma_x} \rho^p \, ds \right)^{1/p} h^{(p-1)/p}$$

by Hölder's inequality. Fubini's theorem then implies that

$$\mu(E) \leq h^{p-1} \int_G \rho^p \, d\mu,$$

as required. This proves (5.3.12).

Lower semicontinuous admissible functions

As previously remarked, estimating the modulus from above is usually substantially easier than estimating it from below. The problem is that for a lower bound one has to show an estimate for all admissible functions. To such end, it would be desirable to know whether the pool of admissible metrics could somehow be restricted. In general, one cannot restrict the admissible metrics to continuous or bounded functions. For example, it follows from the formulas in (5.3.8), and from the subadditivity, that the p-modulus for $1 \leq p \leq n$ of the family of all nonconstant curves in $\mathbb{R}^n$ that pass through a given point is zero. Obviously, there are no admissible bounded metrics for this family.

A useful observation is that it suffices to consider lower semicontinuous functions as admissible metrics. This follows directly from the definitions and from the Vitali–Carathéodory theorem (see Section 4.2).

Proposition 5.3.13 *For every curve family Γ in X we have that*

$$\mathrm{Mod}_p(\Gamma) = \inf \int_X \rho^p \, d\mu \,,$$

where the infimum is taken over all lower semicontinuous functions $\rho : X \to [0, \infty]$ that are admissible for Γ.

The following proposition links the p-modulus of curve families for various ranges of p. It is easily proved by an application of Hölder's inequality and is left to the reader to verify.

Proposition 5.3.14 *Let Γ be a family of curves all of which lie in a measurable set $A \subset X$ with $\mu(A) < \infty$. Then, for $1 \leq q < p$,*

$$\mathrm{Mod}_q(\Gamma)^p \leq \mu(A)^{p-q} \, \mathrm{Mod}_p(\Gamma)^q.$$

In particular, if such a family Γ is p-exceptional for some $p > 1$ then it is q-exceptional for each $1 \leq q \leq p$.

Later, in Chapter 9, we link Poincaré inequalities to modulus, and provide some nontrivial lower bounds.

5.4 Notes to Chapter 5

The modulus was first called the *extremal length* or *distance* (for $p = 2$). This terminology still appears in the function theory literature for curve families in the complex plane. As defined in this book, the modulus is the reciprocal of the extremal length. Beurling made systematic use of the modulus in his 1933 thesis [27, pp. 1–107]. Later the method was bolstered by Beurling, Ahlfors, and many others. Notable early applications appeared also in the works of Grötzsch and Teichmüller. The modulus is a particularly important tool in higher-dimensional function theory, in the theory of quasiconformal and quasiregular mappings where other tricks of complex analysis are often lost. See [273], [279], [234], and [120].

The first abstract treatment of modulus was the seminal 1957 paper [89] by Fuglede. This elegant work is highly recommended reading. Later, the modulus was used in connection with Sobolev functions, especially by Ohtsuka [220], who called it the extremal length and first considered the concept of p-exceptional sets in the context of Euclidean domains for all $p \geq 1$.

An integration theory with respect to the modulus was developed in [196], but so far there have been no applications of this theory.

The definitions in Chapter 5.1 are standard adaptations of the classical $\mathbb{R}^n$-valued case. In particular, Väisälä's monograph [273] has a careful discussion of both curves and modulus in $\mathbb{R}^n$. Systematic uses of moduli in abstract metric measure spaces, especially in connection with Sobolev spaces and quasiconformal mappings, can be found in for example [125], [168], [247], [248], [271], and [272]. An important predecessor was [223]. The basic estimates here all originate from well-known results in $\mathbb{R}^n$.

The notion of the p-modulus of curve families and its more abstract versions (due originally to Fuglede) are still under active study; see for example [4], [11], [16], [30], [115], and [227].

Proposition 5.2.11 is due to Ziemer and Ohtsuka; see [288, Lemma 2.3]. Their proof employs Clarkson's inequalities instead of Mazur's lemma. The fact that one can restrict to lower semicontinuous functions in the definition of the modulus (Proposition 5.3.13) had already been pointed out by Fuglede [89, p. 173].

The discussion relating to Proposition 5.3.3 and Remark 5.3.5 is linked to the classification of metric spaces according to whether the collection of unbounded rectifiable curves that start from a fixed ball has positive p-modulus; a metric measure space for which such a collection of curves has positive p-modulus is called p-hyperbolic. Euclidean spaces of dimension n are p-hyperbolic precisely when $1 \leq p < n$, whereas the standard hyperbolic space $\mathbb{H}^n$ is p-hyperbolic for all $p \geq 1$. A discussion of p-hyperbolicity can be found in the work of Grigor'yan [103], [104] and the dissertation [137] of Holopainen.

6

Upper gradients

We begin this chapter with a brief introduction to classical Sobolev spaces on Euclidean open sets. Our treatment of the topic is standard, and some routine details are left (with references) to the reader. The initial discussion serves as a motivation for the ensuing abstract definitions.

The main bulk of this chapter is a thorough analysis of upper gradients, on which the later definition for Sobolev spaces in metric measure spaces is based; this is given in Chapter 7. We discuss upper gradients here in the general framework of functions that are defined in metric measure spaces and have values in an arbitrary metric space.

6.1 Classical first-order Sobolev spaces

We denote by Ω an open subset of $\mathbb{R}^n$, $n \geq 1$, and by $C_c^\infty(\Omega)$ the vector space of all infinitely many times differentiable functions with compact support in Ω. We typically abbreviate $dm_n(x)$ to dx for the Lebesgue n-measure in $\mathbb{R}^n$. Unless otherwise specifically stated, all measure theoretic notions in this section are understood with respect to Lebesgue measure. Standard basis vectors in $\mathbb{R}^n$ are denoted by $e_1, \ldots, e_n$.

Dirichlet and Sobolev spaces

Let u be a real-valued locally integrable function on Ω and let $i \in \{1, \ldots, n\}$. A real-valued locally integrable function v_i on Ω is said to be the *weak ith partial derivative* of u on Ω if

$$\int_\Omega v_i(x)\varphi(x)\, dx = -\int_\Omega u(x)\frac{\partial \varphi}{\partial x_i}(x)\, dx \tag{6.1.1}$$

whenever $\varphi \in C_c^\infty(\Omega)$.

It is easy to see that each weak derivative of a function, whenever it exists, *is uniquely defined as an element of* $L^1_{\mathrm{loc}}(\Omega)$. Indeed, for this it suffices to observe the following elementary fact: if $v \in L^1_{\mathrm{loc}}(\Omega)$ satisfies

$$\int_\Omega v(x)\,\varphi(x)\,dx = 0$$

for every $\varphi \in C^\infty_c(\Omega)$ then $v = 0$ almost everywhere. It is also obvious that the existence of a weak derivative is independent of the Lebesgue representative of the given function u in $L^1_{\mathrm{loc}}(\Omega)$. We use the notation

$$\partial_i u = \frac{\partial u}{\partial x_i}$$

for the weak ith partial derivative of a function u. If u has a weak ith partial derivative for each $i = 1, \dots, n$, we call the $\mathbb{R}^n$-valued locally integrable function

$$\nabla u := (\partial_1 u, \dots, \partial_n u)$$

the *weak derivative* or *gradient* of u. Recall that the standard Euclidean norm is used in $\mathbb{R}^n$ unless otherwise stated, so that

$$|\nabla u|^2 = (\partial_1 u)^2 + \cdots + (\partial u_n)^2 .$$

Remarks 6.1.2 (a) If u is continuously differentiable then, by integration by parts, the weak derivative agrees with the classical derivative and we use the same notation, $\nabla u = (\partial_1 u, \dots, \partial_n u)$, for the gradient. It is not true, however, that a locally integrable function that is almost everywhere differentiable with a locally integrable almost everywhere defined (classical) gradient necessarily possesses a weak gradient. The function may miss the key property of being absolutely continuous on lines; see Theorem 6.1.17. For example, the standard Cantor function $c : [0, 1] \to [0, 1]$ is a continuous increasing surjection, with $c'(x) = 0$ for almost every $x \in [0, 1]$, but c does not possess weak derivatives. See also Theorem 6.1.13.

(b) In the language of distribution theory, the weak partial derivative $\partial_i u$ is precisely the distributional (ith partial) derivative of the distribution $u \in L^1_{\mathrm{loc}}(\Omega)$, assumed to be representable by an element in $L^1_{\mathrm{loc}}(\Omega)$. We will not discuss distributions further in this book.

The *Dirichlet space* $L^{1,p}(\Omega)$, $1 \le p < \infty$, consists of all real-valued functions u in $L^1_{\mathrm{loc}}(\Omega)$ with weak derivative ∇u in $L^p(\Omega : \mathbb{R}^n)$. The expression

$$\|u\|_{L^{1,p}(\Omega)} := \left(\int_\Omega |\nabla u(x)|^p\,dx\right)^{1/p} \tag{6.1.3}$$

defines a seminorm on $L^{1,p}(\Omega)$; it is not a norm because it vanishes for (locally) constant functions which may not be identically zero.

The *Sobolev space* $W^{1,p}(\Omega)$, $1 \le p < \infty$, is the normed space of functions u in $L^{1,p}(\Omega) \cap L^p(\Omega)$ equipped with the norm

$$\|u\|_{W^{1,p}(\Omega)} := \|u\|_{L^p(\Omega)} + \|u\|_{L^{1,p}(\Omega)}. \tag{6.1.4}$$

Because the equality of the integrals in (6.1.1) is seen to persist under L^p-limits, we have that $W^{1,p}(\Omega)$ *is a Banach space for every* $1 \le p < \infty$.

As in the case of Lebesgue spaces L^p we call the members of the spaces $L^{1,p}(\Omega)$ and $W^{1,p}(\Omega)$ functions, even though strictly speaking they are equivalence classes of functions, with two functions identified if and only if they agree outside a set of measure zero. Later, in Proposition 7.1.31, we will consider a finer equivalence relation, defined in terms of capacity. This finer relation leads to a more precise description of the pointwise behavior of Sobolev functions and is important for the geometric development in this book.

We next demonstrate the important fact that smooth functions are dense in $W^{1,p}(\Omega)$. In particular, this leads to an alternate description of the Sobolev space as the closure of smooth functions under the norm (6.1.4). To conform with standard notation, we make the following definition.

Definition 6.1.5 Denote by $H^{1,p}(\Omega)$ the norm closure of the linear subspace $C^\infty(\Omega) \cap W^{1,p}(\Omega)$ in $W^{1,p}(\Omega)$.

Theorem 6.1.6 *We have that*

$$H^{1,p}(\Omega) = W^{1,p}(\Omega). \tag{6.1.7}$$

Moreover, the linear subspace $C_c^\infty(\mathbb{R}^n)$ *of* $H^{1,p}(\mathbb{R}^n)$ *is dense in* $W^{1,p}(\mathbb{R}^n)$.

For the proof, we recall that the *convolution* between a locally integrable function v and a bounded compactly supported measurable function η is given by

$$v * \eta(x) = \int_{\mathbb{R}^n} v(x-y)\,\eta(y)\,dy = \int_{\mathbb{R}^n} v(y)\,\eta(x-y)\,dy. \tag{6.1.8}$$

Proof of Theorem 6.1.6 It suffices to show that every function $u \in W^{1,p}(\Omega)$ can be approximated in the Sobolev norm by functions in $C^\infty(\Omega)$. We prove this in the case $\Omega = \mathbb{R}^n$ only, with the smaller space $C_c^\infty(\mathbb{R}^n)$. The general case requires only routine technical modifications using partitions of unity. (See for example [81, Theorem 2, p. 125] or [193, Theorem 1.45, p. 25].) Note that in general $C_c^\infty(\Omega)$ fails to be dense in $W^{1,p}(\Omega)$; cf. [2, p. 57, Theorem 3.23].

Thus, let $u \in \mathrm{W}^{1,p}(\mathbb{R}^n)$. We first claim that u can be approximated in $W^{1,p}(\mathbb{R}^n)$ by compactly supported functions in $\mathrm{W}^{1,p}(\mathbb{R}^n)$. To this end, fix $R > 0$ and let $\varphi \in C_c^\infty(\mathbb{R}^n)$ be such that $0 \le \varphi \le 1$, $\varphi = 1$ on $B(0, R)$, and $|\nabla\varphi| \le 2$ on $\mathbb{R}^n$. It easily follows from the definitions that $u(1-\varphi)$ belongs to $W^{1,p}(\mathbb{R}^n)$, with

$$\nabla(u(1-\varphi)) = (1-\varphi)\nabla u - u\nabla\varphi.$$

Moreover, the Sobolev norm of $u(1-\varphi)$ can be estimated from above by the expression

$$3\left(\int_{\mathbb{R}^n\setminus B(0,R)} |u|^p\,dx\right)^{1/p} + \left(\int_{\mathbb{R}^n\setminus B(0,R)} |\nabla u|^p\,dx\right)^{1/p},$$

which tends to zero as $R \to \infty$. This proves the claim, and it follows that we can assume initially that u vanishes outside a compact set.

Next, consider the convolution $u * \eta$ as in (6.1.8), where $\eta \in C_c^\infty(\mathbb{R}^n)$ is a fixed function. Given a standard basis vector e_i, we can estimate the expression

$$\left|\frac{u*\eta(x+re_i) - u*\eta(x)}{r} - u*\partial_i\eta(x)\right|$$

from above by

$$\int_{\mathbb{R}^n} |u(y)| \left|\frac{\eta(x+re_i-y)-\eta(x-y)}{r} - \partial_i\eta(x-y)\right| dy$$
$$= \int_{\mathbb{R}^n} |u(y)|\,|\partial_i\eta(x_{r,y}-y) - \partial_i\eta(x-y)|\,dy\,,$$

for some $x_{r,y} \in \mathbb{R}^n$ such that $|x_{r,y} - x| \le r$. From the uniform continuity of $\partial_i\eta$ and the compactness of the support of u, we obtain that $u * \eta$ is differentiable on $\mathbb{R}^n$ with derivative $u * \nabla\eta$. We repeat the argument with η replaced by $\partial_j\eta$ for any $1 \le j \le n$. Continuing inductively, we conclude that $u * \eta$ is a smooth function and hence its derivative is the same as its weak derivative. If we choose η to be in addition symmetric (that is, $\eta(z) = \eta(-z)$, which can be ensured by choosing η in such a way that $\eta(z) = \eta(|z|)$), then we see by the fact that u has a weak derivative that

$$\partial_i(u*\eta) = u*\partial_i\eta = \partial_i u*\eta\,. \tag{6.1.9}$$

To produce a sequence of such smooth convolutions approximating u in the Sobolev norm, fix a nonnegative symmetric function $\eta \in C_c^\infty(\mathbb{R}^n)$ such that $\int_{\mathbb{R}^n}\eta(y)\,dy = 1$, set

$$\eta_\epsilon(x) := \epsilon^{-n}\eta(x/\epsilon)\,, \qquad \epsilon > 0\,,$$

and let $u_\epsilon = u * \eta_\epsilon$. By the preceding argument, $u_\epsilon \in C_c^\infty(\mathbb{R}^n)$ and $\partial_i u_\epsilon = (\partial_i u)_\epsilon$ for each $i = 1, \ldots, n$. It therefore suffices to show that, given

$v \in L^p(\mathbb{R}^n)$, we have that $v_\epsilon \to v$ in $L^p(\mathbb{R}^n)$ as $\epsilon \to 0$. Since continuous functions are dense in $L^p(\mathbb{R}^n)$, and since convolution with η_ϵ is a linear operation which does not increase the L^p-norm, it follows that we may assume, initially, that v is (uniformly) continuous with compact support. The preceding understood, we use Hölder's inequality to obtain

$$\begin{aligned} |v_\epsilon(x) - v(x)|^p &= \left| \int_{\mathbb{R}^n} (v(x-y) - v(x))\eta_\epsilon(y)\, dy \right|^p \\ &\le \left(\int_{\mathbb{R}^n} |v(x-y) - v(x)|^p |\eta_\epsilon(y)|\, dy \right) \left(\int_{\mathbb{R}^n} |\eta_\epsilon(y)|\, dy \right)^{p-1} \\ &= \int_{\mathbb{R}^n} |v(x-y) - v(x)|^p |\eta_\epsilon(y)|\, dy\,. \end{aligned}$$

Then an application of Fubini's theorem shows that

$$\int_{\mathbb{R}^n} |v_\epsilon(x) - v(x)|^p\, dx \le \int_{\mathbb{R}^n} \int_{\mathbb{R}^n} |v(x-y) - v(x)|^p |\eta_\epsilon(y)|\, dx\, dy\,. \tag{6.1.10}$$

Since $\mathrm{spt}(\eta_\epsilon) \subset B(0,\epsilon)$, it follows that the right-hand side of (6.1.10) converges to zero as $\epsilon \to 0$, by the continuity of translations on $L^p(\mathbb{R}^n : V)$ (see Remark 3.3.54).

This completes the proof of Theorem 6.1.6. □

Remark 6.1.11 The proof of (6.1.9) shows, more generally, that

$$\partial_\nu(u * \varphi) = u * \partial_\nu \varphi = u * (\nabla\varphi \cdot \nu) \tag{6.1.12}$$

for each unit vector $\nu \in \mathbb{R}^n$, where ∂_ν signifies the derivative of a function $\varphi \in C_c^\infty(\mathbb{R}^n)$ in the direction of ν. We will later show, see Theorem 6.1.13, that each $u \in W^{1,p}(\Omega)$ has a representative such that the partial derivatives $\partial_i u$ exist almost everywhere.

Local Sobolev spaces

The definitions of the *local Dirichlet* and *Sobolev spaces* $L^{1,p}_{\mathrm{loc}}(\Omega)$ and $W^{1,p}_{\mathrm{loc}}(\Omega)$ are obvious. To wit, a function belongs to $L^{1,p}_{\mathrm{loc}}(\Omega)$ if it belongs to $L^{1,p}(\Omega')$ for each open set Ω' with compact closure in Ω, and similarly for $W^{1,p}_{\mathrm{loc}}(\Omega)$.

A modification to the proof of Theorem 6.1.6 yields smooth approximations to Sobolev functions in the class $\mathrm{W}^{1,p}_{\mathrm{loc}}(\Omega)$, with convergence of the approximants and their partial derivatives in $L^p_{\mathrm{loc}}(\Omega)$. More precisely, given $u \in \mathrm{W}^{1,p}_{\mathrm{loc}}(\Omega)$ and given that Ω' is compactly contained in Ω, there exists a sequence of smooth functions on Ω' converging to u in $W^{1,p}(\Omega')$. See [81, p. 123, Theorem 1].

There are obvious obstacles in attempting to abstract the preceding definition of Sobolev classes $W^{1,p}(\Omega)$ to general metric spaces. There are no smooth functions in general metric spaces, and in any event one cannot integrate by parts. We next give more promising, alternative definitions for the classical Sobolev spaces. Recall the definition of absolutely continuous functions given in Section 4.4.

Absolute continuity on lines

A real-valued function u on Ω is said to be *absolutely continuous on lines* in Ω if the restriction of u to almost every compact line segment in Ω, parallel to the coordinate axes of $\mathbb{R}^n$, is absolutely continuous. (See Section 4.4.) More precisely, we consider the following definition of absolute continuity on lines.

Definition A function $u : \Omega \to \mathbb{R}$ is absolutely continuous on lines in Ω if, for every subspace $V_i := \{x = (x_1, \ldots, x_n) : x_i = 0\} \subset \mathbb{R}^n$, $i = 1, \ldots, n$, there is a set of Lebesgue $(n-1)$-measure zero, $N_i \subset V_i$, with the following property: for every $y \in V_i \setminus N_i$ the function $t \mapsto u(y + te_i)$ is absolutely continuous on every compact interval $[a, b]$ such that $y + te_i \in \Omega$ for $a \leq t \leq b$.

An elementary but important observation based on (5.3.12) is that a function u as above is absolutely continuous on lines in Ω if and only if it is absolutely continuous on p-a.e. compact line segment that is parallel to a coordinate axis of $\mathbb{R}^n$ and contained in Ω. The value of $p \geq 1$ is immaterial. See Section 5.3.

An absolutely continuous real-valued function on an interval $[a, b]$ is differentiable almost everywhere on $[a, b]$. Thus, by Fubini's theorem, a real-valued function u that is absolutely continuous on lines in Ω has classical partial derivatives m_n-almost everywhere on Ω. If u is in addition measurable then the partial derivatives can be shown to be measurable as well, by noting that for each rational number $h \neq 0$ the map $x \mapsto h^{-1}(u(x + he_i) - u(x))$ is measurable and that the partial derivatives are limits (as $h \to 0$) of these maps almost everywhere in Ω; see Corollary 3.1.5. If these derivatives belong in addition to $L^p(\Omega)$, $1 \leq p < \infty$, then u is said to be *of the class* $ACL^p(\Omega)$.

By using Fubini's theorem and integration by parts (which is valid for absolutely continuous functions of one variable), it is easy to see that a locally integrable function of the class $ACL^p(\Omega)$ belongs to the Dirichlet space $L^{1,p}(\Omega)$. The weak derivatives in this case are the classical partial derivatives. The following theorem shows the converse, giving an alternative characterization of Sobolev classes on open sets in Euclidean space. Recall that by a Lebesgue representative of u we mean a function $\tilde{u}$ that agrees with u almost everywhere in Ω.

Theorem 6.1.13 *A function* $u \in L^1_{\mathrm{loc}}(\Omega)$ *belongs to the Dirichlet space* $L^{1,p}(\Omega)$ *if and only if there is a Lebesgue representative of* u *that belongs to the class* $ACL^p(\Omega)$. *In particular,*

$$W^{1,p}(\Omega) = L^p(\Omega) \cap ACL^p(\Omega).$$

Proof As explained before the statement of the theorem, only the necessity part requires a proof. Thus, let $u \in L^{1,p}(\Omega)$. The family of all compact line segments in Ω can be expressed as a countable union of families of line segments, each contained in a relatively compact subdomain of Ω. It therefore suffices to consider segments that lie in a fixed relatively compact subdomain Ω' of Ω.

From the proof of Theorem 6.1.6, we know that, since $\nabla u \in L^p(\Omega, \mathbb{R}^n)$, the convolutions u_ϵ have the property that $\partial_i u_\epsilon = (\partial_i u)_\epsilon$ converges in $L^p(\Omega')$ to $\partial_i u$. This does not require u itself to be in $L^p(\Omega')$.

Fuglede's lemma 5.2, together with Proposition 3.3.23, guarantees that, for a subsequence,

$$\int_\gamma |\partial_i u_{\epsilon_k} - \partial_i u| \, ds \to 0 \tag{6.1.14}$$

for p-almost every line segment γ in Ω' and every $i = 1, \dots, n$, provided that we choose appropriate (Borel) representatives of $\partial_i u$ and and a sequence $\epsilon_k \to 0$. Moreover, by Lemma 5.2.8 we have that

$$\int_\gamma |\partial_i u| \, ds < \infty \tag{6.1.15}$$

for p-almost every curve γ in Ω', for every $i = 1, \dots, n$.

We continue to work with the subsequence chosen above. Since, as pointed out in the proof of Theorem 6.1.6, the u_ϵ are smooth in Ω', they are absolutely continuous on *all* line segments parallel to the direction of the vector e_i, $i = 1, \dots, n$. Furthermore, since $u \in L^1(\Omega')$ we know that u_ϵ converges to u in $L^1(\Omega)$ and hence, by passing to a subsequence if necessary, we can also ensure that $u_{\epsilon_k} = u_k$ converges almost everywhere in Ω' to u. We redefine u in such a way that $u(x) = \lim_k u_k(x)$ whenever this limit exists. It follows by using Fubini's theorem that, for almost every line segment in Ω' that is parallel to the direction of e_i, we have $u_k \to u$ pointwise outside a set of $\mathcal{H}_1$-measure zero.

Next, let α be a compact line segment on Ω' parallel to the ith coordinate axis such that (6.1.14) and (6.1.15) hold for every compact subsegment γ of α and, in addition, $u_j \to u$ $\mathcal{H}_1$-almost everywhere on α. By the preceding discussion and by (5.2.5), almost every such compact line segment α in Ω' has this property. We assume that α as a curve is (isometrically) parametrized by its arc length. Let $s, r \in [0, \text{length}(\alpha)]$ such that $r < s$ and $u_k(\alpha(r)) \to u(\alpha(r))$ and $u_k(\alpha(s)) \to u(\alpha(s))$; almost every r and s in $[0, \text{length}(\alpha)]$ satisfy this condition. By the fundamental theorem of calculus applied to the smooth functions u_k, we see that

$$\begin{aligned} |u(\alpha(r)) - u(\alpha(s))| &= \lim_k |u_k(\alpha(r)) - u_k(\alpha(s))| \\ &\le \lim_k \int_{\alpha|[r,s]} |\partial_i u_k| \, ds = \int_{\alpha|[r,s]} |\partial_i u| \, ds. \end{aligned}$$

Given that $\int_\alpha |\partial_i u| \, ds$ is finite, the absolute continuity of u on α follows from the absolute continuity of the integral, provided that we can show that the limit $\lim_k u_k(\alpha(s))$ exists for *all* $s \in [0, \text{length}(\alpha)]$; recall that u coincides with the above limit whenever the latter exists. To this end, notice from (6.1.14) that

$$\lim_k \int_{\alpha|[r,s]} \partial_i u_k \, ds = \int_{\alpha|[r,s]} \partial_i u \, ds \tag{6.1.16}$$

for all $0 \le r \le s \le \text{length}(\alpha)$. Fixing r to be such that $\lim_k u_k(\alpha(r))$ exists, it immediately follows from the fundamental theorem of calculus applied to the smooth functions u_k, together with the absolute continuity of the integral and (6.1.15), that the limit also exists at every $r \le s \le \text{length}(\alpha)$. In particular, the limit exists at $\text{length}(\alpha)$, and an analogous argument then gives the existence of the limit for all s.

This completes the proof of Theorem 6.1.13. □

A function $u : \Omega \to \mathbb{R}$ is said to be *absolutely continuous on a curve* γ in Ω if $\gamma : [0, \text{length}(\gamma)] \to \Omega$ is rectifiable and parametrized by the arc length and the function $u \circ \gamma : [0, \text{length}(\gamma)] \to \mathbb{R}$ is absolutely continuous; cf. Section 4.4.

Using the preceding concept we can formulate an important generalization of Theorem 6.1.13.

Theorem 6.1.17 *A function $u \in L^1_{\text{loc}}(\Omega)$ belongs to the Dirichlet space $L^{1,p}(\Omega)$ if and only if there is a Lebesgue representative of u that is absolutely continuous on p-a.e. compact curve in Ω and the partial derivatives of u belong to $L^p(\Omega)$.*

The formulation requires an explanation. If the function u is absolutely continuous on p-a.e. compact curve in Ω then in particular u is absolutely continuous on lines and hence possesses classical partial derivatives almost everywhere. This follows from (5.3.12).

Proof The sufficiency in Theorem 6.1.17 follows from Theorem 6.1.13. Toward the necessity, we follow the line of argument in the proof of Theorem 6.1.13. Instead of line segments we consider general rectifiable curves and replace ∂u_k with ∇u_k; similarly for u. The almost everywhere convergence of our subsequence to u guarantees that, for p-a.e. curve, we have m_1-almost everywere convergence along the curve in question, see Lemma 5.2.15. □

Remarks 6.1.18 (a) Changing the value of a function $u \in W^{1,p}(\Omega)$ arbitrarily on a hyperplane does not change the representative of u in $W^{1,p}(\Omega)$. However, after such a change u may not be absolutely continuous on lines. Thus the representative of a Sobolev function offered by Theorem 6.1.13 is "better" than an arbitrary representative. Although it is not clear at this point, Theorem 6.1.17 offers an even better representative called a quasicontinuous representative. Representatives in $ACL^p(\Omega)$ can be changed in sets whose projections to the coordinate hyperplanes have zero m_{n-1}-measure, whereas quasicontinuous representatives can be changed only in sets of capacity zero, which is a more stringent condition when $p \geq n-1$. For the development of the theory offered in this book, it is important to understand these distinctions in general. The theory of capacity will be developed fully in Chapter 7.

(b) The characterization of Sobolev functions as functions that are absolutely continuous on lines, with appropriately integrable partial derivatives, immediately yields a removability result. If $F \subset \mathbb{R}^n$ is a closed set whose projections to each coordinate hyperplane has m_{n-1}-measure zero, and if u is a member of $W^{1,p}(\Omega \setminus F)$ for some open neighborhood Ω of F, then u belongs to $W^{1,p}(\Omega)$. Note that such an F must have Lebesgue n-measure zero, so that the extension of u is automatic; what is nontrivial is that this extension is in the Sobolev space $W^{1,p}(\Omega)$.

6.2 Upper gradients

It follows from Theorem 6.1.17 that each function $u \in L^{1,p}(\Omega)$ has a representative that satisfies the condition

$$|u(\gamma(a)) - u(\gamma(b))| \leq \int_\gamma |\nabla u| \, ds$$

for p-a.e. compact curve $\gamma : [a, b] \to \Omega$. We use this inequality as a starting point for defining a substitute for the derivative in general metric measure spaces. In fact, we only have a substitute for the *size of the derivative*.

For the rest of this chapter, we will assume that $X = (X, d_X, \mu)$ is a metric measure space as defined in Section 3.3 (that is, (X, d) is separable and μ is a locally finite Borel regular measure on X) and that $Z = (Z, d_Z)$ is a metric space. We also assume that $1 \leq p < \infty$ unless otherwise specifically stated.

Definition A Borel function $\rho : X \to [0, \infty]$ is said to be an *upper gradient* of a map $u : X \to Z$ if

$$d_Z\big(u(\gamma(a)), u(\gamma(b))\big) \leq \int_\gamma \rho \, ds \tag{6.2.1}$$

for every rectifiable curve $\gamma : [a, b] \to X$. We say that ρ is a *p-weak upper gradient* of u if (6.2.1) holds for p-a.e. rectifiable curve in X. If $u : A \to Z$ is defined only on a subset A of X then we speak with self-explanatory meaning about upper gradients or p-weak upper gradients of u in A, where A is considered as a metric measure space itself.

The concept of an upper gradient is purely metric and makes sense for functions defined on an arbitrary metric space. The concept of a p-weak upper gradient depends on the underlying measure structure as well as on p. Pure upper gradient theory does not seem to be very fruitful because this concept is not stable under limiting processes, and so the main emphasis in this book is on weak upper gradients. Moreover, p-weak upper gradients have richer properties if they are in addition p-integrable. This extra assumption is imposed in most of the ensuing discussion, and remarks will be made that illuminate to what extent it is needed.

The following lemma shows that in considering p-weak upper gradients we do not move very far from upper gradients.

Lemma 6.2.2 *If g is a nonnegative Borel measurable function on X that is finite-valued μ-almost everywhere and if $u : X \to Z$ has g as a p-weak upper gradient in X then there is a sequence (g_k) of upper gradients of u such that $g \leq g_{k+1} \leq g_k$ and $\|g - g_k\|_{L^p(X)} \to 0$ as $k \to \infty$.*

Proof Let Γ be the collection of all nonconstant compact rectifiable curves in X for which the upper gradient inequality (6.2.1) fails for the pair (u, g). Then, by assumption, $\mathrm{Mod}_p(\Gamma) = 0$ and so by Lemma 5.2.8 we have a nonnegative Borel measurable function ρ on X such that $\int_\gamma \rho \, ds = \infty$ for each $\gamma \in \Gamma$ and $\rho \in L^p(X)$. Now, the functions $g_k = g + 2^{-k}\rho$ form a sequence of upper gradients of u that satisfies our requirements. This completes the proof of this lemma. □

Equation (6.2.1) is the *upper gradient inequality for the pair (u, ρ) on γ*. If ρ is an upper gradient, or a p-weak upper gradient, of u then the pair (u, ρ) is called a *function–upper-gradient pair.* Note that the potential dependence on p is suppressed in the latter terminology, such matters being usually clear from the context.

Examples and basic properties of upper gradients

In general, a given function has infinitely many distinct upper gradients: if ρ is an upper gradient of a function then so is $\rho + \sigma$ for every nonnegative Borel function σ. Moreover, the pointwise maximum of two upper gradients is an upper gradient.

The function $\rho \equiv \infty$ is an upper gradient of every function.

If X has no nonconstant rectifiable curves then $\rho \equiv 0$ is an upper gradient of every function on X.

If u is L-Lipschitz then $\rho \equiv L$ is an upper gradient of u.

If ρ is an upper gradient of a function $u : X \to Z$ and if $z_0 \in Z$ then ρ is also an upper gradient of the real-valued function $x \mapsto d_Z(u(x), z_0)$. More generally, if W is a metric space and $f : Z \to W$ is L-Lipschitz then $L\rho$ is an upper gradient of the function $f \circ u : X \to W$ whenever ρ is an upper gradient of u.

If ρ is an upper gradient of a function $u : X \to Z$ and if $A \subset X$ then the restriction $\rho|A : A \to [0, \infty]$ is an upper gradient of the restriction $u|A : A \to Z$.

The collection of upper gradients of a given function u is a convex set: if ρ and σ are two upper gradients of u and if $0 \leq \lambda \leq 1$ then $(1 - \lambda)\rho + \lambda\sigma$ is an upper gradient of u.

If $Z = (Z, |\cdot|)$ is a normed space, if two functions $u_1, u_2 : X \to Z$ have respective upper gradients ρ_1 and ρ_2, and if $\lambda \in \mathbb{R}$ then $|\lambda|\rho_1 + \rho_2$ is an upper gradient of the function $\lambda u_1 + u_2$ and ρ_1 is an upper gradient of the function $|u_1|$.

Analogous statements hold for p-weak upper gradients. Furthermore, because of Lemma 6.2.2, the L^p-closure of the convex set of all p-integrable upper gradients of u is the convex set of all p-integrable p-weak upper gradients of u; see Corollary 6.3.12 below.

The fundamental theorem of calculus implies that if u is a smooth real- or Banach space-valued function on an open set Ω in $\mathbb{R}^n$ then $|\nabla u|$ is an upper gradient of u. Moreover, the proof of Theorem 6.1.17 implies that $|\nabla u|$ is a p-weak upper gradient of a function $u \in L^{1,p}(\Omega)$, provided that both u and $|\nabla u|$ are pointwise defined. Here we need to take a good representative for u, as guaranteed by Theorem 6.1.17, while any Borel representative for $|\nabla u|$ is a p-weak upper gradient by Lemma 5.2.16 (cf. Lemma 6.2.8 below). Analogous remarks hold if Ω is replaced by a Riemannian manifold and $|\nabla u|$ is the length of the Riemannian gradient of u.

The *pointwise lower Lipschitz-constant function* of a map $u : X \to Z$ is given by

$$\operatorname{lip} u(x) := \liminf_{r \to 0} \sup_{y \in B(x,r)} \frac{d_Z(u(x), u(y))}{r}. \tag{6.2.3}$$

Similarly, the *pointwise upper Lipschitz-constant function* of $u : X \to Z$ is given by

$$\operatorname{Lip} u(x) := \limsup_{r \to 0} \sup_{y \in B(x,r)} \frac{d_Z(u(x), u(y))}{r}. \tag{6.2.4}$$

Lemma 6.2.5 *Let $u : X \to Z$ be continuous. Then* lip u *and* Lip u *are Borel functions.*

Proof Define, for $x \in X$ and $r > 0$,

$$l_r u(x) := \inf_{s<r} \sup_{y \in B(x,s)} \frac{d_Z(u(x), u(y))}{s}$$

and

$$L_r u(x) := \sup_{s<r} \sup_{y \in B(x,s)} \frac{d_Z(u(x), u(y))}{s}.$$

Because $\operatorname{lip} u(x) = \lim_{r \to 0} l_r u(x)$ and $\operatorname{Lip} u(x) = \lim_{r \to 0} L_r u(x)$ for every $x \in X$, it suffices to show that the functions $l_r u$ and $L_r u$ are Borel for every $r > 0$ (Proposition 3.3.22). We will do this for l_r, the proof for L_r being similar.

Thus, fix $r > 0$. The function

$$u_s^y(x) := \frac{d_Z(u(x), u(y))}{s} \chi_{B(x,s)}(y) = \frac{d_Z(u(x), u(y))}{s} \chi_{B(y,s)}(x)$$

is lower semicontinuous for every $y \in X$ and $s > 0$. Therefore, the function

$$u_s(x) := \sup_{y \in X} u_s^y(x) = \sup_{y \in B(x,s)} \frac{d_Z(u(x), u(y))}{s}$$

is lower semicontinuous and hence Borel for every $s > 0$. Recall that lower semicontinuity is characterized by the condition that the super level sets $\{u > \tau\}$ for $\tau \in \mathbb{R}$ are open. Hence the supremum of a collection of lower semicontinuous functions is lower semicontinuous. Let (s_n) be a countable dense set in $(0, \infty)$. We claim that

$$l_r u(x) = \inf_{s_n < r} u_{s_n}(x)$$

for every $x \in X$. Because the infimum of a countable set of Borel functions is Borel (consider the preimage of $[a, \infty)$ for each $a \in \mathbb{R}$), the assertion follows from this. To prove the claim, fix $x \in X$. Then fix $\epsilon > 0$ and choose $s < r$ such that

$$\frac{d_Z(u(x), u(y))}{s} < l_r u(x) + \epsilon$$

for every $y \in B(x, s)$. There exists $s_n \in (\max\{s - \epsilon, s/2\}, s)$ such that

$$\frac{d_Z(u(x), u(y))}{s_n} = \frac{s}{s_n} \frac{d_Z(u(x), u(y))}{s} < \frac{s}{\max\{s - \epsilon, s/2\}} (l_r u(x) + \epsilon)$$

for every $y \in B(x, s_n)$. Thus

$$\inf_{s_n < r} u_{s_n}(x) \leq \frac{s}{\max\{s - \epsilon, s/2\}} (l_r u(x) + \epsilon) .$$

Obviously $l_r u(x) \leq \inf_{s_n < r} u_{s_n}(x)$, and so the claim follows on letting $\epsilon \to 0$ in the preceding inequality. The lemma is proved. □

Lemma 6.2.6 *If $u : X \to Z$ is locally Lipschitz then* lip u *is an upper gradient of u.*

Proof By the preceding discussion, lip u is a nonnegative Borel function. Let $\gamma : [0, \text{length}(\gamma)] \to X$ be a nonconstant rectifiable curve parametrized by its arc length. Because u is locally Lipschitz the map $u \circ \gamma : [0, \text{length}(\gamma)] \to Z$ is absolutely continuous, and we find from (4.4.26) that on the one hand

$$d_Z\big(u(\gamma(a)), u(\gamma(b))\big) \leq \int_0^{\text{length}(\gamma)} |(u \circ \gamma)'(t)|\, dt , \qquad (6.2.7)$$

where we have the metric differential of $u \circ \gamma$ on the right-hand side. On the other hand, for $t \in (0, \text{length}(\gamma))$ and for $h \in \mathbb{R}$ with $|h|$ small enough, we have

$$\frac{d_Z((u \circ \gamma)(t), (u \circ \gamma)(t + h))}{|h|} \leq \sup_{y \in B(\gamma(t), |h|))} \frac{d_Z(u(\gamma(t)), u(y))}{|h|} .$$

Because the left-hand side of this inequality tends to $|(u \circ \gamma)'(t)|$ as $|h| \to 0$ for m_1-a.e. t (Theorem 4.4.8), we may infer that $|(u \circ \gamma)'(t)| \leq \text{lip}\, u(\gamma(t))$ for m_1-a.e. $t \in (0, \text{length}(\gamma))$. The lemma follows from this and from (6.2.7). □

Next we record an important robustness property of weak upper gradients. The lemma follows directly from Lemma 5.2.16.

Lemma 6.2.8 *If ρ is a p-weak upper gradient of a map $u : X \to Z$ and if $\sigma : X \to [0, \infty]$ is a Borel function such that $\sigma = \rho$, almost everywhere in X, then σ is a p-weak upper gradient of u. In particular, if E is a Borel set of* *measure zero and* ρ *is a* p*-weak upper gradient of* u *then* $\rho\chi_{X\setminus E}$ *is a* p*-weak upper gradient of* u.

The following simple lemma will also be used later.

Lemma 6.2.9 *Let* $A \subset X$ *be closed, let* ρ *be an upper gradient of a map* $u : A \to Z$*, and let* $z_0 \in Z$*. Define* $u_0 : X \to Z$ *by* $u_0(x) = u(x)$ *if* $x \in A$ *and* $u(x) = z_0$ *if* $x \in X \setminus A$*, and define* $\rho_0 : X \to [0, \infty]$ *by* $\rho_0(x) = \rho(x)$ *if* $x \in A$ *and* $\rho_0(x) = \infty$ *if* $x \in X \setminus A$*. Then* ρ_0 *is an upper gradient of* u_0 *in* X.

Proof The assertion is clear since for every nonconstant rectifiable curve γ in X that meets $X \setminus A$ we must have $\int_\gamma \rho_0 \, ds = \infty$ because A is closed. □

Remarks 6.2.10 (a) An assertion analogous to Lemma 6.2.8 for upper gradients is false: the function $\rho \equiv 1$ is an upper gradient of every 1-Lipschitz function on $\mathbb{R}^2$ but $\rho\chi_{\mathbb{R}^2\setminus L}$ need not be an upper gradient of such a function if $L \subset \mathbb{R}^2$ is a line.

(b) Lemma 6.2.9 is not true for an arbitrary subset $A \subset X$: consider $A = \mathbb{R} \setminus \{0\} \subset \mathbb{R}$ and the function $\chi_{(0,\infty)} : \mathbb{R} \setminus \{0\} \to \mathbb{R}$ together with an upper gradient $\rho \equiv 0$.

6.3 Maps with p-integrable upper gradients

In this section we study maps that have p-integrable upper gradients. Recall the standing assumptions in this chapter that $X = (X, d_X, \mu)$ is a metric measure space, that $Z = (Z, d_Z)$ is a metric space, and that $1 \leq p < \infty$ unless otherwise specifically stated.

Absolute continuity on curves

By Theorem 6.1.17 a classical Sobolev function in $\mathbb{R}^n$ has a representative that is absolutely continuous on p-almost every curve. We next prove an analogous fact for functions defined on arbitrary metric measure spaces.

We generalize the definition given just before Theorem 6.1.17. A map $u : X \to Z$ is said to be *absolutely continuous on a curve* γ in X if γ is rectifiable and the map $u \circ \gamma_s : [0, \text{length}(\gamma)] \to Z$ is absolutely continuous as defined in Section 4.4. (Recall the notation γ_s for the arc length parametrization of γ, from (5.1.6).)

We begin with a lemma which follows directly from the definitions and from the absolute continuity of the integral. We leave its verification to the reader.

Lemma 6.3.1 *Let* $u : X \to Z$ *be a map and let* γ *be a rectifiable compact curve in* X*. Assume that* $\rho : X \to [0, \infty]$ *is a Borel function such that* ρ *is* *integrable on γ and that the pair (u, ρ) satisfies the upper gradient inequality* (6.2.1) *on γ as well as on each compact subcurve of γ. Then u is absolutely continuous on γ.*

The next proposition is fundamental.

Proposition 6.3.2 *Suppose that ρ is a p-integrable p-weak upper gradient of a map $u : X \to Z$. Then p-almost every compact rectifiable curve γ in X has the following property: ρ is integrable on γ and the pair (u, ρ) satisfies the upper gradient inequality* (6.2.1) *on γ and each of its compact subcurves. In particular, every map $u : X \to Z$ that has a p-integrable p-weak upper gradient is absolutely continuous on p-a.e. compact curve in X.*

Proof Denote by Γ_0 the family of all compact rectifiable curves γ such that either the upper gradient inequality (6.2.1) does not hold or ρ is not integrable on γ. Then $\text{Mod}_p(\Gamma_0) = 0$ by the definition of weak upper gradients, by (5.2.6), and by Lemma 5.2.8. Next, if Γ is the family of all compact curves in X that have a subcurve in Γ_0 then $\text{Mod}_p(\Gamma) = 0$ by (5.2.5). The first assertion follows.

The second assertion follows from the first and from Lemma 6.3.1. $\square$

The argument in the proof of the preceding proposition is typical in the theory of modulus, and will be repeatedly used in this book.

Next we show how the metric differential along curves is related to the upper gradient. Recall the definition of a metric differential from Theorem 4.4.8.

Proposition 6.3.3 *Let $u : X \to Z$ be a map and let $\gamma : [0, \text{length}(\gamma)] \to X$ be a rectifiable curve parametrized by its arc length. Assume that $\rho : X \to [0, \infty]$ is a Borel function such that ρ is integrable on γ and the pair (u, ρ) satisfies the upper gradient inequality* (6.2.1) *on γ and each of its compact subcurves. Then u is absolutely continuous on γ, and the inequality*

$$|(u \circ \gamma)'(t)| \leq (\rho \circ \gamma)(t) \tag{6.3.4}$$

holds for almost every $t \in [0, \text{length}(\gamma)]$; $|(u \circ \gamma)'(t)|$ denotes the metric differential. In particular, if ρ is a p-integrable p-weak upper gradient of a function $u : X \to Z$ then (6.3.4) *holds for p-a.e. curve $\gamma : [0, \text{length}(\gamma)] \to X$ parametrized by its arc length. Furthermore, if u has a p-integrable p-weak upper gradient in X and ρ is a nonnegative p-integrable Borel measurable function on X such that* (6.3.4) *holds for p-a.e. absolutely continuous rectifiable curve γ in X then ρ is a p-weak upper gradient of u.*

Proof The absolute continuity of u on γ follows from Lemma 6.3.1. Next, we observe that the hypotheses give

$$\frac{d_Z\big((u \circ \gamma)(t), (u \circ \gamma)(t+h)\big)}{h} \leq \frac{1}{h} \int_t^{t+h} (\rho \circ \gamma)(s)\, ds$$

for every $t \in [0, \text{length}(\gamma))$ and for every $h \in (0, \text{length}(\gamma) - t)$. For almost every t, as $h \to 0$ the left-hand side of this inequality tends to $|(u \circ \gamma)'(t)|$ (by Theorem 4.4.8) and the right-hand side tends to $(\rho \circ \gamma)(t)$ (by the Lebesgue differentiation theorem; see (3.4.10) and (3.4.11)). Hence we conclude that (6.3.4) holds. The penultimate assertion follows from the first assertion and from Proposition 6.3.2. Towards the final claim, recall that if u has a p-integrable p-weak upper gradient on X then it is absolutely continuous on p-almost every curve γ. The proof is completed by an application of Proposition 4.4.25 to $u \circ \gamma$. □

Localization

A reasonable notion of a derivative must satisfy the condition that every locally constant map has the zero function as derivative. We now demonstrate that a p-integrable p-weak upper gradient has this property. The main result in this regard is Corollary 6.3.16. We also establish the important lattice property of upper gradients (Corollary 6.3.12).

Lemma 6.3.5 *Suppose that $u : X \to Z$ is absolutely continuous on p-almost every compact rectifiable curve in X. If there exists $c \in Z$ such that $u \equiv c$, μ-almost everywhere in X, then the collection Γ_E of all nonconstant curves in X that intersect the set*

$$E := \{x \in X \,:\, u(x) \neq c\}$$

has p-modulus zero. In particular, $\rho \equiv 0$ is a p-weak upper gradient of u.

Proof We have that p-almost every nonconstant rectifiable curve γ in X has the following two properties: u is absolutely continuous on γ and the length of γ in E is zero (Lemma 5.2.15). In the present situation, therefore, we must have that $u \circ \gamma \equiv c$ and that, in particular γ cannot meet E at all. However, every curve in Γ_E does meets E. This gives $\text{Mod}_p(\Gamma_E) = 0$, by (5.2.5), as required. The second assertion is clear from the first. □

Proposition 6.3.2 and Lemma 6.3.5 yield the following corollary.

Corollary 6.3.6 *Suppose that $u : X \to Z$ has a p-integrable p-weak upper gradient and that there exists $c \in V$ such that $u \equiv c$ μ-almost everywhere in X. Then $\rho \equiv 0$ is a p-weak upper gradient of u.*

Remark 6.3.7 Lemma 6.3.5 does not hold if the hypothesis of absolute continuity along p-a.e. curve is relaxed. Consider, for example, the characteristic function of a point in $\mathbb{R}$; the collection of all nonconstant curves in $\mathbb{R}$ that meet the point has nonzero 1-modulus. The converse to the second assertion is not true in general. If the metric space X contains no rectifiable curves then the zero function is an upper gradient of any function on X. In Section 7.5 we will study a condition on X which suffices for the following assertion: the only maps on X which have the zero function as a p-weak upper gradient are the (essentially) constant functions.

Next we explain further techniques for constructing new upper gradients from old.

Lemma 6.3.8 *Suppose that σ and τ are two p-integrable p-weak upper gradients of a map $u : X \to Z$. If E is a Borel subset of X then the function*

$$\rho = \sigma \cdot \chi_E + \tau \cdot \chi_{X \setminus E} \tag{6.3.9}$$

is a p-weak upper gradient of u.

Proof From the hypotheses we have that p-almost every compact rectifiable curve γ in X enjoys the following properties: u is absolutely continuous on γ, the function $\sigma + \tau$ is integrable on γ, and the upper gradient inequality (6.2.1) holds for both pairs (u, σ) and (u, τ) on γ as well as on each of its compact subcurves. (See Proposition 6.3.2 and Lemma 5.2.8.) Let $\gamma : [0, \ell] \to X$ be such a curve, parametrized by arc length. We will prove that the upper gradient inequality (6.2.1) holds for the pair (u, ρ) on γ, where ρ is as in (6.3.9).

Because $E' := \gamma^{-1}(E)$ is a Borel subset of $[0, \ell]$, there is a sequence of compact sets (F_n) such that $F_1 \subset F_2 \subset \cdots \subset E'$ and $m_1(E' \setminus F_n) \to 0$ as $n \to \infty$ (see (3.3.38)). Set $U_n := [0, \ell] \setminus F_n$. Then U_n is relatively open in $[0, \ell]$, and so it is the union of a pairwise disjoint collection of intervals. Enumerate the intervals $I_1 = (a_1, b_1)$, $I_2 = (a_2, b_2), \ldots$ Denote the restrictions of γ to the closures of these intervals by $\gamma_1, \gamma_2, \ldots$ We then have that

$$\begin{aligned} d_Z\big(u(\gamma(0)), u(\gamma(\ell))\big) &\le d_Z\big(u(\gamma(0)), u(\gamma(a_1))\big) + d_Z\big(u(\gamma(a_1)), u(\gamma(b_1))\big) \\ &\quad + d_Z\big(u(\gamma(b_1)), u(\gamma(\ell))\big) \\ &\le \int_{\gamma \setminus \gamma_1} \sigma \, ds + \int_{\gamma_1} \tau \, ds , \end{aligned}$$

with an obvious notation $\gamma \setminus \gamma_1$. Proceeding by induction we obtain that

$$d_Z\big(u(\gamma(0)), u(\gamma(\ell))\big) \le \int_{\gamma \setminus \bigcup_{1 \le i \le j} \gamma_i} \sigma \, ds + \int_{\bigcup_{1 \le i \le j} \gamma_i} \tau \, ds$$

for each positive integer j, which becomes

$$d_Z\big(u(\gamma(0)), u(\gamma(\ell))\big) \le \int_{F_n} \sigma(\gamma(t))\,dt + \int_{U_n} \tau(\gamma(t))\,dt \tag{6.3.10}$$

upon application of the Lebesgue dominated convergence theorem. Finally, another application of the dominated convergence theorem yields

$$d_Z\big(u(\gamma(0)), u(\gamma(\ell))\big) \le \int_{E'} \sigma(\gamma(t))\,dt + \int_{[0,\ell]\setminus E'} \tau(\gamma(t))\,dt = \int_\gamma \rho\,ds$$

as required. The proof of the lemma is complete. □

Corollary 6.3.11 *Suppose that σ and τ are two p-integrable p-weak upper gradients of a map $u : X \to Z$. Then the function* $\min(\sigma, \tau)$ *is a p-integrable p-weak upper gradient of u.*

Proof Apply Lemma 6.3.8 to the Borel set $E = \{\sigma < \tau\}$. □

A *lattice* in $L^p(X)$ is a collection of functions that is closed under pointwise minimum and maximum operations. Corollary 6.3.11, together with Fuglede's lemma (see Section 5.1) and the remarks in Section 6.2, implies the following result.

Corollary 6.3.12 *The collection $\mathcal{U}$ of all p-integrable p-weak upper gradients of a map $u : X \to Z$ is a closed convex lattice inside $L^p(X)$.*

It will be shown below, in Theorem 6.3.20, that the set of p-weak upper gradients as in the preceding corollary, if nonempty, has a unique element of smallest L^p-norm. This follows from Corollary 2.4.16 for $p > 1$ by the uniform convexity of $L^p(X)$, but there is a proof that works for $p = 1$ as well; see Theorem 6.3.20.

Remark 6.3.13 Divide [0, 1] into two disjoint Borel sets A and B in such a way that every open subinterval of [0, 1] meets both A and B on a set of positive m_1-measure. Then $\sigma = \infty \cdot \chi_A$ and $\tau = \infty \cdot \chi_B$ are upper gradients of every function $u : [0, 1] \to V$, but $\min(\sigma, \tau) \equiv 0$. Thus Corollary 6.3.11 is not true for arbitrary upper gradients.

Lemma 6.3.14 *Suppose that $u : X \to Z$ is absolutely continuous on p-a.e. compact rectifiable curve. Let E be a Borel subset of X and assume that there are maps $v, w : X \to Z$ such that $u = v$ μ-almost everywhere on E and $u = w$ μ-almost everywhere on $X \setminus E$. If v and w possess p-integrable p-weak upper gradients σ and τ, respectively, then*

$$\rho = \sigma\chi_E + \tau\chi_{X\setminus E} \tag{6.3.15}$$

is a p-integrable p-weak upper gradient of u.

Proof We will show that both $\rho_1 = \sigma + \tau\chi_{X\setminus E}$ and $\rho_2 = \sigma\chi_E + \tau$ are p-weak upper gradients of u. Because both ρ_1 and ρ_2 are p-integrable, Lemma 6.3.8 then gives the desired conclusion. By symmetry it suffices to show that ρ_1 is a p-weak upper gradient of u.

As earlier, it follows from the definitions, from the basic properties of the modulus, from Lemma 5.2.15, and from Proposition 6.3.2 that p-a.e. compact rectifiable curve γ in X enjoys the following three properties: the maps u, v, and w are all absolutely continuous on γ; the two map–upper-gradient pairs (v, σ) and (w, τ) satisfy the upper gradient inequality (6.2.1) on γ and each of its compact subcurves; and the length of γ is zero in the sets

$$E' := \{x \in E : u(x) \neq v(x)\}, \quad F' := \{x \in X \setminus E : u(x) \neq w(x)\}.$$

Let $\gamma : [0, \ell] \to X$ be such a curve parametrized by its arc length. We will show that (6.2.1) holds on γ for the pair (u, ρ_1).

Since u, v, w are continuous on γ, we have that $\gamma^{-1}(\{u \neq v\} \cap \{u \neq w\}) = \gamma^{-1}(E' \cap F')$ must be open. Since the length of γ inside $E' \cup F'$ is zero, it follows that γ does not intersect $E' \cap F'$. Therefore, at each point of γ either $u = v$ or $u = w$. Since $\gamma^{-1}(\{u = v\})$ is closed, its complement in $[0, \ell]$ is a pairwise disjoint union of relatively open intervals that are mapped by γ into $X \setminus (E \cup F')$. Now a repetition of the argument in the proof of Lemma 6.3.8 tells us that (u, ρ_1) satisfies (6.2.1).

This completes the proof of Lemma 6.3.14. □

Lemma 6.3.14 implies an important localization property for integrable upper gradients:

Corollary 6.3.16 *Let ρ be a p-integrable p-weak upper gradient of a map $u : X \to Z$, and let $c \in Z$. Then the function $\rho\chi_{X\setminus E}$ is a p-integrable p-weak upper gradient of u for every Borel set $E \subset \{x \in X : u(x) = c\}$.*

Proof The constant map $v \equiv c$ has $\sigma \equiv 0$ as a p-integrable upper gradient and $u|E = v$. The assertion therefore follows from Proposition 6.3.2 and Lemma 6.3.14. □

Remark 6.3.17 As earlier (cf. Remark 6.3.13), the assumption that ρ is p-integrable in Corollary 6.3.16 is necessary. If u is the characteristic function of the origin in $\mathbb{R}$ then the function $\rho_1 = \infty \cdot \chi_{\mathbb{R}\setminus\{0\}}$ is an upper gradient of u, but $\rho_2 \equiv 0$ is not.

Minimal upper gradients

A p-weak upper gradient ρ of a map $u : X \to Z$ is said to be a *minimal p-weak upper gradient* if it is p-integrable and if $\rho \leq \sigma$ almost everywhere in X whenever σ is a p-integrable p-weak upper gradient of u. Obviously, such a p-weak upper gradient, if it exists, is unique up to a set of measure zero (courtesy of Lemma 6.2.8). Also note that the minimal p-weak upper gradient has the smallest L^p-norm amongst all the upper gradients of u.

We denote the minimal p-weak upper gradient of a map u by ρ_u. The existence of minimal upper gradients is guaranteed under rather general circumstances by Theorem 6.3.20 below.

If we consider functions with values in a normed space then minimal upper gradients enjoy the following simple properties:

$$\rho_u = \rho_{-u}\,, \quad \rho_{u+v} \le \rho_u + \rho_v\,, \quad \rho_{|u|} \le \rho_u\,, \quad \rho_{\lambda u} = |\lambda|\rho_u\,, \tag{6.3.18}$$

where $\lambda \in \mathbb{R}$. We will prove later that $\rho_u = \rho_{|u|}$ for measurable functions $u : X \to \mathbb{R}$ (Corollary 6.3.27). Moreover, if W is a metric space and $f : Z \to W$ is L-Lipschitz then

$$\rho_{f\circ u} \le L\,\rho_u\,. \tag{6.3.19}$$

Note that the definition of a minimal p-weak upper gradient depends on p. To keep matters simple, this a priori dependence on p is suppressed in the notation. It turns out that in a large class of metric measure spaces the minimal p-weak upper gradient of a real-valued function is independent of p, although this is not true in general. See Section 13.5.

The minimal p-weak upper gradient ρ_u should be thought of as a substitute for $|\nabla u|$, the length of a gradient, for functions defined in metric measure spaces.

Theorem 6.3.20 *The collection of all p-integrable p-weak upper gradients of a map $u : X \to Z$ is a closed convex lattice inside $L^p(X)$ and, if nonempty, contains a unique element of smallest L^p-norm. In particular, if a map has a p-integrable p-weak upper gradient then it has a minimal p-weak upper gradient.*

Proof It was recorded in Corollary 6.3.12 that the collection $\mathcal{U}$ of all p-integrable p-weak upper gradients of u is a closed convex lattice inside $L^p(X)$. It is a general fact that every such set, if nonempty, contains a unique element of smallest L^p-norm. Indeed, assume that $(\rho_i) \subset \mathcal{U}$ is a sequence such that

$$\lim_{i\to\infty} \|\rho_i\|_p = \inf\{\|\rho\|_p : \rho \in \mathcal{U}\}.$$

By the lattice property (Corollary 6.3.12), replacing ρ_i with $\min_{1\le j\le i} \rho_j$ if necessary we can assume that the sequence (ρ_i) is pointwise decreasing. The limit function

$$\rho_u := \lim_{i\to\infty} \rho_i$$

is Borel by Proposition 3.3.22. By the Lebesgue monotone convergence theorem we have that $\rho_i \to \rho_u$ in $L^p(X)$, so that by Fuglede's lemma we know that $\rho_u \in \mathcal{U}$. It is clear that ρ_u has the required property of minimality. □

Remarks 6.3.21 (a) In the definition of minimal upper gradients, it is important to restrict to p-integrable upper gradients, for otherwise existence cannot be guaranteed even in simple situations. The function $u(x) = x$ on $X = [0, 1]$ has $\rho_u(x) \equiv 1$ as minimal weak upper gradient (for all $p \geq 1$) as defined above, but the function $\sigma = \infty \cdot \chi_O$ is an upper gradient of u for every dense open set $O \subset [0, 1]$; in particular, by choosing O appropriately we have that $\sigma < \rho_u$ on a set of positive measure.

(b) As mentioned in the proof of Theorem 6.3.20, for $p > 1$ we could have invoked the uniform convexity of $L^p(X)$ together with Corollary 2.4.16 to conclude that there is a unique element ρ_u of smallest L^p-norm $\mathcal{U}$. The fact that $\mathcal{U}$ is a lattice allows us to give an argument which works for all $p \geq 1$.

Proposition 6.3.22 *Let $u, v : X \to Z$ be two maps with respective minimal p-weak upper gradients ρ_u and ρ_v. If $u = v$ almost everywhere in a Borel set E then $\rho_u = \rho_v$ almost everywhere in E.*

Proof By the existence of p-integrable p-weak upper gradients, both u and v are absolutely continuous on p-a.e. compact curve (Proposition 6.3.2). We thus have from Lemma 6.3.14 that $\rho_u \cdot \chi_E + \rho_v \cdot \chi_{X\setminus E}$ is a p-integrable p-weak upper gradient of v. The minimality of ρ_v implies that $\rho_v \leq \rho_u$ almost everywhere in E, and the lemma follows by symmetry. □

We now use Proposition 6.3.22 to show that minimal upper gradients behave well under the truncation of real-valued functions. The next proposition also shows that the class of functions with p-integrable upper gradients forms a lattice.

Proposition 6.3.23 *Let $u_1, u_2 : X \to \mathbb{R}$ be two measurable functions with respective minimal p-weak upper gradients ρ_{u_1} and ρ_{u_2}. Then the following equalities are valid pointwise almost everywhere in X:*

$$\rho_{\min\{u_1,u_2\}} = \rho_{u_1} \cdot \chi_{\{u_1 \leq u_2\}} + \rho_{u_2} \cdot \chi_{\{u_2 < u_1\}}, \tag{6.3.24}$$

$$\rho_{\max\{u_1,u_2\}} = \rho_{u_1} \cdot \chi_{\{u_1 > u_2\}} + \rho_{u_2} \cdot \chi_{\{u_2 \geq u_1\}}. \tag{6.3.25}$$

Remark 6.3.26 Proposition 6.3.23 does not assert that the functions on the right in equalities (6.3.24) and (6.3.25) are the minimal upper gradients of $\min(u_1, u_2)$ and $\max(u_1, u_2)$, respectively. The proposition implies only that such an assertion holds for Borel representatives of these functions.

Proof of Proposition 6.3.23 First let ρ_1, ρ_2 be a choice of p-integrable upper gradients of u_1, u_2 respectively. Let γ be a nonconstant compact rectifiable curve in X, and consider $u = \min\{u_1, u_2\}$. Denoting the two end points of γ by x and y, without loss of generality let $u(x) \geq u(y)$. Then, without loss of generality, suppose that $u(y) = u_1(y) \leq u_2(y)$. Consequently $u(x) \leq u_1(x)$, and so

$$|u(x) - u(y)| = u(x) - u(y) \leq u_1(x) - u_1(y) \leq \int_\gamma \rho_1 \, ds \leq \int_\gamma (\rho_1 + \rho_2) \, ds.$$

It follows that $\rho_1 + \rho_2$ is a p-integrable upper gradient of u, and so u is absolutely continuous on p-almost every curve. Thus the triple u, u_1, u_2 of functions satisfies the hypotheses of Lemma 6.3.14, with E a Borel set containing $\{x \in X : u_1(x) \leq u_2(x)\}$ with equal measures. Thus (6.3.24) is verified for u by combining Lemma 6.3.14 and Proposition 6.3.22. A similar argument also verifies (6.3.25) for $\max\{u_1, u_2\}$. This completes the proof of Proposition 6.3.23. □

Because $|u| = \max\{u, -u\}$ and because $\rho_u = \rho_{-u}$, we have the following corollary to Proposition 6.3.23.

Corollary 6.3.27 *If $u : X \to \mathbb{R}$ is measurable and has a p-integrable p-weak upper gradient then $\rho_u = \rho_{|u|}$.*

The next proposition shows that a product rule is valid for real-valued functions with p-integrable upper gradients.

Proposition 6.3.28 *Let V be a Banach space. Assume that $u : X \to V$ and $m : X \to \mathbb{R}$ are measurable functions that are absolutely continuous on p-a.e. compact rectifiable curve in X. Assume further that ρ and σ are p-weak upper gradients of u and m respectively. Then every Borel representative of $|m| \cdot \rho + |u| \cdot \sigma$ is a p-weak upper gradient of $mu : X \to V$. In particular, if $|m| \cdot \rho + |u| \cdot \sigma$ is in $L^p(X)$ then*

$$\rho_{m \cdot u} \leq |m| \cdot \rho + |u| \cdot \sigma$$

almost everywhere in X.

Proof Let τ be a Borel representative of the function $|m| \cdot \rho + |u| \cdot \sigma$ (Proposition 3.3.23). Let $\gamma : [0, \text{length}(\gamma)] \to X$ be a rectifiable curve parametrized by its arc length so that the length of γ in the set

$$\{\tau \neq |m| \cdot \rho + |u| \cdot \sigma\}$$

is zero, the upper gradient inequality holds for both pairs (u, ρ) and (m, σ) on γ as well as on each of its compact subcurves, and the sum $\rho + \sigma$ is integrable on γ. In particular, both u and m, and hence the product mu, are absolutely continuous on γ. By Lemma 5.2.15 and Proposition 6.3.2, p-a.e. compact curve γ for which $\int_\gamma \rho+\sigma < \infty$ has these properties. Next, using the definition of the metric differential in (4.4.9) and Proposition 6.3.3 we deduce that

$$\begin{aligned} |((m \cdot u) \circ \gamma)'(t)| &\leq |(m \circ \gamma)(t)|\,|(u \circ \gamma)'(t)| + |(m \circ \gamma)'(t)|\,|(u \circ \gamma)(t)| \\ &\leq |(m \circ \gamma)(t)|\,(\rho \circ \gamma)(t) + (\sigma \circ \gamma)(t)\,|(u \circ \gamma)(t)| \\ &= \tau(t) \end{aligned}$$

for almost every $t \in [0, \text{length}(\gamma)]$. The proposition follows from these remarks together with the inequality (6.3.4) of Proposition 6.3.3. □

Upper gradients and p-exceptional sets

Recall the definition of a p-exceptional set from Section 5.2. The first assertion in the following proposition is just a rephrasing of Lemma 6.3.5. The second follows from the first and from Proposition 6.3.2.

Proposition 6.3.29 *Suppose that $u : X \to Z$ is absolutely continuous on p-a.e. rectifiable curve in X. If there exists $c \in Z$ such that $u \equiv c$ almost everywhere in X then the set $\{x \in X : u(x) \neq c\}$ is p-exceptional. In particular, if $u : X \to Z$ has a p-integrable p-weak upper gradient and if for some $c \in Z$ we have that $u \equiv c$, almost everywhere in X, then the set $\{x \in X : u(x) \neq c\}$ is p-exceptional.*

Thus p-exceptional sets can be used, for example, to measure the ambiguity in pointwise defined representatives for maps with p-integrable p-weak upper gradients. The next proposition illustrates this remark. (See also Proposition 7.1.31 below).

Proposition 6.3.30 *Let (u_i) be a sequence of maps $u_i : X \to Z$ with (ρ_i) a corresponding sequence of p-integrable p-weak upper gradients. Assume that there exists a map $u : X \to Z$ together with a p-exceptional set $E \subset X$ such that $\lim_{i\to\infty} u_i(x) = u(x)$ for every $x \in X \setminus E$. Assume further that there exists a Borel function $\rho : X \to [0, \infty]$ such that $\rho_i \to \rho$ in $L^p(X)$ as $i \to \infty$. Then ρ is a p-weak upper gradient of u.*

Proof By the hypotheses and by Fuglede's lemma, we find that p-a.e. compact rectifiable curve γ in X satisfies the following properties: the upper gradient inequality (6.2.1) holds for each pair (u_i, ρ_i) on γ; γ does not meet E; and $\lim_{i\to\infty} \int_\gamma \rho_i\, ds = \int_\gamma \rho\, ds$. Let $\gamma : [a, b] \to X$ be such a curve. Then

$$d_Z\big(u(\gamma(a)), u(\gamma(b))\big) = \lim_{i\to\infty} d_Z\big(u_i(\gamma(a)), u_i(\gamma(b))\big) \le \lim_{i\to\infty} \int_\gamma \rho_i \, ds,$$

which equals $\int_\gamma \rho \, ds$. The proof is complete. □

6.4 Notes to Chapter 6

The classical theory of Sobolev spaces permeates modern mathematics. The theory was systematically developed by Sobolev [253], [254] in the mid 1930s; his influential book [255], [256] was published in 1950. Sobolev's theory had important predecessors, and in particular the idea of functions that are absolutely continuous on lines evolved in the papers by Levi [183], Tonelli [269], and Nikodym [219]. A treatment of Sobolev functions (in the Euclidean setting) in terms of their absolute continuity on line segments can also be found in [273], where the terminology ACL_p was used. The notion of p-exceptional sets in the Euclidean setting was studied in [220]. See [212, Chapter 1.8, p. 19] and [202, p. 29] for further comments and references to important early works.

The approximation of Sobolev functions by smooth functions (Theorem 6.1.6) is a version of the classical results of Deny and Lyons [73], and Meyers and Serrin [207]. A Lipschitz density result for vector-valued Sobolev functions on certain classes of metric measure spaces will be proven in Theorem 8.2.1.

There exist several comprehensive monographs on classical Sobolev spaces, e.g., [1], [2], [81], [193], [202], [212], [290].

Upper gradients were introduced in [124], [125]. They were initially called "very weak gradients", but the more appropriate term "upper gradient" was soon suggested. Functions with p-integrable p-weak upper gradients were subsequently studied in [168], while the theory of Sobolev spaces based on upper gradients was systematically developed in [247], [248], and by Cheeger in [53]. We will begin to discuss this approach to Sobolev spaces in Chapter 7. For a recent use of the machinery of upper gradients in the study of geometric integration theory à la Whitney in a metric space context, see [228].

The elegant proof of Theorem 6.3.20 was pointed out to us by Piotr Hajłasz. It avoids the use of the uniform convexity of L^p-spaces for $p > 1$, as pointed out in Remark 6.3.21(b). For the existence of minimal weak upper gradients relating to more general Banach lattices than L^p, see Malý [194].

7
Sobolev spaces

In this chapter we introduce and study Sobolev spaces of functions defined on arbitrary metric measure spaces with values in a Banach space. Central to this discussion is the theory of upper gradients developed in the previous chapter. We also discuss capacity. Capacity is an outer measure on a given metric measure space, defined with the aid of the Sobolev norm, and is used in this book to describe the pointwise behavior of functions in Sobolev and Dirichlet classes.

We assume throughout this chapter that $X = (X, d, \mu)$ is a metric measure space as defined in Section 3.3, and that V is a Banach space. We also assume that $1 \leq p < \infty$ unless otherwise specifically stated.

7.1 Vector-valued Sobolev functions on metric spaces

The theory of weak upper gradients developed in Sections 6.2 and 6.3 replaces the theory of weak or distributional derivatives in the construction of Sobolev classes of functions on metric measure spaces. Note that in Sections 6.2 and 6.3 we mostly studied maps with values in an arbitrary metric space. To obtain a linear function space, we need to have a linear target; moreover, for reasons of measurability and Bochner integration, we need to assume that we have a Banach space.

Dirichlet classes

In Section 6.1 we introduced the Dirichlet space $L^{1,p}(\Omega)$ as the space of those locally integrable functions on an open set Ω in $\mathbb{R}^n$ that have distributional derivatives in $L^p(\Omega)$. The importance of Dirichlet spaces lies in the fact that imposing the p-integrability condition for a function, in addition to its gradient, is sometimes an unnecessarily strong requirement. We next discuss analogs of Dirichlet spaces of functions defined on metric measure spaces.

The *Dirichlet space*, or *Dirichlet class*, $D^{1,p}(X : V)$ consists of all measurable functions $u : X \to V$ that possess a p-integrable p-weak upper gradient in X. For brevity, we set $\mathrm{D}^{1,p}(X) := \mathrm{D}^{1,p}(X : \mathbb{R})$.

A measurable function $u : X \to V$ belongs to $D^{1,p}(X : V)$ if and only if it possesses a p-integrable upper gradient (Lemma 6.2.2).

In contrast with Section 6.1, the definition of a function in a Dirichlet space *does not include the requirement of local integrability*. It does include the requirement of measurability, albeit not Borel measurability. However, it will be shown in Proposition 7.1.2 that for functions in open sets in $\mathbb{R}^n$ the two definitions are equivalent. Moreover, in Chapter 9 we will show that, for real-valued functions, even the a priori requirement of measurability can be removed if X supports a Poincaré inequality; using the Pettis measurability theorem in Section 3.1 one can even remove the requirement of measurability for more general functions on such X.

Dirichlet classes $D^{1,p}(X : V)$ are vector spaces, and we can equip them with the seminorm

$$\|u\|_{D^{1,p}(X\,:\,V)} := \|\rho_u\|_{L^p(X)}, \tag{7.1.1}$$

where ρ_u is the minimal p-weak upper gradient of u. However, in this book we rarely emphasize the seminorm structure on $D^{1,p}(X : V)$. We consider $D^{1,p}(X : V)$ simply as a collection, or a vector space, of pointwise defined functions.

Proposition 7.1.2 *Let $\Omega \subset \mathbb{R}^n$ be open and let $u \in D^{1,p}(\Omega)$. Then u is locally integrable in Ω. In particular, we have that*

$$D^{1,p}(\Omega) = L^{1,p}(\Omega) \tag{7.1.3}$$

in the following sense: if $u \in D^{1,p}(\Omega)$ then $u \in L^{1,p}(\Omega)$; if $u \in L^{1,p}(\Omega)$ is a function then u has a Lebesgue representative in $D^{1,p}(\Omega)$.

Note that the equality (7.1.3) has to be interpreted properly, as the members of $L^{1,p}(\Omega)$ are (Lebesgue) equivalence classes of functions whereas the members of $D^{1,p}(\Omega)$ are functions that are necessarily absolutely continuous on p-a.e. curve.

Proof We wish to show that $u \in L^1_{\mathrm{loc}}(\Omega)$. To this end we fix $R > 0$, and we would like to show that $u \in L^1(\Omega_R)$, where Ω_R consists of points in Ω that are a distance at least $4R$ away from $\partial\Omega$ and are at most a distance $4/R$ from a fixed point $x_0 \in \Omega$. We consider a p-integrable upper gradient g of u (such a function g exists by Lemma 6.2.2). It follows from the existence of such g that, necessarily, $|u(x)| < \infty$ for almost every $x \in \Omega$. Fix such a point $x \in \Omega_R$. Points $y \in B(x, R)$ can be represented by polar coordinates (τ, θ) based at x, with θ in the unit sphere S^{n-1}. Let L_y be the (radial) line segment connecting x to y. Then, by the upper gradient inequality (6.2.1),

$$|u(x) - u(y)| \le \int_{L_y} g\, ds = \int_0^\tau g(s, \theta)\, ds.$$

Integrating with respect to the variable y in $B(x, R)$, we obtain the estimate

$$\begin{aligned}\int_{B(x,R)} |u(y) - u(x)|\, dy &\le \int_{S^{n-1}} \int_0^R \tau^{n-1} \int_0^\tau g(s, \theta)\, ds\, d\tau\, d\theta \\ &\le \int_{S^{n-1}} \int_0^R \tau^{n-1} \int_0^R g(s, \theta)\, ds\, d\tau\, d\theta.\end{aligned}$$

An application of Tonelli's theorem (see for example [236, p. 309]) yields

$$\begin{aligned}\int_{B(x,R)} |u(y) - u(x)|\, dy &\le \int_0^R \tau^{n-1} \left(\int_{S^{n-1}} \int_0^R \frac{g(s, \theta)}{s^{n-1}}\, s^{n-1}\, ds\, d\theta \right) d\tau \\ &= \frac{R^n}{n} \int_{B(x,R)} \frac{g(y)}{|x - y|^{n-1}}\, dy.\end{aligned}$$

Thus, to conclude that $u \in L^1(B(x, R))$ it suffices to know that the last term in the above sequence of inequalities is finite. This may not be the case for every $x \in \Omega_R$, but it is enough for us to show that it holds for almost every $x \in \Omega_R$, for then we can cover Ω_R by balls $B(x, R)$, and the compactness of Ω_R then yields the desired result.

The map

$$x \to \int_{\mathbb{R}^n} \frac{g(y)}{|x - y|^{n-1}}\, dy$$

is called the Riesz potential of g. This potential is closely connected with the maximal function of g, as we show now. For nonnegative integers i we set $B_i = B(x, 2^{-i} R)$. Then

$$\begin{aligned}\int_{B(x,R)} \frac{g(y)}{|x - y|^{n-1}}\, dy &= \sum_{i=0}^\infty \int_{B_i \setminus B_{i+1}} \frac{g(y)}{|x - y|^{n-1}}\, dy \\ &\le \sum_{i=0}^\infty \int_{B_i} \frac{g(y)}{(2^{-(i+1)} R)^{n-1}}\, dy\end{aligned}$$

$$\leq C_n R \sum_{i=0}^{\infty} 2^{-(i+1)} \fint_{B_i} g(y)\, dy$$
$$\leq C_n R \sum_{i=0}^{\infty} 2^{-(i+1)} Mg_R(x) = CRMg_R(x),$$

where g_R is the zero extension of g outside Ω_R. We are now in a position to use Theorem 3.5.6, obtaining

$$\left|\left\{x \in \Omega_R : \int_{B(x,R)} \frac{g(y)}{|x-y|^{n-1}}\, dy > t\right\}\right| \leq \left|\left\{x \in \Omega_R\ :\ Mg_R(x) > \frac{t}{CR}\right\}\right|$$
$$\leq \frac{CR}{t} \int_{\Omega_R} g(y)\, dy$$

for $t > 0$. This immediately yields that the integral

$$\int_{B(x,R)} \frac{g(y)}{|x-y|^{n-1}}\, dy$$

is finite almost everywhere in Ω_R, completing the proof. □

Remark 7.1.4 The above proof yields as a side product an important relation known as a Poincaré inequality. Let us fix a ball $B \subset \Omega$. Then the above proof also tells us that, when $x \in B$ and R is the radius of B,

$$\fint_B |u(y) - u(x)|\, dy \leq CRM\tilde{g}(x),$$

where $\tilde{g} = g\chi_B$. Now integrating over $x \in B$, we obtain

$$\fint_B \fint_B |u(y) - u(x)|\, dy\, dx \leq CR \fint_B M\tilde{g}(x)\, dx$$
$$\leq CR \left(\fint_B M\tilde{g}(x)^p\, dx\right)^{1/p}.$$

Hence, if $p > 1$ we may apply the second part of Theorem 3.5.6 to obtain

$$\fint_B |u(x) - u_B|\, dx \leq \fint_B \fint_B |u(y) - u(x)|\, dy\, dx \leq CR \left(\fint_B g(x)^p\, dx\right)^{1/p}.$$

The relationship

$$\fint_B |u(x) - u_B|\, dx \leq CR \left(\fint_B g(x)^p\, dx\right)^{1/p} \tag{7.1.5}$$

is generally called a $(1, p)$-Poincaré inequality. We will study this inequality in Chapter 8 in the setting of metric measure spaces. In the Euclidean setting considered here, the above inequality is valid even for $p = 1$;

see (8.1.2) below for a proof formulated in the special case of a smooth function and the modulus of its gradient.

Lemma 7.1.6 *If $u \in D^{1,p}(X : V)$ and if $v : X \to V$ is a function that agrees with u outside a p-exceptional set, then v belongs to $D^{1,p}(X : V)$. Conversely, if two functions in $D^{1,p}(X : V)$ agree almost everywhere, then they agree outside a p-exceptional set.*

Proof Let $E := \{x \in X : u(x) \neq v(x)\}$. Then, by assumption, the collection of all curves γ in X that meet E has p-modulus zero. In particular, every p-integrable p-weak upper gradient of u is also a p-integrable p-weak upper gradient of v, and the first assertion follows.

Next, assume that $u, v \in D^{1,p}(X : V)$ are such that the set $E := \{u \neq v\}$ has measure zero. Since $u - v$ has a p-integrable p-weak upper gradient, we have that E is p-exceptional by Proposition 6.3.29. □

Truncation properties of Dirichlet functions

An important feature of classical first-order Sobolev spaces is that they are closed under the truncation of functions. The Dirichlet classes on metric measure spaces bear the same hallmark.

The following simple fact follows from the basic definitions by Corollary 6.3.27.

Lemma 7.1.7 *If u is a function in the class $D^{1,p}(X : V)$ then $|u|$ is in the class $D^{1,p}(X)$ and*

$$\| \, |u| \, \|_{D^{1,p}(X)} = \|u\|_{D^{1,p}(X\,:\,V)} \,.$$

The next result follows from Proposition 6.3.23.

Proposition 7.1.8 *Let u_1 and u_2 be two functions in $D^{1,p}(X)$ with respective minimal p-weak upper gradients ρ_{u_1} and ρ_{u_2}. Then the following equalities are valid pointwise almost everywhere in X:*

$$\rho_{\min\{u_1,u_2\}} = \rho_{u_1} \cdot \chi_{\{u_1 \le u_2\}} + \rho_{u_2} \cdot \chi_{\{u_2 < u_1\}}\,, \tag{7.1.9}$$

$$\rho_{\max\{u_1,u_2\}} = \rho_{u_1} \cdot \chi_{\{u_1 > u_2\}} + \rho_{u_2} \cdot \chi_{\{u_2 \ge u_1\}}\,. \tag{7.1.10}$$

In particular, if $u \in D^{1,p}(X)$ and $t \in \mathbb{R}$, then

$$\rho_{u_t} \le \rho_u, \tag{7.1.11}$$

where $u_t \in D^{1,p}(X)$ is either of the two functions $\min\{u, t\}$ *or* $\max\{u, t\}$.

In the same way as in Remark 6.3.26, we cannot assert that the functions on the right in (7.1.9) and (7.1.10) are p-weak upper gradients, for they may not be Borel. However, by modifying them on sets of measure zero, we obtain Borel representatives for which the above identity holds.

The last claim in Proposition 7.1.8 has a version for Banach space-valued functions as well. We also discover that truncated functions are dense in the Sobolev and Dirichlet classes. To explain this, we first define what is meant by a truncation of Banach space-valued functions.

For $t > 0$, consider the mapping $r_t : V \to V$ defined by

$$r_t(v) := \frac{t}{|v|}\, v \ \text{ if } |v| > t \qquad \text{and} \qquad r_t(v) := v \ \text{ if } |v| \le t\,. \tag{7.1.12}$$

Lemma 7.1.13 *The mapping r_t is a Lipschitz retraction from V to the closed ball $B_t := \{v \in V : |v| \le t\}$. More precisely, $r_t : V \to B_t$ is a 3-Lipschitz surjection fixing B_t pointwise.*

Proof It suffices to verify the Lipschitz property of r_t. Assume first that $a, b \in V$ with $|a|, |b| \ge t$. Then

$$\begin{aligned} |r_t(a) - r_t(b)| &= t\left|\frac{a}{|a|} - \frac{b}{|b|}\right| = t\left|\frac{|b|a - |b|b + |b|b - |a|b}{|a||b|}\right| \\ &\le 2t\,\frac{|a-b|}{|a|} \le 2\,|a-b|\,. \end{aligned}$$

If $|a| \ge t$ and $|b| < t$, there is a point b' on the line segment $[a, b]$ (the collection of points $a + t(b-a) \in V, 0 \le t \le 1$) such that $|b'| = t$. Then, by the preceding computation,

$$\begin{aligned} |r_t(a) - r_t(b)| &\le |r_t(a) - r_t(b')| + |r_t(b') - r_t(b)| \\ &\le 2|a - b'| + |b - b'| \le 3|a-b|\,. \end{aligned}$$

By symmetry, and by the fact that $r_t(v) = v$ for $|v| \le t$, we have that the map $r_t : V \to B_t$ is 3-Lipschitz as asserted. □

We define a *two-sided truncation* of a function $u : X \to V$ to be

$$T_t u := r_t \circ u : X \to V\,, \quad t > 0\,. \tag{7.1.14}$$

Note that $|T_t u(x)| \le t$ for every $x \in X$ and that

$$T_t u = \max\{\min\{u, t\}, -t\} \tag{7.1.15}$$

if $V = \mathbb{R}$. Furthermore,

$$|T_t u(x)| \le |u(x)| \qquad \text{and} \qquad \lim_{t\to\infty} T_t u(x) = u(x) \tag{7.1.16}$$

for every $x \in X$. It follows that

$$\lim_{t\to\infty} \|T_t u - u\|_{L^p(X:V)} = 0 \tag{7.1.17}$$

if $u \in L^p(X:V)$ for some $1 \le p < \infty$, by the dominated convergence theorem applied to $|T_t u - u|$ dominated by $2|u|$.

Proposition 7.1.18 *Let $u \in D^{1,p}(X:V)$. Then*

$$\lim_{t\to\infty} \|\rho_{(u-T_t u)}\|_{L^p(X)} = 0\,. \tag{7.1.19}$$

Proof From the basic properties of minimal upper gradients (6.3.18) and (6.3.19) and from Lemma 7.1.13, we obtain that

$$\rho_{u-T_t u} \le \rho_u + \rho_{T_t u} \le \rho_u + 3\rho_u = 4\rho_u\,.$$

On the other hand, using Proposition 6.3.22 and the Borel regularity of the measure, we infer that $\rho_{u-T_t u} = 0$ almost everywhere in $\{|u| \le t\}$. Hence

$$\int_X \rho^p_{u-T_t u}\, d\mu \le 4^p \int_{\{|u|>t\}} \rho^p_u\, d\mu \to 0$$

as $t \to \infty$, proving (7.1.19). □

Characterizations of Dirichlet functions

We now present some useful characterizations of functions in the Dirichlet classes. With the aid of these characterizations, many problems regarding Banach space-valued functions can be reduced to the real-valued case.

Theorem 7.1.20 *Let $u : X \to V$ be a measurable function. Then the following four conditions are equivalent:*

(i) *u has a μ-representative in the Dirichlet class $D^{1,p}(X:V)$.*
(ii) *There exists a p-integrable Borel function $\rho : X \to [0,\infty]$ with the following property: for each 1-Lipschitz function $\varphi : V \to \mathbb{R}$ there exists a μ-representative u_φ of the function $\varphi \circ u$ in $D^{1,p}(X)$ such that the minimal upper gradient ρ_{u_φ} of u_φ satisfies $\rho_{u_\varphi} \le \rho$ almost everywhere.*
(iii) *There exists a p-integrable Borel function $\rho : X \to [0,\infty]$ with the following property: for each v^* in the dual space V^* with dual norm $|v^*| \le 1$ there exists a μ-representative u_{v^*} of the function $\langle v^*, u\rangle$ in $D^{1,p}(X)$ such that the minimal upper gradient $\rho_{u_{v^*}}$ of u_{v^*} satisfies $\rho_{u_{v^*}} \le \rho$ almost everywhere.*
(iv) *There exists a p-integrable Borel function $\rho : X \to [0,\infty]$ with the following property: for each $z \in u(X)$ there exists a μ-representative u_z*

of the function $x \mapsto |u(x) - z|$ in $D^{1,p}(X)$ such that the minimal upper gradient ρ_{u_z} of u_z satisfies $\rho_{u_z} \leq \rho$ almost everywhere.

Moreover, if $u : X \to V$ is a function in $D^{1,p}(X : V)$, then there exists a countable set of linear functionals (v_i^*) with $|v_i^*| \leq 1$ and a countable set of points (z_i) in $u(X)$ such that the equalities

$$\rho_u(x) = \sup_i \rho_{u_{v_i^*}}(x) \tag{7.1.21}$$

and

$$\rho_u(x) = \sup_i \rho_{u_{z_i}}(x) \tag{7.1.22}$$

hold for almost every x in X.

Proof We first prove the implication (i) $\Rightarrow$ (ii). Let $u_0 \in D^{1,p}(X : V)$ be a μ-representative of u, and let $\rho : X \to [0, \infty]$ be any p-integrable upper gradient of u_0. We claim that ρ satisfies the requirement in (ii). Indeed, let $\varphi : V \to \mathbb{R}$ be 1-Lipschitz. Then, by (6.3.19), ρ is an upper gradient of the function $u_\varphi := \varphi \circ u_0$. Because u_φ is also measurable (Proposition 3.1.8), we have that u_φ belongs to $D^{1,p}(X)$. Obviously u_φ agrees with $\varphi \circ u$ almost everywhere, and the minimal upper gradient ρ_{u_φ} is essentially majorized by ρ. This proves the implication.

The implications (ii) $\Rightarrow$ (iii) and (ii) $\Rightarrow$ (iv) are clear, since the functions $v^* \in V^*$, $|v^*| \leq 1$, and the functions d_z, $z \in V$, given by $d_z(v) = |v - z|$, are 1-Lipschitz and real-valued.

Next we prove the implication (iii) $\Rightarrow$ (i). Because $u : X \to V$ is measurable, there exists a set Z in X of measure zero such that $u(X \setminus Z)$ is a separable subset of V (by Egoroff's theorem; see Section 3.1). Hence we can choose a countable set $(v_i) \subset V$ whose closure in V contains the difference set

$$u(X \setminus Z) - u(X \setminus Z) \subset V\,.$$

Now we select a countable subset (v_i^*) of V^* such that $\langle v_i^*, v_i \rangle = |v_i|$ and $|v_i^*| = 1$ for each i. Such a set exists, by the Hahn–Banach theorem. Let Z_i be the collection of all the points in X at which $u_{v_i^*} \in D^{1,p}(X)$ differs from $\langle v_i^*, u \rangle$. Then the measure of the set $Z_0 = Z \cup \bigcup_i Z_i$ is zero, and $u_{v_i^*}(x) = \langle v_i^*, u(x) \rangle$ for every i and for every $x \in X \setminus Z_0$.

Let $\rho : X \to [0, \infty]$ be the p-integrable Borel function guaranteed by the hypotheses. Then the Borel function ρ_u^*, given by

$$\rho_u^*(x) := \sup_i \rho_{u_{v_i^*}}(x)$$

for $x \in X$, is in $L^p(X)$. Furthermore, it follows from Proposition 6.3.2, from Lemmas 5.2.16 and 5.2.15, and from the subadditivity of the modulus (5.2.6) that p-a.e. compact rectifiable curve $\gamma : [a, b] \to X$ satisfies the following three properties: ρ_u^* is integrable on γ; the upper gradient inequality (6.2.1) holds for each pair $(u_{v_i^*}, \rho_u^*)$ on γ as well as on each of its subcurves; and the length of γ in Z_0 is zero. Denote the collection of all such nonconstant curves by Γ.

Assume now that $\gamma \in \Gamma$ is a curve with both end points $\gamma(a), \gamma(b)$ outside Z_0. Then we can pick a subsequence (v_{i_j}) converging to $u(\gamma(a)) - u(\gamma(b))$. Consequently, with an obvious notation $v_{i_j}^*$, we have that

$$\begin{aligned} |u(\gamma(a)) - u(\gamma(b))| &= \lim_{j\to\infty} |v_{i_j}| \\ &\le \limsup_{j\to\infty}\Big(\big|\langle v_{i_j}^*, v_{i_j} - u(\gamma(a)) + u(\gamma(b))\rangle\big| \\ &\qquad + \big|\langle v_{i_j}^*, u(\gamma(a)) - u(\gamma(b))\rangle\big|\Big) \\ &\le \limsup_{j\to\infty} |v_{i_j} - u(\gamma(a)) + u(\gamma(b))| \\ &\qquad + |u_{v_{i_j}^*}(\gamma(a)) - u_{v_{i_j}^*}(\gamma(b))| \\ &\le \int_\gamma \rho_u^*\, ds\,. \end{aligned} \tag{7.1.23}$$

Next, assume that $\gamma \in \Gamma$ is a curve with at least one end point in Z_0, say $\gamma(a) \in Z_0$. Choose a sequence $(t_k) \subset [a, b]$ such that $t_k \to a$, as $k \to \infty$, and $\gamma(t_k) \notin Z_0$ for each k. Then, by manipulaions like those in (7.1.23), we find that

$$|u(\gamma(t_k)) - u(\gamma(t_l))| \le \int_{\gamma|[t_k, t_l]} \rho_u^*\, ds\,,$$

where $\gamma|[t_k, t_l]$ is the restriction of γ to $[t_k, t_l]$. Because ρ is integrable on γ, it follows that the sequence $\big(u(\gamma(t_k))\big)$ is convergent in V and that the limit is independent of the sequence (t_k). In fact, the limit is independent of the curve γ as well, for if $\gamma_1 : [a_1, b_1] \to X$ is another curve in Γ with $\gamma_1(a_1) = \gamma(a)$ and if $(s_m) \subset [a_1, b_1]$ is a sequence converging to a_1 with $\gamma_1(s_m) \notin Z_0$ then we have that

$$|u(\gamma_1(s_m)) - u(\gamma(t_k))| \le \int_{\gamma_1|[a_1, s_m]} \rho_u^*\, ds + \int_{\gamma|[a, t_k]} \rho_u^*\, ds \to 0$$

as $m, k \to \infty$.

We can now define a function u_0 by first setting $u_0(x) = u(x)$ if $x \notin Z_0$. If $x \in Z_0$ and if there is a nonconstant rectifiable curve $\gamma : [a, b] \to X$ in Γ with $\gamma(a) = x$, then $u_0(x)$ is defined as the limit in the preceding argument. Thus u_0 is defined outside a p-exceptional set of measure zero; this is the subset of

Z_0 that none of the curves in Γ meets and u_0 can be defined arbitrarily there. The upper gradient inequality (6.2.1) holds for the pair (u_0, ρ_u^*) on each curve in Γ, and hence we can conclude that $u_0 \in D^{1,p}(X : V)$ and that the minimal p-weak upper gradient of u_0 is bounded above by ρ_u^*.

The proof for the implication (iv) $\Rightarrow$ (i) is similar, using the fact that

$$|u(x) - v_{i_j}| - |u(y) - v_{i_j}| \to |u(x) - u(y)|$$

for a subsequence (v_{i_j}), of a dense set (v_i) in $u(X \setminus Z_0)$, converging to $u(y)$, where Z_0 is a set of measure zero as in the proof of (iii) $\Rightarrow$ (i).

Finally, let us prove (7.1.21) and (7.1.22). As above, we set

$$\rho_u^*(x) := \sup_i \rho_{u_{v_i^*}}(x)$$

for $x \in X$, where the functionals v_i^* are chosen as in the proof of the implication (iii) $\Rightarrow$ (i). From the above argument, we know that $\rho_u \leq \rho_u^*$ almost everywhere in X. On the other hand, because the maps $v \mapsto \langle v_i^*, v\rangle$ are 1-Lipschitz, it is clear that $\rho_u^* \leq \rho_u$ almost everywhere.

The proof for the statement involving the functions d_z, $z \in u(X)$, is similar, and Theorem 7.1.20 is now completely verified. □

Sobolev classes

Now we will define the Sobolev classes. The definition is akin to that for functions in the Dirichlet classes, but an integrability requirement is placed on the functions as well, and not only on the upper gradients. Moreover, in contrast with $D^{1,p}$, the Sobolev classes are equipped with a norm. Recall that X is an arbitrary metric measure space as in Section 3.3, V is a Banach space, and $1 \leq p < \infty$.

Let $\tilde{N}^{1,p}(X : V)$ denote the collection of all p-integrable functions u that have an upper gradient in $L^p(X)$. We emphasize that this definition is framed in terms of genuine *functions*, as opposed to equivalence classes. Thus,

$$\tilde{N}^{1,p}(X : V) = D^{1,p}(X : V) \cap L^p(X : V).$$

In the above equality $D^{1,p}(X : V)$ consists of *functions*, and so $L^p(X : V)$ stands for the collection of functions that are p-integrable. With the help of Proposition 6.3.29, we will soon remove this ambiguity in notation.

It follows from the definitions that $\tilde{N}^{1,p}(X : V)$ is a vector space. We equip it with the seminorm

$$\|u\|_{\tilde{N}^{1,p}(X : V)} = \|u\|_{L^p(X : V)} + \|\rho_u\|_{L^p(X)}, \tag{7.1.24}$$

where ρ_u is the minimal p-weak upper gradient of u guaranteed by Theorem 6.3.20. We recall again that a function u has an upper gradient in $L^p(X)$

if and only if it has a p-weak upper gradient in $L^p(X)$, as demonstrated by Lemma 6.2.2. Moreover, Lemma 6.2.2 implies that the expression (7.1.24) can be written equivalently as

$$\|u\|_{\tilde{N}^{1,p}(X:V)} := \|u\|_{L^p(X:V)} + \inf \|\rho\|_{L^p(X)}, \tag{7.1.25}$$

where the infimum is taken over all upper gradients ρ of u.

The seminorm in (7.1.24) is not a norm in general. For example, if E is a nonempty p-exceptional subset of X of zero measure, and if $c \in V \setminus \{0\}$, then the function $u = c\chi_E$ is a nonzero function from X to V with $\|u\|_{\tilde{N}^{1,p}(X:V)} = 0$.

We obtain a normed space $N^{1,p}(X:V)$ by passing to equivalence classes of functions in $\tilde{N}^{1,p}(X:V)$, where $u_1 \sim u_2$ if and only if $\|u_1 - u_2\|_{\tilde{N}^{1,p}(X:V)} = 0$. (Compare Remark 2.1.16.) Thus,

$$N^{1,p}(X:V) := \tilde{N}^{1,p}(X:V)/\{u \in \tilde{N}^{1,p}(X:V) : \|u\|_{\tilde{N}^{1,p}(X:V)} = 0\}. \tag{7.1.26}$$

The normed space $N^{1,p}(X:V)$ of equivalence classes of functions in $\tilde{N}^{1,p}(X:V)$ is called the *Sobolev space* of V-valued functions on X. We write

$$\|u\|_{N^{1,p}(X:V)}$$

for the (quotient) norm of $u \in N^{1,p}(X:V)$. If $V = \mathbb{R}$, we make the abbreviation

$$N^{1,p}(X) := N^{1,p}(X:\mathbb{R}). \tag{7.1.27}$$

It will be shown later, in Section 7.3, that $N^{1,p}(X:V)$ is a Banach space. Before this, we explore some basic properties of the Sobolev spaces $N^{1,p}(X:V)$.

If $Y \subset X$ is an arbitrary open subset then we have the metric measure space (Y, d, μ_Y) (see Section 3.3), and it follows from the definitions that the restriction map $u \mapsto u|_Y$ yields a bounded operator $\tilde{N}^{1,p}(X:V) \to \tilde{N}^{1,p}(Y:V)$,

$$\|u|_Y\|_{\tilde{N}^{1,p}(Y:V)} \leq \|u\|_{\tilde{N}^{1,p}(X:V)}. \tag{7.1.28}$$

Moreover, (7.1.28) and the definition for the quotient norm in (2.1.17) give that also

$$\|u|_Y\|_{N^{1,p}(Y:V)} \leq \|u\|_{N^{1,p}(X:V)}, \tag{7.1.29}$$

where the inequality is naturally interpreted in terms of equivalence classes.

We next define local Sobolev spaces. Let $\tilde{N}^{1,p}_{\mathrm{loc}}(X : V)$ be the vector space of functions $u : X \to V$ with the property that every point $x \in X$ has a neighborhood U_x in X such that $u \in \tilde{N}^{1,p}(U_x : V)$. Two functions u_1 and u_2 in $\tilde{N}^{1,p}_{\mathrm{loc}}(X : V)$ are said to be equivalent if every point $x \in X$ has a neighborhood U_x in X such that the restrictions $u_1|_{U_x}$ and $u_2|_{U_x}$ determine the same element in $N^{1,p}(U_x : V)$. It follows from (7.1.28) that such a neighborhood can be assumed to be open. The *local Sobolev space* $N^{1,p}_{\mathrm{loc}}(X : V)$ is the vector space of equivalence classes of functions in $\tilde{N}^{1,p}_{\mathrm{loc}}(X : V)$ under the preceding equivalence relation.

As in the Lebesgue theory, we speak of functions rather than equivalence classes of elements in $N^{1,p}(X : V)$. To be able to do this carefully, it is important to understand the amount of ambiguity in the chosen representatives. A good rule of thumb is given in Lemma 7.1.6. For further information see Corollary 7.2.10 and Section 7.5 below, especially Proposition 7.5.2.

We next clarify the above equivalence relation in the general context of local Sobolev spaces.

Lemma 7.1.30 *Two functions u_1, u_2 in $\tilde{N}^{1,p}_{\mathrm{loc}}(X : V)$ determine the same element in $N^{1,p}_{\mathrm{loc}}(X : V)$ if and only if $u_1 - u_2 = 0$ in $N^{1,p}(X : V)$.*

Proof It suffices to prove the following: a function $u : X \to V$ determines the zero element in $N^{1,p}_{\mathrm{loc}}(X : V)$ if and only if $\|u\|_{\tilde{N}^{1,p}(X : V)} = 0$.

If $\|u\|_{\tilde{N}^{1,p}(X : V)} = 0$, then, by (7.1.28), we know that u determines the zero element in $N^{1,p}_{\mathrm{loc}}(X : V)$.

Assume next that $u = 0$ in $N^{1,p}_{\mathrm{loc}}(X : V)$. Then $E := \{x \in X : u(x) \neq 0\}$ has measure zero (Lemma 3.3.31). We will show in addition that E is p-exceptional. To this end, fix $x \in X$ and let U_x be a neighborhood of x such that $u|_{U_x} = 0$ in $N^{1,p}(U_x : V)$. It follows from Lemma 7.1.6 that $E \cap U_x$ is p-exceptional. Consequently, Lemma 5.2.10 gives that E is p-exceptional. The preceding understood, we have that $\rho \equiv 0$ is a p-weak upper gradient of u and hence that $\|u\|_{\tilde{N}^{1,p}(X : V)} = 0$ as required. The lemma follows. □

We summarize the preceding discussion in the following proposition.

Proposition 7.1.31 *If $u \in \tilde{N}^{1,p}_{\mathrm{loc}}(X : V)$ and if $v : X \to V$ is a function that agrees with u outside a p-exceptional set of measure zero, then v belongs to $\tilde{N}^{1,p}_{\mathrm{loc}}(X : V)$ and the two functions determine the same element in $N^{1,p}_{\mathrm{loc}}(X : V)$. In particular, if in addition $u \in \tilde{N}^{1,p}(X : V)$, then we also have $v \in \tilde{N}^{1,p}(X : V)$, and the two functions determine the same element in $N^{1,p}(X : V)$.*

Conversely, if two functions in $\tilde{N}^{1,p}_{\mathrm{loc}}(X : V)$ agree almost everywhere, then they agree outside a p-exceptional set. In particular, if two μ-representatives *of a function in an equivalence class in* $N^{1,p}_{\text{loc}}(X : V)$ *both lie in* $\tilde{N}^{1,p}_{\text{loc}}(X : V)$*, then they differ only in a* p*-exceptional set of measure zero.*

Proof The assertions in the first paragraph of the proposition follow directly from the definitions and from Lemmas 7.1.6 and 7.1.30. The first assertion in the second paragraph follows from Lemma 7.1.6. Finally, the last assertion follows from the assertion before it and from Lemma 3.3.31. The proposition is proved. □

Proposition 7.1.31 implies that a function in $N^{1,p}_{\text{loc}}(X : V)$, and in particular a function in $N^{1,p}(X : V)$, is well defined outside a p-exceptional subset of measure zero. While p-exceptionality is an easily defined condition for "small sets", an alternate description intrinsically in terms of the Sobolev space $N^{1,p}(X : V)$ would also be desirable. Such a characterization is proven later, in Corollary 7.2.10, in terms of capacity. This characterization is also used in the proof that $N^{1,p}(X : V)$ is a Banach space.

Nontriviality of Sobolev classes

Every function in $N^{1,p}(X : V)$ belongs to the Lebesgue space $L^p(X : V)$ by definition, and the inclusion $N^{1,p}(X : V) \subset L^p(X : V)$ is a bounded embedding. (Note that this inclusion is indeed an injection, by Proposition 7.1.31.) Sometimes the Sobolev space reduces to the Lebesgue space; in other words, the equality

$$N^{1,p}(X : V) = L^p(X : V) \tag{7.1.32}$$

may hold. The precise meaning of this equality of spaces is that every Lebesgue equivalence class in $L^p(X : V)$ determines a unique equivalence class in $N^{1,p}(X : V)$.

Equality (7.1.32) holds for all spaces X without nonconstant rectifiable curves; thus it holds for all totally disconnected spaces and for all snowflake spaces X for example. A metric space (X, d) is called a *snowflake space* if there exists $\epsilon > 0$ such that $d^{1+\epsilon}$ is a metric on X. Note that $(X, d^{1-\delta})$ is a snowflake space for every $0 < \delta < 1$.

More generally, equality in (7.1.32) occurs if the p-modulus of the collection of all nonconstant curves in X is zero. It turns out that the converse also holds.

We say that the Sobolev space $N^{1,p}(X : V)$ is *nontrivial* if it is strictly contained in $L^p(X : V)$.

Proposition 7.1.33 *The Sobolev space* $N^{1,p}(X : V)$ *is nontrivial if and only if the* p*-modulus of the collection of all nonconstant curves in* X *is positive.*

Proof As mentioned just before the proposition, the necessity part of the assertion is clear from the definitions.

To prove sufficiency, we use the subadditivity of the modulus and the countable covering of X by open balls of the form $B(x_i, q)$, where (x_i) is a fixed countable dense subset of X and q is a positive rational number, to conclude that there is an open ball B in X with the following property: the p-modulus of the family Γ of all curves in X with one end point in B and the other in $X \setminus \overline{B}$ is positive.

Given the preceding, we claim that the L^p-function $c\chi_B$, where c is a fixed nonzero vector in V, cannot have a representative in $N^{1,p}(X : V)$. Towards a contradiction, suppose that u is such a representative. By Borel regularity, there exists a Borel set E in X of measure zero such that $u|B \setminus E \equiv c$ and $u|X \setminus (B \cup E) \equiv 0$. It follows that there is a curve γ in X, parametrized by its arc length, that intersects both $B \setminus E$ and $X \setminus (\overline{B} \cup E)$ such that u is absolutely continuous on γ and γ meets E in a set of zero length (Proposition 6.3.2 and Lemma 5.2.15). This is a contradiction because, on the dense set $\gamma \setminus E$, the absolutely continuous function u takes only two vector values, 0 and c, and it takes both values on sets of positive length, violating absolute continuity. The proposition follows. □

Remark 7.1.34 If the Sobolev space $N^{1,p}(X : V)$ is nontrivial then the canonical embedding $N^{1,p}(X : V) \to L^p(X : V)$ is never isometric. Indeed, by the proof of Proposition 7.1.33, by Lemma 3.3.28, and by (5.2.5), we can find two concentric balls of finite measure, $B(x, r) \subset B(x, R) \subset X$, $0 < r < R$, such that the p-modulus of all the curves that start in $B(x, r)$ and end in $X \setminus B(x, R)$ is positive. This implies that the p-integrable Lipschitz function $u(z) = \operatorname{dist}(z, X \setminus B(x, R))$ cannot have the zero function as its minimal p-weak upper gradient. (Here we ensure that u takes values in V by fixing an isometric embedding $\mathbb{R} \to V$.) We leave the details of this argument to the reader.

By Theorem 6.1.6, functions with compact support are dense in the Sobolev space $W^{1,p}(\mathbb{R}^n)$. An analogous fact holds true in general for mappings into a normed vector space V. For the purposes of the next proposition we say that a function $u : X \to V$ has *bounded support* if $u = 0$ outside a bounded set in X.

Proposition 7.1.35 *The vector subspace of $N^{1,p}(X : V)$ consisting of bounded functions with bounded support is dense in $N^{1,p}(X : V)$.*

Proof Let $u \in N^{1,p}(X : V)$. By Proposition 7.1.18 it suffices to show that u can be approximated in $N^{1,p}(X : V)$ by functions with bounded support. To this end, we may assume that X is unbounded. Fix $x_0 \in X$ and, for each $i = 1, 2, \ldots$, fix a 1-Lipschitz function $\varphi_i : X \to [0, 1]$ such that $\varphi_i|_{B(x_0,i)} = 1$ and $\varphi_i|_{X \setminus B(x_0,2i)} = 0$. We claim that the functions $u_i := \varphi_i \cdot u$ yield an approximating sequence as desired. Indeed, every Borel representative of the function $\varphi_i \rho_u + |u| \rho_{\varphi_i}$ is a p-integrable p-weak upper gradient of u_i (Proposition 6.3.28). Hence, in particular, $u_i \in N^{1,p}(X : V)$ for every i. Similarly, by (6.3.18) and by Proposition 6.3.22 we find that the minimal p-weak upper gradient of $u - u_i$ satisfies $\rho_{u-u_i}|_{B(x_0,i)} = 0$, and on $X \setminus B(x_0, i)$ we have

$$\rho_{u-u_i} \leq 2\rho_u + |u| .$$

It follows that

$$\begin{aligned}\int_X \rho_{u-u_i}^p \, d\mu &= \int_{X \setminus B(x_0,i)} \rho_{u-u_i}^p \, d\mu \\ &\leq 2^{p+1} \left(\int_{X \setminus B(x_0,i)} \rho_u^p + |u|^p \, d\mu \right) \to 0\end{aligned}$$

as $i \to \infty$. Because we also have $u_i \to u$ in $L^p(X : V)$, the proposition follows. □

Metric space-valued Sobolev maps

By the embedding theorems of Chapter 3, one can consider metric space-valued Sobolev mappings on X in the present framework. Assume that $Y = (Y, d_Y)$ is a metric space, $y_0 \in Y$, and consider the Kuratowski embedding $Y \subset l^\infty(Y)$ as in Section 4.1.

We define the Sobolev and Dirichlet classes of mappings from X to Y as follows:

$$N^{1,p}(X : Y) := \{u \in N^{1,p}(X : l^\infty(Y)) \, : \, u(x) \in Y \text{ for a.e. } x \in X\}$$

and similarly,

$$D^{1,p}(X : Y) := \{u \in D^{1,p}(X : l^\infty(Y)) \, : \, u(x) \in Y \text{ for a.e. } x \in X\}.$$

The class $N^{1,p}(X : Y)$ depends on the choice of the base point y_0, but for simplicity this dependence is suppressed from the notation. If $\mu(X)$ is finite then there is a natural bijection between the classes corresponding to different base points, which is an isometry between subsets of the normed space $N^{1,p}(X : l^\infty(Y))$.

Analogously we can consider the local version of $N^{1,p}(X : Y)$, which we denote $N^{1,p}_{\mathrm{loc}}(X : Y)$. This is the class of all maps $f : X \to l^\infty(Y)$ such that each $x \in X$ has a bounded open neighborhood U_x for which $f|_{U_x} \in N^{1,p}(U_x : Y)$. Given that X is a metric measure space and hence μ is locally finite, we can also require $\mu(U_x)$ to be finite. Hence it is clear that membership in the local class is independent of the base point $y_0 \in Y$.

Because the elements in $N^{1,p}(X : l^\infty(Y))$ and $N^{1,p}_{\mathrm{loc}}(X : l^\infty(Y))$ are equivalence classes of functions, we should think of the elements in the classes $N^{1,p}(X : Y)$ and $N^{1,p}_{\mathrm{loc}}(X : Y)$ in the same way. Because two equivalent functions agree almost everywhere, the given definitions are consistent.

The next result describes the classes $N^{1,p}(X : Y)$ and $D^{1,p}(X : Y)$ in more precise terms. Kuratowski embedding with respect to y_0 is still understood to apply, and we let $\overline{Y}$ denote the metric completion of Y.

Proposition 7.1.36 *A function $u : X \to l^\infty(Y)$ is in the class $N^{1,p}(X : Y)$ if and only if $u : X \to l^\infty(Y)$ is p-integrable, $u(x) \in Y$ for almost every x in X, and there exists a p-integrable Borel function $\rho : X \to [0, \infty]$ such that for every nonconstant rectifiable curve $\gamma : [a, b] \to X$ we have*

$$|u(\gamma(a)) - u(\gamma(b))| \le \int_\gamma \rho \, ds. \tag{7.1.37}$$

Similarly, a function $u : X \to l^\infty(Y)$ is in the class $D^{1,p}(X : Y)$ if and only if $u : X \to l^\infty(Y)$ is measurable, $u(x) \in Y$ for almost every x in X, and there exists a p-integrable Borel function $\rho : X \to [0, \infty]$ such that (7.1.37) *holds for every nonconstant rectifiable curve $\gamma : [a, b] \to X$.*

Moreover, in both cases, if these two equivalent conditions hold then $u(x) \in \overline{Y}$ for all x outside a p-exceptional subset in X.

Proof The first two claims are simple reformulations of the definitions as given in Section 7.1. To prove the last claim, we may assume that $u \in D^{1,p}(X : Y)$ and that Y is complete. Let E denote the set of points $x \in X$ such that $u(x) \notin Y$. Now, p-a.e. curve $\gamma : [a, b] \to X$ enjoys the following two properties: γ meets E on a set of zero length (Lemma 5.2.15) and u is absolutely continuous on γ. In particular, such a curve γ cannot meet E at all, for else u would map a nondegenerate subcurve of γ outside $\overline{Y}$, which is impossible. This proves the proposition. □

In what follows, $\ell(\gamma)$ denotes the length of a rectifiable curve γ.

Proposition 7.1.38 *Suppose that (Y, d_Y) is a complete metric space. A map $f : X \to Y$ is in $D^{1,p}(X : Y)$ if and only if there is a nonnegative p-integrable Borel function ρ on X such that, whenever $\gamma : [0, \ell(\gamma)] \to X$ is a nonconstant rectifiable curve in X,*

$$d_Y\big(f(\gamma(\ell(\gamma))), f(\gamma(0))\big) \le \int_\gamma \rho \, ds. \tag{7.1.39}$$

Moreover, a map in $D^{1,p}(X : Y)$ can be modified on a p-exceptional set of measure zero to obtain a map into Y. Furthermore, if $\mu(X)$ is finite then

$f \in N^{1,p}(X:Y)$ if and only if $f \in D^{1,p}(X:Y)$ and the function $x \mapsto d_Y(f(x), y_0)$ is p-integrable for some (and hence every) $y_0 \in Y$.

Proof It is clear from the definition of $D^{1,p}(X:Y)$ and the proof of Proposition 6.3.28 that if f, ρ satisfies (7.1.39) then $f \in D^{1,p}(X:Y)$. Note that this does not require the completeness of (Y, d_Y), but the next argument does.

Now suppose that $f \in D^{1,p}(X:Y)$ and that ρ is a p-integrable upper gradient of $f : X \to l^\infty(Y)$. Let $E \subset X$ be the set of points in X that are mapped by f into $l^\infty(Y)\setminus Y$. Then, by the definition of $D^{1,p}(X:Y)$, we know that $\mu(E) = 0$ and so p-a.e. nonconstant rectifiable curve $\gamma : [a,b] \to X$ has $\text{length}(E \cap \gamma) = 0$. By discarding a further collection of nonconstant compact rectifiable curves, of p-modulus zero, we can also ensure that f is absolutely continuous on γ. It follows that (because Y is complete) $f \circ \gamma([a,b]) \subset Y$; that is, p-a.e. nonconstant compact rectifiable curve in X does not intersect E. It follows that E is p-exceptional and so, redefining f on E if necessary, we obtain a map from X into Y that satisfies (7.1.39).

The last assertion of the proposition follows from the first, since the only difference between $D^{1,p}(X:Y)$ and $N^{1,p}(X:Y)$ is the integrability of f. □

Proposition 7.1.38 shows that there is an intrinsic way of determining membership of $D^{1,p}(X:Y)$ and $N^{1,p}_{\text{loc}}(X:Y)$ that is independent of the embedding of Y into a Banach space. One should, however, keep in mind that the metric on $N^{1,p}(X:Y)$ does indeed depend on the embedding of Y into a Banach space (just as the space $L^p(X:Y)$ depends on the embedding of Y); see the discussion in Section 7.6. However, if Y is complete, then given $N^{1,p}(X:Y)$, the metric imposed on this function space, via an isometric embedding of Y into a Banach space, makes $N^{1,p}(X:Y)$ complete. Furthermore, the topology on $N^{1,p}(X:Y)$ is independent of the embedding of Y; this fact is useful in considering related variational problems.

7.2 The Sobolev p-capacity

We recall that throughout this chapter, $X = (X, d, \mu)$ is an arbitrary metric measure space and $1 \le p < \infty$.

The *p-capacity* of a set $E \subset X$ is defined to be the (possibly infinite) number

$$\text{Cap}_p(E) := \inf\left(\int_X |u|^p \, d\mu + \int_X \rho_u^p \, d\mu\right), \tag{7.2.1}$$

where the infimum is taken over all functions $u \in N^{1,p}(X)$ such that $u \ge 1$ on E outside a p-exceptional set of measure zero.

The functions u as in the preceding paragraph are called *p-admissible*, or sometimes just *admissible*, for the set E. Recall that functions in $N^{1,p}(X)$ are well defined up to p-exceptional sets of measure zero (Proposition 7.1.31), and so the preceding definition makes sense: two equivalent functions in $N^{1,p}(X)$ are simultaneously admissible. It is convenient to assume that admissible functions always satisfy $u \geq 1$ everywhere on E; obviously this assumption can be made without loss of generality. Note that we can alternatively use the following equivalent definition: *the infimum in* (7.2.1) *is taken over all functions $u \in \tilde{N}^{1,p}(X)$ such that $u \geq 1$ on E.*

If no admissible functions exist, we set $\mathrm{Cap}_p(E) = \infty$. If $\mathrm{Cap}_p(E) = 0$, we say that E is a *set of zero p-capacity* or that *E has zero p-capacity*. Every p-exceptional set of zero measure has zero p-capacity since the characteristic function of that set is an admissible function. Proposition 7.2.8 below shows that the converse is also true.

In the classical theory of Sobolev spaces in $\mathbb{R}^n$, in the definition of capacity it is customary to require that the admissible test functions u as above satisfy $u \geq 1$ in a neighborhood of E. In the theory of $N^{1,p}(X)$ spaces, even when $X = \mathbb{R}^n$ with the usual distance and measure, such a requirement is not needed. This advantage is due to the better pointwise behavior of functions in $N^{1,p}(X)$. In Chapter 8 we will show that, under the assumption that X supports a $(1, p)$-Poincaré inequality (as we saw in (7.1.5), $X = \mathbb{R}^n$ has this property), the two approaches give the same value for Cap_p.

The capacity satisfies

$$\mathrm{Cap}_p(\emptyset) = 0\,, \tag{7.2.2}$$

moreover, *monotonicity* holds:

$$\mathrm{Cap}_p(E_1) \leq \mathrm{Cap}_p(E_2) \tag{7.2.3}$$

if $E_1 \subset E_2$.

Lemma 7.2.4 *The p-capacity* Cap_p *is a countably subadditive set function, hence an outer measure on X. That is,*

$$\mathrm{Cap}_p\left(\bigcup_{i=1}^{\infty} E_i\right) \leq \sum_{i=1}^{\infty} \mathrm{Cap}_p(E_i) \tag{7.2.5}$$

whenever (E_i) is a sequence of sets in X.

The proof of Lemma 7.2.4 is less straightforward than one might expect, for we have not yet proven that $N^{1,p}(X)$ is a Banach space. Indeed, the subadditivity condition (7.2.5) is needed later for our proof of Theorem 7.3.6.

First, we require a simple but useful lemma.

Lemma 7.2.6 *The p-capacity* $\mathrm{Cap}_p(E)$ *is equal to the infimum of*

$$\int_X |u|^p \, d\mu + \int_X \rho_u^p \, d\mu$$

over all functions $u \in N^{1,p}(X)$ *such that* $0 \le u \le 1$ *on* X *and* $u = 1$ *on* E.

Proof Given a function $u \in N^{1,p}(X)$ such that $u \ge 1$ in E, we apply Proposition 7.1.8 and infer that the function $\max\{0, \min\{u, 1\}\}$ is in $N^{1,p}(X)$ and is admissible with norm not exceeding that of u. □

Proof of Lemma 7.2.4 We may assume that the sum on the right in (7.2.5) is finite. Fix $\epsilon > 0$. Then, for each $i = 1, 2, \ldots$, pick a (pointwise defined) function $u_i \in N^{1,p}(X)$ such that $0 \le u_i \le 1$, $u_i = 1$ on E_i, and

$$\int_X u_i^p \, d\mu + \int_X \rho_{u_i}^p \, d\mu \le \mathrm{Cap}_p(E_i) + 2^{-i}\epsilon \, .$$

By Proposition 7.1.8, the functions $v_j = \max\{u_i : 1 \le i \le j\}$ are in $N^{1,p}(X)$ with the limit $v(x) := \lim_{j\to\infty} v_j(x)$ well defined at every point x in X. Note that $v = 1$ on $\bigcup_{i=1}^{\infty} E_i$. Next, let $\rho_i = \rho_{u_i}$ be the minimal p-weak upper gradient of u_i. Then, by Proposition 7.1.8, the function $\sigma_j = \max\{\rho_i : 1 \le i \le j\}$ is a p-weak upper gradient of v_j. Furthermore, because

$$0 \le v_j^p \le \sum_{i=1}^{j} u_i^p$$

and because

$$0 \le \sigma_j^p \le \sum_{i=1}^{j} \rho_i^p \, ,$$

we have that

$$\begin{aligned} \|v_j\|_{L^p(X)}^p + \|\rho_{v_j}\|_{L^p(X)}^p &\le \sum_{i=1}^{\infty} \left(\|u_i\|_{L^p(X)}^p + \|\rho_i\|_{L^p(X)}^p \right) \\ &\le \sum_{i=1}^{\infty} \mathrm{Cap}_p(E_i) + \epsilon \end{aligned} \tag{7.2.7}$$

for each j. In particular, the limit $\sigma(x) := \lim_{j\to\infty} \sigma_j(x)$ is a Borel function and belongs to $L^p(X)$ and $\sigma_j \to \sigma$ in $L^p(X)$ by the monotone convergence theorem. It therefore follows from Proposition 6.3.30 that v is in $N^{1,p}(X)$ with σ its p-weak upper gradient. It is immediate that v is admissible for $\bigcup_{i=1}^{\infty} E_i$.

Finally, because we also have $v_j \to v$ in $L^p(X)$, (7.2.7) gives that

$$\operatorname{Cap}_p\left(\bigcup_{i=1}^{\infty} E_i\right) \le \|v\|_{L^p(X)}^p + \|\sigma\|_{L^p(X)}^p \le \sum_{i=1}^{\infty} \operatorname{Cap}_p(E_i) + \epsilon .$$

By letting $\epsilon \to 0$ we obtain that (7.2.5) holds, and the proposition is proved. □

The next proposition characterizes p-exceptional sets in terms of capacity.

Proposition 7.2.8 *A subset $E \subset X$ is of zero p-capacity if and only if $\mu(E) = 0$ and E is p-exceptional.*

Proof As mentioned earlier, a p-exceptional set of measure zero has zero p-capacity essentially by definition. Namely, the characteristic function χ_E is both p-integrable and has the zero function as a p-weak upper gradient; that is, $\chi_E \in N^{1,p}(X)$ with norm zero.

Assume next that $\operatorname{Cap}_p(E) = 0$. For each positive integer i we can find a function $u_i \in N^{1,p}(X)$ such that $0 \le u_i \le 1$, $u_i = 1$ on E, and

$$\|u_i\|_{L^p(X)}^p + \|\rho_{u_i}\|_{L^p(X)}^p \le 2^{-i} . \tag{7.2.9}$$

As in the proof of (7.2.5), we find that for positive integers $j \ge j_0$ the functions $v_j = \min\{u_i : j_0 \le i \le j\}$ are in $N^{1,p}(X)$ with p-weak upper gradients $\sigma_j = \max\{\rho_i : j_0 \le i \le j\}$ by Proposition 7.1.8. Moreover, the limit function $w_{j_0}(x) := \lim_{j\to\infty} v_j(x)$ belongs to $N^{1,p}(X)$ with $g_{j_0} := \lim_{j\to\infty} \sigma_j$ as a p-weak upper gradient (Proposition 6.3.30). In particular $\|w_{j_0}\|_{N^{1,p}(X)} \le 2^{-j_0/p}$ by (7.2.9). A further application of the monotone convergence theorem gives a limit function $v \in N^{1,p}(X)$ for which $\|v\|_{N^{1,p}(X)} = 0$; notice that (w_j) is monotone increasing and (g_j) is decreasing. Since $v|_E = 1$, we obtain from Proposition 7.1.31 that E is p-exceptional and of measure zero. This completes the proof. □

The following corollary rephrases part of Proposition 7.1.31 in terms of capacity.

Corollary 7.2.10 *Two functions in $\tilde{N}^{1,p}(X : V)$ determine the same element in $N^{1,p}(X : V)$ if and only if they agree outside a set of zero p-capacity. Moreover, if two functions in $\tilde{N}^{1,p}(X : V)$ agree almost everywhere then they agree outside a set of zero p-capacity.*

Remark 7.2.11 The condition that E has measure zero in Proposition 7.2.8 is essential. If X has no rectifiable curves then X itself is p-exceptional but not of measure zero and hence is not of capacity zero. Also, consider the example

$X = (\mathbb{R}^n, |x-y|, m_n + \delta_0)$ as in Section 5.2; then the origin is p-exceptional for $1 \le p \le n$ but not of p-capacity zero.

Later, in Section 7.5, we will explore a condition on X that guarantees the conclusion in Proposition 7.2.8 without the requirement that the measure of E be zero.

As pointed out in the comment before (7.2.2), when X satisfies a Poincaré inequality as discussed in Chapter 8, one may require that admissible functions u for Cap_p also satisfy $u \ge 1$ on a *neighborhood* of E. However, even without these additional conditions on X the following proposition tells us that if X is proper then this additional assumption on test functions is legitimate for sets of capacity zero. Recall that X is proper if the closed and bounded subsets of X are compact. The proposition below is a key ingredient in proving the quasicontinuity of Sobolev functions (Section 7.4). The properness condition on X can be relaxed to local compactness, with appropriate modifications. For example, in a locally compact setting we can reduce the situation of Proposition 7.2.12 below (by the countable subadditivity of capacity) to sets E that lie in an open subset of X with compact closure; then in Lemma 7.2.13 we may assume that F contains the complement of this open set. To keep the notation in the proof simple, we will assume that X is proper.

Proposition 7.2.12 *Let X be a proper metric measure space. Then $\inf\{\mathrm{Cap}_p(U) : E \subset U \text{ and } U \text{ an open subset of } X\} = 0$ for every $E \subset X$ such that* $\mathrm{Cap}_p(E) = 0$.

To prove Proposition 7.2.12 we need the following lemma.

Lemma 7.2.13 *Let X be a proper metric measure space and ρ be a nonnegative lower semicontinuous function on X with $\rho \in L^p_{\mathrm{loc}}(X)$. Then, given a nonempty closed set $F \subset X$ and $\tau > 0$ such that $\rho \ge \tau$ on $X \setminus F$, the map $u : X \to \mathbb{R}$ defined by*

$$u(x) = \min\left\{1, \inf_\gamma \int_\gamma \rho\, ds\right\},$$

where the infimum is taken over compact rectifiable curves γ with an end point in F and an end point x, is lower semicontinuous and hence is measurable. In addition, ρ is an upper gradient of u and so u belongs to $N^{1,p}_{\mathrm{loc}}(X)$.

Proof To prove that u is lower semicontinuous, we need to show that for each $a > 0$ the sublevel set $F_a := \{x \in X : u(x) \le a\}$ is closed. Observe that $F \subset F_a$. To prove that F_a is closed, we proceed by considering a sequence (x_j) in F_a that converges to $x \in X$. If the tail end of the sequence lies in F or if $x \in F$, we are done; so we assume that, for each j, $x_j \notin F$ and that $x \notin F$.

If there is no rectifiable curve connecting x_j, to F for some j, then $u(x_j) = 1$ for that choice of j and so $a \geq 1$ because $x_j \in F_a$, from which it would follow that $x \in F_a = X$. Hence we may assume without loss of generality that for each j there is a compact rectifiable curve γ_j, parametrized by the arc length, such that $\gamma_j(0) = x_j$, $\gamma_j(\ell_j) \in F$ with ℓ_j the length of γ_j, $\gamma_j([0, \ell_j)) \subset X \setminus F$, and $\int_{\gamma_j} \rho\, ds \leq a + j^{-1}$. Since $\rho \geq \tau$, it follows that

$$\ell_j \leq \frac{a}{\tau} + \frac{1}{\tau j} \leq \frac{1+a}{\tau} =: M.$$

We can extend the domain of definition of each γ_j to $[0, M]$ by setting $\gamma_j(t) = \gamma_j(\ell_j)$ for $t \in [\ell_j, M]$. Now, by the Arzelà–Ascoli theorem 5.1.10 there is a subsequence, also denoted γ_j, that converges uniformly on $[0, M]$ to a 1-Lipschitz map γ_∞, which satisfies $\gamma_\infty(0) = x$, $\gamma_\infty(M) \in F$ (because F is closed), and length$(\gamma_\infty) \leq a/\tau$. By (5.1.15) we have $\int_{\gamma_\infty} \rho\, ds \leq \int_0^m \rho \circ \gamma_\infty(t)\, dt$, where $m = \liminf_j \ell_j$. On the other hand,

$$\liminf_{j\to\infty} \int_{\gamma_j} \rho\, ds = \liminf_{j\to\infty} \int_0^{\ell_j} \rho \circ \gamma_j(t)\, dt$$

and, for fixed $\epsilon > 0$ and for sufficiently large j, we have $\ell_j \geq m - \epsilon$. Hence by Fatou's lemma, by the lower semicontinuity of ρ, and by the fact that the curves γ_j are arc length parametrized,

$$\begin{aligned} a \geq \liminf_{j\to\infty} \int_{\gamma_j} \rho\, ds &\geq \liminf_{j\to\infty} \int_0^{m-\epsilon} \rho \circ \gamma_j(t)\, dt \\ &\geq \int_0^{m-\epsilon} \liminf_{j\to\infty} \rho \circ \gamma_j(t)\, dt \geq \int_0^{m-\epsilon} \rho \circ \gamma_\infty(t)\, dt. \end{aligned}$$

Letting $\epsilon \to 0$, we see that $\int_{\gamma_\infty} \rho\, ds \leq \int_0^m \rho \circ \gamma_\infty(t)\, dt \leq a$, that is, $u(x) \leq a$, from which we conclude that $x \in F_a$, that is, F_a is closed.

It remains to show that ρ is an upper gradient of u. For this, we fix a compact rectifiable curve γ in X and let x, y denote the end points of γ. If both $u(x)$ and $u(y)$ equal 1 then clearly $|u(x) - u(y)| \leq \int_\gamma \rho\, ds$. So, we now assume that $u(y) < 1$. For $\epsilon > 0$ let β_ϵ be a rectifiable curve with one end point in F and the other end point y such that $1 > u(y) \geq \int_{\beta_\epsilon} \rho\, ds - \epsilon$; then the concatenation $\gamma + \beta_\epsilon$ of γ and β_ϵ is a rectifiable path in X connecting F to x. First, suppose that $u(x) = 1$. Then $|u(x) - u(y)| = u(x) - u(y)$. Since $u(x) \leq \int_{\gamma+\beta_\epsilon} \rho\, ds$, we see that

$$|u(x) - u(y)| = u(x) - u(y) \leq \int_{\gamma+\beta_\epsilon} \rho \, ds + \epsilon - \int_{\beta_\epsilon} \rho \, ds.$$

Because $\int_{\beta_\epsilon} \rho \, ds$ is finite, we can subtract it from $\int_{\gamma+\beta_\epsilon} \rho \, ds$ to obtain $|u(x) - u(y)| = u(x) - u(y) \leq \int_\gamma \rho \, ds + \epsilon$. Letting $\epsilon \to 0$ yields the upper gradient inequality in the case $u(x) = 1$. If both $u(x) < 1$ and $u(y) < 1$ then a repetition of the argument above yields $u(x) - u(y) \leq \int_\gamma \rho \, ds$ and, reversing the roles of x and y in the argument, yields $u(y) - u(x) \leq \int_\gamma \rho \, ds$. Together, these two inequalities verify the upper gradient property in the case $u(x), u(y) < 1$.

This completes the proof. □

Now we are ready to prove Proposition 7.2.12.

Proof of Proposition 7.2.12 By the subadditivity property of the p-capacity Cap_p (see Lemma 7.2.4), we may assume that E is bounded. Since $\mathrm{Cap}_p(E) = 0$, it follows that $\mu(E) = 0$. Hence, for each positive integer j we can find a bounded open set $U_j \supset E$ such that $\mu(U_j) \leq 1/j$; see Lemma 3.3.37. Again, since $\mathrm{Cap}_p(E) = 0$, by Proposition 7.2.8 we know that $\chi_E \in N^{1,p}(X)$. In particular, there is a nonnegative Borel measurable function ρ on X such that $\rho \in L^p(X)$ and, whenever γ is a nonconstant compact rectifiable curve intersecting E, we have $\int_\gamma \rho \, ds = \infty$. By the Vitali–Carathéodory theorem, see Section 4.2, we may assume that ρ is lower semicontinuous. Note that $X \setminus U_j$ is a closed set and that $\rho + \chi_{U_j}$ is a lower semicontinuous function.

An application of Lemma 7.2.13 with $F = X \setminus U_j$ and the lower semicontinuous function $\rho + \chi_{U_j}$, and setting $\tau = 1$, gives a function $u_j \in N^{1,p}(X)$ with upper gradient $\rho + \chi_{U_j}$ such that u_j is also lower semicontinuous. Since for every nonconstant compact rectifiable curve γ that intersects E we have $\int_\gamma \rho \, ds = \infty$, we see that $u_j = 1$ on E. Furthermore, for $x \notin U_j$ we have $u_j(x) = 0$; hence, by Proposition 6.3.22, $(1 + \rho)\chi_{U_j}$ is a p-weak upper gradient of u_j. By the lower semicontinuity of u_j, the level set $V_j := \{x \in X : u_j(x) > 3/4\}$ is an open set containing E, and the function $\frac{4}{3}u_j$ is admissible for computing the p-capacity of V_j. Therefore

$$\begin{aligned}\left(\frac{3}{4}\right)^p \mathrm{Cap}_p(V_j) &\leq \int_X u_j^p \, d\mu + \int_{U_j} (1+\rho)^p \, d\mu \\ &\leq (1+2^p)\mu(U_j) + \int_{U_j} \rho^p \, d\mu \leq \frac{1+2^p}{j} + \int_{U_j} \rho^p \, d\mu.\end{aligned}$$

Since $\rho \in L^p(X)$, it follows that as $j \to \infty$ the last term above tends to zero. This completes the proof of the proposition. □

7.3 $N^{1,p}(X : V)$ is a Banach space

The goal of this section is to demonstrate that the normed space $N^{1,p}(X : V)$ as defined before (7.1.26) is a Banach space.

We first introduce the concept of quasiuniform convergence (compare Egoroff's theorem, to be found in Section 3.1). A sequence of functions $u_i : X \to V$ converges *p-quasiuniformly* to a function $u : X \to V$ if for every $\epsilon > 0$ there exists a set $F_\epsilon \subset X$ such that $\mathrm{Cap}_p(F_\epsilon) < \epsilon$ and $u_i \to u$ uniformly in $X \setminus F_\epsilon$. Obviously, a p-quasiuniformly convergent sequence of functions converges pointwise outside a set of p-capacity zero or, equivalently, outside a p-exceptional set of measure zero (Proposition 7.2.8).

Next we formulate and prove the following crucial result. The reader should compare the ensuing proof with that of Proposition 2.3.13.

Proposition 7.3.1 *Every Cauchy sequence of functions in $N^{1,p}(X : V)$ contains a p-quasiuniformly convergent subsequence. Moreover, the pointwise limit function belongs to $N^{1,p}(X : V)$ and is independent of the chosen subsequence.*

Proof Every Cauchy sequence in $N^{1,p}(X : V)$ is a Cauchy sequence in $L^p(X : V)$. Therefore any two limit functions agree almost everywhere, and so the independence assertion of the proposition follows from the first two assertions and Proposition 7.1.31.

Now choose a subsequence (u_i) of a given Cauchy sequence in $N^{1,p}(X : V)$ such that (u_i) converges pointwise almost everywhere to its L^p-limit $\tilde{u}$ and

$$\|u_i - u_{i+1}\|^p_{L^p(X : V)} + \|\rho_{i+1,i}\|^p_{L^p(X)} \leq 2^{-i(p+1)}, \tag{7.3.2}$$

where we denote by $\rho_{i,j}$ the minimal p-weak upper gradient of $u_i - u_j$. In general,

$$u_i = u_1 + \sum_{k=1}^{i-1}(u_{k+1} - u_k)$$

has

$$\rho_i = \rho_1 + \sum_{k=1}^{i-1} \rho_{k+1,k}$$

as a p-weak upper gradient. Moreover,

$$\|\rho_j - \rho_{j+i}\|_{L^p(X)} \leq \sum_{k=j}^{j+i-1} \|\rho_{k+1,k}\|_{L^p(X)} \leq \sum_{k=j}^{\infty} 2^{-k} \to 0$$

as $j \to \infty$. It follows that (ρ_i) is a Cauchy sequence in $L^p(X)$ and hence converges in $L^p(X)$ to a nonnegative Borel function ρ. Define a function u by

$$u(x) = \lim_{i\to\infty} u_i(x), \tag{7.3.3}$$

wherever this limit exists. Since $u_i \to \tilde{u}$ almost everywhere, the limit exists and satisfies $u(x) = \tilde{u}(x)$ for almost every x. In particular, $u \in L^p(X : V)$. (For this membership to hold, it is immaterial how u is defined on the set where the limit in (7.3.3) does not exist.)

We next show that the sequence (u_i) converges to u in a p-quasiuniform way. This implies in particular that the limit in (7.3.3) exists outside a set of p-capacity zero, and so the second claim of the proposition (that the limit function belongs to $N^{1,p}(X : V)$) follows from Propositions 6.3.30 and 7.2.8. To this end, define

$$E_i = \{x \in X : |u_i(x) - u_{i+1}(x)| > 2^{-i}\}$$

and

$$F_j = \bigcup_{i=j}^{\infty} E_i .$$

If $x \notin F_j$ then we have $|u_i(x) - u_{i+1}(x)| \le 2^{-i}$ for all $i \ge j$. This implies that the sequence $(u_i(x))$ is a Cauchy sequence in V and hence has a limit; by (7.3.3) this limit is $u(x)$. Moreover, on the one hand

$$|u(x) - u_i(x)| \le 2^{-i+1} \tag{7.3.4}$$

whenever $i \ge j$ and $x \notin F_j$; that is, $u_i \to u$ uniformly in $X \setminus F_j$. On the other hand, the function $2^i|u_i - u_{i+1}|$ belongs to $N^{1,p}(X)$ (Lemma 7.1.7) and satisfies $2^i|u_i - u_{i+1}| \ge 1$ on E_i by the definition of E_i. Therefore, by inequality (7.3.2),

$$\mathrm{Cap}_p(E_i) \le 2^{ip}\|u_i - u_{i+1}\|^p_{L^p(X)} + 2^{ip}\|\rho_{i+1,i}\|^p_{L^p(X)} \le 2^{-i} .$$

Thus, by the subadditivity property (7.2.5) we find that

$$\mathrm{Cap}_p(F_j) \le \sum_{i=j}^{\infty} 2^{-i} = 2^{-j+1}$$

for $j = 1, 2, \ldots$ The proposition follows from this and from (7.3.4). □

Remark 7.3.5 If, in the preceding proof of Proposition 7.3.1, the sequence (u_i) consists of continuous functions, then the sets F_j are open. In particular, it follows that the pointwise limit function u is continuous outside an open set of arbitrarily small (prescribed) capacity. This observation will be used later in Section 7.4.

Theorem 7.3.6 *The normed space $N^{1,p}(X:V)$ is a Banach space.*

Proof Let (u_i) be a Cauchy sequence in $N^{1,p}(X:V)$. By considering a subsequence if necessary, we may assume that the sequence (u_i) satisfies the conclusions of Proposition 7.3.1 and the condition in (7.3.2) with appropriate notation. In particular, the functions u_i converge pointwise to a function $u \in N^{1,p}(X:V)$ outside a p-exceptional set E of measure zero. Because

$$u(x) - u_i(x) = \sum_{k=i}^{\infty} u_{k+1}(x) - u_k(x)$$

for $x \in X \setminus E$, and because

$$\sum_{k=i}^{n} \rho_{k+1,k} \to \sum_{k=i}^{\infty} \rho_{k+1,k}$$

in $L^p(X)$ by our assumption that (7.3.2) holds, we deduce from Proposition 6.3.30 that

$$\sum_{k=i}^{\infty} \rho_{k+1,k}$$

is a p-weak upper gradient of $u - u_i$. Moreover, inequality (7.3.2) further gives that

$$\begin{aligned} \|u - u_i\|_{N^{1,p}(X:V)} &\le \sum_{k=i}^{\infty} \left(\|u_k - u_{k+1}\|_{L^p(X:V)} + \|\rho_{k+1,k}\|_{L^p(X)}\right) \\ &\le 2\sum_{k=i}^{\infty} \left(\|u_k - u_{k+1}\|^p_{L^p(X:V)} + \|\rho_{k+1,k}\|^p_{L^p(X)}\right)^{1/p} \\ &\le 2\sum_{k=i}^{\infty} 2^{-k(1+1/p)} \le 4 \times 2^{-i}. \end{aligned}$$

Therefore $u_i \to u$ in $N^{1,p}(X:V)$ as $i \to \infty$, and the theorem follows. □

An important consequence of reflexivity is the weak (pre-)compactness of bounded sequences, as in Theorem 2.4.1. The following proposition and its corollaries show that, even without reflexivity, a version of weak compactness holds on $N^{1,p}(X)$. In many applications, this suffices.

Proposition 7.3.7 *Let (u_i) be a sequence of functions in $N^{1,p}(X : V)$ with (ρ_i) a corresponding p-weak upper gradient sequence. If $u_i \to u$ in $L^p(X : V)$ and if $\rho_i \to \rho$ in $L^p(X)$ then u has a representative in $N^{1,p}(X : V)$ with each Borel representative of ρ as its p-weak upper gradient. Moreover, a subsequence of (u_i) converges pointwise to this representative of u outside a set of p-capacity zero.*

We will prove this proposition after Remark 7.3.14 below.

We can single out three consequences of Proposition 7.3.7; these are Theorems 7.3.8, 7.3.9, and 7.3.12. They are essentially variants of each other but deserve separate formulations.

Recall that $L^p(X : V)$, $1 < p < \infty$, is reflexive if V is reflexive (see the discussion of the duality of spaces in Section 3.2). In particular, we can have $V = \mathbb{R}$ in the following.

Theorem 7.3.8 *Assume that $1 < p < \infty$ and that $L^p(X : V)$ is reflexive. Let (u_i) be a bounded sequence in $N^{1,p}(X : V)$ with (ρ_i), a corresponding sequence of p-weak upper gradients, bounded in $L^p(X)$.*

Then there exists a function $u \in N^{1,p}(X : V)$ together with a p-weak upper gradient $\rho \in L^p(X)$ such that u belongs to the closure of the convex hull of the sequence (u_i) in $L^p(X : V)$ and ρ belongs to the closure of the convex hull of the sequence (ρ_i) in $L^p(X)$. More precisely, there is a sequence $(\tilde{u}_j)$ whose members are convex combinations of the functions u_i and there is a sequence $(\tilde{\rho}_j)$ whose members are convex combinations of the functions ρ_i such that $(\tilde{u}_j, \tilde{\rho}_j)$ is a function–upper-gradient pair for each j, $\tilde{u}_j \to u$ in $L^p(X : V)$, and $\tilde{\rho}_j \to \rho$ in $L^p(X)$.

Proof Because the relevant L^p-spaces are reflexive, we can apply Theorem 2.4.1 together with Mazur's lemma (see Section 2.3) and infer that a sequence of convex combinations of the functions u_i converges to a function u in $L^p(X : V)$. By passing to another subsequence, we may likewise assume that the corresponding convex-combination sequence of weak upper gradients converges in $L^p(X)$ to some function ρ. (Recall that the upper gradient property is preserved under convex combinations.) The assertion, therefore, follows from Proposition 7.3.7. □

In many situations we do not wish to insist on the reflexivity of V (which means that we cannot insist on the reflexivity of $L^p(X : V)$). However, in such situations we often deal with a bounded sequence of functions in $N^{1,p}(X : V)$ that converges in $L^p(X : V)$. This is the case when one considers, for example, discrete convolutions in $N^{1,p}(X : V)$, as in the proofs of Theorems 10.3.4 and 10.4.3. Thus the following theorem comes in handy.

Theorem 7.3.9 *Assume that $1 < p < \infty$ and that (u_i) is a bounded sequence in $N^{1,p}(X : V)$ that converges weakly in $L^p(X : V)$ to a function u. Then u has a representative in $N^{1,p}(X : V)$ such that*

$$\|u\|_{L^p(X : V)} \le \liminf_{i\to\infty} \|u_i\|_{L^p(X : V)}, \qquad \|\rho_u\|_{L^p(X)} \le \liminf_{i\to\infty} \|\rho_{u_i}\|_{L^p(X)}. \tag{7.3.10}$$

In particular,

$$\|u\|_{N^{1,p}(X : V)} \le \liminf_{i\to\infty} \|u_i\|_{N^{1,p}(X : V)}. \tag{7.3.11}$$

Proof By passing to subsequences if necessary, we may assume that

$$\liminf_{i\to\infty} \|\rho_{u_i}\|_{L^p(X)} = \lim_{i\to\infty} \|\rho_{u_i}\|_{L^p(X)}$$

and that ρ_{u_i} converges weakly in $L^p(X)$ to a function ρ (by Theorem 2.4.1 together with Theorem 2.4.9 and Proposition 2.4.19). By Mazur's lemma (see Section 2.3) we can form sequences $(\tilde{u}_j)$ and $(\tilde{\rho}_j)$ of convex combinations of the sequences (u_i) and (ρ_{u_i}), respectively, such that $(\tilde{u}_j, \tilde{\rho}_j)$ is a function–upper-gradient pair for each j, $\tilde{u}_j \to u$ in $L^p(X : V)$, and $\tilde{\rho}_j \to \rho$ in $L^p(X)$. The assertion now follows from Proposition 7.3.7 and from the lower semicontinuity of norms under weak convergence (Proposition 2.3.5). □

For further reference, we single out the following direct consequence of Theorems 7.3.9 and 2.4.1.

Theorem 7.3.12 *Assume that $1 < p < \infty$ and that $L^p(X : V)$ is reflexive. Then every bounded sequence (u_i) in $N^{1,p}(X : V)$ has a subsequence that converges weakly in $L^p(X : V)$ to a function u such that $u \in N^{1,p}(X : V)$ and*

$$\|u\|_{N^{1,p}(X : V)} \le \liminf_{i\to\infty} \|u_i\|_{N^{1,p}(X : V)}. \tag{7.3.13}$$

Remark 7.3.14 In the situation of Theorem 7.3.12, it is not true in general that every bounded sequence in $N^{1,p}(X : V)$ contains a subsequence that converges weakly in $N^{1,p}(X : V)$. In fact, such an assertion is tantamount to the reflexivity of $N^{1,p}(X : V)$, by the Eberlein–Šmulian theorem [74, p. 17].

Proof of Proposition 7.3.7 To simplify our notation, we write ρ for the given Borel representative of the limit of the sequence (ρ_i). We may assume that $u_i \to u$ pointwise almost everywhere. By Proposition 6.3.30 it suffices to prove that there exists a representative $\tilde{u}$ of u such that

$$\lim_{i\to\infty} u_i(x) = \tilde{u}(x) \tag{7.3.15}$$

for x outside a p-exceptional set. Let E denote the set where the limit on the left in (7.3.15) does not exist, and define $\tilde{u}(x)$ accordingly for $x \notin E$. Then E has measure zero. We claim that E is in addition p-exceptional.

To prove this, observe first that p-a.e. compact rectifiable curve γ in X has the following properties: the length of γ in E is zero, the upper gradient inequality (6.2.1) holds for each pair (u_i, ρ_i) on γ and all its subcurves, each function u_i is absolutely continuous on γ, each function ρ_i is integrable on γ, ρ is integrable on γ, and

$$\lim_{i\to\infty} \int_{\gamma'} \rho_i \, ds = \int_{\gamma'} \rho \, ds \tag{7.3.16}$$

for every subcurve γ' of γ. (This assertion follows from the basic properties of the modulus, such as (5.2.5) and Lemma 5.2.8, and from Proposition 6.3.2, Lemma 5.2.15, and Fuglede's lemma of Section 5.2.) Let $\gamma : [a, b] \to X$ be such a curve. We claim that γ does not meet E.

Towards proving this claim, we will establish that the family (u_i) satisfies the following equicontinuity property on γ: given $t \in [a, b]$ and given $\epsilon > 0$, there exists $\delta > 0$ such that $|u_i(\gamma(s)) - u_i(\gamma(t))| < \epsilon$ for every i whenever $s \in [a, b]$ satisfies $|s - t| < \delta$. To prove this, fix $t \in [a, b]$ and $\epsilon > 0$. Then observe that, for every $\delta > 0$, we have

$$\limsup_{i\to\infty} \sup_{s\in[a,b],|s-t|<\delta} |u_i(\gamma(s)) - u_i(\gamma(t))| \le \limsup_{i\to\infty} \int_{\gamma_{t,\delta}} \rho_i \, ds = \int_{\gamma_{t,\delta}} \rho \, ds\,, \tag{7.3.17}$$

where $\gamma_{t,\delta}$ denotes the restriction of γ to $\{r \in [a, b] : |r - t| \le \delta\}$. In particular, by the integrability of ρ on γ, we can find $i_0 \ge 1$ and $\delta > 0$ such that

$$|u_i(\gamma(s)) - u_i(\gamma(t))| < \epsilon$$

whenever $i \ge i_0$ and $s \in [a, b]$ satisfies $|s - t| < \delta$. The desired equicontinuity property follows. (Note that each u_i is continuous on γ.) Now we show that the limit in (7.3.15) exists for every point $x = \gamma(t)$. Indeed, fix $t \in [a, b]$ and fix $\epsilon > 0$. Then fix $\delta > 0$ as in the equicontinuity requirement corresponding to t and ϵ. Because the length of γ in E is zero, we can find $s \in [a, b]$ such that $|s - t| < \delta$ and $\gamma(s) \notin E$. In particular, we have that

$$\begin{aligned} |u_i(\gamma(t)) - u_j(\gamma(t))| &\le |u_i(\gamma(t)) - u_i(\gamma(s))| + |u_i(\gamma(s)) - u_j(\gamma(s))| \\ &\quad + |u_j(\gamma(s)) - u_j(\gamma(t))| \\ &\le 2\epsilon + |u_i(\gamma(s)) - u_j(\gamma(s))|\,. \end{aligned}$$

Because $u_i(\gamma(s)) \to u(\gamma(s))$, this gives that $(u_i(\gamma(t)))$ is a Cauchy sequence in V and hence has a limit; that is, $\gamma \cap E$ is empty.

We have thus established that p-almost every nonconstant compact rectifiable curve in X avoids E. In particular E is p-exceptional, and the theorem follows. □

Remark 7.3.18 By using the argument at the beginning of the proof of Proposition 7.3.1 and the argument in the proof of Proposition 7.3.7, one obtains a proof of the fact that $N^{1,p}(X:V)$ is a Banach space without any discussion of capacities. However, the conclusion that every Cauchy sequence in $N^{1,p}(X:V)$ subconverges p-quasiuniformly is useful extra information.

The proof of the following result illustrates the use of Proposition 7.3.7 and its consequences.

Proposition 7.3.19 *Assume that* $1 < p < \infty$. *If* $E_1 \subset E_2 \subset \cdots$ *is an increasing sequence of subsets of* X *then*

$$\operatorname{Cap}_p\left(\bigcup_{i=1}^{\infty} E_i\right) = \lim_{i\to\infty} \operatorname{Cap}_p(E_i). \tag{7.3.20}$$

Proof By the monotonicity property (7.2.3), it suffices to show that the limit on the right in (7.3.20) is at least as great as the expression on the left. We may assume that the limit is finite.

Fix $\epsilon > 0$. For each i we can find a function $u_i \in N^{1,p}(X)$ such that $0 \le u_i \le 1$, $u_i = 1$ on E_i, and

$$\|u_i\|^p_{L^p(X)} + \|\rho_{u_i}\|^p_{L^p(X)} \le \operatorname{Cap}_p(E_i) + \epsilon \tag{7.3.21}$$

(see Lemma 7.2.6). The sequence (u_i) is bounded in $N^{1,p}(X)$; in particular it is bounded in $L^p(X)$. By passing to a subsequence, we may assume that u_i converges weakly in $L^p(X)$ to a function u. We obtain from the proof of Theorem 7.3.9 that a sequence of convex combinations $(\tilde{u}_j)$ of the sequence (u_i) converges to u in $L^p(X)$ and that (7.3.10) holds. It follows from Proposition 7.3.7 that $u = 1$ on $E := \bigcup_{i=1}^{\infty} E_i$, except perhaps on a set of p-capacity zero. Therefore, by (7.3.10) and (7.3.21),

$$\operatorname{Cap}_p(E) \le \|u\|^p_{L^p(X)} + \|\rho_u\|^p_{L^p(X)} \le \lim_{i\to\infty} \operatorname{Cap}_p(E_i) + \epsilon\,.$$

By letting $\epsilon \to 0$ we complete the proof. □

We did not exclude the case $p = 1$ in Proposition 7.3.19 by accident. Indeed, the sets $E_i = B(0, 1 - 1/i) \subset \mathbb{R}^n$ form an increasing sequence whose union is $B(0, 1)$. If we equip $\mathbb{R}^n$ with the Euclidean distance and the weighted Lebesgue measure corresponding to the weight $\omega(x) = 1+\chi_{\mathbb{R}^n\setminus B(0,1)}(x)$ then $\lim_{i\to\infty} \mathrm{Cap}_1(E_i) = \omega_{n-1}$ but $\mathrm{Cap}_1(\bigcup_i E_i) = 2\omega_{n-1}$; see the reasoning in Section 5.3.

The following lemma demonstrates the local nature of Sobolev functions.

Lemma 7.3.22 *Let $1 < p < \infty$. Suppose that (Ω_n) is a sequence of open subsets of X, with $\Omega_n \subset \Omega_{n+1}$ for each positive integer n. Furthermore, suppose that, with $\Omega = \bigcup_n \Omega_n$, $f \in L^p(\Omega)$ and for each positive integer n there is a function $f_n \in N^{1,p}(\Omega_n : V)$ with an upper gradient $g_n \in L^p(\Omega_n)$ such that $f_n \to f$ in $L^p(\Omega_{n_0} : V)$ for each $n_0 \in \mathbb{N}$; finally, suppose that $\int_{\Omega_n} g_n^p \, d\mu \le 1$. Then there is a function $\hat{f} \in N^{1,p}(\Omega : V)$ such that $\hat{f} = f$ almost everywhere in Ω and $\hat{f}$ has a p-weak upper gradient g in Ω with $\int_\Omega g^p \, d\mu \le 1$.*

Proof An application of Lemma 3.3.19 yields a subsequence (g_{n_k}) and a function $g_\infty \in L^p(\Omega)$ such that (g_{n_k}) converges weakly in $L^p(\Omega)$ to g_∞. By a modification of g_∞ on a set of μ-measure zero if necessary, we may also assume that g_∞ is Borel measurable.

Next, for each positive integer k_0 we apply Theorem 7.3.9 to the sequence $(f_{n_{k+k_0}})$ with respect to the open set Ω_{k_0} to obtain a function $\hat{f}_{k_0} \in N^{1,p}(\Omega_{k_0} : V)$ such that $\hat{f}_{k_0}$ is an $L^p(\Omega_{k_0} : V)$-representative of $f|_{\Omega_{k_0}}$.

Next, observe that, given two positive integers k_0 and k_1 with $k_0 \le k_1$, the two functions $\hat{f}_{k_0}, \hat{f}_{k_1}$ coincide μ-almost everywhere in Ω_{k_0}, and so from Proposition 7.1.31 we know that $\hat{f}_{k_0}$ and $\hat{f}_{k_1}$ belong to the same equivalence class in $N^{1,p}(\Omega_{k_0} : V)$. Thus the function $\hat{f}$ defined by $\hat{f}(x) = \hat{f}_{k_0}(x)$ when $x \in \Omega_{k_0}$ is well defined and equals f μ-almost everywhere in Ω. Furthermore, from the second inequality in (7.3.10) we know that the mimimal p-weak upper gradient g_∞ of $\hat{f}$ satisfies

$$\int_\Omega g_\infty^p \, d\mu \le 1.$$

Since $\Omega = \bigcup_n \Omega_n$ and for each positive integer n we have $\Omega_n \subset \Omega_{n+1}$, it follows that g_∞ is a p-weak upper gradient of $\hat{f}$ in Ω. Note that by Lemma 3.3.19 the function g_∞ is measurable on Ω. A careful examination of the proof of that lemma reveals that g_∞ is also Borel, because each Ω_n is open. Because $\int_\Omega g_\infty^p \, d\mu \le \limsup_{k\to\infty} \int_\Omega G_k^p \, d\mu$, with G_k the zero-extension of g_k to Ω, we see that $\hat{f}$ has a p-weak upper gradient in $L^p(\Omega)$ with norm at most 1. This completes the proof of the lemma. □

7.4 The space $HN^{1,p}(X : V)$ and quasicontinuity

The Sobolev space $N^{1,p}(X : V)$ contains a distinguished closed subspace that is isomorphic to the completion $HN^{1,p}(X : V)$ of locally Lipschitz functions in the norm of $N^{1,p}(X : V)$. This subspace is also denoted by $HN^{1,p}(X : V)$. We make the abbreviation

$$HN^{1,p}(X) := HN^{1,p}(X : \mathbb{R}) .$$

Recall that a function $u : X \to V$ is locally Lipschitz if every point in X has a neighborhood where u is Lipschitz. There are always nonconstant locally Lipschitz functions in $N^{1,p}(X : V)$. For example, choose a ball $B \subset X$ and put $u(x) := \operatorname{dist}(x, X \setminus B)$.

A priori the elements of $HN^{1,p}(X : V)$ are Cauchy sequences (φ_i) in $N^{1,p}(X : V)$ of locally Lipschitz functions $\varphi_i : X \to V$. The following, more concrete, description of $HN^{1,p}(X : V)$ follows from Proposition 7.3.1.

Proposition 7.4.1 *With each element (φ_i) in $HN^{1,p}(X : V)$ there is a uniquely associated function $u \in N^{1,p}(X : V)$ that is a representative of the $L^p(X : V)$-limit of the sequence (φ_i). Moreover, a function $u : X \to V$ is in $HN^{1,p}(X : V)$ if and only if there exists a Cauchy sequence (φ_i) in $N^{1,p}(X : V)$ of locally Lipschitz functions that converges to u both in $N^{1,p}(X : V)$ and p-quasiuniformly.*

A function $u : X \to V$ is said to be *p-quasicontinuous* if, for every $\epsilon > 0$, there exists an open set $G_\epsilon \subset X$ with $\operatorname{Cap}_p(G_\epsilon) < \epsilon$ such that the restriction $u|_{X \setminus G_\epsilon}$ is continuous.

By Remark 7.3.5 we obtain the following result.

Theorem 7.4.2 *The functions in $HN^{1,p}(X : V)$ are p-quasicontinuous.*

Remark 7.4.3 In Theorem 8.2.1 we will prove that, under favorable assumptions, locally Lipschitz continuous functions are dense in $N^{1,p}(X : V)$; thus, in such cases we have the equality

$$N^{1,p}(X : V) = HN^{1,p}(X : V) . \tag{7.4.4}$$

In particular, if (7.4.4) holds and X is locally compact (in particular, if X is complete and μ is doubling), then the functions in $N^{1,p}(X : V)$ are p-quasicontinuous; see [16].

We are now in a position to establish the equality between classical Sobolev spaces and the spaces $N^{1,p}$ for open sets in Euclidean spaces. In the following, we write $N^{1,p}(\Omega)$ for an open set $\Omega \subset \mathbb{R}^n$ equipped with the Euclidean distance and Lebesgue n-measure. Recall the definition of $H^{1,p}(\Omega)$ from Definition 6.1.5.

Theorem 7.4.5 *Let $\Omega \subset \mathbb{R}^n$ be open. Then*

$$N^{1,p}(\Omega) = W^{1,p}(\Omega) = H^{1,p}(\Omega) \tag{7.4.6}$$

in the following precise sense: a function from any of these three Banach spaces has a representative in the other two and this correspondence is a linear isometry between the three spaces. The representatives that belong to $N^{1,p}(\Omega)$ are precisely those functions that are p-quasicontinuous; these functions are pointwise uniquely defined up to a set of p-capacity zero.

Proof The equality $W^{1,p}(\Omega) = H^{1,p}(\Omega)$ follows from Theorem 6.1.6. Next, for the inclusion $H^{1,p}(\Omega) \subset N^{1,p}(\Omega)$ it suffices to show that, for all smooth $\varphi \in H^{1,p}(\Omega)$, we have $\|\varphi\|_{N^{1,p}(\Omega)} \le \|\varphi\|_{H^{1,p}(\Omega)}$. Since $|\nabla\varphi|$ is an upper gradient of φ by the fundamental theorem of calculus, this inequality does indeed hold true. Now an application of Proposition 7.3.7 together with Theorem 7.3.8 shows that $H^{1,p}(\Omega) \subset N^{1,p}(\Omega)$ in the sense of the second paragraph of the above theorem.

Finally, to show that $N^{1,p}(\Omega) \subset W^{1,p}(\Omega)$ with $\|\varphi\|_{W^{1,p}(\Omega)} \le \|\varphi\|_{N^{1,p}(\Omega)}$, we note by the absolute continuity of the integral (see Remark 2.3.16) and by Propositions 6.3.2 and 6.3.3 that functions in $N^{1,p}(\Omega)$ are absolutely continuous on almost every line segment in Ω parallel to coordinate axes with partial derivatives majorized by any p-integrable upper gradient of the function. It follows that $N^{1,p}(\Omega) \subset L^p(\Omega) \cap ACL^p(\Omega)$. Thus, by Theorem 6.1.13 we have $N^{1,p}(\Omega) \subset W^{1,p}(\Omega)$.

For $u \in N^{1,p}(\Omega)$, Theorem 6.1.6 thus gives us a sequence of smooth functions φ_i that converges to u in $W^{1,p}(\Omega)$. By Lemma 6.2.6, $\rho_i(x) = |\nabla\varphi_i(x)|$ is an upper gradient of φ_i; hence this sequence is Cauchy in $N^{1,p}(\Omega)$ also. If we knew that ρ_i were the minimal p-weak upper gradient of φ_i, we could conclude that $\|\varphi\|_{W^{1,p}(\Omega)} \le \|\varphi\|_{N^{1,p}(\Omega)}$.

It remains to be verified that $|\nabla\varphi|$ is the minimal p-weak upper gradient of a smooth function $\varphi \in H^{1,p}(\Omega)$. To see this, let ρ be an arbitrary p-integrable upper gradient of such a function φ. By Fubini's theorem, we can easily infer that almost every point $x \in \Omega$ satisfies, on the one hand,

$$\lim_{\epsilon\to 0} \frac{1}{\epsilon} \int_{[x,x+\epsilon e_i]} \rho \, ds = \rho(x)$$

for every $i = 1, \dots, n$. On the other hand,

$$\lim_{\epsilon\to 0} \frac{|\varphi(x) - \varphi(x+\epsilon e_i)|}{\epsilon} = |\partial_i\varphi(x)|$$

for every $x \in \Omega$. This implies that $|\partial_i \varphi(x)| \le \rho(x)$ for almost every $x \in \Omega$ and every $i = 1, \dots, n$. We can repeat the above argument in a rotated coordinate system to conclude that, given a unit vector ν in $\mathbb{R}^n$,

$$|\nabla \varphi(x) \cdot \nu| = |\partial_\nu \varphi(x)| \le \rho(x)$$

for almost every $x \in \Omega$, where $\partial_\nu \varphi(x)$ is the derivative of φ at x in the direction ν. By employing a countable dense collection of unit vectors (ν_i) we find that

$$|\nabla \varphi(x) \cdot \nu_i| \le \rho(x)$$

for almost every $x \in \Omega$ and for every ν_i. The density of the collection (ν_i) then gives that $|\nabla \varphi| \le \rho$ almost everywhere in Ω, as required. □

A discussion about Sobolev spaces in the Euclidean setting is not complete without a consideration of the weighted Euclidean setting, as given in [128]. A weight ω on $\mathbb{R}^n$ is a nonnegative measurable function that is positive almost everywhere. Such a weight is said to be a p-admissible weight if the weighted measure $\mu_\omega = \omega\, dm_n$ is doubling and there are constants $C, \lambda \ge 1$ such that whenever u is a measurable function on $\mathbb{R}^n$ with a weak derivative ∇u, and B is a ball in $\mathbb{R}^n$, we have

$$\omega(B)^{-1} \int_B |u - u_B|\, \omega\, dm_n \le C \operatorname{rad}(B) \left(\omega(\lambda B)^{-1} \int_{\lambda B} |\nabla u|^p\, \omega\, dm_n \right)^{1/p}.$$

Here, given a measurable set $A \subset \mathbb{R}^n$, the measure $\omega(A)$ is given by $\int_A \omega\, dm_n$. The potential theory associated with such a weight was considered in [128]. The corresponding weighted Sobolev space is the completion of smooth functions on $\mathbb{R}^n$ under the norm

$$\|u\|_{W^{1,p}(\mathbb{R}^n,\omega)} := \left(\int_{\mathbb{R}^n} |u|^p \omega\, dm_n \right)^{1/p} + \left(\int_{\mathbb{R}^n} |\nabla u|^p \omega\, dm_n \right)^{1/p}.$$

For a general weight ω, it is not clear that if v is in the weighted Sobolev class then it necessarily has a weak derivative. However, it turns out that, when ω is a p-admissible weight in the above sense, each function in the weighted Sobolev class does have a weak derivative that belongs to $L^p(\mathbb{R}^n, \mu_\omega)$. The monograph [31] discusses the relationship between the weighted Sobolev class and the upper gradients for p-admissible weights; see [31, Appendix A]. Combining Proposition A.12 and Proposition A.13 of [31] shows that $W^{1,p}(\mathbb{R}^n, \omega) = N^{1,p}(\mathbb{R}^n, \mu_\omega)$.

7.5 Main equivalence classes and the MEC$_p$ property

Given a nonnegative Borel function $\rho : X \to [0, \infty]$ and two points $x, y \in X$, we write $x \sim_\rho y$ if either $x = y$ or there exists a rectifiable curve $\gamma : [a, b] \to X$ such that $\gamma(a) = x$ and $\gamma(b) = y$ and ρ is integrable on γ.

This definition yields an equivalence relation $\sim_\rho$ among the points in X, and we have a corresponding collection of $\sim_\rho$-equivalence classes. If there exists an equivalence class $G_\rho \subset X$ such that $\mu(X \setminus G_\rho) = 0$, then this class G_ρ is said to be a *main equivalence class for* ρ. In fact, given ρ there is at most one such class.

A metric measure space X is said to have the *p-main equivalence class property*, or to satisfy *MEC$_p$*, if X has at least two points, if every open set in X has positive measure, and if there exists a main equivalence class for every nonnegative p-integrable Borel function ρ on X.

The following result shows that MEC$_p$ is a local property.

Proposition 7.5.1 *A connected metric space satisfies MEC$_p$ if and only if every point in the space has an open neighborhood that satisfies MEC$_p$.*

Proof The necessity part of the claim is obvious. To prove sufficiency, denote by U_x an open neighborhood of a point $x \in X$ such that U_x satisfies MEC$_p$. In particular, U_x has at least two points and $\mu(U_x) > 0$. Let ρ be a nonnegative p-integrable Borel function on X. The restriction of ρ to U_x is also p-integrable and hence has a measurable set $G_x \subset U_x$ as its main equivalence class. As X is separable, there is a countable cover $\{U_i\}$ of X by sets of the type $U_i := U_{x_i}$ (Lemma 3.3.27). Set $G_\rho := \bigcup_i G_{x_i}$. We claim that G_ρ is the main equivalence class for the function ρ.

It follows from the definitions that $\mu(X \setminus G_\rho) = 0$. Let $x, y \in G_\rho$ and choose open sets U_{i_x} and U_{i_y} from our fixed countable cover containing x and y, respectively. By the connectedness of X, we can find a finite sequence of open sets $U_{i_x} = U_{z_1}, \ldots, U_{z_k} = U_{i_y}$ from the countable cover such that $U_{z_i} \cap U_{z_{i+1}} \neq \emptyset$ for $i = 1, \ldots, k-1$. Note that as $\mu(U_{z_2} \setminus G_{z_2}) = 0 = \mu(U_{z_3} \setminus G_{z_3})$, and $\mu(U_{z_2} \cap U_{z_3}) > 0$, there is a point $z_2' \in G_{z_2} \cap G_{z_3}$. By the definition of the p-main-equivalence-class property, there must be a curve γ_1 joining x and z_2' such that ρ is integrable on γ_1. Similarly, there must be a point $z_3' \in G_{z_3} \cap G_{z_4}$ and a curve γ_2 joining z_2' and z_3' in G_{z_2} such that ρ is integrable on γ_2. Continuing similarly, we obtain a curve $\gamma = \gamma_1 \cup \cdots \cup \gamma_k$ joining x and y such that ρ is integrable on γ. The proposition follows. $\square$

The following proposition shows that spaces with the MEC$_p$ property have many of the properties expected for Sobolev functions.

Proposition 7.5.2 *Suppose that X has the p-main equivalence class property. Then the following three assertions hold.*

(i) *A subset of X has zero p-capacity if and only if it is p-exceptional.*
(ii) *If a function $u : X \to V$ in the Dirichlet class $D^{1,p}(X : V)$ has the zero function as a p-weak upper gradient then there exists $c \in V$ such that $u \equiv c$ outside a set of zero p-capacity.*
(iii) *If a measurable function $u : X \to V$ has the property that $u \circ \gamma \equiv c$ for p-almost every curve γ in X, for some vector $c \in V$, then $u \equiv c$ outside a set of zero p-capacity.*

Proof To prove (i) it suffices to show that every p-exceptional subset of X has measure zero (Proposition 7.2.8). To this end, suppose that $E \subset X$ is p-exceptional. Let ρ be a p-integrable nonnegative Borel function such that

$$\int_\gamma \rho \, ds = \infty$$

for every rectifiable nonconstant curve γ that meets E (Lemma 5.2.8). This means that each point in E belongs to a singleton $\sim_\rho$-equivalence class all by itself. Recall that, because X satisfies MEC_p, nonempty open sets have positive measure and X has at least two points. It follows that isolated points of X must carry positive measure. Therefore, if a singleton set were the main equivalence class of ρ then X could have no other point (because singleton sets are closed sets), which is not possible. Consequently no singleton set can be the main equivalence class of ρ. It follows in particular that no point in E can be equivalent to a point in a main equivalence class G_ρ. Hence E must have measure zero, as required.

To prove (ii) we show that u must take on a constant value almost everywhere in X. The claim then follows from Lemma 7.1.6 and from (i). To this end, choose a nonnegative p-integrable Borel function ρ on X such that

$$\int_\gamma \rho \, ds = \infty$$

for every nonconstant rectifiable curve γ on which the upper gradient inequality (6.2.1) does not hold for the pair $(u, 0)$. Then the main equivalence class G_ρ satisfies $\mu(X \setminus G_\rho) = 0$ and, for every pair of distinct points $x, y \in G_\rho$, there exists a rectifiable curve γ joining x and y such that

$$\int_\gamma \rho \, ds < \infty .$$

The upper gradient inequality is valid for the pair $(u, 0)$ on every such γ. In particular we can infer that u is constant on G_ρ, as desired.

Finally, under the hypotheses of (iii), the zero function is a p-weak upper gradient of u and hence $u \in D^{1,p}(X : V)$. Now we obtain from (ii) that $u \equiv c'$ outside a set of zero p-capacity for some constant $c' \in V$. The set $E := \{x \in X : u(x) \neq c'\}$ must have measure zero, and hence p-almost every curve γ in X that meets E does so in a set of zero length (Lemma 5.2.15). (Note that we have such curves by (i).) We find therefore that $c = c'$.

The proposition is proved. □

Examples (a) It is not hard to see using Fubini's theorem that $\mathbb{R}^n$, equipped with the Euclidean metric and the Lebesgue measure, satisfies MEC_p for every $n \geq 1$ and every $p \geq 1$. In particular, we obtain from Proposition 7.5.2 that *a subset $E \subset \mathbb{R}^n$ has zero p-capacity if and only if it is p-exceptional.*

(b) The subset $\overline{B}(e_1, 1) \cup \overline{B}(-e_1, 1)$ in $\mathbb{R}^n$, where $e_1 = (1, 0, \ldots, 0)$, satisfies MEC_p if and only if $p > n$.

7.6 Notes to Chapter 7

The Sobolev spaces $N^{1,p}$ were called *Newtonian spaces* in [247], [248], [129], and much of the recent literature. This nomenclature is motivated by the fact that the definition is based on upper gradients and, therefore, ultimately on the fundamental theorem of calculus. We have opted to return to the more familiar term, "Sobolev space". There is one caveat, however. Various competing definitions appear in the literature for Sobolev spaces of functions defined on metric spaces, and it is not always true that these spaces are the same. To emphasize this difference, we have used the symbol $N^{1,p}$ instead of, say, $W^{1,p}$. Some other possible definitions will be discussed in Chapter 10; see Section 10.6 for further references. The index 1 in $N^{1,p}(X : V)$ refers to the fact that the functions in question are associated with first-order Sobolev calculus. At the time of writing, the theory of higher-order Sobolev calculus in metric spaces was less developed; see, however, [187] and the references therein.

Theorem 7.1.20, appearing first in [129], was inspired by the works of Ambrosio [6] and Reshetnyak [233], [232], [231].

For Sobolev spaces on weighted Euclidean spaces, see [128]. A comparison of weighted Sobolev spaces and the corresponding upper-gradient-based Sobolev spaces $N^{1,p}$ associated with the weighted Lebesgue measure for p-admissible weights, can be found in Appendix A of [31].

The term "Dirichlet space" for the class $D^{1,p}$ studied in Section 7.1 refers to the fact that functions in this space are not required to be p-integrable; one assumes only that their Dirichlet energy (the integral of the pth power of the gradient) is finite. The careful treatment of measurability issues in Proposition 7.1.8 is new. There are other observations, such as Proposition 7.1.33, that were also previously unpublished.

Proposition 7.2.12 first appeared in [34]. An exposition of the potential theory associated with $N^{1,p}$-spaces can be found in the monograph [31] of A. and J. Björn. The outer capacity property given in Proposition 7.2.12 and the quasicontinuity property studied in Section 7.4 are used extensively in the study of Perron solutions and the resolutivity of continuous functions; see for example [33], [32], and [31].

The example following the proof of Proposition 7.3.19 is due to Riikka Korte.

The MEC_p property was first identified by Ohtsuka in the Euclidean setting [220]. Although [220] was published in 2003, versions of it were in circulation much earlier. The thesis [247], containing a systematic study of real-valued $N^{1,p}$-spaces, benefited from Ohtsuka's work.

Ambrosio, Gigli, and Savaré demonstrated in [16] that, for $p > 1$, continuous functions are dense in $N^{1,p}(X)$ when μ is locally doubling and X is locally complete. The reflexivity of $N^{1,p}(X)$ under these conditions was established by Ambrosio, Colombo, and Di Marino in [12].

The Sobolev space $N^{1,p}(X)$, $1 < p < \infty$, is not always reflexive. Examples to this effect were given in [122, p. 212].

In [111] and [110] Hajłasz proved that the property of density of Lipschitz mappings in $N^{1,p}(X : Y)$ is not invariant under bilipschitz changes in the metric of the target space Y and that the norm on $N^{1,p}(X : Y)$ depends on the embedding of Y into a Banach space.

8

Poincaré inequalities

In this chapter we introduce and discuss Poincaré inequalities in metric measure spaces. These inequalities are formulated using upper gradients for real- or Banach-space-valued functions in arbitrary metric measure spaces. After the definition, we prove fundamental pointwise estimates in doubling metric measure spaces supporting a Poincaré inequality. These estimates involve maximal functions and they constitute alternative, useful, descriptions of Poincaré inequalities. The pointwise estimates can be used to show, for example, that the validity of a Poincaré inequality in doubling metric measure spaces is independent of the target Banach space. We also establish the quasiconvexity of complete and doubling metric measure spaces that support a Poincaré inequality. This fact is applied to show that for the validity of a Poincaré inequality in such spaces it suffices to consider Lipschitz functions with continuous upper gradients. The density of Lipschitz functions in a Sobolev space is also established.

Throughout this chapter we let $X = (X, d, \mu)$ be a metric measure space, as defined in Section 3.3, and V a Banach space. Unless otherwise stipulated, we assume that $1 \leq p < \infty$. This latter assumption is often repeated for emphasis.

8.1 Poincaré inequality and pointwise inequalities

In this section we define a Poincaré inequality for real-valued as well as vector-valued functions. Then we show how in doubling metric measure spaces this inequality can be expressed in the form of various pointwise estimates.

A measurable function $u : X \to V$ is said to be *integrable on balls* if $u \in L^1(B : V)$ for every ball $B \subset X$.

Poincaré inequality for real-valued functions

We first define the Poincaré inequality for real-valued functions.

Definition We say that X *supports a p-Poincaré inequality* if every ball in X has positive and finite measure and if there exist constants $C > 0$ and $\lambda \geq 1$ such that

$$\fint_B |u - u_B| \, d\mu \leq C \operatorname{diam}(B) \left(\fint_{\lambda B} \rho^p \, d\mu \right)^{1/p} \tag{8.1.1}$$

for every open ball B in X, for every function $u : X \to \mathbb{R}$ that is integrable on balls, and for every upper gradient ρ of u in X. The parameters p, C, and λ are called the *data* of the Poincaré inequality (8.1.1). We sometimes say that (X, d, μ) supports a p-Poincaré inequality, to signify our metric and measure; notice that the definition of an upper gradient depends on the metric in question. We also say that X *supports a Poincaré inequality* if it supports a p-Poincaré inequality for some $p \geq 1$.

Recall that the barred integral sign denotes an integral average, that $u_B = \fint_B u \, d\mu$ stands for the mean value of u over the ball B (see the text after (1.2)), and that λB is a ball with the same center as B but with radius that is λ times the radius of B. Recall that balls may have more than one center and more than one radius; hence, strictly speaking, a ball denotes the ball as a set together with its preassigned center and radius. Also, recall that it follows from our assumptions that $\mu(U) > 0$ whenever U is a nonempty open set.

It is immediate from Hölder's inequality that if a space supports a p-Poincaré inequality for some p then it supports a q-Poincaré inequality for all $q \geq p$.

If u is a smooth function in an open ball $B \subset \mathbb{R}^n$ then

$$\fint_B |u - u_B| \, dx \leq C(n) \operatorname{diam}(B) \fint_B |\nabla u| \, dx \,, \tag{8.1.2}$$

where $C(n) > 0$ is a constant that depends only on n. In some texts on Sobolev spaces the radius of B is used instead of the diameter, $\operatorname{diam}(B)$. Since in the metric setting the radius of a ball is not uniquely determined, we use the diameter instead. Observe, however, that if $\operatorname{rad}(B)$ is a pre-determined radius of B then $\operatorname{diam}(B) \leq 2 \operatorname{rad}(B)$.

We have already seen a proof of a weaker version of the above statement in Remark 7.1.4. A modification of that proof yields (8.1.2). For the convenience of the reader, we now give a complete proof of (8.1.2).

As in the proof of Proposition 7.1.2, fix $x \in B$ and use polar coordinates $y = (\tau, \theta)$ based at x (that is, $\tau = |x - y|$ and $\theta = |x - y|^{-1}(x - y)$). Then for smooth functions u we have

$$|u(x) - u(y)| \leq \int_0^{\tau} |\nabla u(s, \theta)| \, ds.$$

Let $|\widetilde{\nabla u}|$ be the zero-extension of $|\nabla u|$ to $\mathbb{R}^n \setminus B$; $|\widetilde{\nabla u}| = |\nabla u| \, \chi_B$. Let R be the radius of B. Then

$$|u(x) - u(y)| \leq \int_0^{2R} |\widetilde{\nabla u}|(s, \theta) \, ds.$$

Integrating with respect to $y \in B \subset B(x, 2R)$, we obtain

$$\begin{aligned}
\int_B |u(x) - u(y)| \, dy &\leq \int_B \int_0^{2R} |\widetilde{\nabla u}|(s, \theta) \, ds \, dy \\
&\leq \int_{B(x,2R)} \int_0^{2R} |\widetilde{\nabla u}|(s, \theta) \, ds \, dy \\
&\leq \int_{S^{n-1}} \int_0^{2R} \tau^{n-1} \int_0^{2R} |\widetilde{\nabla u}|(s, \theta) \, ds \, d\tau \, d\theta \\
&\leq CR^n \int_{S^{n-1}} \int_0^{2R} \frac{|\widetilde{\nabla u}|(s, \theta)}{s^{n-1}} s^{n-1} \, ds \, d\theta \\
&= CR^n \int_{B(x,2R)} \frac{|\widetilde{\nabla u}|(z)}{|x - z|^{n-1}} \, dz \\
&\leq CR^n \int_{4B} \frac{|\widetilde{\nabla u}|(z)}{|x - z|^{n-1}} \, dz.
\end{aligned}$$

An application of Tonelli's theorem now yields

$$\begin{aligned}
\int_B \int_B |u(x) - u(y)| \, dy \, dx &\leq CR^n \int_B \int_{4B} \frac{|\widetilde{\nabla u}|(z)}{|x - z|^{n-1}} \, dz \, dx \\
&= CR^n \int_{4B} \left(\int_B \frac{1}{|x - z|^{n-1}} \, dx \right) |\widetilde{\nabla u}|(z) \, dz \\
&\leq CR^n \int_{4B} \left(\int_{B(z,2R)} \frac{1}{|x - z|^{n-1}} \, dx \right) |\widetilde{\nabla u}|(z) \, dz.
\end{aligned}$$

A polar coordinate integration yields

$$\int_{B(z,2R)} \frac{1}{|x - z|^{n-1}} \, dx = C_n R,$$

and so we have

$$\int_B \int_B |u(x) - u(y)|\, dy\, dx \le CR^{n+1} \int_{4B} |\widetilde{\nabla u}|(z)\, dz$$
$$= CR^{n+1} \int_B |\nabla u(z)|\, dz,$$

from which (8.1.2) follows.

We emphasize that the preceding argument depends strongly upon several distinctive features of Euclidean space, specifically polar coordinates and the convexity of balls (the latter property was used in the proof of Proposition 7.1.2).

On the one hand, as a consequence of (8.1.2) and Theorems 7.4.5 and 6.1.6, it follows that $\mathbb{R}^n$ supports a p-Poincaré inequality for all function–upper-gradient pairs and all $p \geq 1$. On the other hand, if X contains no nonconstant rectifiable curves and has more than one point then it cannot support a Poincaré inequality. Between these two extreme cases, there is a rich supply of spaces that do support a Poincaré inequality. We will not present further examples at this juncture, but refer the reader to Chapter 14.

It follows easily that every space that supports a Poincaré inequality must be connected (see Proposition 8.1.6 below). We will prove in Theorem 8.3.2 the more difficult result that every complete and doubling metric measure space that supports a Poincaré inequality is quasiconvex. Spaces that support a Poincaré inequality, and satisfy some additional geometric measure theoretic conditions, admit a remarkable amount of first-order differential analysis. This will become evident later.

The Poincaré inequality in (8.1.1) differs from the Euclidean inequality (8.1.2) in two respects. First, the right-hand side is the averaged L^p-integral instead of the averaged L^1-integral. Second, the integration on the right-hand side is over a larger ball than that on the left-hand side. Generally speaking these two differences constitute a weakening of inequality (8.1.2). Under suitable conditions on the underlying metric measure space, (8.1.1) can be shown to be equivalent to certain ostensibly stronger inequalities. This self-improving aspect of the left-hand side of inequality (8.1.1) will be discussed in more detail later; see Theorem 9.1.15.

The strongest inequality is the 1-Poincaré inequality. Different exponents present genuinely distinct cases. One can show that, given $1 \leq p < q$, there exist (compact and doubling) metric measure spaces that support a q-Poincaré inequality but not a p-Poincaré inequality. It is a deep fact that if a complete and doubling metric measure space supports a p-Poincaré inequality for some $p > 1$ then it also supports a q-Poincaré inequality for some $q < p$. We will prove this result in Chapter 12. Examples of spaces supporting a Poincaré inequality appear in Chapter 14.

One could axiomatize a Poincaré inequality in many ways similar to that in the preceding discussion. Here we require that inequality (8.1.1) holds for all (open) balls and for all function–upper-gradient pairs (u, ρ), where u is integrable on balls. Alternatively, one could require that (8.1.1) holds for functions that are not necessarily globally defined (see Theorem 8.1.53) or that it holds for continuous or Lipschitz functions only (see Theorems 8.1.53, 8.4.1, and 8.4.2).

The following is a direct consequence of the definition and Lemma 6.2.2.

Proposition 8.1.3 *Suppose that X supports a p-Poincaré inequality. If $u : X \to \mathbb{R}$ is a function that is integrable on balls and possesses a p-integrable p-weak upper gradient in X then*

$$\fint_B |u - u_B|\, d\mu \leq C \operatorname{diam}(B) \left(\fint_{\lambda B} \rho_u^p \, d\mu \right)^{1/p} \tag{8.1.4}$$

for every open ball B in X, where ρ_u is the minimal p-weak upper gradient of u and C, λ are as in (8.1.1).

The requirement that u be integrable on balls can be replaced with the weaker requirement that u be measurable on balls, as the following lemma shows.

Lemma 8.1.5 *Suppose that X supports a p-Poincaré inequality. Then each measurable function $u : X \to \mathbb{R}$ with an upper gradient $\rho \in L^p_{\mathrm{loc}}(X)$ is integrable over a given ball.*

Proof For each positive integer n we consider the truncated function

$$u_n = \max\{-n, \min\{n, u\}\}.$$

Then, by Proposition 6.3.23, ρ is a p-weak upper gradient of u_n. Since u_n is bounded and measurable, it is integrable on balls; hence we can apply the Poincaré inequality to the pair (u_n, ρ) to obtain

$$\begin{aligned}\fint_B \fint_B |u_n(x) - u_n(y)|\, d\mu(x)\, d\mu(y) &\leq 2 \fint_B |u_n - (u_n)_B|\, d\mu \\ &\leq 2C \operatorname{diam}(B) \left(\fint_{\lambda B} \rho^p \, d\mu \right)^{1/p}.\end{aligned}$$

Note that the sequence of functions $\varphi_n : B \times B \to \mathbb{R}$ given by $\varphi_n(x, y) = |u_n(x) - u_n(y)|$ is monotone increasing. Hence we can invoke the monotone convergence theorem to conclude that

$$\fint_B \fint_B |u(x) - u(y)|\, d\mu(x)\, d\mu(y) \leq 2C\, \mathrm{diam}(B) \left(\fint_{\lambda B} \rho^p\, d\mu \right)^{1/p} < \infty.$$

Hence it follows that u is integrable on the ball B. □

As a simple topological consequence of the Poincaré inequality, we establish the following.

Proposition 8.1.6 *Every space that supports a Poincaré inequality is connected.*

Proof Suppose that X supports a Poincaré inequality. Assume that X has two disjoint nonempty open subsets U, V such that $X = U \cup V$. Then the function $u = \chi_U$ has the zero function $\rho \equiv 0$ as an upper gradient. It is clear that (8.1.1) cannot hold for any ball B in X that meets both U and V. The proposition follows. □

Pointwise inequalities

The Poincaré inequality (8.1.1) can be characterized in terms of pointwise inequalities between functions and their upper gradients, at least when the underlying metric measure space is doubling. Many such characterizations express relationships between two functions and as such have nothing to do with upper gradients. Moreover, the proofs for general Banach space-valued functions are no harder.

Recall that for $R > 0$ the restricted maximal function is given by

$$M_R u(x) = \sup_{0<r<R} \fint_{B(x,r)} |u(y)|\, d\mu(y),$$

whenever $u : X \to V$ is a locally integrable function. Recall also the definition of a doubling metric measure space from Section 3.4.

Theorem 8.1.7 *Suppose that X is a doubling metric measure space and that $1 \leq p < \infty$. Let $u : X \to V$ be integrable on balls and let $g : X \to [0, \infty]$ be measurable. Then the following three conditions are equivalent.*

(i) *There exist constants $C > 0$ and $\lambda \geq 1$ such that*

$$\fint_B |u - u_B|\, d\mu \leq C\, \mathrm{diam}(B) \left(\fint_{\lambda B} g^p\, d\mu \right)^{1/p} \qquad (8.1.8)$$

for every open ball B in X.

(ii) *There exist constants* $C > 0$ *and* $\lambda \geq 1$ *such that*

$$|u(x) - u_B| \leq C \operatorname{diam}(B) \left(M_{\lambda \operatorname{diam}(B)}\, g^p(x)\right)^{1/p} \tag{8.1.9}$$

for every open ball B *in* X *and for almost every* $x \in B$.

(iii) *There exist constants* $C > 0$ *and* $\lambda \geq 1$ *and* $A \subset X$ *with* $\mu(A) = 0$ *such that*

$$|u(x) - u(y)| \leq Cd(x, y) \left(M_{\lambda d(x,y)}\, g^p(x) + M_{\lambda d(x,y)}\, g^p(y)\right)^{1/p} \tag{8.1.10}$$

for every $x, y \in X \setminus A$.

The constants C *and* λ *are not necessarily the same for* (8.1.8), (8.1.9) *and* (8.1.10) *but they depend only on each other and on the doubling constant of* μ.

While we have tried to be careful about notation in this book, the reader should be aware that in some existing literature a more relaxed approach is taken; for instance it may be asserted that (8.1.10) is required to hold for almost every $x, y \in X$. We would like to point out here that this statement regarding almost every $x, y \in X$ is to be taken in the sense we have specified above, where the exceptional set A is a subset of X with $\mu(A) = 0$ rather than a set K in $X \times X$ with $(\mu \times \mu)(K) = 0$.

Remark 8.1.11 The proof of Theorem 8.1.7 will show that if (i) holds then (8.1.9) and (8.1.10) hold for all Lebesgue points x and y of u.

Proof of Theorem 8.1.7 First we prove the implication (i) ⇒ (ii). Let $B \subset X$ be an open ball, and let $x \in B$ be a Lebesgue point of u. By Theorem 3.4.3 and the Lebesgue differentiation theorem, almost every point is such a point. Write $r = \operatorname{diam}(B)$ and $B_i = B(x, 2^{-i} r)$ for each nonnegative integer i; clearly we may assume that $r > 0$. Then $\lim_{i\to\infty} u_{B_i} = u(x)$. Using the doubling property of μ together with (8.1.8), we obtain

$$\begin{aligned}
|u(x) - u_{B_0}| &\leq \sum_{i=0}^{\infty} |u_{B_i} - u_{B_{i+1}}| \leq \sum_{i=0}^{\infty} \fint_{B_{i+1}} |u - u_{B_i}|\, d\mu \\
&\leq C \sum_{i=0}^{\infty} \fint_{B_i} |u - u_{B_i}|\, d\mu \\
&\leq C \sum_{i=0}^{\infty} 2^{-i}\, r \left(\fint_{\lambda B_i} g^p\, d\mu\right)^{1/p} = Cr\, (M_{\lambda r}\, g^p(x))^{1/p}.
\end{aligned}$$

Here C depends only on the doubling constant of μ and on the constants in inequality (8.1.8). It follows that

$$\begin{aligned}|u(x) - u_B| &\leq |u(x) - u_{B_0}| + |u_{B_0} - u_B| \\ &\leq Cr\,(M_{\lambda r}\, g^p(x))^{1/p} + |u_{B_0} - u_B|\,.\end{aligned}$$

The last term in the preceding line can be estimated as earlier, by using the doubling property of μ and noting that $B \subset 2B_0$:

$$\begin{aligned}|u_B - u_{B_0}| &\leq |u_B - u_{2B_0}| + |u_{2B_0} - u_{B_0}| \\ &\leq ⨍_B |u - u_{2B_0}|\, d\mu + ⨍_{B_0} |u - u_{2B_0}|\, d\mu \\ &\leq 2C ⨍_{2B_0} |u - u_{2B_0}|\, d\mu \\ &\leq Cr\left(⨍_{2\lambda B_0} g^p\, d\mu\right)^{1/p} \leq Cr\,(M_{2\lambda r}\, g^p(x))^{1/p}.\end{aligned}$$

This concludes the proof of (8.1.9).

To verify the implication (ii) $\Rightarrow$ (iii), consider a countable collection of open balls $B_1, B_2, \ldots$ whose centers run through a fixed countable dense set in X and whose radii run through the positive rationals. Let $A_i \subset B_i$ be the set of points such that $\mu(B_i \setminus A_i) = 0$ and (8.1.9) holds for every $x \in A_i$ (with $B = B_i$). Next, set $C_i = B_i \setminus A_i$ and set $A = X \setminus \bigcup_i C_i$. We have that $\mu(X \setminus A) = 0$. Moreover, if $x, y \in A$ then it follows from the definitions that $x, y \in A_i$ for some B_i such that $\operatorname{diam}(B_i) \leq 3d(x, y)$. Thus

$$|u(x) - u(y)| \leq |u(x) - u_{B_i}| + |u(y) - u_{B_i}|\,,$$

and the claim follows from this on applying (8.1.9).

It remains to prove the implication (iii) $\Rightarrow$ (i). Assume first that $p > 1$. Fix a ball B in X. Without loss of generality we may assume that $g^p \in L^1(3\lambda B)$, where $\lambda \geq 1$ is as in (iii). By applying Cavalieri's principle (3.5.5) and the weak-type estimate (3.5.7) for the maximal function, we obtain

$$\begin{aligned}⨍_B |u - u_B|\, d\mu &\leq ⨍_B ⨍_B |u(x) - u(y)|\, d\mu(x)\, d\mu(y) \\ &\leq C \operatorname{diam}(B) ⨍_B \left(M(g^p \chi_{3\lambda B})(x)\right)^{1/p}\, d\mu(x) \\ &= C \operatorname{diam}(B) \mu(B)^{-1} \int_0^\infty \mu\left(\{x \in B : M(g^p \chi_{3\lambda B})(x) > t^p\}\right) dt\end{aligned}$$

$$\leq C \operatorname{diam}(B)\mu(B)^{-1}\left(\int_0^{t_0} \mu(B)\,dt + \int_{t_0}^{\infty}\left(\frac{C}{t^p}\int_{3\lambda B} g^p\,d\mu\right)dt\right)$$
$$= C \operatorname{diam}(B)\mu(B)^{-1}\left(t_0\mu(B) + Ct_0^{1-p}\int_{3\lambda B} g^p\,d\mu\right).$$

The claim follows upon choosing

$$t_0 = \left(\mu(B)^{-1}\int_{3\lambda B} g^p\,d\mu\right)^{1/p}.$$

It remains to consider the case $p = 1$ in the implication (iii) $\Rightarrow$ (i). This case is more involved, and will be derived from a more general result, Theorem 8.1.18. (See, however, Theorem 8.1.29 and the comment preceding it.)

For later purposes also, we introduce the following concept. A metric measure space $X = (X, d, \mu)$ is said to satisfy a *relative lower volume decay of order* $Q \geq 0$ if there is a constant $C_0 \geq 1$ such that

$$\left(\frac{s}{r}\right)^Q \leq C_0 \frac{\mu(B(x,s))}{\mu(B(a,r))} \tag{8.1.12}$$

whenever $a \in X$, $x \in B(a,r)$, and $0 < s \leq r$. Note that $Q = 0$ only if $\mu(\{x\}) > 0$ for every $x \in X$.

Lemma 8.1.13 *Every doubling metric measure space satisfies a relative lower volume decay of order* $\log_2 C_\mu$. *More precisely, we have that*

$$\left(\frac{s}{r}\right)^{\log_2 C_\mu} \leq 4^{\log_2 C_\mu} \frac{\mu(B(x,s))}{\mu(B(a,r))} \tag{8.1.14}$$

whenever $a \in X$, $x \in B(a,r)$, *and* $0 < s \leq r$.

Here, as usual, C_μ denotes the doubling constant of μ from (3.4.8).

Proof Let a, x, s, and r be as in the statement of the lemma, with $2^{-k-1}r < s \leq 2^{-k}r$ for some nonnegative integer k. Then $B(a,r) \subset B(x, 2^{k+2}s)$, which gives

$$\mu(B(a,r)) \leq C_\mu^{k+2}\mu(B(x,s)) \leq C_\mu^2\left(\frac{r}{s}\right)^{\log_2 C_\mu}\mu(B(x,s))$$

as required. □

Remark 8.1.15 A doubling metric measure space may satisfy a relative lower volume decay of order $Q < \log_2 C_\mu$. For example, if X consists of two points, both of positive measure, then (8.1.12) holds with $Q = 0$ but $C_\mu \geq 2$.

The above lemma gives us a useful *reverse doubling* condition for μ when X is connected. When $x \in X$ and $0 < r < \operatorname{diam}(X)/2$, there is a point

$z \in X \setminus B(x, 2r)$, and so $\partial B(x, 3r/2)$ is nonempty. Hence we can find a point $w \in B(x, 2r)$ such that $d(x, w) = 3r/2$ and $B(w, r/2) \subset B(x, 2r) \setminus B(x, r)$. Therefore, by the above relative lower decay property,

$$1 - \frac{\mu(B(x, r))}{\mu(B(x, 2r))} = \frac{\mu(B(x, 2r) \setminus B(x, r))}{\mu(B(x, 2r))} \geq \frac{1}{C}. \tag{8.1.16}$$

Thus with $0 < c = 1 - (C)^{-1} < 1$, we see that

$$\mu(B(x, r)) \leq c\mu(B(x, 2r)). \tag{8.1.17}$$

Theorem 8.1.18 *Suppose that X is a doubling metric measure space satisfying a relative lower volume decay of order $Q > 0$. Let $u : X \to V$ be integrable on balls and let $h : X \to [0, \infty]$ be measurable. Let B be an open ball in X and assume that there is a set $A \subset X$ with $\mu(A) = 0$ and*

$$|u(x) - u(y)| \leq d(x, y)\,(h(x) + h(y)) \tag{8.1.19}$$

for every $x, y \in 2B \setminus A$. Then there exists a constant $C > 0$ depending only on the constants in (8.1.12) *such that*

$$⨍_B |u - u_B|\, d\mu \leq C\, \mathrm{diam}(B) \left(⨍_{2B} h^q\, d\mu \right)^{1/q}, \tag{8.1.20}$$

where $q = Q/(Q+1)$.

Now let us assume Theorem 8.1.18 and show how the proof of Theorem 8.1.7 can be completed with it. Thus, suppose that (iii) holds for $p = 1$. We use the fact (Lemma 8.1.13) that X satisfies a relative lower volume decay estimate with constants C_0 and Q depending only on the doubling constant of μ. Fix an open ball B in X and let q be as in (8.1.20). We may assume that $g \in L^1(6\lambda B)$. Then Lemma 3.5.10 gives that

$$\left(⨍_{2\lambda B} (M_{4\lambda\, \mathrm{rad}(B)} g)^q\, d\mu \right)^{1/q} \leq C ⨍_{6\lambda B} g\, d\mu, \tag{8.1.21}$$

where $C > 0$ depends only on the doubling constant of μ. Since the statement (8.1.10) implies that (8.1.19) holds for the function $h = C\, M_{4\lambda\, \mathrm{rad}(B)} g$, we obtain from (8.1.20) and (8.1.21) that

$$⨍_B |u - u_B|\, d\mu \leq C\, \mathrm{diam}(B) ⨍_{6\lambda B} g\, d\mu$$

for some constant $C > 0$ that depends only on the data in (8.1.10) and the doubling constant of μ; the constant $\lambda \geq 1$ is as in (8.1.10). This gives (8.1.8) as desired, and hence the remaining implication (iii) $\Rightarrow$ (i) is proved in all cases.

Therefore, assuming Theorem 8.1.18, the proof of Theorem 8.1.7 is complete. □

Proof of Theorem 8.1.18 Let Q, u, h, and B be as in the hypotheses, and put $q = Q/(Q+1)$. We first make several simplifying reductions. If $\text{rad}(B) > \text{diam}(B)$ then, writing x for the assigned center of B, we would have $B = B(x, \text{rad}(B)) = B(x, \text{diam}(B))$; thus we may assume that $\text{rad}(B) \leq \text{diam}(B)$. Furthermore, by simultaneously replacing the metric d by $(\text{rad}(B))^{-1} d$ and the function h by $\text{rad}(B)\, h$, we see that we can assume that $\text{rad}(B) = 1$. Similarly, by replacing the measure μ by $(\mu(2B))^{-1}\mu$, we see that we can assume that $\mu(2B) = 1$.

Next, we may assume that

$$0 < \left(\int_{2B} h^q \, d\mu\right)^{1/q} \leq 2^{1/q}\, h(x) \tag{8.1.22}$$

for every $x \in 2B$. Indeed, if the first inequality fails then $u = u_B$ almost everywhere in B, and if the second inequality fails then we may replace h by the function

$$h(x) + \left(\int_{2B} h^q \, d\mu\right)^{1/q}.$$

Finally, we may assume that $h < \infty$ almost everywhere in $2B$.

The preceding understood, we now consider the sets

$$E_k := \{x \in G : h(x) \leq 2^k\}$$

for every integer k, where $G \subset 2B$ is a set such that $\mu(2B \setminus G) = 0$ and (8.1.19) holds whenever $x, y \in G$. Then $E_k \subset E_{k+1}$, $\mu(E_k) \to \mu(2B)$ as $k \to \infty$, $\mu(E_k) \to 0$ as $k \to -\infty$, and $u|_{E_k}$ is 2^{k+1}-Lipschitz. Let k_0 be the integer for which

$$\mu(E_{k_0 - 1}) < \frac{\mu(2B)}{2} = \frac{1}{2} \leq \mu(E_{k_0}). \tag{8.1.23}$$

By multiplying both u and h by 2^{-k_0}, we may assume that $k_0 = 0$. It follows from (8.1.23) and the fact that $\mu(2B) = 1$ that

$$\frac{1}{2} \geq \mu(E_{-1}) = 1 - \mu(\{x \in G : h(x) > 2^{-1}\}),$$

and so, by Chebyshev's inequality

$$\mu\left(\{x \in G : h(x) > 2^{-1}\}\right) = \mu(\{x \in G : h(x)^q > 2^{-q}\}) \leq 2^q \int_{2B} h^q \, d\mu,$$

we obtain

$$2^{-(1+1/q)} \leq \left(\int_{2B} h^q \, d\mu \right)^{1/q} .$$

Furthermore, because $\mu(E_0) \geq 1/2 > 0$, there is a point $x \in G$ with $h(x) \leq 1$; hence, by (8.1.22), we have

$$\left(\int_{2B} h^q \, d\mu \right)^{1/q} \leq 2^{1/q}.$$

Combining these two inequalities, we conclude that

$$2^{-(1+1/q)} \leq \left(\int_{2B} h^q \, d\mu \right)^{1/q} \leq 2^{1/q} . \tag{8.1.24}$$

We will fix an integer $k_1 > 0 = k_0$, to be determined later, that depends only on the doubling constant of μ.

We have $\mu(E_{k_1}) > 0$ by (8.1.23), and hence there is a vector $v_0 \in V$ such that

$$\operatorname{ess\,inf}_{x \in E_{k_1}} |u(x) - v_0| = 0 .$$

(For example, we may choose $v_0 = u(z)$ for any Lebesgue point $z \in E_{k_1}$ of u by the Lebesgue differentiation theorem; see Section 3.4.) We may replace u by $u - v_0$ and assume to begin with that

$$\operatorname{ess\,inf}_{x \in E_{k_1}} |u(x)| = 0 . \tag{8.1.25}$$

Since

$$⨍_B |u - u_B| \, d\mu \leq 2 ⨍_B |u - v_0| \, d\mu ,$$

the preceding reductions being understood, it suffices to prove that

$$\int_B |u| \, d\mu \leq C \tag{8.1.26}$$

for some constant $C > 0$ that depends only on the doubling constant of μ. (Since we have the doubling property of μ and since $\mu(2B) = 1$, we know that $C_\mu^{-1} \leq \mu(B) \leq 1$.)

If $\mu(B \setminus E_{k_1+1}) = 0$ then (8.1.26) follows from (8.1.25) by the fact that $u|_{E_{k_1+1}}$ is 2^{k_1+2}-Lipschitz, for we also have $\operatorname{rad}(B) = 1$.

We may assume, therefore, that $\mu(B \setminus E_{k_1+1}) > 0$. For each $k \geq k_1 + 1$, set

$$r_k := 16 \, C_0^{1/Q} \, \mu(2B \setminus E_{k-1})^{1/Q} . \tag{8.1.27}$$

Here C_0 is the constant from (8.1.12). By combining Chebyshev's inequality and (8.1.24), we find that

$$r_{k+1} \le 16 C_0^{1/Q} 2^{-qk/Q} \left(\int_{2B} h^q \, d\mu \right)^{1/Q} \le 16 C_0^{1/Q} 2^{-qk/Q} \, 2^{1/Q}$$

for every $k \geq k_1$. An appropriate choice for k_1 then gives the estimate

$$r_{k+1} \le b \, 2^{-(k+1-k_1)q/Q}, \qquad b = \frac{1}{2}(2^{q/Q} - 1). \tag{8.1.28}$$

Such a choice of k_1 depends solely on C_0, q, and Q, not on u, h, or B.

Suppose now that $k \geq k_1 + 1$ is such that $\mu((E_{k+1} \setminus E_k) \cap B)) > 0$ (if such a k does not exist then $\mu(B \setminus E_{k_1+1}) = 0$, contradicting our assumption). Then in particular $r_{k+1} > 0$. Pick a point $x_{k+1} \in (E_{k+1} \setminus E_k) \cap B$. From (8.1.28) we find that the ball $B(x_{k+1}, r_{k+1})$ lies entirely in $2B$. Hence the inequalities (8.1.12) and (8.1.27) give

$$\mu(B(x_{k+1}, r_{k+1})) \ge 2^Q \mu(2B \setminus E_k),$$

which implies that $\mu(B(x_{k+1}, r_{k+1}) \cap E_k) > 0$. Therefore, observing (8.1.28), we can pick a point $x_k \in B(x_{k+1}, r_{k+1}) \cap E_k$ such that

$$d(x_{k+1}, x_k) \le r_{k+1} \le b \, 2^{-(k+1-k_1)q/Q}.$$

We claim that we can continue in a similar fashion to find points

$$\begin{aligned} x_{k+1} &\in (E_{k+1} \setminus E_k) \cap B, \\ x_k &\in B(x_{k+1}, r_{k+1}) \cap E_k, \\ &\vdots \\ x_{k+1-i} &\in B(x_{k+1-(i-1)}, r_{k+1-(i-1)}) \cap E_{k+1-i}, \\ &\vdots \\ x_{k_1} &\in B(x_{k_1+1}, r_{k_1+1}) \cap E_{k_1}, \end{aligned}$$

for every $i = 1, \ldots, k+1-k_1$. To prove the claim, suppose that x_{k+1-i} has been chosen for some $1 \le i \le k - k_1$. If $a \in X$ is such that $B = B(a, 1)$ then, by (8.1.28),

$$\begin{aligned} d(a, x_{k+1-i}) &\le d(a, x_{k+1}) + \cdots + d(x_{k+1-(i-1)}, x_{k+1-i}) \\ &\le 1 + r_{k+1} + \cdots + r_{k+1-(i-1)} \\ &\le 1 + b \sum_{m=0}^{i-1} 2^{-(k+1-k_1-m)q/Q} \le 1 + \frac{1}{2}. \end{aligned}$$

Noting that $r_{k+1-i} \leq b < 1/2$, this implies that $B(x_{k+1-i}, r_{k+1-i}) \subset 2B$. Using Lemma 8.1.13 again, and the definition for r_k, we thus obtain

$$\mu(B(x_{k+1-i}, r_{k+1-i})) \geq 2^Q \mu(2B \setminus E_{k-i}).$$

Hence a point x_{k-i} can be chosen as required. This finishes the proof of the claim.

Because u is 2^{k_1+1}-Lipschitz continuous on E_{k_1}, it follows from (8.1.25) that $|u(x_{k_1})| \leq 2^{k_1+1} \operatorname{diam}(E_{k_1}) \leq 4 \times 2^{k_1+1}$. By using the sequence $x_{k+1}, x_k, \ldots, x_{k_1}$ as in the preceding claim, we now obtain the estimate

$$\begin{aligned}
|u(x_{k+1})| &\leq \sum_{i=0}^{k-k_1} |u(x_{k+1-i}) - u(x_{k-i})| + |u(x_{k_1})| \\
&\leq 2 \sum_{i=0}^{k-k_1} 2^{k+1-i} d(x_{k+1-i}, x_{k-i}) + 4 \times 2^{k_1+1} \\
&\leq 2 \sum_{i=0}^{k-k_1} 2^{k+1-i} r_{k+1-i} + 4 \times 2^{k_1+1} \\
&\leq 2b \sum_{i=0}^{k-k_1} 2^{k+1-i} 2^{-(k+1-k_1-i)q/Q} + 4 \times 2^{k_1+1} \\
&\leq C\, 2^{(k+1-k_1)(1-q/Q)},
\end{aligned}$$

where we have used (8.1.28), (8.1.25), and the fact that $u|E_k$ is 2^{k+1}-Lipschitz. Because this estimate holds for every point $x_{k+1} \in (E_{k+2} \setminus E_{k+1}) \cap B$, if the latter set has positive measure then we can deduce that

$$\begin{aligned}
\int_B |u|\, d\mu &\leq \sum_{j=1}^{\infty} \int_{(E_{k_1+j+1} \setminus E_{k_1+j}) \cap B} |u|\, d\mu + \int_{E_{k_1+1}} |u|\, d\mu \\
&\leq C \sum_{j=1}^{\infty} 2^{j(1-q/Q)} \mu(E_{k_1+j+1} \setminus E_{k_1+j}) + C \\
&\leq C \sum_{j=1}^{\infty} 2^{j(1-q/Q)} 2^{-jq} \int_{E_{k_1+j+1} \setminus E_{k_1+j}} |h|^q\, d\mu + C \\
&= C \sum_{j=1}^{\infty} \int_{E_{k_1+j+1} \setminus E_{k_1+j}} |h|^q\, d\mu + C \\
&\leq C \int_{2B} |h|^q\, d\mu + C \leq C,
\end{aligned}$$

where we have also used Chebyshev's inequality, the choice $q = Q/(Q+1)$, and (8.1.24).

This completes the proof of (8.1.26) and hence that of Theorem 8.1.18. □

If we restrict ourselves to real-valued functions and their upper gradients then there is a more direct alternative proof for the implication (iii) ⇒ (i) in Theorem 8.1.7, for $p = 1$. We will present this proof also, as it involves important techniques. For completeness, we formulate the statement as a separate theorem although it is a special case of Theorem 8.1.7.

Theorem 8.1.29 *Suppose that X is a doubling metric measure space. Assume that there exist constants $C > 0$ and $\lambda \geq 1$ such that*

$$|u(x) - u(y)| \leq Cd(x, y)\big(M_{\lambda d(x,y)}\, \rho(x) + M_{\lambda d(x,y)}\, \rho(y)\big) \qquad (8.1.30)$$

for every $x, y \in X \setminus A$, whenever $u : X \to \mathbb{R}$ is integrable on balls and $\rho : X \to [0, \infty]$ is an upper gradient of u; $A \subset X$ with $\mu(A) = 0$ depends on u (and a priori could depend on ρ as well). Then X supports a 1-Poincaré inequality with data depending only on C, λ, and the doubling constant of μ.

The following lemma is crucial; it expresses the interesting fact that a weak-type estimate implies a strong-type estimate when function–upper-gradient pairs are considered (see Theorem 3.5.6).

Lemma 8.1.31 *Let X be a metric measure space, let $1 \leq p \leq q < \infty$, and let $B \subset X$ be a ball such that $0 < \mu(B) < \infty$. Assume that there exist constants $C_1 > 0$ and $\lambda \geq 1$ such that*

$$\mu(\{x \in B : |u(x) - u_B| > s\}) \leq C_1 s^{-q} \left(\int_{\lambda B} \rho^p \, d\mu \right)^{q/p} \qquad (8.1.32)$$

for each $s > 0$, for each measurable function $u : \lambda B \to \mathbb{R}$ that is integrable in B, and for each upper gradient ρ of u in λB. Then there exists a constant $C_2 > 0$ such that

$$\left(\int_B |u - u_B|^q \, d\mu \right)^{1/q} \leq C_2 \left(\int_{\lambda B} \rho^p \, d\mu \right)^{1/p} \qquad (8.1.33)$$

for each measurable function $u : \lambda B \to \mathbb{R}$ that is integrable in B and for each upper gradient ρ of u in λB; we recall that u_B is the mean value of u on B (see (1.2)). We can choose $C_2 = 32 C_1^{1/q}$.

Remark 8.1.34 An examination of the proof of the above lemma yields a stronger result. We can actually obtain that (8.1.33) holds under the following weaker assumption: whenever $u : \lambda B \to [0,\infty]$ is integrable, has an upper gradient ρ in λB, and satisfies

$$\mu(\{x \in B : u(x) = 0\}) \geq \tfrac{1}{2}\mu(B),$$

then (8.1.32) holds.

Proof of Lemma 8.1.31 Let $u : \lambda B \to \mathbb{R}$ be a measurable function that is integrable in B. We may obviously assume that u possesses a p-integrable upper gradient in λB. It then follows from Theorem 6.3.20 that u has a minimal p-weak upper gradient ρ_u in the ball λB. If u is constant almost everywhere on B then there is nothing to be proved. Therefore, assume that u is nonconstant on B and pick a real number t such that

$$\mu(\{x \in B : u(x) \geq t\}) \geq \tfrac{1}{2}\mu(B)$$

and

$$\mu(\{x \in B : u(x) \leq t\}) \geq \tfrac{1}{2}\mu(B).$$

Because

$$\left(\fint_B |u - u_B|^q \, d\mu\right)^{1/q} \leq 2\left(\fint_B |u - t|^q \, d\mu\right)^{1/q}, \tag{8.1.35}$$

it suffices to verify inequality (8.1.33) with u_B replaced by t.

By considering the function $u - t$, we may assume that $t = 0$. We will estimate the integral of only the positive part u^+ of u, as the reasoning for the negative part u^- is identical. Note that, by Proposition 6.3.23, the minimal p-weak upper gradient ρ_{u^+} agrees almost everywhere with the function $\rho_u \chi_{\{u>0\}}$. To simplify our notation, we write u for u^+ and assume thus that $u \geq 0$ and that $\mu(\{x \in B : u(x) = 0\}) \geq \frac{1}{2}\mu(B)$.

For each integer j we define a function v_j by setting

$$v_j(x) := \max\{0, \min\{u(x) - 2^j, 2^j\}\}.$$

Notice that

$$\mu(\{x \in B : v_j(x) = 0\}) \geq \tfrac{1}{2}\mu(B). \tag{8.1.36}$$

It follows from Proposition 6.3.23 that the minimal p-weak upper gradient ρ_{v_j} satisfies

$$\rho_{v_j} = \rho_u \chi_{L_j} \tag{8.1.37}$$

almost everywhere in λB, where $L_j := \{x \in \lambda B : 2^j \leq u(x) < 2^{j+1}\}$. Recalling that $v_j \leq 2^j$ we have by (8.1.36) that $(v_j)_B \leq 2^{j-1}$, which implies that

$$\mu(\{x \in B : u(x) \geq 2^{j+1}\}) = \mu(\{x \in B : \; v_j(x) \geq 2^j\})$$
$$\leq \mu(\{x \in B : \; |v_j(x) - (v_j)_B| \geq 2^{j-1}\}).$$

Consequently, we deduce from our weak-type assumption (8.1.32) that

$$\mu(L_{j+1} \cap B) \leq \mu(\{x \in B : \; |v_j(x) - (v_j)_B| \geq 2^{j-1}\})$$
$$\leq C_1 2^q \, 2^{-qj} \left(\int_{\lambda B} \rho^p \, d\mu \right)^{q/p}$$

for every upper gradient ρ of v_j in λB. By Lemma 6.2.2 and (8.1.37), we can conclude therefore that

$$\mu(L_{j+1} \cap B) \leq C_1 2^q \, 2^{-qj} \left(\int_{L_j} \rho_u^p \, d\mu \right)^{q/p}.$$

Finally, this gives

$$\int_B u^q \, d\mu \leq \sum_{j \in \mathbb{Z}} 2^{q(j+2)} \mu(L_{j+1} \cap B) \leq 2^{3q} C_1 \sum_{j \in \mathbb{Z}} \left(\int_{L_j} \rho_u^p \, d\mu \right)^{q/p}$$
$$\leq 2^{3q} C_1 \left(\int_{\lambda B} \rho_u^p \, d\mu \right)^{q/p},$$

since $q \geq p$. The proof is completed upon observing that ρ_u is almost everywhere less than or equal to any p-integrable upper gradient of u in λB and that for the constant C_2 we pick up two additional factors of 2 from (8.1.35) and from the splitting $u = u^+ - u^-$. The lemma follows. □

Proof of Theorem 8.1.29 Let $u : X \to \mathbb{R}$ be a function that is integrable on balls and let ρ be an upper gradient of u in X. Then fix an open ball B in X.

From Remark 8.1.34 we infer that we need only to verify the weak-type inequality

$$\mu(\{x \in B : u(x) > s\}) \leq C \operatorname{diam}(B) \, s^{-1} \int_{\lambda B} \rho \, d\mu, \tag{8.1.38}$$

for $s > 0$, under the assumption that $u \geq 0$ and $\mu(\{x \in B : \; u(x) = 0\}) \geq \mu(B)/2$. The constants $C > 0$ and $\lambda \geq 1$ in (8.1.38) are allowed to depend only on the constants appearing in (8.1.30) and on the doubling constant of μ. As usual, we let C and λ denote any such constants.

To prove (8.1.38), fix $s > 0$ and let $x \in B \setminus A$ be such that $u(x) > s$. By our assumption (8.1.30), for every such x and for every $y \in B \setminus A$ such that $u(y) = 0$ we have

$$\mu(B(w, r_w)) \leq C d(x, y) s^{-1} \int_{B(w, r_w)} \rho \, d\mu \tag{8.1.39}$$

for some $0 < r_w < \lambda d(x, y)$, where $w = x$ or $w = y$. Assume first that for every $y \in B \setminus A$ with $u(y) = 0$ we can find an $x \in B$ such that inequality (8.1.39) holds for $w = y$. By the $5B$-covering lemma (see Section 3.3), we may cover the set $\{y \in B : u(y) = 0\}$ by countably many balls $B(y_i, 5r_{y_i})$ in such a way that the balls $B(y_i, r_{y_i})$ are pairwise disjoint and satisfy (8.1.39) for $w = y_i$. Then, by the doubling property of μ,

$$\mu(B) \le 2\mu(\{y \in B : u(y) = 0\}) \le 2\sum_i \mu(B(y_i, 5r_{y_i}))$$
$$\le C\sum_i \mu(B(y_i, r_{y_i})) \le C \operatorname{diam}(B)\, s^{-1} \int_{\lambda B} \rho \, d\mu \, .$$

Hence inequality (8.1.38) follows in this case.

If the above assumption fails then there is some $y \in B \setminus A$ with $u(y) = 0$ such that, for each $x \in B \setminus A$ with $u(x) > s$, we know that inequality (8.1.39) holds with $w = x$. By again using the $5B$-covering lemma, we obtain a countable collection of balls $B(x_i, r_{x_i})$ such that

$$\mu(\{x \in B : u(x) > s\}) \le \sum_i \mu(B(x_i, 5r_{x_i})) \le C \sum_i \mu(B(x_i, r_{x_i}))$$
$$\le C \operatorname{diam}(B)\, s^{-1} \int_{\lambda B} \rho \, d\mu.$$

Thus inequality (8.1.38) always holds, and the proof is complete. □

Poincaré inequality for Banach space-valued functions

The Poincaré inequality for Banach space-valued functions can be defined analogously to the definition at the start of Section 8.1.

Definition 8.1.40 We say that X *supports a p-Poincaré inequality for V-valued functions*, or *for functions valued in V*, if every ball in X has positive and finite measure and if there exist constants $C > 0$ and $\lambda \ge 1$ such that

$$⨍_B |u - u_B| \, d\mu \le C \operatorname{diam}(B) \left(⨍_{\lambda B} \rho^p \, d\mu \right)^{1/p} \tag{8.1.41}$$

for every open ball B in X, for every function $u : X \to V$ that is integrable on balls, and for every upper gradient ρ of u in X. As before, the *data* of inequality (8.1.41) consist of the triple (p, C, λ).

If $V = \mathbb{R}$, we will generally retain the abbreviated terminology of Section 8.1. Alternatively, we say that X *supports a Poincaré inequality for real-valued functions*.

It turns out that, for doubling metric measure spaces, the validity of a Poincaré inequality is independent of the target Banach space. The proof of this result relies on the fundamental pointwise estimates of Theorem 8.1.7.

Theorem 8.1.42 *Let $1 \leq p < \infty$. If a doubling metric measure space supports a p-Poincaré inequality for functions valued in some Banach space then it supports a p-Poincaré inequality for functions valued in an arbitrary Banach space. The data in the conclusion depend only on the data in the assumption and on the doubling constant of the underlying measure.*

Proof Suppose that X supports a p-Poincaré inequality for functions valued in a fixed Banach space. Because every Banach space contains a subspace isometric to $\mathbb{R}$, it follows that X supports a p-Poincaré inequality for real-valued functions with the same data. This understood, we will show that if V is a Banach space and $u : X \to V$ is a function that is integrable on balls, and if $\rho : X \to [0, \infty]$ is an upper gradient of u in X, then

$$|u(x) - u(y)| \leq Cd(x, y)\big(M_{\lambda d(x,y)}\rho^p(x) + M_{\lambda d(x,y)}\rho^p(y)\big)^{1/p} \tag{8.1.43}$$

for every $x, y \in X \setminus A$, and for some $C > 0$ and $\lambda \geq 1$ as required, where $A \subset X$ with $\mu(A) = 0$ depends on u. This suffices, by Theorem 8.1.7.

We argue in a similar way to that in the proof of the implication (iii)⇒(i) in Proposition 7.1.20. Thus, we first single out a set $Z \subset X$ of measure zero such that $u(X \setminus Z)$ is separable (via the Pettis measurability theorem; see Section 3.1), and then let (v_i) be a countable set in V whose closure contains the difference set

$$u(X \setminus Z) - u(X \setminus Z) \subset V\,.$$

Next, we pick a countable subset (v_i^*) of V^* such that $\langle v_i^*, v_i\rangle = |v_i|$ and $|v_i^*| \leq 1$ for each i (the Hahn–Banach theorem). Set $u_i := \langle v_i^*, u\rangle$. Then ρ is an upper gradient of every u_i (see Section 6.2). Because u_i is integrable on balls and because X supports a p-Poincaré inequality, it follows from Theorem 8.1.7 that

$$|u_i(x) - u_i(y)| \leq Cd(x, y)\big(M_{\lambda d(x,y)}\rho^p(x) + M_{\lambda d(x,y)}\rho^p(y)\big)^{1/p} \tag{8.1.44}$$

for all x, y outside a set $Z_i \subset X$ of measure zero, for some $C > 0$ and $\lambda \geq 1$ as required.

We now set $A := Z \cup \bigcup_{i=1}^\infty Z_i$ and fix $x, y \in X \setminus A$. Upon choosing a subsequence (v_{i_k}) converging to $u(x) - u(y)$, we find that

$$\big|u_{i_k}(x) - u_{i_k}(y) - |v_{i_k}|\big| = |\langle v_{i_k}^*, u(x) - u(y) - v_{i_k}\rangle| \to 0$$

as $i_k \to \infty$. In particular, $|u_{i_k}(x) - u_{i_k}(y)| \to |u(x) - u(y)|$ as $i_k \to \infty$. This together with (8.1.44) proves (8.1.43) and the theorem follows. $\square$

In contrast with Banach space-valued functions with upper gradients, the situation for metric space-valued Sobolev maps is more delicate. By embedding the target metric space into a Banach space one can consider instead the corresponding Poincaré inequality; however, the mean value f_B of a metric-space-valued map on a ball $B \subset X$ need not be in the embedded image of the target metric space, even if the target space is complete. A more suitable version of the Poincaré inequality in this setting is the following:

$$\fint_B \fint_B d_Y(f(x), f(y))\, d\mu(y)\, d\mu(x) \le C \operatorname{diam}(B) \left(\fint_{\lambda B} \rho^p \, d\mu \right)^{1/p} \tag{8.1.45}$$

whenever $f : \lambda B \to Y$ has ρ as a (p-weak) upper gradient in λB. Note that, because

$$\begin{aligned} \frac{1}{2} \fint_B \fint_B d_Y(f(x), f(y))\, d\mu(y)\, d\mu(x) &\le \fint_B |f - f_B|\, d\mu \\ &\le \fint_B \fint_B d_Y(f(x), f(y))\, d\mu(y)\, d\mu(x), \end{aligned}$$

in the event that the target space itself happens to be a Banach space, the above version of Poincaré inequality is equivalent to that given in (8.1.41). Now the geometry of the target space Y plays a role in determining whether X supports such a Poincaré inequality. If Y happens to be a singleton set then, regardless of the structure of X, the above version of the Poincaré inequality holds. Furthermore, if balls in X have the MEC_p property in the sense of Section 7.5 and the target space Y is a discrete space then a function $f : \lambda B \to Y$, with B a ball in X, is constant μ-almost everywhere in B if it has an upper gradient $\rho \in L^p(\lambda B)$; in this case (8.1.45) holds also. Thus the validity of (8.1.45) is not equivalent to the situation where X supports a Poincaré inequality for all real-valued functions. However, we have the following result.

Proposition 8.1.46 *Suppose that Y is a metric space that contains a nontrivial quasiconvex curve. Then X supports the Poincaré-type inequality (8.1.45) for Y-valued functions if and only if X supports a p-Poincaré inequality for real-valued functions.*

A curve $\gamma : [a, b] \to Y$ in a metric space is said to be *quasiconvex* (with constant $c > 0$) if γ is absolutely continuous and

$$c\, d_Y(\gamma(t), \gamma(s)) \ge \operatorname{length}(\gamma|_{[t,s]}) = |t - s| \tag{8.1.47}$$

whenever $a \le t < s \le b$.

Proof Let γ be a nontrivial quasiconvex curve in Y; to simplify the notation, we assume that $\gamma : [0, 1] \to Y$, with appropriate changes for more general curves left to the reader. Since we can always embed Y into a Banach space isometrically, the comments above together with Theorem 8.1.42 show that if X supports a p-Poincaré inequality for real-valued functions then it supports (8.1.45). This does not require any condition on Y.

To prove the converse, suppose that X supports (8.1.45) for Y-valued functions. Let $B \subset X$ be a ball and $f : \lambda B \to \mathbb{R}$ have an upper gradient $\rho \in L^p(\lambda B)$. For a positive integer n we consider the truncated function $f_n : \lambda B \to [-n, n]$ given by $f_n = \max\{-n, \min\{f, n\}\}$. Then ρ is still an upper gradient of f_n. Let $F_n : \lambda B \to Y$ be given by

$$F_n(x) = \gamma\left(\frac{1}{2n} f_n(x) + \frac{1}{2}\right).$$

From (6.3.19) and (6.3.18), it follows that $\rho_n = (C/2n)\rho$ is a p-weak upper gradient of F_n on λB. Now an application of (8.1.45) yields

$$\fint_B \fint_B d_Y(F_n(x), F_n(y))\, d\mu(x)\, d\mu(y) \le \frac{C\ \mathrm{diam}(B)}{2n} \left(\fint_{\lambda B} \rho^p\, d\mu\right)^{1/p}. \tag{8.1.48}$$

However, since γ is quasiconvex we know that

$$d_Y(F_n(x), F_n(y)) \ge C^{-1} \frac{1}{2n} |f_n(x) - f_n(y)|,$$

which, together with (8.1.48) gives

$$\fint_B \fint_B |f_n(x) - f_n(y)|\, d\mu(x)\, d\mu(y) \le C\ \mathrm{diam}(B) \left(\fint_{\lambda B} \rho^p\, d\mu\right)^{1/p}.$$

By the monotone convergence theorem we conclude that

$$\fint_B \fint_B |f(x) - f(y)|\, d\mu(x)\, d\mu(y) \le C\ \mathrm{diam}(B) \left(\fint_{\lambda B} \rho^p\, d\mu\right)^{1/p},$$

which, by the comments above, implies the p-Poincaré inequality (8.1.1). □

The following theorem summarizes various characterizations of the Poincaré inequality. It is a direct consequence of Theorems 8.1.7 and 8.1.42, Remark 8.1.11, Proposition 8.1.46 and the definitions.

Theorem 8.1.49 *Suppose that X is a doubling metric measure space, that V is a Banach space, and that $1 \le p < \infty$. Then the following four conditions are equivalent.*

(i) *X supports a p-Poincaré inequality for real-valued functions.*
(ii) *X supports a p-Poincaré inequality for V-valued functions.*

(iii) *There exist constants $C > 0$ and $\lambda \geq 1$ such that*

$$|u(x) - u_B| \leq C \operatorname{diam}(B) \left(M_{\lambda \operatorname{diam}(B)} \rho^p(x)\right)^{1/p} \tag{8.1.50}$$

for every open ball B in X and for almost every $x \in B$ whenever $u : X \to V$ is integrable on balls and ρ is an upper gradient of u in X.

(iv) *There exist constants $C > 0$ and $\lambda \geq 1$ such that*

$$|u(x) - u(y)| \leq Cd(x, y)\left(M_{\lambda d(x,y)} \rho^p(x) + M_{\lambda d(x,y)} \rho^p(y)\right)^{1/p} \tag{8.1.51}$$

for every $x, y \in X \setminus A$ whenever $u : X \to V$ is integrable on balls and ρ is an upper gradient of u in X; here $A \subset X$, such that $\mu(A) = 0$, is a set whose choice depends on u.

(v) *If Y is a metric space that contains a nontrivial quasiconvex curve then there exist constants $C > 0$ and $\lambda \geq 1$ such that whenever $u : X \to Y$ is integrable on balls and ρ is an upper gradient of u in X we have*

$$⨍_B ⨍_B d_Y(f(x), f(y))\, d\mu(y)\, d\mu(x) \leq C \operatorname{diam}(B) \left(⨍_{\lambda B} \rho^p \, d\mu\right)^{1/p}. \tag{8.1.52}$$

Moreover, (8.1.50) and (8.1.51) hold for all Lebesgue points x and y of u.

The various constants are not necessarily the same in (8.1.50), (8.1.51), *and* (8.1.52) *but they depend only on each other and on the doubling constant of the underlying measure; the constants associated with* (8.1.52) *also depend on the constant c in* (8.1.47).

We close this section by showing that there is some flexibility in Definition 8.1.40 regarding the Poincaré inequality: the functions need not be globally defined.

Theorem 8.1.53 *Suppose that X supports a p-Poincaré inequality for V-valued functions for some $1 \leq p < \infty$. Then there exist constants $C \geq 1$ and $\lambda \geq 1$, depending only on the data associated with the Poincaré inequality* (8.1.41), *such that, for every open ball $B \subset X$,*

$$⨍_B |u - u_B| \, d\mu \leq C \operatorname{diam}(B) \left(⨍_{\lambda B} \rho^p \, d\mu\right)^{1/p} \tag{8.1.54}$$

whenever $u : \lambda B \to V$ is integrable in B and ρ is an upper gradient of u in λB.

Similarly, we have local versions of inequalities (8.1.50) and (8.1.51).

Theorem 8.1.55 *Suppose that X is a doubling metric measure space that supports a p-Poincaré inequality for V-valued functions for some $1 \le p < \infty$. Then there exist constants $C \ge 1$ and $\lambda \ge 1$, depending only on the data associated with the Poincaré inequality* (8.1.41) *and on the doubling constant of the underlying measure, such that, for every open ball $B \subset X$,*

$$|u(x) - u_B| \le C \operatorname{diam}(B)\big(M_{\lambda \operatorname{diam}(B)}\, \rho^p(x)\big)^{1/p} \tag{8.1.56}$$

and

$$|u(x) - u(y)| \le Cd(x, y)\big(M_{\lambda d(x,y)}\, \rho^p(x) + M_{\lambda d(x,y)}\, \rho^p(y)\big)^{1/p} \tag{8.1.57}$$

whenever $u : \lambda B \to V$ is integrable in B, ρ is an upper gradient of u in λB, and $x, y \in B$ are Lebesgue points of u. In particular, given such a u, inequalities (8.1.56) *and* (8.1.57) *hold for every $x, y \in B \setminus A$, where A is the set of all points in X that are not Lebesgue points of u; note that $\mu(A) = 0$.*

For the proofs of these two theorems, we require the following simple lemma.

Lemma 8.1.58 *Let $u : X \to Z$ be a map between two metric spaces, let $A \subset X$ be closed, and let $\rho : A \to [0, \infty]$ be an upper gradient of the restriction $u|_A : A \to Z$. Then the function $\bar\rho : X \to [0, \infty]$ defined by $\bar\rho(x) = \rho(x)$ if $x \in A$ and $\bar\rho(x) = \infty$ if $x \in X \setminus A$ is an upper gradient of u in X.*

Proof The assertion follows from the fact that for every nonconstant rectifiable curve γ in X that intersects the open set $X \setminus A$ we must have $\int_\gamma \bar\rho\, ds = \infty$. □

Remark 8.1.59 Lemma 8.1.58 is not true for an arbitrary subset $A \subset X$. Consider $A = \mathbb{R} \setminus \{0\} \subset \mathbb{R}$ and the function $u = \chi_{(0,\infty)} : \mathbb{R} \to \mathbb{R}$ together with an upper gradient $\rho \equiv 0$ on A.

Proof of Theorem 8.1.53 Let $B = B(x, r)$ and write $B_j = \overline{B}(x, (1-2^{-j})r)$, $u_j = u|_{B_j}$, for $j = 1, 2, \ldots$ Then

$$|u_j - (u_j)_{B_j}| \le |u| + \frac{1}{\mu(B_1)} \int_B |u|\, d\mu\,,$$

which gives, by the dominated convergence theorem,

$$\lim_{j\to\infty} \fint_{B_j} |u_j - (u_j)_{B_j}|\, d\mu = \fint_B |u - u_B|\, d\mu\,. \tag{8.1.60}$$

Similarly, by monotone convergence,

$$\lim_{j\to\infty} \fint_{\lambda B_j} \rho^p \, d\mu = \fint_{\lambda B} \rho^p \, d\mu . \tag{8.1.61}$$

However, because B_j is closed and because ρ is an upper gradient of u in λB_j, we obtain from Lemma 8.1.58 that the pair $(\bar{u}_j, \bar{\rho}_j)$ is a function–upper-gradient pair in X, where $\bar{u}_j(x) = u(x)$ if $x \in \lambda B_j$ and $\bar{u}_j(x) = 0$ if $x \in X \setminus \lambda B_j$ and $\bar{\rho}_j(x) = \rho(x)$ if $x \in \lambda B_j$ and $\bar{\rho}_j(x) = \infty$ if $x \in X \setminus \lambda B_j$. By assumption we have, therefore, that

$$\begin{aligned}\fint_{B_j} |u_j - (u_j)_{B_j}| \, d\mu &= \fint_{B_j} |\bar{u}_j - (\bar{u}_j)_{B_j}| \, d\mu \\ &\le C \operatorname{diam}(B_j) \left(\fint_{\lambda B_j} \bar{\rho}_j^p \, d\mu \right)^{1/p} \\ &= C \operatorname{diam}(B_j) \left(\fint_{\lambda B_j} \rho^p \, d\mu \right)^{1/p} .\end{aligned}$$

The assertion now follows from (8.1.60) and (8.1.61). □

Proof of Theorem 8.1.55 Let $B = B(x_0, r)$ be an open ball in X, let $u : 3\lambda B \to V$ be integrable in B, and let ρ be an upper gradient of u in $3\lambda B$, where $\lambda \ge 1$ is a constant such that inequalities (8.1.50) and (8.1.51) hold. We consider balls $B_j = \overline{B}(x_0, (1 - 2^{-j})r)$ and functions $u_j = u|B_j$, for $j = 1, 2, \ldots$ Arguing as in the proof of Theorem 8.1.53, we find function–upper-gradient pairs $(\bar{u}_j, \bar{\rho}_j)$ in X such that $\bar{u}_j|_{3\lambda B_j} = u|_{3\lambda B_j}$ and $\bar{\rho}_j|_{3\lambda B_j} = \rho|_{3\lambda B_j}$. The assertion now follows from Theorem 8.1.49 when one observes the following three facts: every Lebesgue point of u in B belongs to B_j for j sufficiently large, $u_{B_j} \to u_B$ as $j \to \infty$, and $B(x, \lambda \operatorname{diam}(B_j)) \subset 3\lambda B_j$ for $x \in B_j$. □

8.2 Density of Lipschitz functions

In Section 7.4 we defined the space $HN^{1,p}(X : V)$ as the closure of locally Lipschitz functions in $N^{1,p}(X : V)$. According to Proposition 7.4.1, a function $u : X \to V$ belongs to $HN^{1,p}(X : V)$ if and only if there exists a Cauchy sequence (φ_i) of locally Lipschitz functions in $N^{1,p}(X : V)$ converging to u both in $N^{1,p}(X : V)$ and pointwise p-quasiuniformly. Recall from Section 7.3 that convergence is p-quasiuniform if, for every $\epsilon > 0$, there is a measurable set of p-capacity no more than ϵ such that the sequence converges uniformly to the limit function outside this set. By Remark 7.3.5 we can in addition assume that the set of small p-capacity is also open if the functions in the approximating sequence are continuous (see also Theorem 7.4.2.).

We next prove the important fact that in doubling metric measure spaces X with a p-Poincaré inequality the identity $HN^{1,p}(X : V) = N^{1,p}(X : V)$ holds.

Theorem 8.2.1 *Let $1 \leq p < \infty$. Suppose that X is a doubling metric measure space that supports a p-Poincaré inequality. Then Lipschitz functions are dense in $N^{1,p}(X : V)$ and $HN^{1,p}(X : V) = N^{1,p}(X : V)$. If X is in addition locally compact then every function in $N^{1,p}(X : V)$ is p-quasicontinuous.*

We do not know to what extent the doubling assumption is necessary for the concluding statement of Theorem 8.2.1, even when $V = \mathbb{R}$ (see Remark 7.4.3 and Section 8.5 below).

Proof The second claim in the theorem follows from the first and from Theorem 7.4.2.

To prove the first claim, let u be a function in $N^{1,p}(X : V)$. We need to find a sequence of Lipschitz functions converging to u in $\mathrm{N}^{1,p}(X : V)$. For this, we may assume that u vanishes outside some ball (Proposition 7.1.35). By picking a representative, we may also assume that u is pointwise defined everywhere. Let ρ_u be the minimal p-weak upper gradient of u. By Theorem 8.1.49 there is a set $E \subset X$ of measure zero such that

$$|u(x) - u(y)| \leq Cd(x, y)\big(M\rho_u^p(x) + M\rho_u^p(y)\big)^{1/p} \tag{8.2.2}$$

whenever $x, y \in X \setminus E$. For $t > 0$ write

$$E_t = \{x \in X : \; M\rho_u^p(x) > t^p\}.$$

Then (8.2.2) implies that the restriction of u to $X \setminus (E_t \cup E)$ is Lipschitz continuous with constant $C_1 t$, where $C_1 > 0$ is independent of t. By Theorem 4.1.21 we can find a $C_2\,t$-Lipschitz function $u_t : X \to V$ such that $u_t(x) = u(x)$ for $x \in X \setminus (E_t \cup E)$, where $C_2 > 0$ is independent of t.

Let B_0 be a ball in X such that $u(x) = 0$ for every $x \in X \setminus B_0$. We claim that $E_t \subset 2B_0$ for every large enough $t > 0$. To see this, suppose that $x \in E_t \cap (X \setminus 2B_0)$, and let B be a ball centered at x such that

$$t^p < \fint_B \rho_u^p \, d\mu .$$

Because $\rho_u = 0$ in $X \setminus B_0$ (Proposition 6.3.22), B must meet B_0. It follows that $B_0 \subset 3B$, which implies that

$$t^p < C \fint_{3B} \rho_u^p \, d\mu \le C \frac{1}{\mu(B_0)} \int_X \rho_u^p \, d\mu := t_0^p .$$

This proves the claim.

It follows that $u_t = u = 0$ almost everywhere in $X \setminus 2B_0$ for $t \ge t_0$. Because u_t is $C_2 t$-Lipschitz on X, we can deduce that $|u_t|$, for $t \ge t_0$, is bounded in X by a constant $C_3 t$, where $C_3 > 0$ is independent of t. Hence

$$\begin{aligned}\int_X |u - u_t|^p \, d\mu &= \int_{\{u \neq u_t\}} |u - u_t|^p \, d\mu \\ &\le C \int_{\{u \neq u_t\}} |u|^p \, d\mu + C \, t^p \mu(\{u \neq u_t\}) ,\end{aligned}$$

where $C > 0$ is independent of $t \ge t_0$. Since $\mu(\{u \neq u_t\}) \le \mu(E_t)$, it follows from the preceding inequality and from Proposition 3.5.15 that $u_t \to u$ in $L^p(X)$ as $t \to \infty$.

Next, let F be a Borel set containing $E_t \cup E$ such that $\mu(F) = \mu(E_t \cup E)$. Then $u - u_t = 0$ in $X \setminus F$, and it follows from (6.3.18) and from Proposition 6.3.22 that

$$\rho_{u - u_t}(x) \le (\rho_u(x) + C_2 \, t) \, \chi_F(x)$$

for almost every $x \in X$. In particular,

$$\int_X \rho_{u-u_t}^p \, d\mu \le C \int_F \rho_u^p \, d\mu + C \, t^p \, \mu(E_t) ,$$

where $C > 0$ is independent of $t \ge t_0$. As earlier, we find that $\rho_{u-u_t} \to 0$ in $L^p(X)$ as $t \to \infty$. We have thus proved that

$$\lim_{t \to \infty} \|u - u_t\|_{N^{1,p}(X:V)} = 0$$

and the theorem follows. □

At various points in this book we have assumed that the metric space X is complete. As we demonstrate next, in many of these situations the assumption of completeness (which is equivalent to properness in the presence of the doubling property) is not very restrictive. Recall that a measure μ defined on a metric subspace X of Z has a null extension $\overline{\mu}$ to Z given by (3.3.14).

Lemma 8.2.3 *Suppose that (X, d, μ) is a locally compact metric space and μ is a doubling measure supporting a p-Poincaré inequality. Then the metric completion $\widehat{X}$ of X also supports a p-Poincaré inequality when equipped* *with the null extension of μ. Furthermore, the null extension $\overline{\mu}$ is also doubling. The constants relating to the doubling and Poincaré inequality properties of $(\widehat{X}, d, \overline{\mu})$ depend only on the doubling and Poincaré constants of (X, d, μ). Moreover, every $u \in N^{1,p}(X)$ has an extension $\widehat{u} \in N^{1,p}(\widehat{X})$ with $\|\widehat{u}\|_{N^{1,p}(\widehat{X})} = \|u\|_{N^{1,p}(X)}$.*

Proof To check that $\overline{\mu}$ is doubling on $\widehat{X}$, we fix $x_0 \in \widehat{X}$ and $r > 0$. Choose $x_1 \in X$ such that $d(x_1, x_0) < r/4$ (if $x_0 \in X$ then we can choose $x_1 = x_0$). Then,

$$\begin{aligned}
\overline{\mu}(\widehat{B}(x_0, 2r)) &= \mu(X \cap \widehat{B}(x_0, 2r)) \le \mu(B(x_1, 3r)) \\
&\le C_\mu^2\, \mu(B(x_1, 3r/4)) \le C_\mu^2 \mu(X \cap \widehat{B}(x_0, r)) \\
&= C_\mu^2 \overline{\mu}(\widehat{B}(x_0, r)).
\end{aligned}$$

Here $\widehat{B}$ denotes a ball in $\widehat{X}$ centered at a point in $\widehat{X}$, while B denotes a ball in X with center in X. We can therefore conclude that $\overline{\mu}$ is doubling with constant C_μ^2, where C_μ is the doubling constant of μ.

We now use the doubling property of $\overline{\mu}$ to verify the Poincaré inequality for $\widehat{X}$. Indeed, because we assume X to be locally compact, X is an open subset of $\widehat{X}$. So, if $\widehat{u}$ is a $\overline{\mu}$-measurable function on $\widehat{X}$ with upper gradient $\widehat{\rho}$ then the μ-measurable function $u = \widehat{u}|_X$ has the Borel function $\rho = \widehat{\rho}|_X$ as an upper gradient in X. Hence, if $\widehat{B}(x_0, r)$ is a ball in $\widehat{X}$ then, by the p-Poincaré inequality on X applied to the ball $B(x_1, 2r)$ with $x_1 \in X$ such that $d(x_1, x_0) < r/4$, we have

$$\begin{aligned}
\frac{1}{2} \fint_{\widehat{B}(x_0,r)} |\widehat{u} - (\widehat{u})_{\widehat{B}(x_0,r)}| \, d\overline{\mu} &\le \fint_{\widehat{B}(x_0,r)} |\widehat{u} - u_{B(x_1,2r)}| \, d\overline{\mu} \\
&\le C_\mu^2 \fint_{B(x_1,2r)} |u - u_{B(x_1,2r)}| \, d\mu \\
&\le 2C_\mu^2 C_P r \left(\fint_{B(x_1,2\lambda r)} \rho^p \, d\mu \right)^{1/p} \\
&\le 2C_\mu^2 C_P r \left(\fint_{\widehat{B}(x_0,3\lambda r)} \widehat{\rho}^p \, d\overline{\mu} \right)^{1/p}.
\end{aligned}$$

To verify the final claim, note that if $u \in N^{1,p}(X)$ then u can be approximated in $N^{1,p}(X)$ by Lipschitz functions in X; see Theorem 8.2.1. Lipschitz functions on X uniquely extend to $\widehat{X}$, and these extensions lie in $N^{1,p}_{\mathrm{loc}}(\widehat{X})$. Since X is an open subset of $\widehat{X}$, it follows from Lemma 6.3.8 that minimal p-weak upper gradients of the Lipschitz functions on X are also minimal p-weak upper gradients on $\widehat{X}$ of the extended Lipschitz functions. It follows that, for Lipschitz functions in $N^{1,p}(X)$, the last claim of Lemma 8.2.3 holds with the $N^{1,p}$-norm of the extended function equals the $N^{1,p}$-norm of the original Lipschitz function. The density of Lipschitz functions in $N^{1,p}(X)$ together with Proposition 7.3.1 completes the proof provided we know that $\mathrm{Cap}_p^X(E) = 0$ implies $\mathrm{Cap}_p^{\widehat{X}}(E) = 0$ for $E \subset X$. Here Cap_p^X and $\mathrm{Cap}_p^{\widehat{X}}$ are p-capacities with respect to the Sobolev spaces $N^{1,p}(X)$ and $N^{1,p}(\widehat{X})$ respectively.

Let $E \subset X$ be such a set. Since X is locally compact and separable, we can find an increasing sequence (K_n) of compact sets with $X = \bigcup_n K_n$. Set $E_n = K_n \cap E$. Then for each n we have $\mathrm{Cap}_p^X(E_n) = 0$ and hence, by Proposition 7.2.8, E_n is p-exceptional with respect to X. It follows that there is a nonnegative Borel-measurable (with respect to X) function $\rho \in L^p(X)$ such that $\int_\gamma \rho\, ds = \infty$ whenever γ is a nonconstant rectifiable curve in X that intersects E; see Lemma 6.2.2. We now show that E_n is p-exceptional with respect to $\widehat{X}$ as well. If β is a nonconstant rectifiable curve in $\widehat{X}$ that intersects E_n then, since $E_n \subset K_n$ and K_n is a compact subset of X with X open in $\widehat{X}$, β must have a subcurve in X that intersects E_n. The line integral of ρ on this subcurve is therefore infinite. We extend ρ by zero to $\widehat{X}$; since ρ is Borel measurable in X, it follows that this extension, also denoted ρ, is Borel measurable in $\widehat{X}$. Thus $\int_\beta \rho\, ds$ makes sense, and we have $\int_\beta \rho\, ds = \infty$. Because $\rho \in L^p(X)$, we also have $\rho \in L^p(\widehat{X})$ because $\overline{\mu}(\widehat{X} \setminus X) = 0$. Thus by Lemma 6.2.2 we know that E_n is p-exceptional in $\widehat{X}$ and, because $\overline{\mu}(E_n) = \mu(E_n) = 0$, we may use Proposition 7.2.8 again to conclude that $\mathrm{Cap}_p^{\widehat{X}}(E_n) = 0$. Now the subadditivity of $\mathrm{Cap}_p^{\widehat{X}}$ guarantees that $\mathrm{Cap}_p^{\widehat{X}}(E) = 0$.

The proof is now complete. □

Note that in the above lemma we needed the doubling and Poincaré inequality properties of X itself in order to extend functions in $N^{1,p}(X)$ to $N^{1,p}(\widehat{X})$ without increasing the norm. Without this assumption such an extension is impossible, as demonstrated by the metric measure space that is the Euclidean (planar) slit disk. We use Lemma 8.2.3 to extend the outer capacity property for null-capacity sets from the context of proper metric spaces to that of locally compact metric spaces; see Proposition 7.2.12.

Lemma 8.2.4 *Suppose that X is locally compact and μ is a doubling measure supporting a p-Poincaré inequality. If* $\mathrm{Cap}_p(K) = 0$ *then for each* $\epsilon > 0$ *we can find an open set* $U_\epsilon \supset K$ *such that* $\mathrm{Cap}_p(U_\epsilon) < \epsilon$.

Proof By Lemma 8.2.3 we know that $\mathrm{Cap}_p(K) = \mathrm{Cap}_p^{\widehat{X}}(K)$, and so $K \subset \widehat{X}$ where $(\widehat{X}, d, \overline{\mu})$ satisfy the hypotheses of Proposition 7.2.12. It follows that for every $\epsilon > 0$ we can find an open set $\widehat{U}_\epsilon \subset \widehat{X}$ with $K \subset \widehat{U}_\epsilon$ such that $\mathrm{Cap}_p(\widehat{U}_\epsilon \cap X) \le \mathrm{Cap}_p^{\widehat{X}}(\widehat{U}_\epsilon) < \epsilon$. The first inequality follows from the fact that for each $A \subset \widehat{X}$ we have $\mathrm{Cap}_p(A \cap X) \leq \mathrm{Cap}_p^{\widehat{X}}(A)$. Since $\widehat{U}_\epsilon \cap X$ is open in X, the proof of the lemma is complete. □

We now use Theorem 8.2.1 and Lemma 8.2.4 to extend the outer capacity property of Proposition 7.2.12 beyond sets of zero p-capacity.

Corollary 8.2.5 *Let $1 \leq p < \infty$. Let X be a locally compact doubling metric measure space that supports a p-Poincaré inequality. Then* $\mathrm{Cap}_p(A) = \inf\{\mathrm{Cap}_p(U) : X \supset U$ *open,* $A \subset U\}$ *whenever* $A \subset X$.

Proof Denote the desired infimum by α_A. If $\mathrm{Cap}_p(A)$ is infinite there is nothing to prove, hence we may assume that $\mathrm{Cap}_p(A) < \infty$. Let $u \in N^{1,p}(X)$ be admissible for $\mathrm{Cap}_p(A)$; $u \geq 1$ on A except for a set of zero p-capacity. Truncating if necessary, we also assume that $u \geq 0$ on X. By Theorem 8.2.1 we know that u is equivalent to a p-quasicontinuous function. Hence, by Lemma 8.2.4, for a fixed $\eta > 0$ there is an open set $K_\eta \subset X$ with $\mathrm{Cap}_p(K_\eta) < \eta$ such that u is equal to its p-quasicontinuous representative outside K_η. For fixed $0 < \epsilon < 1$, the set $U_\epsilon := \{x \in X : u(x) > 1 - \epsilon\}$ satisfies $A \subset U_\epsilon \cup K_\eta$. Moreover, there is an open set $V_{\epsilon,\eta}$ with $\mathrm{Cap}_p(V_{\epsilon,\eta}) < \eta$ such that $U_\epsilon \cup K_\eta \cup V_{\epsilon,\eta}$ is open. Note that $\mathrm{Cap}_p(K_\eta \cup V_{\epsilon,\eta}) < 2\eta$; so we can find a nonnegative function $w \in N^{1,p}(X)$ such that $w \geq 1$ on $K_\eta \cup V_{\epsilon,\eta}$ and $\int_X (w^p + g_w^p)\, d\mu < 2\eta$.

The function $\max\{(1-\epsilon)^{-1}u, w\}$ satisfies $\max\{(1-\epsilon)^{-1}u, w\} \in N^{1,p}(X)$, $\max\{(1-\epsilon)^{-1}u, w\} \geq 1$, on the open set $U_\epsilon \cup K_\eta \cup V_{\epsilon,\eta}$ and so

$$\alpha_A \leq \mathrm{Cap}_p(U_\epsilon \cup K_\eta \cup V_{\epsilon,\eta}) \leq (1-\epsilon)^{-p} \int_X (u^p + g_u^p)\, d\mu + \int_X (w^p + g_w^p)\, d\mu,$$

that is, $\alpha_A \leq (1-\epsilon)^{-p} \int_X (u^p + g_u^p)\, d\mu + 2\eta$. Letting $\epsilon \to 0$, then $\eta \to 0$, and after that taking the infimum over all admissible functions u for A yields $\alpha_A \leq \mathrm{Cap}_p(A)$. The reverse inequality follows from the monotonicity of the p-capacity. □

8.3 Quasiconvexity and the Poincaré inequality

A metric space Z is said to be C-*quasiconvex*, $C \geq 1$, if each pair of points $x, y \in Z$ can be joined by a rectifiable curve γ in Z such that

$$\mathrm{length}(\gamma) \leq C d_Z(x, y)\,. \tag{8.3.1}$$

We also say that Z is quasiconvex if it is C-quasiconvex for some C. The self-explanatory term *quasiconvex metric* is also used. The least C such that (8.3.1) holds is called the *quasiconvexity constant* of the metric.

Quasiconvexity is an important geometric consequence of the Poincaré inequality for complete, doubling, metric measure spaces. For example, in quasiconvex metric measure spaces every function with a bounded upper gradient is necessarily Lipschitz continuous.

Theorem 8.3.2 *Every complete and doubling metric measure space that supports a Poincaré inequality is quasiconvex. The quasiconvexity constant depends only on the doubling constant of the measure and the data associated with the Poincaré inequality.*

A related but stronger notion is *annular quasiconvexity*. A metric space Z is *annularly C-quasiconvex* for some constant $C \geq 1$ if, whenever $z \in Z$ and $r > 0$, each pair of points $x, y \in B(z, r) \setminus B(z, r/2)$ can be connected in $B(z, Cr) \setminus B(z, r/C)$ by a C-quasiconvex curve. The association between annular quasiconvexity and Poincaré inequalities will be explored in the next chapter; see Theorem 9.4.1.

Remark 8.3.3 The ensuing proof of Theorem 8.3.2 shows that the conclusion of the theorem holds under weaker hypotheses. It suffices to assume that the Poincaré inequality (8.1.1) holds for every Lipschitz function $u : X \to V$ and for every Lipschitz continuous upper gradient $\rho : X \to [0, \infty)$ of u (with constants C and λ independent of u, ρ, and B, of course). Alternatively, it suffices to assume that there are constants $C > 0$ and $\lambda \geq 1$ such that

$$\fint_B |u - u_B| \, d\mu \leq C \operatorname{diam}(B) \left(\fint_{\lambda B} (\operatorname{Lip} u)^p \, d\mu \right)^{1/p} \tag{8.3.4}$$

for every open ball B in X and for every Lipschitz function $u : X \to \mathbb{R}$, where we recall (see (6.2.4)) that the pointwise upper Lipschitz-constant function is defined by

$$\operatorname{Lip} u(x) = \limsup_{r \to 0} \sup_{y \in B(x,r)} \frac{|u(x) - u(y)|}{r}. \tag{8.3.5}$$

The conclusions are similarly quantitative in that the quasiconvexity constant depends only on the doubling constant of the measure and the data associated with the various Poincaré inequalities. These remarks will be used later, in Section 8.4.

Proof of Theorem 8.3.2 Let X be a complete and doubling metric measure space that supports a p-Poincaré inequality; evidently, we may assume that $p > 1$. Fix a point $x \in X$. Given an integer $k \geq 1$, a finite sequence $x = x_0$, $x_1, \ldots, x_l = y$ of points in X is said to be a *k-chain from x to y* if $d(x_{i+1}, x_i) \leq 1/k$ for each $0 \leq i \leq l - 1$. We claim that, for each point $y \in X$ and for each $k \geq 1$, there is a k-chain from x to y. To see this, observe that the set U_x consisting of all those points $y \in X$ such that there exists a k-chain from x to y is open. Because the complement $X \setminus U_x$ is open as well, and because X is connected (Proposition 8.1.6), we must have that $U_x = X$ as claimed. If we assume the weaker hypotheses of Remark 8.3.3 then we do not a priori have the connectivity of X but we notice that $\operatorname{dist}(U_x, X \setminus U_x) \geq 1/k$, and this implies that the characteristic function χ_{U_x} is Lipschitz and has $\rho \equiv 0$ as an upper gradient, which in turn yields $U_x = X$.

Next, define a function $u_k : X \to [0, \infty)$ by setting

$$u_k(y) := \inf \sum_{i=0}^{l-1} d(x_{i+1}, x_i),$$

where the infimum is taken over all k-chains from x to y. Then u_k is 1-Lipschitz in every ball of radius at most $1/k$, so the constant function $\rho \equiv 1$ is an upper gradient of u_k (Lemma 6.2.6). Since we also have $u_k(x) = 0$, it follows from Theorem 8.1.7 that

$$u_k(y) \leq Cd(x, y)\left(M_{\lambda d(x,y)}\rho^p(x) + M_{\lambda d(x,y)}\rho^p(y)\right)^{1/p} \leq Cd(x, y) \tag{8.3.6}$$

whenever $y \in X$, where $C > 0$ depends only on the data in the hypotheses. Note that, because u_k is continuous, every point is a Lebesgue point of u_k.

Now fix a point $y \in X$ with $y \neq x$. Choose, for each integer $k \geq \max\{1, 2/d(x, y)\}$, a k-chain $x = x_{k,0}, x_{k,1}, \ldots, x_{k,l(k)} = y$ such that

$$\sum_{i=0}^{l(k)-1} d(x_{k,i+1}, x_{k,i}) \leq 2u_k(y) \leq 2Cd(x, y), \tag{8.3.7}$$

where $C > 0$ is as in (8.3.6), and

$$\max\{d(x_{k,i-1}, x_{k,i}), d(x_{k,i}, x_{k,i+1})\} \geq 1/(2k) \tag{8.3.8}$$

for each $i = 1, 2, \ldots, l(k) - 1$. Requirement (8.3.7) can be accomplished by using the definition of u_k and (8.3.6), and requirement (8.3.8) can be accomplished by first taking any k-chain such that (8.3.7) holds and then dispensing with extra points, if necessary. One easily computes from (8.3.7) and (8.3.8) that

$$\frac{l(k) - 1}{4k} \leq 2Cd(x, y).$$

Consequently,

$$d(x, y) \leq \frac{l(k)}{k} \leq 10Cd(x, y) \tag{8.3.9}$$

for all $k \geq k_0$.

Next, set $L := 2Cd(x, y)$ and $\varepsilon_k := L/l(k)$, and let

$$N_k := \{0, \varepsilon_k, 2\,\varepsilon_k, \dots, (l(k) - 1)\,\varepsilon_k, L\} \subset [0, L]$$

be an ε_k-net in $[0, L]$. It follows from (8.3.9) that

$$\frac{1}{5k} \le \varepsilon_k \le \frac{2C}{k} \tag{8.3.10}$$

for all $k \ge k_0$. Therefore, the map $\gamma_k : N_k \to X$, $\gamma_k(i\varepsilon_k) := x_{k,i}$, is 5-Lipschitz for $k \ge k_0$. We will show that the sequence (γ_k) gives rise to a 5-Lipschitz map $\gamma : D \to X$ defined on a dense subset $D \subset [0, L]$ such that $0, L \in D$ with $\gamma(0) = x$ and $\gamma(L) = y$. Such a map extends to a 5-Lipschitz map $[0, L] \to X$, thus providing a curve from x to y of length at most $5L = 10Cd(x, y)$, where $C > 0$ is the constant in (8.3.6) with possible dependence on other data.

It remains to construct the map γ. This is done via a familiar Arzelà–Ascoli type argument, as we next explain. First we note that all the images $\gamma_k(N_k)$ lie in some fixed compact subset K of X; the reason is that X is proper (Lemma 4.1.14). Denote by

$$D_n := \{m2^{-n}L : m = 0, 1, \dots, 2^n\}$$

the set of all dyadic points in $[0, L]$ at the level $n = 0, 1, 2, \dots$ Set $\gamma(0) = x$ and $\gamma(L) = y$. Next, consider the point $z = L/2$. There is a subsequence $\gamma_{k_1}, \gamma_{k_2}, \dots$ of the sequence (γ_k) such that for some points $z_{k_j} \in N_{k_j}$ we have both $z_{k_j} \to z$ and $\gamma_{k_j}(z_{k_j}) \to w_z \in K$. We set $\gamma(z) = w_z$. Because each map γ_k is 5-Lipschitz, the element w_z is independent of the choice of subsequence. From the properties of the maps γ_k we have that $\gamma : D_1 \to X$ is 5-Lipschitz. Using appropriate subsequences of (γ_{k_j}) we argue similarly and define $\gamma(z)$ for points $z \in D_2 \setminus D_1$ in such a way as to obtain a 5-Lipschitz map $\gamma : D_2 \to X$. Continuing in this manner, by passing to further subsequences we obtain a 5-Lipschitz map $\gamma : D \to X$ where $D = \bigcup_n D_n$, as desired. We leave the details to the reader.

The proof of Theorem 8.3.2 is now completed by extending the 5-Lipschitz map on the dense set D to $[0, L]$. □

Length spaces and geodesic spaces

A metric space is said to be a *length space* if the distance between every pair of points in the space is equal to the infimum of the lengths of the curves joining the points.

A metric space is said to be *geodesic* if every pair of points in the space can be joined by a curve whose length is equal to the distance between the points.

Such a curve is called a *geodesic* between the two points. We also say that a metric e on Z is a *geodesic metric* (resp. *length metric*) if (Z, e) is a geodesic metric space (resp. length metric space).

A geodesic space is always a length space. If V is a normed space of dimension at least 2, then $V \setminus \{0\}$ is always a length space but in general is not geodesic. However, we have the following result.

Lemma 8.3.11 *Proper length spaces are geodesic.*

Proof This is a straightforward application of the Arzelà–Ascoli theorem 5.1.10. Indeed, given two points x and y in a proper length space Z, let $\gamma_i : [0, l_i] \to Z$ be curves joining x and y parametrized by the arc length, where the $l_i = \text{length}(\gamma_i)$ satisfy $d_Z(x, y) \le l_i \to d_Z(x, y)$ as $i \to \infty$. We may assume that there is some finite positive number l such that $\gamma_i(0) = x$, $\gamma_i(l_i) = y$, and $l_i \le l$ for all i. By defining $\gamma_i(t) = y$ for $l_i \le t \le l$ we have a sequence of 1-Lipschitz maps $\gamma_i : [0, l] \to Z$. (See Section 5.1.) Now, Theorem 5.1.10 gives a subsequence γ_{i_j} that converges uniformly on $[0, l]$ to a map $\gamma : [0, l] \to Z$. This map is 1-Lipschitz and satisfies $\gamma(0) = x$ and $\gamma(t) = y$ for $d_Z(x, y) \le t \le l$. The restriction $\gamma|_{[0, d_Z(x,y)]}$ is therefore a geodesic from x to y. The lemma follows. □

The geodesic property of a metric space is often useful for technical considerations. We will use the next proposition several times in this book.

Proposition 8.3.12 *Suppose that (Z, d_Z) is a proper and rectifiably path connected metric space. Define, for $x, y \in Z$,*

$$\hat{d}(x, y) := \inf_{\gamma} \text{length}(\gamma), \tag{8.3.13}$$

where the infimum is taken over all curves γ that join x and y in Z. Then $\hat{d}$ is a geodesic metric in Z and if (Z, d_Z) is quasiconvex then

$$d_Z(x, y) \le \hat{d}(x, y) \le C d_Z(x, y) \tag{8.3.14}$$

for all $x, y \in X$, where $C \ge 1$ is the quasiconvexity constant of d_Z.

Proof The proof is similar to that of Lemma 8.3.11. First we observe that $\hat{d}$ is indeed a metric. Second, the first inequality in (8.3.14) holds by (5.1.12) and thus (8.3.14) is immediate if (Z, d_Z) is quasiconvex. Next, given $x, y \in Z$, let $\gamma_i : [0, l_i] \to Z$ be curves parametrized by the arc length such that $\gamma_i(0) = x$, $\gamma_i(l_i) = y$, and that $l_i \to l := \hat{d}(x, y)$ as $i \to \infty$. Because we may assume that $l \le l_i \le C d_Z(x, y)$, arguing as in the proof of Lemma 8.3.11 we obtain a 1-Lipschitz map $\gamma : [0, l] \to (Z, d_Z)$ such that $\gamma(0) = x$ and $\gamma(l) = y$.

Because $d_Z \leq \hat{d}$, we have that $\gamma : [0, l] \to (Z, \hat{d})$ is 1-Lipschitz as well, and hence provides a geodesic as desired. The proposition follows. □

In general, two metrics d_1 and d_2 in a set Z are said to be *bilipschitz equivalent* if there is a constant $C \geq 1$ such that

$$C^{-1} d_2(x, y) \leq d_1(x, y) \leq C d_2(x, y) \tag{8.3.15}$$

for every pair of points $x, y \in Z$.

We have the following corollary to Theorem 8.3.2 and Proposition 8.3.12.

Corollary 8.3.16 *A complete and doubling metric measure space that supports a Poincaré inequality admits a geodesic metric that is bilipschitz equivalent to the underlying metric. The bilipschitz constant depends only on the doubling constant of the measure and on the data of the Poincaré inequality.*

Remark 8.3.17 In light of Remark 8.3.3, the conclusion of Corollary 8.3.16 holds for complete and doubling metric measure spaces that support a Poincaré inequality for Lipschitz functions and their Lipschitz-continuous upper gradients.

Lemma 8.3.18 *Let d_1 and d_2 be two bilipschitz-equivalent metrics on X. If a measure μ on X is doubling with respect to d_1 then it is also doubling with respect to d_2. If in addition (X, d_1, μ) supports a p-Poincaré inequality then also (X, d_2, μ) supports a p-Poincaré inequality.*

Proof The first claim is easily verified from the definitions; we leave this to the reader. Let C be the bilipschitz constant relating d_1 to d_2. If ρ is an upper gradient of a function u with respect to the metric d_2 then $C\rho$ is an upper gradient of u with respect to the metric d_1. Thus, given a function u that is integrable on balls and an upper gradient ρ of u with respect to d_2, we have the Poincaré inequality (8.1.1) with ρ replaced by $C\rho$, for all d_1-balls. By the first claim and Theorem 8.1.7, we conclude by using the pointwise estimate (8.1.10), the maximal function being taken with respect to d_1. Doubling allows us to replace d_1 by d_2 in this estimate, and the claim follows from Theorem 8.1.7. □

8.4 Continuous upper gradients and pointwise Lipschitz constants

This section is devoted to the proof of the following theorem (and its variant Theorem 8.4.2), which will be applied when the stability of the Poincaré inequality under the convergence of metric spaces is studied, in Chapter 11.

Theorem 8.4.1 *Suppose that X is a complete and doubling metric measure space. Let $1 \le p < \infty$. Then X supports a p-Poincaré inequality if and only if there exist constants $C > 0$ and $\lambda \ge 1$ such that* (8.1.1) *holds for every open ball B in X, for every Lipschitz function $u : X \to \mathbb{R}$, and for every Lipschitz-continuous upper gradient $\rho : X \to [0, \infty)$ of u in X. The data of the Poincaré inequalities depend only on each other and on the doubling constant of the measure.*

For the next theorem, recall the definition of the pointwise upper Lipschitz-constant function from (8.3.5).

Theorem 8.4.2 *Suppose that X is a complete and doubling metric measure space. Let $1 \le p < \infty$. Then X supports a p-Poincaré inequality if and only if there exist constants $C > 0$ and $\lambda \ge 1$ such that*

$$\fint_B |u - u_B| \, d\mu \le C \operatorname{diam}(B) \left(\fint_{\lambda B} (\operatorname{Lip} u)^p \, d\mu \right)^{1/p} \tag{8.4.3}$$

for every open ball B in X and for every Lipschitz function $u : X \to \mathbb{R}$. The data of the Poincaré inequalities depend only on each other and on the doubling constant of the measure.

We now first prove a simple lemma which reduces the latter theorem to the former. (Note that necessity is clear in both statements, since every Lipschitz function in a doubling metric measure space is integrable on balls.)

Lemma 8.4.4 *Suppose that Z is a C-quasiconvex metric space and that $u : Z \to V$ is a function. Then*

$$\operatorname{Lip} u(x) \le C\rho(x) \tag{8.4.5}$$

for every $x \in Z$ and for every continuous upper gradient ρ of u in Z.

Proof Let ρ be a continuous upper gradient of u and fix $x \in X$. Let $r > 0$ and $y \in B(x, r)$. Choose a curve γ joining x to y with length $(\gamma) \le Cd(x, y)$. Then

$$|u(x) - u(y)| \le \int_\gamma \rho \, ds \le Cr \max_{z \in \overline{B}(x, Cr)} \rho(z),$$

and the desired inequality follows. □

Suppose now that X supports a Poincaré inequality for all Lipschitz functions and their pointwise upper Lipschitz constants as in (8.4.3). Then, by Remark 8.3.3, X is quasiconvex and we obtain from Lemma 8.4.4 that X

supports a Poincaré inequality for all Lipschitz functions and their Lipschitz-continuous upper gradients. All this is quantitative as well. Thus Theorem 8.4.2 is a consequence of Theorem 8.4.1.

Proof of Theorem 8.4.1 As pointed out earlier, the necessity part of the statement is obvious. The proof of the sufficiency part requires a repeated application of Theorem 8.1.7. Thus, assume that the Poincaré inequality (8.1.1) holds, with some unspecified data, for every Lipschitz function in X and for every Lipschitz continuous upper gradient of that function in X. Then, by Theorem 8.1.7, we have that (8.1.10) holds for every such pair of functions, with constants that depend only on the initial data; we will denote these constants for definiteness by C_1 and λ_1 (instead of C and λ). We also have that X is L-quasiconvex, where $L \geq 1$ depends only on the data (Remark 8.3.3).

Next, fix a function $u : X \to \mathbb{R}$ that is integrable on balls, fix an upper gradient $\rho : X \to [0, \infty]$ of u, and fix an open ball B in X. It involves no loss of generality to assume that $\rho \in L^p(21L\lambda_1 B)$, for if it does not then (8.1.1) obviously holds for $\lambda = 21L\lambda_1$. For each given $\delta > 0$, the Vitali–Carathéodory theorem in Section 4.2 together with the dominated convergence theorem implies that there is an $\epsilon > 0$ and a lower semicontinuous function $\hat{\rho} : 21L\lambda_1 B \to [\epsilon, \infty)$ such that $\rho|_{21L\lambda_1} \leq \hat{\rho}$ in $21L\lambda_1 B$ and

$$\int_{6\lambda_1 B} \hat{\rho}^p \, d\mu \leq \int_{6\lambda_1 B} \rho^p \, d\mu + \delta\,. \tag{8.4.6}$$

Replacing ρ with $\hat{\rho}$ if necessary, we may clearly assume that ρ is lower semicontinuous in $21L\lambda_1 B$ and satisfies $\rho \geq \epsilon$ in $21L\lambda_1 B$ for some $\epsilon > 0$. The preceding understood, we will prove that

$$|u(x) - u(y)| \leq 2LC_1\, d(x, y)\Big(M_{\lambda_1 d(x,y)}\rho^p(x) + M_{\lambda_1 d(x,y)}\rho^p(y)\Big)^{1/p} \tag{8.4.7}$$

for every pair of points $x, y \in E$, where $E \subset 2B$ is such that $\mu(2B \setminus E) = 0$. The claim then follows from Theorem 8.1.7; recall that X can be covered by a countable number of balls.

To this end, we set $\sigma := L\lambda_1$. By the local integrability of u, we know that $|u(x)| < \infty$ almost everywhere. It thus suffices to verify (8.4.7) for each pair $x, y \in 2B$ of points with $|u(x)| < \infty$ and $|u(y)| < \infty$. Fix such a pair x, y. We may clearly assume that $u(x) = 1$ and $u(y) = 0$. Because $\rho|_{21\sigma B}$ is lower semicontinuous, Proposition 4.2.2 implies that there is an increasing sequence of Lipschitz functions $\rho_i : 21\sigma B \to [\epsilon, \infty)$ converging pointwise to

ρ in $21\sigma B$. We define, for each i, a function $u_i : 10\sigma B \to \mathbb{R}$ by setting

$$u_i(z) = \inf_{\gamma} \int_{\gamma} \rho_i \, ds \,,$$

where the infimum is taken over all rectifiable curves that join z to y in $20\sigma LB$. Note that there exists at least one such curve for every $z \in 10\sigma B$ by the L-quasiconvexity of X.

We next show that u_i is Lipschitz for every i. Thus, fix i. Let $z, w \in 10\sigma B$ be two distinct points, and suppose that $u_i(z) \geq u_i(w)$. Fix $\delta > 0$ and pick a rectifiable curve $\gamma_{w,y}$ joining w to y in $20\sigma LB$ such that

$$\int_{\gamma_{w,y}} \rho_i \, ds \leq u_i(w) + \delta \,.$$

Again by the L-quasiconvexity of X, we may join z to w by a rectifiable curve $\gamma_{z,w}$ such that $\text{length}(\gamma_{z,w}) \leq Ld(z, w)$; in particular, $\gamma_{z,w}$ lies in $20\sigma LB$. Let γ be the curve obtained by concatenating $\gamma_{w,y}$ and $\gamma_{z,w}$. Then

$$\begin{aligned} |u_i(z) - u_i(w)| = u_i(z) - u_i(w) &\leq \int_{\gamma} \rho_i \, ds - \int_{\gamma_{w,y}} \rho_i \, ds + \delta \\ &\leq \int_{\gamma_{z,w}} \rho_i \, ds + \delta \leq M_i L d(z, w) + \delta \,, \end{aligned}$$

where

$$M_i := \sup_{20\sigma B} \rho_i < \infty \,.$$

By letting δ tend to zero, we conclude that $u_i : 10\sigma B \to \mathbb{R}$ is Lipschitz with constant $M_i L$. By using the Lipschitz extension lemma, we can extend both ρ_i (from $21\sigma B$) and u_i (from $10\sigma B$) to be Lipschitz functions in the whole of X. These extensions will still be denoted ρ_i and u_i.

As in the proof of Lemma 7.2.13, we see that ρ_i is an upper gradient of u_i in $10\sigma B$. By arguing as in the proof of Lemma 8.4.4, we also obtain from the preceding that

$$\operatorname{Lip} u_i(z) \leq L\rho_i(z) \tag{8.4.8}$$

for every $z \in 10\sigma B$.

Next, put $v_i := \varphi u_i$, where

$$\varphi(z) = \min\{1, (\sigma \operatorname{rad}(B))^{-1} \operatorname{dist}(z, X \setminus 9\sigma B)\} \,.$$

Then $\varphi : X \to [0, 1]$ is a Lipschitz function with $\varphi = 1$ on $8\sigma B$. In particular, $v_i : X \to \mathbb{R}$ is a Lipschitz function that vanishes outside $9\sigma B$ and satisfies $v_i|8\sigma B = u_i|8\sigma B$. Consider the Lipschitz function η given by

$$\eta(z) := \begin{cases} 0 & \text{if } z \in 6\sigma B \cup (X \setminus 11\sigma B), \\ \operatorname{dist}(z, 6\sigma B)/\operatorname{dist}(X \setminus 8\sigma B, 6\sigma B) & \text{if } z \in 8\sigma B \setminus 6\sigma B, \\ 1 & \text{if } z \in 9\sigma B \setminus 8\sigma B, \\ 1 - \operatorname{dist}(z, 9\sigma B)/\operatorname{dist}(X \setminus 11\sigma B, 9\sigma B) & \text{if } z \in 11\sigma B \setminus 9\sigma B. \end{cases}$$

Because $(\sigma \operatorname{rad}(B))^{-1}\chi_{9\sigma B\setminus 8\sigma B}$ is an upper gradient of φ, it follows that $(\sigma \operatorname{rad}(B))^{-1}\eta$ is a Lipschitz upper gradient of φ. We will use this function to obtain a Lipschitz upper gradient of v_i as follows. Set

$$\tau_i(z) := \rho_i(z) + (\sigma \operatorname{rad}(B))^{-1} \sup_{10\sigma B} |u_i| \cdot \eta(z).$$

Then, by (8.4.8) and the definitions of v_i and τ_i, we have

$$\operatorname{Lip} v_i(z) \leq \operatorname{Lip} u_i(z) + \operatorname{Lip} \varphi(z) \sup_{10\sigma B} |u_i| \leq L\tau_i(z)$$

for every $z \in X$. We can deduce from Lemma 6.2.6, therefore, that $L\tau_i$ is a Lipschitz-continuous upper gradient of the Lipschitz function v_i. Hence, by the assumption, by Remark 8.1.11, and by the notational conventions made at the beginning of the proof, we have

$$|v_i(x) - v_i(y)| \leq C_1\, d(x, y)\Big(M_{\lambda_1 d(x,y)}\tau_i^p(x) + M_{\lambda_1 d(x,y)}\tau_i^p(y)\Big)^{1/p}. \tag{8.4.9}$$

Since we have $v_i|_{8\sigma B} = u_i|_{8\sigma B}$, $\tau_i|_{6\sigma B} = \rho_i|_{6\sigma B}$, and $\rho_i \leq \rho$, we obtain from (8.4.9) that in fact

$$|u_i(x) - u_i(y)| \leq LC_1\, d(x, y)\Big(M_{\lambda_1 d(x,y)}\rho^p(x) + M_{\lambda_1 d(x,y)}\rho^p(y)\Big)^{1/p}. \tag{8.4.10}$$

Therefore, to prove (8.4.7) (under our various reductions), it suffices to show that $|u_i(x)| \geq 1/2$ for some i.

Suppose on the contrary that $u_i(x) < 1/2$ for all i. By the definition of u_i, we can find a sequence (γ_i) of rectifiable curves joining x and y in $20\sigma B$ such that

$$\int_{\gamma_i} \rho_i < \frac{1}{2} \tag{8.4.11}$$

for each i. Because $\rho_i \geq \epsilon$, inequality (8.4.11) gives that $\operatorname{length}(\gamma_i) \leq 1/(2\epsilon)$. The preceding being understood, we argue as in the proof of Lemma 8.3.11. We may assume that there is an $H > 0$ such that $\operatorname{length}(\gamma_i) \to H$, each $\gamma_i : [0, H_i] \to X$ is 1-Lipschitz with $\gamma_i(0) = x$ and $\gamma_i(H_i) = y$, where $H_i = \max\{H, \operatorname{length}(\gamma_i)\}$, and $\gamma_i|_{[0,\operatorname{length}(\gamma_i)]}$ is parametrized by the arc length.

By the Arzelà–Ascoli theorem 5.1.10, we can further assume that the maps γ_i converge uniformly in $[0, H]$ to a 1-Lipschitz map $\gamma : [0, H] \to X$. (Note that X is proper by Lemma 4.1.14.)

Fix i_0 and $0 < \delta < H$. Since ρ_{i_0} is continuous, Fatou's lemma gives

$$\begin{aligned}\int_0^{H-\delta} \rho_{i_0}(\gamma(t))\,dt &= \int_0^{H-\delta} \lim_{i\to\infty} \rho_{i_0}(\gamma_i(t))\,dt \le \liminf_{i\to\infty} \int_0^{H-\delta} \rho_{i_0}(\gamma_i(t))\,dt \\ &\le \liminf_{i\to\infty} \int_0^{\operatorname{length}(\gamma_i)} \rho_{i_0}(\gamma_i(t))\,dt = \liminf_{i\to\infty} \int_{\gamma_i} \rho_{i_0}\,ds\,.\end{aligned}$$

Using first the fact that the sequence (ρ_i) is increasing and then letting $\delta \to 0$, we obtain from the preceding and from (8.4.11) that

$$\int_0^{H} \rho_{i_0}(\gamma(t))\,dt \le \liminf_{i\to\infty} \int_{\gamma_i} \rho_i\,ds \le 1/2\,.$$

Combining this with Lemma 5.1.14 yields $\int_\gamma \rho_{i_0}\,ds \le 1/2$. Finally, since the functions ρ_i increase to ρ in $21\sigma B$ and since the image of γ lies in $20\sigma\overline{B}$, we deduce that

$$\int_\gamma \rho\,ds = \lim_{i_0\to\infty} \int_\gamma \rho_{i_0}\,ds \le 1/2\,.$$

Since $|u(x) - u(y)| = 1$, the fact that ρ is an upper gradient of u is contradicted. This completes the proof of (8.4.7), and hence of Theorem 8.4.1. $\square$

Finally, we point out that the requirement of completeness in this section can be relaxed to local completeness; see Lemma 8.2.3.

8.5 Notes to Chapter 8

The argument using concentric balls with dyadically decreasing radii, as in the proof of Theorem 8.1.7, is known in the literature as a *telescoping argument* [120]. The inequality (8.1.10) should be compared with the Hajłasz inequality (10.2.1), which is the basis for the Hajłasz–Sobolev space considered in [108]; see Chapter 10. Theorem 8.1.7 first made its appearance in [129]. The result from Theorem 8.1.7 for real-valued functions can be found in [125] and [114]. Theorem 8.1.18 is found in [108], and is used extensively to develop the theory of Banach-space-valued Sobolev functions. The truncation technique employed in the proof of Theorem 8.1.7 is originally due to Maz'ya [201], [200], [198].

One consequence of the Poincaré inequality, the density of Lipschitz functions (see Theorem 8.2.1), holds for real-valued functions in a larger context.

The proof of Theorem 8.2.1 given here is originally due to Semmes. It was shown in [16] that, for $p > 1$, if the metric space is complete and the measure is doubling then Lipschitz functions are always dense in $N^{1,p}(X)$. Thus, when X is complete and the measure is doubling, the functions in $N^{1,p}(X)$ are always quasicontinuous.

The quasiconvexity of metric measure spaces supporting a Poincaré inequality (Theorem 8.3.2) was first recorded by Cheeger in [53]. Various versions of the proof of quasiconvexity can also be found in [114], [244], [69], [125], [150], and [31].

Theorems 8.4.1 and 8.4.2 are due to Keith [150]. A weaker version of Theorem 8.4.2 appeared in [126], where it was shown that to verify that a doubling metric measure space supports a p-Poincaré inequality it suffices to verify the Poincaré inequality for Lipschitz functions and *all* their upper gradients. It will be shown in Chapter 13 that if the metric measure space is doubling and supports a p-Poincaré inequality, then the minimal p-weak upper gradient of a Lipschitz function is its pointwise Lipschitz-constant function (see Theorem 13.5.1). All these arguments can be viewed as variants of part of an argument due to Ziemer [287], [288], [289] for the coincidence of the variational p-capacity of a compact set E in a Euclidean domain Ω and the p-modulus of the family of curves joining E to the complement of Ω.

9

Consequences of Poincaré inequalities

In this chapter we discuss some further consequences of Poincaré inequalities in metric measure spaces. We show that many Sobolev-type inequalities follow from a basic Poincaré inequality in doubling metric measure spaces. The Lebesgue differentiation theorem tells us that every integrable function has μ-a.e. point as a Lebesgue point. We will strengthen the Lebesgue point property for Sobolev functions and show that p-capacity-a.e. point is a Lebesgue point of a function in $N^{1,p}(X : V)$. Finally, we also demonstrate that a metric space supporting a Poincaré inequality necessarily has the MEC_p property in the sense of Section 7.5.

Throughout this chapter we let $X = (X, d, \mu)$ be a metric measure space as defined in Section 3.3 and V a Banach space and suppose that X is locally compact and supports a p-Poincaré inequality. Unless otherwise stipulated, we assume that $1 \leq p < \infty$.

9.1 Sobolev–Poincaré inequalities

The Poincaré inequality (8.1.1), or its Banach-space-valued counterpart (8.1.41), gives control over the mean oscillation of a function in terms of the p-means of its upper gradient. In many classical situations, for example in Euclidean space $\mathbb{R}^n$, various *Sobolev–Poincaré inequalities* demonstrate that one can similarly control the q-means of the function $|u - u_B|$ for certain values of $q > 1$. Analogous results are valid in metric measure spaces satisfying a Poincaré inequality. This is the topic of the current section.

We recall one of the pointwise estimates (8.1.56) that follows from the p-Poincaré inequality in a doubling metric measure space X. If B is an open ball in X and if $u : \lambda B \to V$ is integrable in B with ρ an upper gradient of u in λB then

$$|u(x) - u_B| \leq C \operatorname{diam}(B) \left(M_{\lambda \operatorname{diam}(B)} \rho^p(x)\right)^{1/p} \qquad (9.1.1)$$

for almost every $x \in B$. Here $\lambda \geq 1$ and $C > 0$ are fixed constants depending only on the data associated with X. It follows from the mapping properties of the maximal operator (Theorem 3.5.6) and from (9.1.1) that $u \in L^q(B)$ for all $q < p$ provided that $\rho \in L^p(\lambda B)$. Moreover, an application of Lemma 8.1.31 shows that u is in fact p-integrable in B in this case and that the following strengthening of the Poincaré inequality holds.

Theorem 9.1.2 *Suppose that X is a doubling metric measure space supporting a p-Poincaré inequality for some $1 \leq p < \infty$. Then there are constants $C > 0$ and $\lambda \geq 1$, depending only on the data associated with the Poincaré inequality and the doubling constant of the underlying measure, such that*

$$\int_B |u - u_B|^p \, d\mu \leq C \operatorname{diam}(B)^p \int_{\lambda B} \rho^p \, d\mu \qquad (9.1.3)$$

whenever B is an open ball in X, u is a measurable real-valued function in λB that is integrable in B, and ρ is an upper gradient of u in λB.

Proof The discussion preceding the statement of the theorem, together with inequality (8.1.32) with $q = p$, gives the following weak-type inequality:

$$\mu\left(\{x \in B : |u(x) - u_B| \geq s\}\right) \leq C s^{-p} \operatorname{diam}(B)^p \int_{\lambda B} \rho^p \, d\mu \, .$$

The claim (9.1.3) then follows from Lemma 8.1.31. □

Recall that if $\rho \in L^p(\lambda B)$ is an upper gradient of $u : \lambda B \to V$ in λB then $|u| : \lambda B \to \mathbb{R}$ also has ρ as an upper gradient in λB (see (6.3.18)).

Corollary 9.1.4 *Suppose that X is a doubling metric measure space supporting a p-Poincaré inequality for some $1 \leq p < \infty$. Let V be a Banach space, B be a ball in X, and $u : \lambda B \to V$ be measurable such that $\rho \in L^p(\lambda B)$ is an upper gradient of u. Then*

$$\left(\int_B |u - u_B|^p \, d\mu\right)^{1/p} \leq 2C \operatorname{diam}(B) \left(\int_{\lambda B} \rho^p \, d\mu\right)^{1/p} .$$

Proof Because of Lemma 8.1.5, we know that $|u| \in L^1(B)$ and so, by Proposition 3.2.4, the Banach space-valued function u is integrable on B. Thus we have

$$⨍_B |u - u_B| \, d\mu \leq C \operatorname{diam}(B) \left(⨍_{\lambda B} \rho^p \, d\mu\right)^{1/p} . \qquad (9.1.5)$$

Observe that $x \mapsto |u(x) - u_B|$ is a composition of a 1-Lipschitz function with u. Hence, by (6.3.19), ρ is an upper gradient of this function as well. Applying Theorem 9.1.2 to this real-valued function, we have

$$\left(\fint_B \left| |u - u_B| - \left(\fint_B |u - u_B| \, d\mu \right) \right|^p d\mu \right)^{1/p} \leq C \operatorname{diam}(B) \left(\fint_{\lambda B} \rho^p \, d\mu \right)^{1/p}.$$

Because

$$\left(\fint_B |u - u_B|^p \, d\mu \right)^{1/p} - \fint_B |u - u_B| \, d\mu \leq \left(\fint_B \left| |u - u_B| - \left(\fint_B |u - u_B| \, d\mu \right) \right|^p d\mu \right)^{1/p},$$

we obtain the desired inequality on applying (9.1.5) to the real-valued function $|u - u_B|$. □

The argument leading to Theorem 9.1.2 can be improved in two respects. First, (9.1.1) was proved by chaining the Poincaré inequality. The use of the maximal function to replace the entire sum obtained from the chaining argument is rather crude. A better estimate can be obtained by splitting the sum into two parts and estimating only one of them by a maximal function. In this way we can improve the integrability of $|u - u_B|$ beyond p. Second, if we know that balls in X are reasonably shaped, say the metric is geodesic, then the chaining can be done more effectively in such a way that one never has to leave the original ball B. In this way we can dispense with the factor λ in the integral limit on the right-hand side of (9.1.3).

We will now embark on establishing these improvements. We split the discussion into two parts, corresponding to the real- and Banach space-valued cases, respectively.

Let us begin with a chaining estimate. Recall the definition of a geodesic metric space from Section 8.3.

Lemma 9.1.6 *Suppose that X is a geodesic and doubling metric measure space. Let $B = B(x_0, r_0)$ be an open ball in X and let $x \in B$. Then, for every $\lambda \geq 1$ and for every $0 < \epsilon < (r_0 - d(x_0, x))/10$, there is a sequence or chain $B_0, B_1, \ldots, B_{k+1}$ of open balls in X with the following properties:*

(i) $B_0 = B(x_0, r_0/2\lambda)$ *and* $B_{k+1} = B(x, \epsilon/\lambda)$*;*
(ii) $\lambda B_i \subset B$ *for every* $i = 0, 1, \ldots, k+1$*;*
(iii) $x \in (2\lambda + 1)B_i$ *for every* $i = 0, 1, \ldots, k+1$*;*

(iv) $\max\{\mu(B_i), \mu(B_{i+1})\} \le C\,\mu(B_i \cap B_{i+1})$ *for every* $i = 0, 1, \ldots, k$;
(v) $\sum_{i=0}^{k+1} \chi_{\lambda B_i}(y) \le C\chi_B(y)$ *for every* $y \in X$;
(vi) $\min\{1/(2\lambda), 2\lambda/(6(2\lambda+1))\}\,\mathrm{rad}(B_i) \le \mathrm{rad}(B_{i+1}) \le 2\,\mathrm{rad}(B_i)$ *for every* $i = 0, 1, \ldots, k$;
(vii) $\mathrm{rad}(B_{i+1}) \le 2\lambda/(2\lambda+1)\,\mathrm{rad}(B_i)$ *for all but at most three indices* $i = 0, 1, \ldots, k$.

The constant $C \ge 1$ *depends only on* λ *and the doubling constant of the underlying measure.*

Proof Fix λ and ϵ. Assume first that $x \in B \setminus B(x_0, 2r_0/3)$. Pick a geodesic γ_x joining x to x_0. Put $s = (2\lambda+1)/(2\lambda)$ and $\widehat{B}_0 = B(x, \epsilon/\lambda)$. Then we trace along γ_x, starting from x towards x_0, until we leave $\widehat{B}_0$ at a point z_1. Set $\widehat{B}_1 = B(z_1, s\epsilon/\lambda)$. Assuming that $\widehat{B}_i = B(z_i, s^i\epsilon/\lambda)$ has been defined for $i \ge 1$ and $B(x_0, r_0/(2\lambda)) \cap \widehat{B}_i = \emptyset$, we trace along γ_x from z_i towards x_0 until we leave $\widehat{B}_i$ at a point z_{i+1}. Then we set $\widehat{B}_{i+1} = B(z_{i+1}, s^{i+1}\epsilon/\lambda)$. The process terminates after a finite number of steps. More precisely, we note that $\widehat{B}_0 \cap B(x_0, r_0/(2\lambda)) = \emptyset$ and write $k - 1 = i \ge 1$ for the smallest integer i such that $B(x_0, r_0/(2\lambda)) \cap \widehat{B}_i \ne \emptyset$. Next, we pick a point z_k from the geodesic γ_x such that z_k lies in $\widehat{B}_{k-1}$ and satisfies $d(x_0, z_k) = r_0/(2\lambda)$; such a point exists because γ_x is a geodesic. Then we set $\widehat{B}_k = B(z_k, r_0/(2\lambda))$ and $\widehat{B}_{k+1} = B(x_0, r_0/(2\lambda))$. Finally, we define $B_i = \widehat{B}_{k+1-i}$ for $0 \le i \le k+1$. It remains to be checked that this chain of balls has the desired properties (i)–(vii).

First, condition (i) is clear. Condition (ii) is also clear for $i = 0, 1, k+1$. To check the remaining cases, fix $1 \le i \le k-1$ and pick $z \in \lambda\widehat{B}_i$. Then

$$
\begin{aligned}
d(x_0, z) \le d(x_0, z_i) + d(z_i, z) &< d(x_0, x) - \sum_{j=0}^{i-1} s^j\epsilon/\lambda + \epsilon s^i \\
&= d(x_0, x) - \epsilon\left(\frac{1}{\lambda}\frac{s^i - 1}{s - 1} - s^i\right) \\
&= d(x_0, x) - \epsilon\,(s^i - 2)\,.
\end{aligned}
$$

So, we have $\lambda\widehat{B}_i \subset B$ if $d(x_0, x) - \epsilon(s^i - 2) \le r_0$. Thus suppose that

$$
d(x_0, x) - \epsilon(s^i - 2) = d(x_0, x) + \epsilon(2 - s^i) > r_0.
$$

In this case, we have $0 < r_0 - d(x_0, x) < \epsilon(2 - s^i)$. By the condition imposed on ϵ, this indicates that $2 - s^i > 10$, which is impossible. Hence we have $\lambda\widehat{B}_i \subset B$ as required.

Thus, (ii) follows.

Next for (iii), we first observe that

$$d(z_i, x) < \frac{\epsilon}{\lambda}\sum_{j=1}^{i} s^j = \frac{\epsilon}{\lambda}\,\frac{s(s^i-1)}{s-1} < (2\lambda+1)\,\frac{s^i\epsilon}{\lambda}$$

for $i = 1, \ldots, k-1$. Furthermore, $d(z_i, x) \le r_0 \le (2\lambda+1)r_0/(2\lambda)$ for $i = k, k+1$. Because we also have $x \in B_{k+1}$, (iii) must hold.

To prove condition (iv), for $i = 0, 1, \ldots, k-2$ pick a point y_i in the geodesic γ_x between z_i and z_{i+1} satisfying

$$d(z_{i+1}, y_i) = d(z_i, y_i) = d(z_i, z_{i+1})/2\,,$$

where we understand that $z_0 = x$. For y_{k-1} pick a point in γ_x between z_{k-1} and z_k that satisfies

$$d(z_k, y_{k-1}) = \min\{r_0/4\lambda,\, d(z_k, z_{k-1})/2\}\,,$$

and for y_k pick a point in γ_x such that $d(x_0, y_k) = r_0/4\lambda$. By putting $t_i' = d(z_{i+1}, y_i)$, we have from the construction that

$$B(y_i, t_i') \subset \widehat{B}_i \cap \widehat{B}_{i+1} \tag{9.1.7}$$

for every $i = 0, \ldots, k$, where we note that $z_{k+1} = x_0$. Now (iv) follows from (9.1.7) on using the doubling condition of the measure, provided that we can show that the ratios of $\operatorname{rad}(\widehat{B}_i)$, $\operatorname{rad}(\widehat{B}_{i+1})$, and t_i' are bounded from above and below by a constant that depends only on λ. Indeed, this is clear for $i = 0, \ldots, k-2$ because $\operatorname{rad}(\widehat{B}_{i+1}) = s\operatorname{rad}(\widehat{B}_i) = 2st_i'$. We also have $\operatorname{rad}(\widehat{B}_{k+1}) = \operatorname{rad}(\widehat{B}_k) = 2t_k'$, and so it remains to study the case $i = k-1$. For this, we have the estimate

$$\frac{\epsilon}{\lambda}\,\frac{1-s^k}{1-s} = \frac{\epsilon}{\lambda}\sum_{i=0}^{k-1} s^i \ge \frac{r_0}{6},$$

which gives

$$r_0 \le \frac{6s}{s-1}\operatorname{rad}(\widehat{B}_{k-1}). \tag{9.1.8}$$

On the other hand, from (ii) and from the definitions we have that

$$\lambda\operatorname{rad}(\widehat{B}_{k-1}) \le r_0\left(1 - \frac{1}{2\lambda}\right). \tag{9.1.9}$$

Thus (iv) follows.

Next we will prove (v). Suppose that $z \in \lambda\widehat{B}_i \cap \lambda\widehat{B}_{i+j}$ for some $i = 0, \ldots, k-2$ and $j \geq 1$. Then

$$2\epsilon s^i(s^j - 1) = \sum_{k=i}^{i+j-1} \frac{\epsilon s^k}{\lambda} = d(z_i, z_{i+j}) \leq \lambda\left(\frac{s^i\epsilon}{\lambda} + \frac{s^{i+j}\epsilon}{\lambda}\right) = \epsilon s^i(1 + s^j),$$

from which we obtain that $j \leq \log(3)/\log((2\lambda+1)/(2\lambda))$, proving (v).

Condition (vi) is clear for $i = 2, \ldots, k$, as well as for $i = 0$. For the remaining case, $i = 1$, we have that $\operatorname{rad}(B_1) = \operatorname{rad}(\widehat{B}_k) = r_0/(2\lambda)$ and $\operatorname{rad}(B_2) = \operatorname{rad}(\widehat{B}_{k-1})$, and (vi) follows from (9.1.8) and (9.1.9) in this case as well.

Finally, it is clear that (vii) holds for every index $i = 2, \ldots, k$.

This finishes the proof in the case $x \in B \setminus B(x_0, 2r_0/3)$.

Assume now that $x \in B(x_0, 2r_0/3)$. Set $r_0' = 3d(x_0, x)/2$, $B' = B(x_0, r_0')$, and $\epsilon' = \min\{\epsilon, (r_0' - d(x_0, x))/20\}$. It follows from the first part of the proof, applied to B' and ϵ', that there are balls $B_0', B_1', \ldots, B_{k'+1}'$ such that (ii)–(vi) hold; in addition $B_{k'+1}' \subset B(x, \epsilon)$ and $B_0' = B(x_0, r_0'/(2\lambda)) \subset B(x_0, r_0/(2\lambda))$, and (vii) holds except for $i = 0, 1$. Next, let l be the smallest nonnegative integer satisfying $(2\lambda)^{l+1} r_0' > r_0$. Put $B_{-i}' = (2\lambda)^i B_0'$ for $i = 0, 1, \ldots, l$ and $B_{-l-1}' = B(x_0, r_0/(2\lambda))$. Now all the assertions (i)–(vii) are valid upon re-indexing the balls as follows: $B_i = B_{i-l-1}'$ for $i = 0, 1, \ldots, k+1$, where $k = k' + l + 1$.

The proof of Lemma 9.1.6 is complete. □

We will use Lemma 9.1.6 to establish the following improvement of estimate (8.1.56) in geodesic spaces. In the following statement, the phrase "the data of the hypotheses" refers to the constants in the Poincaré inequality together with the doubling constant of the underlying measure.

Lemma 9.1.10 *Suppose that X is a geodesic and doubling metric measure space supporting a p-Poincaré inequality for some $1 \leq p < \infty$, and suppose that V is a Banach space. Let $\lambda \geq 1$ be a constant that depends only on the data of the hypotheses and such that the conclusions in Theorems 8.1.53 and 8.1.55 are satisfied. Assume that $B = B(x_0, r_0)$ is an open ball in X, that $u : B \to V$ is an integrable function, and that $\rho : B \to [0, \infty]$ is a p-integrable upper gradient of u in B. Finally, let $x \in B$ and let $0 < \epsilon < (r_0 - d(x_0, x))/10$.*

Then, for a sequence of balls $B_0, B_1, \ldots, B_{k+1} \subset B$, associated with x, ϵ, and λ as in Lemma 9.1.6, we have that

$$|u_{B_{k+1}} - u_{B_0}| \leq C \sum_{i=0}^{k+1} \operatorname{diam}(B_i) \left(⨍_{\lambda B_i} \rho^p \, d\mu \right)^{1/p}. \tag{9.1.11}$$

In particular, if x is a Lebesgue point of u, we have that

$$|u(x) - u_{B_0}| \leq C \operatorname{diam}(B) \left(M\rho^p(x)\right)^{1/p}, \tag{9.1.12}$$

where M denotes the maximal function, as defined in (3.5.1), applied to the zero extension of ρ to the complement of B.

In both the inequalities (9.1.11) and (9.1.12) $C > 0$ is a constant that depends only on the data of the hypotheses.

Note that the existence of a constant $\lambda \geq 1$, as in the statement of the lemma, is made possible by Theorem 8.1.49.

Proof Let $B_0, B_1, \ldots, B_{k+1}$ be a chain of balls in B as in Lemma 9.1.6, corresponding to x, λ, and ϵ. We obtain an estimate as follows:

$$\begin{aligned}
|u_{B_{k+1}} - u_{B_0}| &\leq \sum_{i=0}^{k} |u_{B_{i+1}} - u_{B_i}| \\
&\leq \sum_{i=0}^{k} |u_{B_{i+1}} - u_{B_i \cap B_{i+1}}| + |u_{B_i} - u_{B_i \cap B_{i+1}}| \\
&\leq \sum_{i=0}^{k} \fint_{B_i \cap B_{i+1}} \left(|u - u_{B_{i+1}}| + |u - u_{B_i}|\right) d\mu \\
&\leq C \sum_{i=0}^{k+1} \fint_{B_i} |u - u_{B_i}| \, d\mu \\
&\leq C \sum_{i=0}^{k+1} \operatorname{diam}(B_i) \left(\fint_{\lambda B_i} \rho^p \, d\mu\right)^{1/p},
\end{aligned}$$

where we have used the Poincaré inequality (Theorems 8.1.53 and 8.1.49) together with Lemma 9.1.6(ii), (iii). This establishes (9.1.11). Next, by Lemma 9.1.6(iii) and by doubling, we obtain from (9.1.11) that

$$|u_{B_{k+1}} - u_{B_0}| \leq C \left(M\rho^p(x)\right)^{1/p} \sum_{i=0}^{k+1} \operatorname{diam}(B_i). \tag{9.1.13}$$

Because $u_{B_{k+1}} \to u(x)$ at a Lebesgue point x, as $\epsilon \to 0$, and because the radii of the balls B_i behave geometrically as described in Lemma 9.1.6(vi), (vii), in particular by (vii) we have

$$\operatorname{diam}(B_i) \leq C \left(\frac{2\lambda}{2\lambda + 1}\right)^i r_0/\lambda,$$

we obtain (9.1.12) from (9.1.13). The lemma follows upon noticing that, in a geodesic space, the diameter of a ball with nonempty complement is not smaller than its radius. □

In the classical Sobolev–Poincaré inequalities in $\mathbb{R}^n$, the dimension n plays a special role, particularly in the embedding theorems of Sobolev and Morrey. For our versions of the Sobolev–Poincaré-type estimate, a suitable substitute for this threshold parameter is given by a lower decay order for the measure of balls. We say that X has a *relative lower volume decay* of order $Q > 0$ if

$$\left(\frac{\operatorname{diam}(B')}{\operatorname{diam}(B)}\right)^Q \le C_0 \frac{\mu(B')}{\mu(B)} \tag{9.1.14}$$

holds whenever $B' \subset B$ are balls in X. Note that Lemma 8.1.13 implies that, when μ is doubling, inequality (9.1.14) always holds for some $Q \le \log_2 C_\mu$ whenever the two balls are concentric. By the doubling property of μ, it is easy to see that (9.1.14) is valid for all pairs of balls $B' \subset B$ if and only if it is valid for concentric balls; the constant C_0 may change but the exponent Q remains the same.

We establish the following real-valued case first.

Theorem 9.1.15 *Suppose that X is a geodesic and doubling metric measure space supporting a p-Poincaré inequality for some $1 \le p < \infty$. Assume moreover that X has a relative lower volume decay (9.1.14) of order $Q \ge 1$. Then there are positive constants C and c, depending only on the data of the hypotheses, such that the following three statements hold whenever B is an open ball in X, $u : B \to \mathbb{R}$ is an integrable function, and $\rho : B \to [0, \infty]$ is a p-integrable upper gradient of u in B.*

(i) *If $p < Q$ then*

$$\left(\fint_B |u - u_B|^{p^*}\, d\mu\right)^{1/p^*} \le C \operatorname{diam}(B) \left(\fint_B \rho^p\, d\mu\right)^{1/p}, \tag{9.1.16}$$

where $p^* = pQ/(Q-p)$.

(ii) *If $p = Q$ then*

$$\fint_B \exp\left(\left(\frac{|u - u_B|}{c \operatorname{diam}(B) \left(\fint_B \rho^Q\, d\mu\right)^{1/Q}}\right)^{Q/(Q-1)}\right) d\mu \le C\,. \tag{9.1.17}$$

(iii) *If $p > Q$ then*

$$\|u - u_B\|_{L^\infty(B)} \le C \operatorname{diam}(B) \left(\fint_B \rho^p\, d\mu\right)^{1/p}. \tag{9.1.18}$$

Remark 9.1.19 An application of Hölder's inequality, together with Theorem 9.1.15, reveals that when μ is doubling and X is a geodesic space supporting a p-Poincaré inequality, X actually supports the following version of a Poincaré inequality:

$$\left(\fint_B |u - u_B|^p \, d\mu\right)^{1/p} \le C \operatorname{diam}(B) \left(\fint_B \rho^p \, d\mu\right)^{1/p}.$$

Proof of Theorem 9.1.15 Let $B = B(x_0, r_0)$ be an open ball in X, let $u : B \to \mathbb{R}$ be an integrable function, and let ρ be a p-integrable upper gradient of u in B. In what follows, we let $C > 0$ denote any constant that depends only on the data described in the assertion. Moreover, we let $\lambda \ge 1$ be a constant, depending only on the data, such that the conclusions in Theorems 8.1.53 and 8.1.55 are satisfied.

We note that it suffices to establish (9.1.16) and (9.1.18) with u_B replaced by u_{B_0}, where $B_0 = B(x_0, r_0/(2\lambda))$. This follows from the simple estimate

$$|u - u_B| \le |u - u_{B_0}| + |u_{B_0} - u_B| \le |u - u_{B_0}| + \fint_B |u - u_{B_0}| \, d\mu. \tag{9.1.20}$$

The preceding understood, we now proceed to study the three cases.

Case (i), $p < Q$. Let $x \in B$ be a Lebesgue point of u such that $|u(x) - u_{B_0}| > 0$ (see the Lebesgue differentiation theorem in Section 3.4). If no such point exists, there is nothing to prove since then u is constant. Pick $0 < \epsilon < (r_0 - d(x_0, x))/10$ such that

$$|u(x) - u_{B_0}| \le 2\,|u_{B(x,r)} - u_{B_0}| \tag{9.1.21}$$

for $0 < r \le \epsilon$. Let $B_0, \dots, B_{k+1}$ be a chain of balls in B as in Lemma 9.1.6, corresponding to x, λ, and ϵ. Then (9.1.11) of Lemma 9.1.10 gives

$$|u_{B_{k+1}} - u_{B_0}| \le C \sum_{i=0}^{k+1} \operatorname{diam}(B_i) \left(\fint_{\lambda B_i} \rho^p \, d\mu\right)^{1/p}. \tag{9.1.22}$$

Next, fix $0 < t \le 2r_0$. We have from (9.1.22) that

$$|u_{B_{k+1}} - u_{B_0}| \le C(S_t + S^t), \tag{9.1.23}$$

where

$$S_t := \sum_{\operatorname{diam}(B_i) \le t} \operatorname{diam}(B_i) \left(\fint_{\lambda B_i} \rho^p \, d\mu\right)^{1/p}$$

and

$$S^t := \sum_{\operatorname{diam}(B_i)>t} \operatorname{diam}(B_i) \left(\fint_{\lambda B_i} \rho^p \, d\mu \right)^{1/p} .$$

Condition (vii) of Lemma 9.1.6 tells us that, by starting with the ball B_{i_0} of smallest index i_0 for which $\operatorname{diam}(B_i) \le t$, we actually have

$$\operatorname{diam}(B_i) \le C \left(\frac{2\lambda}{2\lambda+1} \right)^{i-i_0} t$$

for B_i satisfying $\operatorname{diam}(B_i) \le t$, and hence, by Lemma 9.1.6(iii) and the doubling condition, we obtain

$$S_t \le Ct \left(M\rho^p(x) \right)^{1/p} .$$

To estimate S^t, we apply (9.1.14) to obtain

$$\begin{aligned} S^t &= \sum_{\operatorname{diam}(B_i)>t} \operatorname{diam}(B_i) \left(\fint_{\lambda B_i} \rho^p \, d\mu \right)^{1/p} \\ &= \sum_{\operatorname{diam}(B_i)>t} \operatorname{diam}(B_i) \mu(\lambda B_i)^{-1/p} \left(\int_{\lambda B_i} \rho^p \, d\mu \right)^{1/p} \\ &\le C \sum_{\operatorname{diam}(B_i)>t} (\operatorname{diam}(B_i))^{(1-Q/p)} (\operatorname{diam}(B))^{Q/p} \mu(B)^{-1/p} \left(\int_B \rho^p \, d\mu \right)^{1/p} \\ &\le Ct^{(1-Q/p)} (\operatorname{diam}(B))^{Q/p} \left(\fint_B \rho^p \, d\mu \right)^{1/p} , \end{aligned}$$

where in the last step we have used the fact that $1 - Q/p < 0$ and the geometric nature of our series of diameters (Lemma 9.1.6(vi), (vii)); using the largest index i for which the ball B_i satisfies $\operatorname{diam}(B_i) \ge t$, by Lemma 9.1.6(vii) we see that

$$\operatorname{diam}(B_i) \ge C^{-1} \left(\frac{2\lambda+1}{2\lambda} \right)^{i-i_0} t.$$

By combining the preceding estimates for S_t and S^t with (9.1.21) and (9.1.23), we arrive at

$$|u(x) - u_{B_0}| \le C \left(t \left(M\rho^p(x) \right)^{1/p} + t^{(1-Q/p)} (\operatorname{diam}(B))^{Q/p} \left(\fint_B \rho^p \, d\mu \right)^{1/p} \right). \tag{9.1.24}$$

Next, note that

$$t \left(M\rho^p(x) \right)^{1/p} \le t^{(1-Q/p)} (\operatorname{diam}(B))^{Q/p} \left(\fint_B \rho^p \, d\mu \right)^{1/p}$$

if and only if

$$t \le \operatorname{diam}(B) \left(\fint_B \rho^p \, d\mu \right)^{1/Q} (M\rho^p(x))^{-1/Q} . \tag{9.1.25}$$

If the right-hand side of (9.1.25) does not exceed $\operatorname{diam}(B) \le 2r_0$ then we take t equal to that right-hand side and obtain from (9.1.24) that

$$|u(x) - u_{B_0}| \le C \operatorname{diam}(B) \left(\fint_B \rho^p \, d\mu \right)^{1/Q} (M\rho^p(x))^{(Q-p)/pQ},$$

whence

$$\begin{aligned} |u(x) - u_{B_0}|^{pQ/(Q-p)} &\le C\, (\operatorname{diam}(B))^{pQ/(Q-p)} \\ &\quad \times \left(\fint_B \rho^p \, d\mu \right)^{p/(Q-p)} M\rho^p(x) . \end{aligned} \tag{9.1.26}$$

If, however, the right-hand side in (9.1.25) is greater than $\operatorname{diam}(B)$ then we take $t = \operatorname{diam}(B)$ and similarly obtain from (9.1.24) that

$$|u(x) - u_{B_0}| \le C \operatorname{diam}(B) \left(\fint_B \rho^p \, d\mu \right)^{1/p} . \tag{9.1.27}$$

We conclude that, for almost every point $x \in B$, we have that either (9.1.26) or (9.1.27) holds.

Let A_1 denote the set of those points in B for which (9.1.26) holds and let A_2 consist of those points in B for which (9.1.27) holds. For $s > 0$, write $E_s = \{x \in B : |u(x) - u_{B_0}| > s\}$. Applying (9.1.26) and the weak-type estimate for the maximal function (given in Theorem 3.5.6), we arrive at

$$\begin{aligned} \mu(A_1 \cap E_s) &\le \mu \left(\left\{ M\rho^p > C\, (\operatorname{diam}(B))^{-p^*} \left(\fint_B \rho^p \, d\mu \right)^{-p^*/Q} s^{p^*} \right\} \right) \\ &\le C\, (\operatorname{diam}(B))^{p^*} s^{-p^*} \left(\fint_B \rho^p \, d\mu \right)^{p^*/Q} \left(\int_B \rho^p \, d\mu \right) \\ &= C\, (\operatorname{diam}(B))^{p^*} \mu(B) s^{-p^*} \left(\fint_B \rho^p \, d\mu \right)^{p^*/p} , \end{aligned}$$

where we recall the notation $p^* = pQ/(Q-p)$. Applying (9.1.27), we obtain that $A_2 \cap E_s = \emptyset$ whenever $s \ge C \operatorname{diam}(B) \left(\fint_B \rho^p \, d\mu \right)^{1/p}$, while for all smaller values of s we find that

$$\mu(A_2 \cap E_s) \le \mu(B) \le C\, (\operatorname{diam}(B))^{p^*} \mu(B) s^{-p^*} \left(\fint_B \rho^p \, d\mu \right)^{p^*/p} .$$

In conclusion, by combining the preceding inequalities we arrive at the estimate

$$\begin{aligned}\mu(\{|u-u_{B_0}|>s\}) &\le C\,(\mathrm{diam}(B))^{p^*}\mu(B)s^{-p^*}\left(\fint_B \rho^p\,d\mu\right)^{p^*/p}\\ &\le C\,(\mathrm{diam}(B))^{p^*}\mu(B)^{-p^*/Q}s^{-p^*}\left(\int_B \rho^p\,d\mu\right)^{p^*/p}.\end{aligned}$$

The claim now follows from this last estimate and from Lemma 8.1.31. Note that although Lemma 8.1.31 does not directly apply in this situation, the proof of that lemma does apply: because of condition (i) of Lemma 9.1.6, $\mu(B_0)$ is comparable to $\mu(B)$ and one simply chooses the cut off level t using B_0.

Case (ii), $p = Q$. From (9.1.16) and Hölder's inequality we already know that we can bound the averaged L^q-norm of $|u - u_B|$ by the averaged L^Q-norm of ρ for every $q < \infty$. The key in the present case will be to estimate the constant in this inequality and to expand the exponential function as a power series. By (9.1.20) it is direct to verify, using Hölder's inequality, that we may replace u_B by u_{B_0}.

Fix $q > \max\{Q, Q/(Q-1)\}$ and a Lebesgue point $x \in B$ of u. Then fix $0 < \delta < q^{-1}$, to be determined later. Using (9.1.21) and (9.1.22), we find a chain of balls $B_0, \dots, B_{k+1}$ (Lemma 9.1.6) and obtain from Lemma 9.1.10 that

$$\begin{aligned}|u(x)-u_{B_0}| &\le C\sum_{i=0}^{k+1}\mathrm{diam}(B_i)\left(\fint_{\lambda B_i}\rho^Q\,d\mu\right)^{1/Q}\\ &\le C\sum_{i=0}^{k+1}\mathrm{diam}(B_i)^{1-\delta}\mu(B_i)^{1/q-1/Q}\\ &\quad\times\left((\mathrm{diam}(B_i))^{q\delta}\fint_{\lambda B_i}\rho^Q\,d\mu\right)^{1/q}\left(\int_{\lambda B_i}\rho^Q\,d\mu\right)^{1/Q-1/q}.\end{aligned}$$

Since $(Q-1)/Q + 1/q + (1/Q - 1/q) = 1$, we can use Hölder's inequality to estimate the last sum from above by

$$\begin{aligned}&C\left(\sum_{i=0}^{k+1}\left((\mathrm{diam}(B_i))^{1-\delta}\mu(B_i)^{1/q-1/Q}\right)^{Q/(Q-1)}\right)^{(Q-1)/Q}\\ &\quad\times\left(\sum_{i=0}^{k+1}(\mathrm{diam}(B_i))^{q\delta}M\rho^Q(x)\right)^{1/q}\left(\int_B\rho^Q\,d\mu\right)^{1/Q-1/q},\end{aligned}$$

where we have replaced the averaged integrals $\fint_{\lambda B_i} \rho^Q \, d\mu$ by the maximal function $CM\rho^Q(x)$ and used the bounded overlap of the balls λB_i to estimate the last sum (Lemma 9.1.6(ii), (iii), (v)).

We pause here to record the estimate

$$|u(x) - u_{B_0}| \le CT_1 T_2 T_3 , \tag{9.1.28}$$

where the three factors are

$$T_1 = \left(\sum_{i=0}^{k+1} \left((\operatorname{diam}(B_i))^{1-\delta} \mu(B_i)^{1/q-1/Q} \right)^{Q/(Q-1)} \right)^{(Q-1)/(Q)},$$

$$T_2 = \left(\sum_{i=0}^{k+1} (\operatorname{diam}(B_i))^{q\delta} M\rho^Q(x) \right)^{1/q},$$

$$T_3 = \left(\int_B \rho^Q \, d\mu \right)^{1/Q-1/q}.$$

To estimate T_2, we write $a = 2\lambda/(2\lambda + 1)$ and use Lemma 9.1.6(i), (vii) to obtain

$$\sum_{i=0}^{k+1} (\operatorname{diam}(B_i))^{q\delta} M\rho^Q(x) \le C \frac{r_0^{q\delta}}{1 - a^{q\delta}} M\rho^Q(x) \le C(q\delta)^{-1} r_0^{q\delta} M\rho^Q(x).$$

In the last inequality we employed the fact that $q\delta < 1$. So, by the choice of a, we have $1 - a^{q\delta} = (2\lambda + 1)^{-q\delta} ((2\lambda + 1)^{q\delta} - (2\lambda)^{q\delta})$ and hence

$$(1 - a^{q\delta})^{-1} \le C(q\delta)^{-1} (2\lambda)^{-1} (2\lambda(2\lambda + 1))^{q\delta} \le C_\lambda (q\delta)^{-1}.$$

Thus

$$T_2 \le C(q\delta)^{-1/q} r_0^{\delta} (M\rho^Q(x))^{1/q}. \tag{9.1.29}$$

Next, to estimate T_1 we use the lower bound on measures (9.1.14) and argue as earlier, now using the fact that $q\delta < 1 \le Q < q$, to obtain

$$\begin{aligned} &\sum_{i=0}^{k+1} \left((\operatorname{diam}(B_i))^{1-\delta} \mu(B_i)^{1/q-1/Q} \right)^{Q/(Q-1)} \\ &\quad \le C(r_0^{1-Q/q} \mu(B)^{1/q-1/Q})^{Q/(Q-1)} \sum_{i=0}^{k+1} (\operatorname{diam}(B_i))^{(Q/q-\delta)Q/(Q-1)} \\ &\quad \le Cq(Q - \delta q)^{-1} \left(r_0^{1-\delta} \mu(B)^{1/q-1/Q} \right)^{Q/(Q-1)}. \end{aligned}$$

If we let $\delta = Qq^{-2}$ then $q(Q - \delta q)^{-1} = q(Q - Q/q)^{-1} \le q$, since $q \ge Q/(Q - 1)$.

Combining the above estimates, (9.1.28) yields

$$|u(x) - u_{B_0}| \le C r_0\, q^{1/q+(Q-1)/Q} \mu(B)^{1/q-1/Q} \times \left(\int_B \rho^Q \, d\mu\right)^{1/Q-1/q} (M\rho^Q(x))^{1/q}, \tag{9.1.30}$$

where in particular $C > 0$ is independent of $q > \max\{Q, Q/(Q-1)\}$. We proceed to estimate the integral of $|u - u_{B_0}|^{q/2}$. Recalling that (9.1.30) is valid for almost every point $x \in B$, we arrive at

$$\int_B |u - u_{B_0}|^{q/2} \, d\mu \le C^q \, r_0^{q/2} \, q^{1/2+q(Q-1)/(2Q)} \mu(B)^{1/2-q/(2Q)} \times \left(\int_B \rho^Q \, d\mu\right)^{q/(2Q)-1/2} \int_B \left(M\rho^Q\right)^{1/2} \, d\mu.$$

Lemma 3.5.10 gives in turn that

$$\int_B \left(M\rho^Q\right)^{1/2} \, d\mu \le C \left(\mu(B) \int_B \rho^Q \, d\mu\right)^{1/2},$$

and hence we conclude that

$$\fint_B |u - u_{B_0}|^{q/2} \, d\mu \le C^q q^{1/2+q(Q-1)/(2Q)} \left(r_0^Q \fint_B \rho^Q \, d\mu\right)^{q/(2Q)}, \tag{9.1.31}$$

where C does not depend on $q > \max\{Q, Q/(Q-1)\}$. Notice that estimate (9.1.31) holds for $1 \le q \le \max\{Q, Q/(Q-1)\}$ as well, by Hölder's inequality and the first part of the theorem.

Now, for $t > 0$,

$$\exp\left((t|u(x) - u_{B_0}|)^{Q/(Q-1)}\right) = \sum_{m=1}^{\infty} (m!)^{-1} (t|u(x) - u_{B_0}|)^{mQ/(Q-1)}.$$

Integrating over B and using estimate (9.1.31), we obtain

$$\fint_B \exp\left((t|u(x) - u_B|)^{Q/(Q-1)}\right) d\mu \le 1 + C \sum_{m=1}^{\infty} (m!)^{-1} m^{1/2+m} \left(t^Q r_0^Q \fint_B \rho^Q \, d\mu\right)^{m/(Q-1)}.$$

The above series converges, provided that

$$t r_0 \left(\fint_B \rho^Q \, d\mu\right)^{1/Q} < e^{-(Q-1)/Q}$$

and the proof of (9.1.17) is thereby complete.

Case (iii), $p > Q$. As in the beginning of the proof of case (ii), $p = Q$, we find balls $B_0, \ldots, B_{k+1}$ as in Lemma 9.1.6 such that

$$|u(x) - u_{B_0}| \le C \sum_{i=0}^{k+1} \operatorname{diam}(B_i)\, \mu(B_i)^{-1/p} \left(\int_{B_i} \rho^p \, d\mu \right)^{1/p}$$

given a Lebesgue point x of u in B_0. By combining this with the measure decay estimate (9.1.14) and the geometric decay of the radii of B_i (Lemma 9.1.6(vi), (vii)), we deduce that

$$\begin{aligned} |u(x) - u_{B_0}| &\le C \sum_{i=1}^{k+1} (\operatorname{diam}(B_i))^{(1-Q/p)} (\operatorname{diam}(B))^{Q/p} \\ &\quad \times \mu(B)^{-1/p} \left(\int_{B_i} \rho^p \, d\mu \right)^{1/p} \\ &\le C \operatorname{diam}(B) \left(⨍_B \rho^p \, d\mu \right)^{1/p}, \end{aligned}$$

as desired.

The proof of Theorem 9.1.15 is complete. □

Now we turn to the general Banach-space-valued case. The statement remains the same, but the proof will differ in the case $p < Q$.

Theorem 9.1.32 *Suppose that X is a geodesic and doubling metric measure space supporting a p-Poincaré inequality for some $1 \le p < \infty$ and that V is a Banach space. Assume further that X has the relative lower volume decay property* (9.1.14) *of order $Q \ge 1$.*

Then there are positive constants C and c, depending only on the data of the hypotheses, such that the following three statements hold whenever B is an open ball in X, $u : B \to V$ is an integrable function, and $\rho : B \to [0, \infty]$ is a p-integrable upper gradient of u in B.

(i) *If $p < Q$, then*

$$\left(⨍_B |u - u_B|^{p^*} \, d\mu \right)^{1/p^*} \le C \operatorname{diam}(B) \left(⨍_B \rho^p \, d\mu \right)^{1/p}, \quad (9.1.33)$$

where $p^* = pQ/(Q - p)$.

(ii) *If $p = Q$ then*

$$⨍_B \exp\left(\left(\frac{|u - u_B|}{c \operatorname{diam}(B) \left(⨍_B \rho^Q \, d\mu \right)^{1/Q}} \right)^{Q/(Q-1)} \right) d\mu \le C. \quad (9.1.34)$$

(iii) *If $p > Q$ then*

$$\|u - u_B\|_{L^\infty(B:V)} \le C\,\mathrm{diam}(B)\left(\fint_B \rho^p\,d\mu\right)^{1/p}. \tag{9.1.35}$$

Proof An inspection of the proof of Theorem 9.1.15 reveals that both (9.1.34) and (9.1.35) follow from (9.1.33) together with the general arguments given in the proof; the fact that u is real-valued there plays no role. Hence the burden of proof here is to establish (9.1.33) for V-valued functions. This is done using the technique employed in the proof of Corollary 9.1.4 and the real-valued result from Theorem 9.1.15 as follows. As in the proof of Corollary 9.1.4, we note that the function $x \mapsto |u(x) - u_B|$ also has ρ as an upper gradient in B and so, by (9.1.16),

$$\begin{aligned}\left(\fint_B |u - u_B|^{p^*}\,d\mu\right)^{1/p^*} &- \fint_B |u - u_B|\,d\mu \\ &\le \left(\fint_B \left||u - u_B| - \fint_B |u - u_B|\,d\mu\right|^{p^*} d\mu\right)^{1/p^*} \\ &\le C\,\mathrm{diam}(B)\left(\fint_B \rho^p\,d\mu\right)^{1/p}.\end{aligned}$$

It remains to estimate $\fint_B |u - u_B|\,d\mu$. The proof is completed via the estimate (9.1.12) and the argument that we employed to prove Theorem 9.1.2. □

The assumption that X is geodesic allowed us to integrate g over B instead of over the larger ball λB, in Theorems 9.1.15 and 9.1.32. Even without this assumption, one can still obtain Sobolev–Poincaré-type inequalities, provided that we use larger balls. For this we restrict our attention to the quasiconvex case, which is easily handled using our previous results. Indeed, if X is quasiconvex (as is the case if X is complete and supports a p-Poincaré inequality; see Theorem 8.3.2), we may consider the associated geodesic metric $\hat{d}_X$ as in Proposition 8.3.12. Then $\hat{d}_X$ is bilipschitz equivalent to the original metric d of X because of the quasiconvexity of X. Therefore, by Lemma 8.3.18, the metric measure space $(X, \hat{d}_X, \mu)$ is a geodesic space that supports a p-Poincaré inequality (with the same p), with constants that depend on those associated with the original p-Poincaré inequality and the quasiconvexity constant. We may apply Theorem 9.1.32 to $(X, \hat{d}_X, \mu)$, and then use the fact that d and $\hat{d}_X$ are bilipschitz equivalent, to obtain the following corollary to Theorem 9.1.32.

Corollary 9.1.36 *Suppose that X is a quasiconvex and doubling metric measure space supporting a p-Poincaré inequality for some $1 \le p < \infty$ and that V is a Banach space. Assume moreover that X has the relative lower volume* *decay property* (9.1.14) *of order* $Q \geq 1$. *Then there are positive constants* C, τ, *and* c *depending only on the data of the hypotheses such that the following three statements hold whenever* B *is an open ball in* X, $u : B \to V$ *is an integrable function, and* $\rho : B \to [0, \infty]$ *is a* p*-integrable upper gradient of* u *in* B.

(i) *If* $p < Q$ *then*

$$\left(\fint_B |u - u_B|^{p^*}\, d\mu\right)^{1/p^*} \leq C \operatorname{diam}(B)\left(\fint_{\tau B} \rho^p\, d\mu\right)^{1/p}, \tag{9.1.37}$$

where $p^* = pQ/(Q - p)$.

(ii) *If* $p = Q$ *then*

$$\fint_B \exp\left(\left(\frac{|u - u_B|}{c \operatorname{diam}(B)\left(\fint_{\tau B} \rho^Q\, d\mu\right)^{1/Q}}\right)^{Q/(Q-1)}\right) d\mu \leq C. \tag{9.1.38}$$

(iii) *If* $p > Q$, *then*

$$\|u - u_B\|_{L^\infty(B:V)} \leq C \operatorname{diam}(B)\left(\fint_{\tau B} \rho^p\, d\mu\right)^{1/p}. \tag{9.1.39}$$

When $p > Q$, it is possible to obtain a better estimate than (9.1.35) or (9.1.39), namely that given by the Morrey embedding theorem, which states that functions in $N^{1,p}(X : V)$ are Hölder continuous. Notice that this follows for suitable Lebesgue representatives from (9.1.39) and (9.1.14) via the triangle inequality. We will give a more refined statement in Section 9.2 after discussing the Lebesgue points of Sobolev functions.

9.2 Lebesgue points of Sobolev functions

In the Lebesgue differentiation theorem (see Section 3.4) we saw that if a function is locally integrable in X then μ-a.e. point in X is a Lebesgue point of that function. Given that the Sobolev functions as considered in this book are better defined than that (for example, they are quasicontinuous), there should be a refined version of this result for Sobolev functions. The goal of this section is to study such a result.

We again assume that the measure on X is doubling. Hence it follows from the discussion in Section 4.1 that X is a doubling metric space and that by Lemma 4.1.12, for each $K \geq 1$ there is a constant $C = C_K > 0$ such that for every $r > 0$ we can find a countable cover of X of the form $\{B(x_i, r)\}_i$ such that $\sum_i \chi_{B(x_i, Kr)} \leq C$.

Correspondingly, as in Section 4.1, we can find a partition of unity subordinate to the above cover; for every i there is a C/r-Lipschitz function $\varphi_{r,i} : X \to [0, 1]$ such that the support of $\varphi_{r,i}$ lies in $B(x_i, 2r)$ and $\sum_i \varphi_{r,i} \equiv 1$.

Given a measurable function $u : X \to V$, let

$$u_r(x) := \sum_i \varphi_{r,i}(x) u_{B(x_i,r)}.$$

Such a function u_r is called a *discrete convolution* of u. This concept was studied by Coifman and Weiss in [63].

Let (r_j) be an enumeration of the positive rationals, and let $M^*u := \sup_j |u|_{r_j}$. Recall also the maximal function $Mu(x) = \sup_{r>0} \fint_{B(x,r)} |u| \, d\mu$.

Lemma 9.2.1 *There exists a constant $C > 0$ such that, for every measurable function $u : X \to V$,*

$$\frac{1}{C} Mu \le M^*u \le CMu.$$

Proof Let $x \in X$. Then

$$|u|_{r_j}(x) = \sum_i \varphi_{r_j,i}(x) |u|_{B(x_i,r_j)}.$$

If i is such that $\varphi_{r_j,i}(x) \neq 0$ then $x \in B(x_i, 2r_j)$, which in turn implies that $B(x_i, 2r_j) \subset B(x, 4r_j)$. Therefore, by the doubling property of μ,

$$|u|_{r_j}(x) \le \sum_i \varphi_{r_j,i}(x) |u|_{B(x,4r_j)} \frac{\mu(B(x, 4r_j))}{\mu(B(x_i, r_j))} \le C|u|_{B(x,4r_j)} \le CMu(x).$$

Taking the supremum over j yields the second inequality.

Next, if $r > 0$, we can find some r_j such that $r_j/4 \le r < r_j/2$. Let $I_j(x) = \{i \in \mathbb{N} : B(x_i, r_j) \cap B(x, r) \neq \emptyset\}$. By the doubling property of μ this is a nonempty finite set. For each $i \in I_j(x)$ we have $B(x, r) \subset B(x_i, 2r_j)$ and $B(x_i, r_j) \subset B(x, 6r)$. By the doubling property of μ,

$$|u|_{B(x,r)} \le C \, |u|_{B(x_i,2r_j)}.$$

Therefore

$$|u|_{B(x,r)} \le C \sum_{i \in I_j(x)} \varphi_{r_j,i}(x) |u|_{B(x_i,2r_j)} = C \, |u|_{2r_j} \le C \, M^*u(x).$$

Now taking the supremum over all $r > 0$ yields the first inequality. □

In what follows, we assume that X supports a better Poincaré inequality. As will be seen in Chapter 12, this requirement is not at all restrictive when X is complete and μ is doubling.

Proposition 9.2.2 *Suppose that $p > 1$, $u \in N^{1,p}(X : V)$, $g_u \in L^p(X)$ is the minimal p-weak upper gradient of u, and X supports a q-Poincaré inequality for some $1 \le q < p$. Then for every $r > 0$ we have that the discrete convolution $u_r \in N^{1,p}(X : V)$ and that there is a constant $C > 0$, independent of u and r, such that $C(Mg_u^q)^{1/q} \in L^p(X)$ is a p-weak upper gradient of u_r. Moreover, M^*u belongs to $N^{1,p}(X)$ with $C\big(Mg_u^q\big)^{1/q}$ as a p-weak upper gradient.*

Proof Note that

$$u_r(x) = u(x) + \sum_i \varphi_{r,i}(x)(u_{B(x_i,r)} - u(x)).$$

Therefore, by (6.3.18) and Proposition 6.3.28,

$$g_u + \sum_i \left(\frac{C}{r} |u_{B(x_i,r)} - u| + g_u \right) \chi_{B(x_i,2r)}$$

is a p-weak upper gradient of u_r. Note that the sum is locally finite.

If $x \in B(x_i, 2r)$ then

$$|u(x) - u_{B(x_i,r)}| \le |u(x) - u_{B(x,4r)}| + |u_{B(x,4r)} - u_{B(x_i,r)}|$$

and, by the Poincaré inequality,

$$\begin{aligned} |u_{B(x,4r)} - u_{B(x_i,r)}| &\le \fint_{B(x_i,r)} |u - u_{B(x,4r)}| \, d\mu \\ &\le C \fint_{B(x,4r)} |u - u_{B(x,4r)}| \, d\mu \\ &\le Cr \left(\fint_{B(x,4r)} g_u^q \, d\mu \right)^{1/q} \\ &\le Cr \left(Mg_u^q(x) \right)^{1/q}. \end{aligned}$$

Moreover, by (8.1.9), for μ-a.e. $x \in X$,

$$|u(x) - u_{B(x,4r)}| \le Cr\big(Mg_u^q\big)^{1/q}(x).$$

Therefore, for μ-a.e. $x \in B(x_i, 2r)$,

$$|u(x) - u_{B(x_i,r)}| \le Cr\big(Mg_u^q\big)^{1/q}(x).$$

By the Lebesgue differentiation theorem (see Section 3.4), for μ-a.e. $x \in X$ we have $g_u(x) \le \left(Mg_u^q(x)\right)^{1/q}$. So $C\left(Mg_u^q\right)^{1/q}$ is a p-weak upper gradient of u_r. Moreover, by Theorem 3.5.6 and the fact that $q < p$, this function is p-integrable.

Given k, since at most C balls $B(x_i, 2r)$ intersect the ball $B(x_k, r)$, we see by an application of Hölder's inequality that

$$\int_{B(x_k,r)} |u_r|^p \, d\mu \le C \sum_{\substack{i \\ B(x_i,2r)\cap B(x_k,r)\neq\emptyset}} \frac{\mu(B(x_k, r))}{\mu(B(x_i, r))} \int_{B(x_i,r)} |u|^p \, d\mu$$
$$\le C \int_{B(x_k,3r)} |u|^p \, d\mu$$

and, summing over k, we conclude that $u_r \in N^{1,p}(X : V)$.

Towards the last claim, notice that $|u| \in N^{1,p}(X)$ with g_u as a p-weak upper gradient of $|u|$, because of (6.3.18). Again by (6.3.18) and by the first part of our claim, for each j we have that $|u|_{r_j} \in N^{1,p}(X)$ with $C\left(Mg_u^q\right)^{1/q}$ as a p-weak upper gradient. For $k \in \mathbb{N}$, let $v_k = \max_{1\le j\le k} |u|_{r_j}$. Then, by Proposition 6.3.23, $v_k \in N^{1,p}(X)$ with the same p-weak upper gradient.

By Lemma 9.2.1 and Theorem 3.5.6, we know that $M^*u \in L^p(X)$ and hence, by the monotone convergence theorem, $v_k \to M^*u$ in $L^p(X)$. Now, by the second part of Proposition 7.3.7, $M^*u \in N^{1,p}(X)$ with $C\left(Mg_u^q\right)^{1/q}$ as a p-integrable p-weak upper gradient. □

Lemma 9.2.3 *If* $0 < r < 1$ *and* $x \in X$ *then*

$$\mathrm{Cap}_p(B(x, r)) \le C\,\frac{\mu(B(x, r))}{r^p}.$$

Proof Let $u : X \to \mathbb{R}$ be the Lipschitz function given by

$$u(y) = \min\left\{\frac{\mathrm{dist}(y, X \setminus B(x, 2r))}{r}, 1\right\}.$$

Then $u = 1$ on $B(x, r)$, $0 \le u \le 1$ on X, and u is supported on $B(x, 2r)$. Moreover, u is $1/r$-Lipschitz. Hence

$$\mathrm{Cap}_p(B(x, r)) \le \|u\|^p_{N^{1,p}(X)} \le C\mu(B(x, 2r)) + \frac{\mu(B(x, 2r))}{r^p}.$$

Since $r < 1$, the conclusion follows from the doubling property of μ. □

Lemma 9.2.4 *If* f *is a nonnegative function in* $L^1_{\mathrm{loc}}(X)$ *and*

$$E = \left\{x \in X \;:\; \limsup_{r\to 0^+} r^p \fint_{B(x,r)} f \, d\mu > 0\right\}$$

then $\mathrm{Cap}_p(E) = 0$.

Proof We prove this lemma for the case $f \in L^1(X)$, the local version being similar. It suffices to show that the set

$$E_\epsilon := \left\{ x \in X : \limsup_{r \to 0^+} r^p \fint_{B(x,r)} f \, d\mu > \epsilon \right\}$$

has zero p-capacity for each $\epsilon > 0$. Fix $\epsilon > 0$.

Recall that, by the absolute continuity of integrals, for every $\epsilon_1 > 0$ there exists $\tau > 0$ such that whenever $A \subset X$ is a measurable set with $\mu(A) \le \tau$, then $\int_A f \, d\mu < \epsilon_1$. For fixed $\epsilon_1 > 0$, let τ be as above and choose $0 < \delta < 1/5$ such that

$$\frac{\delta^p}{\epsilon} \int_X f \, d\mu < \tau.$$

Note that for every $x \in E_\epsilon$ there is some r_x with $0 < r_x < \delta$ such that

$$r_x^p \fint_{B(x,r_x)} f \, d\mu > \epsilon. \tag{9.2.5}$$

We can cover E_ϵ by such balls and, by the $5B$-covering lemma (see Section 3.3), we can find a countable pairwise disjoint subcollection $\{B_i := B(x_i, r_i)\}_i$ such that $E_\epsilon \subset \bigcup_i B(x_i, 5r_i)$. Now, by Lemma 9.2.3, since $5r_i \le 5\delta < 1$,

$$\mathrm{Cap}_p(E_\epsilon) \le \sum_i \mathrm{Cap}_p(B(x_i, 5r_i)) \le C \sum_i \frac{\mu(B_i)}{r_i^p}$$

and hence, by the choice of such balls in the cover and by (9.2.5), we have on the one hand

$$\mathrm{Cap}_p(E_\epsilon) \le \frac{C}{\epsilon} \sum_i \int_{B_i} f \, d\mu = \frac{C}{\epsilon} \int_{\bigcup_i B_i} f \, d\mu.$$

On the other hand,

$$\mu\left(\bigcup_i B_i\right) = \sum_i \mu(B_i) \le \sum_i \frac{r_i^p}{\epsilon} \int_{B_i} f \, d\mu \le \frac{\delta^p}{\epsilon} \int_X f \, d\mu < \tau.$$

Therefore

$$\mathrm{Cap}_p(E_\epsilon) \le \frac{C}{\epsilon} \epsilon_1,$$

and letting $\epsilon_1 \to 0$, we have $\mathrm{Cap}_p(E_\epsilon) = 0$. The lemma is now proved. □

Lemma 9.2.6 *Suppose that $p > 1$ and that X supports a q-Poincaré inequality for some $1 \le q < p$. If $u \in N^{1,p}(X : V)$ then, for every $\lambda > 0$,*

$$\mathrm{Cap}_p(\{x \in X : Mu(x) > \lambda\}) \le \frac{C}{\lambda^p} \|u\|^p_{N^{1,p}(X:V)}.$$

Proof Let

$$E_\lambda = \{x \in X : CM^*u(x) > \lambda\},$$

where C is the comparison constant from Lemma 9.2.1. Then

$$\{x \in X : Mu(x) > \lambda\} \subset E_\lambda,$$

and hence the desired p-capacity is estimated from above by $\mathrm{Cap}_p(E_\lambda)$. Thus it suffices to prove the above estimate for E_λ.

Observe that $(C/\lambda)M^*u$ is in $N^{1,p}(X)$ by Proposition 9.2.2, and hence is admissible for estimating the p-capacity of E_λ. Again by Proposition 9.2.2 and by Lemma 9.2.1 and the Hardy–Littlewood maximal theorem 3.5.6 with exponents $p > 1$ and $p/q > 1$,

$$\begin{aligned}\mathrm{Cap}_p(E_\lambda) &\le \frac{C}{\lambda^p}\|M^*u\|^p_{N^{1,p}(X)}\\ &\le \frac{C}{\lambda^p}\left(\|M^*u\|^p_{L^p(X)} + \|\left(Mg_u^q\right)^{1/q}\|^p_{L^p(X)}\right)\\ &\le \frac{C}{\lambda^p}\left(\|u\|^p_{L^p(X)} + \|g_u\|^p_{L^p(X)}\right) \le \frac{C}{\lambda^p}\|u\|^p_{N^{1,p}(X:V)}.\end{aligned}$$

Thus the lemma is proved. □

We say that a property holds for *p-a.e. point* in X if the set of points for which the property does not hold has p-capacity zero. As in the Lebesgue differentiation theorem (Section 3.4), a point $x \in X$ is a Lebesgue point of a measurable function u if

$$\lim_{r\to 0} \fint_{B(x,r)} |u - u(x)|\, d\mu = 0.$$

Recall from Corollary 9.1.36 that

$$\left(\fint_B |u - u_B|^{p*}\, d\mu\right)^{1/p*} \le C \,\mathrm{diam}(B) \left(\fint_{\tau B} g_u^p\, d\mu\right)^{1/p} \tag{9.2.7}$$

whenever $u \in N^{1,p}(X : V)$ and B is a ball in X, provided that X supports a p-Poincaré inequality.

We now come to the main theorem of this section.

Theorem 9.2.8 *Suppose that $p > 1$, X supports a q-Poincaré inequality for some $1 \le q < p$, and μ satisfies the relative lower volume decay property* (9.1.14) *of order $Q \ge 1$. Let $u \in N^{1,p}(X : V)$. Then p-a.e. point in X is a Lebesgue point of u. Furthermore, if $p < Q$ then, for p-a.e. $x \in X$,*

$$\lim_{r\to 0} \fint_{B(x,r)} |u - u(x)|^{p^*}\, d\mu = 0 \tag{9.2.9}$$

where

$$p^* = \frac{pQ}{Q-p}.$$

Proof Let

$$A = \left\{ x \in X : \limsup_{r \to 0} r^p \fint_{B(x,r)} g_u^p \, d\mu > 0 \right\}.$$

Since $g_u \in L^p(X)$, we have $g_u^p \in L^1(X)$ and hence, by Lemma 9.2.4, $\mathrm{Cap}_p(A) = 0$. By the Poincaré inequality, if $x \in X \setminus A$ then

$$\begin{aligned}\left(\fint_{B(x,r)} |u - u_{B(x,r)}| \, d\mu \right)^p &\le C r^p \fint_{B(x,\lambda r)} g_u^p \, d\mu \\ &\to 0 \quad \text{as} \quad r \to 0,\end{aligned}$$

that is,

$$\lim_{r \to 0} \fint_{B(x,r)} |u - u_{B(x,r)}| \, d\mu = 0 \tag{9.2.10}$$

whenever $x \in X \setminus A$. Since X supports a p-Poincaré inequality, we know that Lipschitz functions are dense in $N^{1,p}(X : V)$; see Theorem 8.2.1. Let (u_n) be a sequence of Lipschitz functions in $N^{1,p}(X : V)$ such that $\|u - u_n\|^p_{N^{1,p}(X:V)} \le 2^{-n(p+1)}$ for each n and there exists a set K with $\mathrm{Cap}_p(K) = 0$ for which $u_n \to u$ pointwise everywhere in $X \setminus K$. Such a sequence exists because of Proposition 7.3.7. For $n \in \mathbb{N}$, let $A_n = \{x \in X : M(u - u_n)(x) > 2^{-n}\}$ and set $E_n = A \cup K \cup \left(\bigcup_{k \ge n} A_k \right)$. By Lemma 9.2.6,

$$\mathrm{Cap}_p(A_n) \le C \times 2^{np} \|u - u_n\|^p_{N^{1,p}(X:V)} \le C \times 2^{-n}.$$

Then, by the subadditivity of the p-capacity (Lemma 7.2.4),

$$\mathrm{Cap}_p(E_n) \le 2C \times 2^{-n}.$$

Note that

$$\begin{aligned}|u_k(x) - u_{B(x,r)}| &\le \fint_{B(x,r)} |u - u_k(x)| \, d\mu \\ &\le \fint_{B(x,r)} |u_k - u| \, d\mu + \fint_{B(x,r)} |u_k - u_k(x)| \, d\mu \\ &\le M(u_k - u)(x) + \fint_{B(x,r)} |u_k - u_k(x)| \, d\mu.\end{aligned}$$

Hence, if $x \in X \setminus E_n$ and $k \geq n$ then

$$\begin{aligned}\limsup_{r\to 0} |u_k(x) - u_{B(x,r)}| &\leq \limsup_{r\to 0} \fint_{B(x,r)} |u - u_k(x)|\, d\mu \\ &\leq M(u_k - u)(x) \leq 2^{-k}.\end{aligned} \tag{9.2.11}$$

Therefore, for every $x \in X \setminus E_n$ and for every $l \geq k \geq n$,

$$\begin{aligned}|u_k(x) - u_l(x)| &\leq \limsup_{r\to 0} |u_k(x) - u_{B(x,r)}| + \limsup_{r\to 0} |u_l(x) - u_{B(x,r)}| \\ &\leq 2^{1-k},\end{aligned}$$

that is, (u_k) converges uniformly on $X \setminus E_n$ to u. (Note that, as $K \subset E_n$, $u_n \to u$ pointwise on $X \setminus E_n$.) Thus, u is continuous on $X \setminus E_n$. Moreover, by (9.2.11), if $x \in X \setminus E_n$ and $k \geq n$ then, for $l \geq k$,

$$\begin{aligned}&\limsup_{r\to 0} \fint_{B(x,r)} |u - u(x)|\, d\mu \\ &\qquad \leq \limsup_{r\to 0} \fint_{B(x,r)} |u - u_k(x)|\, d\mu + |u_k(x) - u(x)| \\ &\qquad \leq 2^{-k} + |u_k(x) - u(x)|\end{aligned}$$

and since $u_k(x) \to u(x)$ as $k \to \infty$, we see that

$$\limsup_{r\to 0} \fint_{B(x,r)} |u - u(x)|\, d\mu = 0.$$

Thus, each point $x \in X \setminus E_n$ is a Lebesgue point of u.

To obtain equation (9.2.9) in the case where $p < Q$, we can apply (9.2.7) instead of the Poincaré inequality to the estimate just before (9.2.10); this gives

$$\lim_{r\to 0} \fint_{B(x,r)} |u - u_{B(x,r)}|^{p*}\, d\mu = 0.$$

Hence, for $x \in X \setminus E_n$, using the fact that x is a Lebesgue point of u,

$$\begin{aligned}&\limsup_{r\to 0} \fint_{B(x,r)} |u - u(x)|^{p*}\, d\mu \\ &\qquad \leq 2^{p*} \lim_{r\to 0} \fint_{B(x,r)} |u - u_{B(x,r)}|^{p*}\, d\mu + 2^{p*} \lim_{r\to 0} |u(x) - u_{B(x,r)}|^{p*},\end{aligned}$$

which equals zero. By taking $E = \bigcap_n E_n$ we see that $\mathrm{Cap}_p(E) = 0$ and the above discussion holds for each $x \in X \setminus E$. This completes the proof. □

Now we will prepare to prove a Morrey-type embedding theorem.

Lemma 9.2.12 *Suppose that X is a quasiconvex and doubling metric measure space supporting a p-Poincaré inequality for some $1 \le p < \infty$ and that V is a Banach space. Assume moreover that X has relative lower volume decay* (9.1.14) *of order $Q \ge 1$. If $p > Q$ then, given a ball $B \subset X$ and a measurable function $u : 4\lambda B \to V$ with upper gradient $\rho \in L^p(4\lambda B)$, there is a set $E \subset B$ with $\mathrm{Cap}_p(E) = 0$ such that*

$$|u(x) - u(y)| \le C \operatorname{diam}(B)^{Q/p} d(x, y)^{1-Q/p} \left(⨍_{4\lambda B} \rho^p \, d\mu \right)^{1/p}$$

whenever $x, y \in B \setminus E$.

Proof Let E_0 be the collection of all non-Lebesgue points of u. Then, by the Lebesgue differentiation theorem (Section 3.4), we know that $\mu(E_0) = 0$. We will first prove the above inequality for all $x, y \in B \setminus E_0$. To do so, for each $k \in \mathbb{Z}$ we set $B_k = B(x, 2^{-k} d(x, y))$ if $k \ge 0$ and $B_k = B(y, 2^k d(x, y))$ if $k < 0$. Then we note that, for each $k \in \mathbb{Z}$, we have $\frac{1}{2} B_k \subset B_{k+1} \subset 2B_k \subset 4B$. Since x, y are Lebesgue points of u, as before we obtain

$$\begin{aligned}
&|u(y) - u(x)| \\
&\quad \le \sum_{k \in \mathbb{Z}} |u_{B_k} - u_{B_{k+1}}| \\
&\quad \le C \sum_{k \in \mathbb{Z}} ⨍_{2B_k} |u - u_{2B_k}| \, d\mu \\
&\quad \le C \sum_{k \in \mathbb{Z}} 2^{-|k|} d(x, y) \left(⨍_{2\lambda B_k} \rho^p \, d\mu \right)^{1/p} \\
&\quad \le C d(x, y)^{1-Q/p} \sum_{k \in \mathbb{Z}} 2^{-|k|(1-Q/p)} \left(\frac{\operatorname{diam}(B_k)^Q}{\mu(B_k)} \int_{2\lambda B_k} \rho^p \, d\mu \right)^{1/p}.
\end{aligned}$$

An application of (9.1.14), together with the fact that $1 - Q/p > 0$, now gives

$$\begin{aligned}
&|u(y) - u(x)| \\
&\quad \le C d(x, y)^{1-Q/p} \sum_{k \in \mathbb{Z}} 2^{-|k|(1-Q/p)} \left(\frac{\operatorname{diam}(B)^Q}{\mu(B)} \int_{2\lambda B_k} \rho^p \, d\mu \right)^{1/p};
\end{aligned}$$

the desired conclusion follows upon noting that $2\lambda B_k \subset 4\lambda B$ for each $k \in \mathbb{Z}$.

By the above argument we know that $u|_{X \setminus E_0}$ is Hölder continuous and hence admits a Hölder-continuous extension $\tilde{u}$ to X. Let E be the collection of all points in E_0 at which u and $\tilde{u}$ differ. To complete the proof, it suffices to show that $\mathrm{Cap}_p(E) = 0$. Since $u \in N^{1,p}(X : V)$, we know that the collection Γ of nonconstant compact rectifiable curves in X on which u is not continuous has zero p-modulus. Because $\mu(E) \le \mu(E_0) = 0$, we know that the collection Γ_E^+ of nonconstant rectifiable curves γ satisfying $\mathcal{H}_1(\gamma^{-1}(E)) > 0$ has zero p-modulus. Since Γ contains all those nonconstant compact rectifiable curves that intersect E but do not lie in Γ_E^+, we know that the collection Γ_E of all the nonconstant compact rectifiable curves that intersect E is a subcollection of $\Gamma \cup \Gamma_E^+$. It follows that $\mathrm{Mod}_p(\Gamma_E) = 0$ and hence, by Proposition 7.2.8, we know that $\mathrm{Cap}_p(E) = 0$. This completes the proof. □

Lemma 9.2.13 *Suppose that X is locally compact and satisfies the hypotheses of Lemma 9.2.12 for some $p > Q$. Let $E \subset X$. If $\mathrm{Cap}_p(E) = 0$ then E is empty.*

Proof By the countable subadditivity of the p-capacity and the separability of X, we may assume that $E \subset B$ for some ball B in X with $\mathrm{diam}(B) < 1/C$, where C is the constant from the conclusion of Lemma 9.2.12.

Suppose that E is nonempty. By the outer-capacity property of Corollary 8.2.5, for every $\epsilon > 0$ we can find a nonempty open set U with $E \subset U \subset B$ such that $\mathrm{Cap}_p(U) < \epsilon$. By the definition of the p-capacity, now we can find $u \in N^{1,p}(X)$ with $0 \le u \le 1$, $u = 1$ on U, such that

$$\int_X u^p \, d\mu + \int_X \rho_u^p \, d\mu < \epsilon.$$

We fix $0 < \epsilon < 2^{-p}\, \mu(B)$. Since U is nonempty and open, it has a Lebesgue point of u; hence, by Lemma 9.2.12,

$$\begin{aligned}\|1-u\|_{L^\infty(B)} &\le \frac{C\ \mathrm{diam}(B)}{\mu(B)^{1/p}} \left(\int_X \rho^p \, d\mu\right)^{1/p} \\ &\le \frac{C\ \mathrm{diam}(B)}{\mu(B)^{1/p}} \epsilon^{1/p} < \frac{1}{\mu(B)^{1/p}} \epsilon^{1/p} < \frac{1}{2}.\end{aligned}$$

It follows that for μ-a.e. $y \in B$ we have $u(y) \ge 1/2$, which means that

$$\frac{\mu(B)}{2^p} \le \int_X u^p \, d\mu < \epsilon,$$

and this violates the assumption on ϵ above. The above contradiction completes the proof. □

Combining Lemmas 9.2.12 and 9.2.13 yields the *Morrey embedding theorem.*

Theorem 9.2.14 *Suppose that X is a quasiconvex and doubling metric measure space supporting a p_0-Poincaré inequality for some $1 \le p_0 < \infty$ and that V is a Banach space. Assume moreover that there are constants $C_0 \ge 1$ and $Q \ge 1$ such that inequality* (9.1.14) *holds whenever $B' \subset B$ are balls in X.*

If $p > Q$ such that $p \geq p_0$ then, given a ball $B \subset X$ and a measurable function $u : 4\lambda B \to V$ with upper gradient $\rho \in L^p(4\lambda B)$, whenever $x, y \in B$ we have

$$|u(x) - u(y)| \leq C \operatorname{diam}(B)^{Q/p} d(x, y)^{1-Q/p} \left(\fint_{4\lambda B} \rho^p \, d\mu \right)^{1/p}.$$

In particular, the functions $u \in L^{1,p}(X : V)$ are locally $(1 - Q/p)$-Hölder continuous.

9.3 Measurability of equivalence classes and MEC_p

Recall from Section 7.5 that, given a nonnegative Borel-measurable function ρ on X, two points $x, y \in X$ are equivalent with respect to ρ, $x \sim_\rho y$, if there is a compact rectifiable curve γ in X connecting x to y with $\int_\gamma \rho \, ds$ finite. The relation $\sim_\rho$ is an equivalence relation; denote by $[x]_\rho := \{y \in X : y \sim_\rho x\}$ the equivalence class of a point $x \in X$. By Proposition 8.1.6 we know that if X supports a p-Poincaré inequality then X is connected. However, to prove that under this assumption X has to be rectifiably path connected and that it has the MEC_p property in the sense of Section 7.5, we need to know that $[x]_\rho$ is measurable. The goal of this section is to address this issue. In particular we will show that, under the assumption of completeness, separability, and p-Poincaré inequality, X must be rectifiably path connected and must satisfy MEC_p and every function with a p-integrable upper gradient must be measurable. For this we do *not* assume that the measure μ is doubling.

Analytic sets are subsets of complete separable metric spaces that are obtained as images of complete separable metric spaces under continuous maps (see [149]). Borel sets are analytic sets, and analytic sets are measurable with respect to any Borel measure on X ([149, Theorem 14.2]). Furthermore, images of analytic sets under continuous maps into a complete separable metric space are analytic sets.

The standing assumptions for this section are as follows:

X is a complete separable metric space and μ is a locally finite Borel regular measure on X such that nonempty open sets have positive measure.

Theorem 9.3.1 *If $\rho : X \to [0, \infty]$ is a Borel function then for all $x \in X$ the set $[x]_\rho$ is measurable. Furthermore, if $F \subset X$ is a closed set then the function $u : X \to [0, \infty]$ defined by*

$$u(x) = \inf_\gamma \int_\gamma \rho \, ds,$$

with the infimum taken over all rectifiable curves that connect x to F, is measurable.

The following are easy corollaries of the above theorem; note that we do not need to assume that μ is doubling here. If μ is doubling, however, the conclusion of the second corollary is an immediate consequence of Theorem 8.3.2.

Corollary 9.3.2 *Let $E, F \subset X$ be two compact sets and X be complete. Let $\Gamma(E, F; X)$ denote the collection of all paths in X connecting E to F. Then*

$$\mathrm{Mod}_p(\Gamma(E, F; X)) = \inf_u \int_X \rho_u^p \, d\mu,$$

where the infimum is taken over all measurable functions u in X that satisfy $u \geq 1$ on E and $u \leq 0$ on F.

The quantity on the right-hand side of the above equality is known in the literature as the relative p-capacity of the ordered triple $(E, F; X)$.

Corollary 9.3.3 *The rectifiable path components of X are measurable. In particular, if X supports a Poincaré inequality then X is rectifiably path connected.*

The following result is another corollary to Theorem 9.3.1.

Theorem 9.3.4 *If X is proper and supports a p-Poincaré inequality then X has the MEC_p property, and every function on X with a p-integrable upper gradient is measurable.*

Note that if X has no nonconstant rectifiable curves then every function has a p-integrable upper gradient. Thus, functions with p-integrable upper gradients are not necessarily measurable in a general setting.

The remainder of this section is devoted to the proofs of Theorems 9.3.1 and 9.3.4. First we record some preliminary results.

Lemma 9.3.5 *Let Z be a topological space and $\mathcal{Y}$ be a collection of functions $g : Z \to [0, \infty]$. Suppose that $\mathcal{Y}$ satisfies the following four conditions:*

(i) *all continuous functions $g : Z \to [0, \infty]$ are in $\mathcal{Y}$;*
(ii) *if $(g_i)_i$ is a monotone increasing sequence of functions in $\mathcal{Y}$ then $\lim_i g_i \in \mathcal{Y}$;*
(iii) *if $r, s > 0$ and $g, f \in \mathcal{Y}$ then $rg + sf \in \mathcal{Y}$;*
(iv) *if $g \in \mathcal{Y}$ with $0 \leq g \leq 1$ then $1 - g \in \mathcal{Y}$.*

Then every Borel function $g : Z \to [0, \infty]$ is in $\mathcal{Y}$.

Proof Let $\mathcal{A}_1$ denote the collection of all sets $E \subset Z$ for which $\chi_E \in \mathcal{Y}$. Since continuous functions are in $\mathcal{Y}$, from Proposition 4.2.2 and condition (ii) above it follows that lower semicontinuous functions are in $\mathcal{Y}$. If U is an open subset of Z then χ_U is a lower semicontinuous function on Z; it follows that $U \in \mathcal{A}_1$.

By condition (iv), if $E \in \mathcal{A}_1$ then $Z \setminus E \in \mathcal{A}_1$. By condition (ii), if (E_i) is an increasing sequence (that is, $E_i \subset E_{i+1}$) of sets from $\mathcal{A}_1$ then $\bigcup_i E_i \in \mathcal{A}_1$. Furthermore, if $E, F \in \mathcal{A}_1$ with $E \subset F$ then repeated applications of conditions (iii) and (iv) tell us that $\chi_E, 1-\chi_F, 1-\chi_F+\chi_E, 1-(1-\chi_F+\chi_E) = \chi_{F\setminus E}$ are in $\mathcal{Y}$ and so $F \setminus E \in \mathcal{A}_1$. Also, we have that $Z \in \mathcal{A}_1$ (since open sets are in $\mathcal{A}_1$).

Collections $\mathcal{A}_1$ that satisfy the conclusions in the above paragraph are called λ-*classes* [149]. We will now show that $\mathcal{A}_1$ contains the σ-algebra generated by the collection of all open subsets of Z, namely the collection of all Borel subsets of Z (this is called the $(\pi - \lambda)$-theorem in the literature). We would like to prove that $\mathcal{A}_1$ is closed under finite intersections, but we are not able to do so directly. However, this is trivially true for the subcollection of open subsets.

By replacing $\mathcal{A}_1$ with the intersection $\mathcal{A}$ of all λ-classes that contain the collection $\mathcal{O}$ of all open subsets of Z, if necessary, we may assume that $\mathcal{A}$ is the smallest λ-class of subsets of Z that contains $\mathcal{O}$. Let $\mathcal{A}_0$ be the collection of all $E \in \mathcal{A}$ such that whenever U is an open subset of Z we have $U \cap E \in \mathcal{A}$. We now claim that $\mathcal{A}_0$ is a λ-class containing $\mathcal{O}$. To see this, note that $Z \in \mathcal{A}_0$. If (E_j) is an increasing sequence of sets from $\mathcal{A}_0$ and U is open then $\bigcup_j E_j \in \mathcal{A}$ because $\mathcal{A}$ is a λ-class, and from $U \cap \bigcup_j E_j = \bigcup_j (E_j \cap U)$ it follows that the increasing sequence $(E_j \cap U)$ is also in $\mathcal{A}$, by the definition of $\mathcal{A}_0$, whence we conclude that $\bigcup_j E_j$ is also in $\mathcal{A}_0$. If $E, F \in \mathcal{A}_0$ such that $E \subset F$ then we have that $F \setminus E \in \mathcal{A}$ and that $U \cap (F \setminus E) = (U \cap F) \setminus (U \cap E) \in \mathcal{A}$, because $U \cap E \subset U \cap F$ and $U \cap E, U \cap F \in \mathcal{A}$. We have verified that $\mathcal{A}_0$ is a λ-class; hence by the assumed minimality of $\mathcal{A}$ we know that $\mathcal{A}_0 = \mathcal{A}$. Hence, whenever $E \in \mathcal{A}$ and U is an open subset of Z, we have that $E \cap U \in \mathcal{A}$.

We now show that $\mathcal{A}$ is closed under finite intersections. To see this, fix $E \in \mathcal{A}$ and consider the collection $\mathcal{A}(E)$ of all sets $F \in \mathcal{A}$ for which $F \cap E \in \mathcal{A}$. As in the previous paragraph, we can show that $\mathcal{A}(E)$ is a λ-class and, by the previous paragraph, contains $\mathcal{O}$. Hence, by the minimality of $\mathcal{A}$, we see that $\mathcal{A} = \mathcal{A}(E)$, that is, $\mathcal{A}$ is closed under the taking of intersections with E. Since $E \in \mathcal{A}$ is arbitrary, it follows that $\mathcal{A}$ is closed under finite intersections. Since $\mathcal{A}$ is also closed under complementation (that is, if E is in $\mathcal{A}$ then so is $Z \setminus E$), it follows that $\mathcal{A}$ is an algebra that is closed under finite intersections and finite unions. Finally, since the union of an increasing sequence of sets from $\mathcal{A}$ is in $\mathcal{A}$, it follows that $\mathcal{A}$ is a σ-algebra and hence contains all Borel sets. Thus we conclude that $\mathcal{A}_1$ also contains all Borel sets.

The above reasoning, together with condition (iii), tells us that nonnegative simple Borel functions are in $\mathcal{Y}$. Now, an application of the fact that every nonnegative Borel function is an increasing limit of simple Borel functions (see, for example, the proof of Proposition 3.3.22) together with condition (ii) yields the desired conclusion. □

To prove Theorem 9.3.1 it suffices to show that, given a nonnegative Borel-measurable function ρ on X and a point $x_0 \in X$, the set $[x_0]_\rho$ is analytic. We do so by demonstrating that it is the union of the images of a sequence of analytic sets under a continuous map. These analytic sets are subsets of metric spaces that are themselves subspaces of a single complete separable metric space; this is the set Y of all curves $\gamma : [0, 1] \to X$ that satisfy $\gamma(0) = x_0$. The metric on Y is given as follows. If $\gamma, \beta \in Y$ then

$$d_\infty(\gamma, \beta) = \sup_{0 \le t \le 1} d(\gamma(t), \beta(t)).$$

It is easy to see that (Y, d_∞) is complete (because X is complete), that a sequence of curves (γ_i) converges to γ if and only if $\gamma_i \to \gamma$ uniformly on $[0, 1]$, and that (Y, d_∞) is separable (because X is separable). For each $L > 0$ we consider the collection Y_L of all L-Lipschitz maps that belong to Y. Note that Y_L is a complete subspace of Y under the metric d_∞. If X is proper (as is the case if X is complete and doubling) then Y_L is sequentially compact and hence compact. Now we are ready to prove Theorem 9.3.1.

Proof of Theorem 9.3.1 For $L > 0$ we consider the complete metric space Y_L described above, equipped with the metric d_∞, and let $\pi_L : Y_L \to X$ be given by $\pi_L(\gamma) = \gamma(1)$. Then we note that π_L is a continuous map from Y_L into the complete separable metric space X; hence the images of analytic subsets of Y_L under π_L are analytic in X.

Corresponding to each Borel-measurable function $g : X \to [0, \infty]$ there is a function $\varphi_g : Y_L \to [0, \infty]$ given by $\varphi_g(\gamma) = \int_\gamma g\, ds$. Let $\mathcal{Y}_L$ consist of all such functions g for which φ_g is a Borel function. We now show that $\mathcal{Y}_L$ satisfies the four conditions that make up the hypothesis of Lemma 9.3.5. Indeed, if g is a continuous function and (γ_i) is a sequence in Y_L that converges in the metric d_∞ (hence uniformly) to γ then $\int_\gamma g\, ds = \lim_i \int_{\gamma_i} g\, ds$; so φ_g is continuous and hence is a Borel function. Thus condition (i) of Lemma 9.3.5 is satisfied. If (g_i) is a monotone increasing sequence of functions in $\mathcal{Y}_L$ then, writing $g = \lim_i g_i$, an application of the monotone convergence theorem tells us that, for $\gamma \in Y_L$,

$$\varphi_g(\gamma) = \int_\gamma g\, ds = \lim_i \int_\gamma g_i\, ds = \lim_i \varphi_{g_i}(\gamma),$$

and so φ_g is the limit of a sequence of Borel functions and hence is Borel (see, for example, Proposition 3.3.22). Thus condition (ii) of Lemma 9.3.5 is also satisfied. Conditions (iii) and (iv) are immediate consequences of the linearity of line integrals and the fact that a linear combination of two Borel functions into $[0, \infty]$ is Borel. Thus, by the conclusion of Lemma 9.3.5, all Borel functions $g : X \to [0, \infty]$ belong to $\mathcal{Y}_L$. In particular, if $\rho : X \to [0, \infty]$ is a Borel function then φ_ρ is a Borel function from Y_L to $[0, \infty]$, and hence $\varphi_\rho^{-1}([0, \infty))$, the preimage of the Borel set $[0, \infty)$, is a Borel subset of the complete separable space Y_L. Hence $\pi_L(\varphi_\rho^{-1}([0, \infty))$ is an analytic subset of X. Because

$$[x_0]_\rho = \bigcup_{k=1}^\infty \pi_k(\varphi_\rho^{-1}([0, \infty)),$$

it follows that $[x_0]_\rho$ is analytic and hence measurable.

A similar argument applied to Z_L, the collection of all L-Lipschitz curves γ from $[0, 1]$ to X with $\gamma(0) \in F$ (where F is closed), shows that the set $[F]_\rho$ of all points $y \in X$ for which there is a point $x \in F$ with $x \sim_\rho y$ is also measurable. Furthermore, by considering $\varphi_\rho^{-1}([0, t])$ for $t \geq 0$ instead of $\varphi_\rho^{-1}([0, \infty))$, we also see that the level sets of the function u given in the second part of the statement of Theorem 9.3.1 are measurable, and hence u is measurable. This proves the second part of the theorem and hence completes the proof. □

Now we prove the main theorem of this section, Theorem 9.3.4.

Proof of Theorem 9.3.4 Let $\rho \in L^p(X)$ be a nonnegative Borel measurable function. We wish to show that there is some $x_0 \in X$ such that $\mu(X \setminus [x_0]_\rho) = 0$. By the Vitali–Carathéodory theorem (see Section 4.2), there is a lower semicontinuous function $\rho_0 \in L^p(X)$ such that $\rho_0 \geq \rho$ pointwise in X. Since for each $x \in X$ we have $[x]_{\rho_0} \subset [x]_\rho$, it suffices to prove the result under the assumption that ρ itself is lower semicontinuous.

The above reduction understood, for each positive integer m we set

$$E_m := \{x \in X : M(\rho^p)(x) \leq m^p\}.$$

By the Hardy–Littlewood maximal theorem 3.5.6 we know that $\mu(X \setminus \bigcup_m E_m) = 0$. Since $E_m \subset E_{m+1}$, there is a smallest positive integer m_0 for which E_{m_0} is nonempty. We will show that $\bigcup_{m \geq m_0} E_m \subset [x_0]_\rho$, according to which $\mu(X \setminus [x_0]_\rho) = 0$. Thus $[x_0]_\rho$ is the main equivalence class for ρ.

We define $u : X \to [0, \infty]$ by setting

$$u(x) = \inf_{\gamma} \int_{\gamma} (1 + \rho)\, ds$$

and, for each positive integer k,

$$u_k(x) = \inf_{\gamma} \int_{\gamma} (1 + \min\{\rho, k\})\, ds,$$

where the infimum is taken over all rectifiable curves that connect x to x_0. By Theorem 9.3.1, u and u_k are measurable. The goal is to show that u is finite on $\bigcup_m E_m$. Set $\rho_k = \min\{k, \rho\}$. As in the proof of Lemma 7.2.13, we see that $1 + \rho_k$ is an upper gradient of u_k; by the quasiconvexity of X (see Theorem 8.3.2), u_k is $C(1 + k)$-Lipschitz continuous on X, and hence every point is a Lebesgue point of u_k. Now we are set to apply the p-Poincaré inequality. We obtain from (8.1.10) that, for $x, y \in X$,

$$|u_k(x) - u_k(y)| \le Cd(x, y) \left(M(1 + \rho_k)^p(x) + M(1 + \rho_k)^p(y) \right)^{1/p}.$$

Since $M(1 + \rho_k)^p \le 2^p M \rho_k^p + 2^p \le 2^p M \rho^p + 2^p$, we see that u_k is $C\, 2^{1+1/p}(m^p + 1)^{1/p}$-Lipschitz continuous on E_m. Since $u_k \le u_{k+1}$ for $k \in \mathbb{N}$, we set $v = \lim_{k \to \infty} u_k$ and note that v is also $8\,Cm$-Lipschitz continuous on E_m. Note also that $v(x_0) = 0$; it follows that for all $x \in \bigcup_m E_m$ we have $v(x) < \infty$. Hence, to show that u is finite on $\bigcup_m E_m$, it suffices to show that $u \le v$ on $\bigcup_m E_m$. To this end, fix $x \in \bigcup_m E_m$. For each $k \in \mathbb{N}$ there is a rectifiable curve γ_k, connecting x_0 to x, such that

$$v(x) \ge u_k(x) \ge \int_{\gamma_k} (1 + \rho_k)\, ds \; - \; 2^{-k}.$$

First note that $\operatorname{length}(\gamma_k) \le v(x) + 2^{-k}$ and so, by the properness of X, we may employ the Arzelà–Ascoli theorem to extract a subsequence of these curves, (γ_{k_j}), and a rectifiable curve γ connecting x_0 to x such that $\gamma_k \to \gamma$ uniformly. Recall that ρ is lower semicontinuous. We argue as in the proof of Lemma 7.2.13 to show that $\int_\gamma (1 + \rho)\, ds \le v(x)$. For $k_0 \in \mathbb{N}$, by the lower semicontinuity of $1 + \rho_{k_0}$, we have

$$\begin{aligned} \int_\gamma (1 + \rho_{k_0})\, ds &\le \liminf_{k \to \infty} \int_{\gamma_k} (1 + \rho_{k_0})\, ds \\ &\le \lim_{k \to \infty} \int_{\gamma_k} (1 + \rho_k)\, ds \le v(x), \end{aligned}$$

where we have used the fact that $\rho_k \ge \rho_{k_0}$ for $k \ge k_0$. Now an application of the monotone convergence theorem tells us that $\int_\gamma (1 + \rho)\, ds \le v(x)$.

Thus, $u(x) \leq v(x) < \infty$ for every $x \in \bigcup_m E_m$ and this in turn implies that $\bigcup_m E_m \subset [x_0]_\rho$, completing the proof that X satisfies MEC_p.

Now, to prove that a function u on X with a p-integrable upper gradient ρ is measurable on X, we may again assume that ρ is lower semicontinuous. We now proceed as in the proof of the MEC_p property above to consider the measurable sets $E_m = \{x \in X : M(\rho^p)(x) \leq m^p\}$. We define $f : X \times X \to [0, \infty]$ as follows:

$$f(x, y) = \inf_\gamma \int_\gamma (1 + \rho)\, ds,$$

where the infimum is taken over all rectifiable curves γ connecting x to y in X. By the proof of the MEC_p property we know that, for $x, y \in E_m$,

$$f(x, y) \leq 8Cmd(x, y)$$

and, in particular, $f(x, y)$ is finite for $x, y \in \bigcup_m E_m$. Furthermore, for $x, y \in E_m$, because ρ is an upper gradient of u we have

$$|u(x) - u(y)| \leq \inf_\gamma \int_\gamma \rho\, ds \leq \inf_\gamma \int_\gamma (1 + \rho)\, ds = f(x, y) \leq 8Cmd(x, y).$$

Here again the infimum is over all rectifiable curves connecting x to y. It follows that u is $8Cm$-Lipschitz continuous on E_m, and so $u\,\chi_{E_m}$ is measurable on X. For $k \geq m_0$, set $F_k = E_k \setminus \bigcup_{m<k} E_m$. Then the F_k are measurable, and

$$u = \sum_{k=m_0}^{\infty} u\chi_{F_k}$$

is measurable. This completes the proof of the theorem. □

Theorem 9.3.4 is quite useful, for it also tells us that the requirement of measurability on functions in the results of this chapter and in those of Chapter 8 can be removed.

We now use Theorem 9.3.4 to provide estimates on the p-modulus of the family of all rectifiable curves that connect $B(x, r)$ to $X \setminus B(x, R)$.

Lemma 9.3.6 *Let X be a complete doubling metric measure space supporting a p-Poincaré inequality. Suppose also that μ satisfies (9.1.14). If $x \in X$ and $0 < r < R$ are such that $X \setminus \overline{B}(x, 3R/2)$ is nonempty, we denote the collection of all curves connecting $B(x, r)$ to $X \setminus B(x, R)$ by $\Gamma(x, r, R)$.*

(i) *If $p < Q$ then*

$$\mathrm{Mod}_p(\Gamma(x, r, R)) \geq \frac{1}{C} \frac{\mu(B(x, r))^{1-p/Q} \mu(B(x, R))^{p/Q}}{R^p}.$$

(ii) *If* $p = Q$ *then*

$$\mathrm{Mod}_Q(\Gamma(x, r, R)) \geq \frac{1}{C}\frac{\mu(B(x, R))}{R^Q}\left(\log\left(\frac{C\mu(B(x, R))}{\mu(B(x, r))}\right)\right)^{1-Q}.$$

(iii) *If* $p > Q$ *then*

$$\mathrm{Mod}_p(\Gamma(x, r, R)) \geq \frac{1}{C}\mu(B(x, R))R^{-p}.$$

Proof As mentioned in Section 5.3, lower bounds for the modulus are obtained by checking each admissible function for the family of curves. So, let ρ be an admissible function for computing $\mathrm{Mod}_p(\Gamma(x, r, R))$; that is, ρ is a nonnegative Borel-measurable function on X such that $\int_\gamma \rho\, ds \geq 1$ for each $\gamma \in \Gamma(x, r, R)$. In order to apply the Poincaré inequality, we need a function–upper-gradient pair. Thanks to Lemma 8.1.5 and Theorem 9.3.4, we know that the function u given by

$$u(x) = \min\left\{1, \inf_\gamma \int_\gamma \rho\, ds\right\},$$

where the infimum is taken over all rectifiable curves γ connecting x to $X \setminus B(x, R)$, is integrable on balls; as in Lemma 7.2.13 it can be directly checked that ρ is an upper gradient of u. Hence we may apply the conclusions of Theorem 9.1.15 to the pair (u, ρ).

By assumption, $X \setminus \overline{B}(x, 3R/2)$ is nonempty; hence, by the path connectedness of X (which follows from the p-Poincaré inequality), there is a point $y \in X$ with $d(x, y) = 3R/2$. Therefore $B(y, R/2) \subset B(x, 2R) \setminus B(x, R)$, and an application of the doubling property of μ yields

$$\mu(B(x, 2R) \setminus B(x, R)) \geq \mu(B(y, R/2)) \geq \frac{1}{C_\mu^4}\mu(B(x, R)).$$

Note that $u = 1$ on $B(x, r)$ and $u = 0$ on $X \setminus B(x, R)$. Hence, with $B = B(x, 2R)$,

$$u_B = \frac{1}{\mu(B(x, 2R))}\int_{B(x,2R)} u\, d\mu \leq \frac{\mu(B(x, R))}{\mu(B(x, 2R))}$$
$$\leq \left(1 + \frac{1}{C_\mu^4}\right)^{-1} =: c < 1.$$

Hence $|u - u_B| \geq 1 - c$ on $B(x, r)$. So, an application of Corollary 9.1.36(i) gives part (i) of our statement and Corollary 9.1.36(ii) gives part (ii) of our statement, while Corollary 9.1.36(iii) yields part (iii) of our statement. Note that the completeness of X, together with Theorem 8.3.2, permits us to use Corollary 9.1.36. □

Recall that if μ is doubling then there is some $Q > 0$ for which (9.1.14) holds. We now explore a stronger condition on μ. We say that μ is *Ahlfors Q-regular* if there is a constant $C > 0$ such that, whenever $x \in X$ and $0 < r < \operatorname{diam}(X)$,

$$\frac{1}{C} r^Q \leq \mu(B(x, r)) \leq Cr^Q. \tag{9.3.7}$$

If X satisfies the hypotheses of Lemma 9.3.6 and in addition μ is Ahlfors Q-regular then the upper bound from Proposition 5.3.9 for $p = Q$ and the upper bound estimate found at the end of the proof of Corollary 5.3.11 for $p < Q$ tell us that the lower bound estimates given in Lemma 9.3.6 for $p \leq Q$ are the best possible. Finally, when $p > Q$ we have the estimate

$$\operatorname{Mod}_p(\Gamma(x, r, R)) \geq cR^{Q-p} > 0,$$

which reinforces the conclusion of Lemma 9.2.13.

9.4 Annular quasiconvexity

In this section we give an analog of Theorem 8.3.2 for annular quasiconvexity.

Theorem 9.4.1 *Every complete metric measure space that is Ahlfors Q-regular and supports a p-Poincaré inequality for some $p < Q$ is annularly quasiconvex. The annular quasiconvexity constant depends only on the Ahlfors regularity constants of the measure and on the data associated with the Poincaré inequality.*

Proof Let (X, d, μ) be a complete metric measure space which is Ahlfors Q-regular and supports a p-Poincaré inequality for some $p < Q$. By Theorem 8.3.2, X is C_q-quasiconvex for some $C_q \geq 1$.

We wish to show that there is a constant $C_a \geq 1$, depending only on the doubling and Poincaré data, such that X is annularly C_a-quasiconvex. In other words, we wish to show that, for each $r > 0$ such that $B(z, r) \setminus B(z, r/2)$ is nonempty and $x, y \in B(z, r) \setminus B(z, r/2)$, there is a C_a-quasiconvex curve in $B(z, C_a r) \setminus B(z, r/C_a)$ connecting x to y. To this end, consider $C \geq \max\{4, C_q\}$. We may assume without loss of generality that $d(x, y) > r/(2C)$. Let Γ be the collection of all rectifiable curves in $B(z, C^2\lambda r)$ connecting $B(x, r/(2C^2))$ to $B(y, r/(2C^2))$. Note that these two balls are contained in $B(z, C^2 r)$.

By Theorem 9.3.1, if ρ is a nonnegative Borel-measurable function on X such that $\int_\gamma \rho\, ds \geq 1$ for each $\gamma \in \Gamma$ then, defining a function u on $B(z, C^2 r)$ by

$$u(w) = \min\left\{1, \int_\beta \rho\, ds\right\},$$

where the infimum is taken over all compact rectifiable curves β in $B(z, C^2\lambda r)$ connecting $B(x, r/(2C^2))$ to w, we have that u is measurable with $u = 0$ on $B(x, r/(2C^2))$ and $u = 1$ on $B(y, r/(2C^2))$. Note that $r/(2C) < d(x, y) < 2r$. Let $B_{-1} = B(x, r/(2C))$, $B_0 = B(z, Cr)$, and $B_1 = B(y, r/(2C))$. Then $u_{B_{-1}} = 0$ and $u_{B_1} = 1$ and so, by the Poincaré inequality and the doubling property of μ,

$$\begin{aligned}
1 = |u_{B_{-1}} - u_{B_1}| &\le |u_{B_{-1}} - u_{B_0}| + |u_{B_0} - u_{B_1}| \\
&\le \frac{1}{\mu(B_{-1})}\int_{B_{-1}} |u - u_{B_0}|\, d\mu + \frac{1}{\mu(B_1)}\int_{B_1} |u - u_{B_0}|\, d\mu \\
&\le \frac{2C}{\mu(B_0)}\int_{B_0} |u - u_{B_0}|\, d\mu \\
&\le Cr\left(\fint_{\lambda B_0} \rho^p\, d\mu\right)^{1/p}.
\end{aligned}$$

It follows that

$$\frac{\mu(B_0)}{Cr^p} \le \int_{\lambda B_0} \rho^p\, d\mu \le \int_{B(z, C^2\lambda r)} \rho^p\, d\mu,$$

that is,

$$\operatorname{Mod}_p(\Gamma) \ge \frac{\mu(B_0)}{C_0 r^p}.$$

We now fix a positive integer $m \ge 4$ and split Γ into three subfamilies Γ_1, Γ_2, and Γ_3, where Γ_1 is the collection of all curves in Γ that intersect $B(z, r/m)$, Γ_2 is the collection of all the curves in Γ that remain in $B(z, C^2\lambda r) \setminus B(z, r/m)$ and have length at least $m\, d(x, y)$, and Γ_3 consists of all the curves in Γ that remain in $B(z, C^2\lambda r) \setminus B(z, r/m)$ and have length at most $m\, d(x, y)$. We wish to show that Γ_3 is nonempty. To do so, we show that the p-moduli of Γ_1 and Γ_2 are small in comparison with the p-modulus of Γ.

Since $\rho_2 := (md(x, y))^{-1}\chi_{B(z, C^2\lambda r)}$ is admissible for Γ_2, we have

$$\operatorname{Mod}_p(\Gamma_2) \le \frac{1}{m^p\, d(x, y)^p}\mu(B(z, C^2\lambda r)) \le \frac{C_2\, \mu(B_0)}{m^p\, r^p}.$$

The fact that each curve in Γ_1 contains a subcurve in $\Gamma(z, r/m, r/2)$ implies that

$$\operatorname{Mod}_p(\Gamma_1) \le \operatorname{Mod}_p(\Gamma(z, r/m, r/2)),$$

where $\Gamma(z, r/m, r/2)$ is as in Lemma 9.3.6. Let i_0 be a positive integer such that $2^{i_0} \le m < 2^{i_0+1}$ and, for $i = 0, 1, \ldots, i_0$, let $\widehat{B}_i = B(z, 2^i r/m)$. Define a function g on $B(z, r)$ by

$$g(w) := \frac{8}{\log(m/2)\, d(z, w)} \chi_{B(z,r/2)\setminus B(z,r/m)}(w).$$

Then g is admissible for $\Gamma(z, r/m, r/2)$ since, for any $\gamma \in \Gamma(z, r/m, r/2)$, we can find subcurves γ_i of γ, $i = 1, \ldots, i_0$, that lie in the annuli $\widehat{B}_i \setminus \widehat{B}_{i-1}$ and connect the inner sphere $\{w \in X : d(z, w) = 2^{i-1} m\}$ to the outer sphere $\{w \in X : d(w, z) = 2^i m\}$. We see that

$$\begin{aligned}
\int_\gamma g\, ds &\ge \sum_{i=1}^{i_0-1} \int_{\gamma_i} g\, ds \ge 8 \sum_{i=1}^{i_0-1} \frac{1}{\log(m/2)} \frac{\operatorname{length}(\gamma_i)}{2^{i+1} r/m} \\
&\ge 8 \frac{1}{\log(m/2)} \sum_{i=1}^{i_0-1} \frac{2^{i-1} r/m}{2^{i+1} r/m} \ge 8 \frac{i_0 - 1}{4 \log(m/2)} \\
&\ge 8 \frac{i_0}{8 \log(m/2)} \ge 1.
\end{aligned}$$

It follows from the Ahlfors Q-regularity of μ that

$$\begin{aligned}
\operatorname{Mod}_p(\Gamma_1) \le \int_X g^p\, d\mu &= \sum_{i=1}^{i_0} \int_{\widehat{B}_i \setminus \widehat{B}_{i-1}} g^p\, d\mu \\
&\le \frac{C}{(\log(m/2))^p} \sum_{i=1}^{i_0} \frac{\mu(\widehat{B}_i \setminus \widehat{B}_{i-1})}{(2^i r/m)^p} \\
&\le \frac{C r^{-p}}{(\log(m/2))^p} \sum_{i=1}^{i_0} \left(2^i/m\right)^{Q-p} \mu(B_0);
\end{aligned}$$

recall that $B_0 = B(z, 2Cr)$. By the choice of i_0 we have

$$\operatorname{Mod}_p(\Gamma_1) \le \frac{C\mu(B_0)}{r^p (\log(m/2))^p} \frac{2^{i_0(Q-p)}}{m^{Q-p}} \le \frac{C_1 \mu(B_0)}{r^p (\log(m/2))^p}.$$

Thus,

$$\begin{aligned}
\operatorname{Mod}_p(\Gamma_1 \cup \Gamma_2) &\le \frac{\mu(B_0)}{r^p} \left(\frac{1}{C_2\, m^p} + \frac{C_1}{(\log(m/2))^p} \right) \\
&\le \frac{(C_1 + C_2)\mu(B_0)}{r^p} (\log(m/2))^{-p}.
\end{aligned}$$

Hence, if $m > 2e^{(2C_0(C_1+C_2))^{1/p}}$ then

$$\mathrm{Mod}_p(\Gamma_3) \geq \frac{\mu(B_0)}{2C_0\, r^p} > 0,$$

and it follows that in $B(z, C^2\lambda r) \setminus B(z, r/(C^2\lambda))$ there is a rectifiable curve with length at most $md(x, y)$, connecting a point x_1 in $B(x, r/(2C))$ to a point y_1 in $B(y, r/(2C))$. The quasiconvexity of X ensures that we can connect x to x_1 and y to y_1 by quasiconvex curves, and the concatenation of these three curves yields the desired curve, verifying the stated annular quasiconvexity. □

The above proof can be adapted to more general doubling measures, but one needs a relative upper volume decay property (an inequality that is the reverse of (8.1.14)) with exponent $Q > p \geq 1$. Such an upper volume decay property is known to hold when μ is doubling and X is connected. Note that the connectivity of X is guaranteed by the support of a Poincaré inequality; hence the condition $Q > p$ is the crucial assumption. Also note that the Euclidean line $\mathbb{R}$ is Ahlfors 1-regular and supports a 1-Poincaré inequality but is not annularly quasiconvex.

9.5 Notes to Chapter 9

The fact that a p-Poincaré inequality on a doubling metric space implies improved integrability was first established by Grigor'yan [19], [103] in the Riemannian manifold setting for $p = 2$. The corresponding results for real-valued functions in this chapter were proven in [113] and [114]. In [114], one can also find a Rellich–Kondrachov compact embedding theorem and a version of the Sobolev embedding theorem on spheres. It is easy to check that the compact embedding theorem can hold only for finite-dimensional target spaces V and thus we have opted not to include it here. Also, we have not included an embedding theorem on spheres as the formulation is somewhat complicated. Towards the noncompactness of embeddings, let us simply consider the special case $V = \ell^\infty$; this argument can easily be generalized to other infinite-dimensional Banach spaces V. We first fix a ball B in X and an L-Lipschitz function η supported on $2B$ with $\eta = 1$ on B and $0 \leq \eta \leq 1$ on X. We choose the maps $v_k : X \to V$ by setting $v_k(x) = \eta(x)\vec{e}_k$ for every $x \in X$, where $\vec{e}_k$ is the sequence in V whose kth entry is 1 and all other entries of which are zero. It is clear that $L\chi_{2B}$ is an upper gradient for v_k and the norm of v_k is at most $(1 + L)\mu(2B)$ for each k. Hence the sequence $\{v_k\}$ is bounded. However, it has no convergent subsequence because the norm of $v_k - v_j$ is at least $\mu(B)$ for each k, j with $k \neq j$.

The proof of the capacitary Lebesgue point property, Theorem 9.2.8, in the metric setting is due to Kinnunen and Latvala [157]. While their result is stated for another Sobolev-type space, called the Hajłasz–Sobolev space (see Chapter 10), their technique is quite versatile. The classical (Euclidean) proof of the capacitary Lebesgue point property relied on a Besicovitch-type covering theorem to verify the capacitary weak-type inequality for maximal functions (as in Lemma 9.2.6). A local version of the Besicovitch covering theorem would suffice here, since the argument is essentially localizable. However, it is not known whether there are any non-Riemannian metric spaces in which a local version of the Besicovitch covering theorem holds; see for example [290, p. 119, Lemma 3.3.1] and [193, p. 89, Theorem 2.54]. The proof in the Euclidean setting is due to Federer and Ziemer [84]. Earlier work by Giusti [95] indicated that the set of non-Lebesgue points of a function in $W^{1,p}(\Omega)$ has Hausdorff dimension at most $n - p$; this is weaker than the statement that the set has null p-capacity.

In the metric setting Kinnunen and Latvala avoided the need for a Besicovitch-type covering by proving that the super level set $\{Mu > t\}$ of a Sobolev function u is contained in the super level set of another related Sobolev function (see Lemma 9.2.1 together with Proposition 9.2.2). Then it is a straightforward computation to prove the needed capacitary weak-type estimate as in Lemma 9.2.6. The proof of this lemma also shows that if u is a nonnegative function in $N^{1,p}(X)$ and $t > 0$ then the p-capacity of the super level set $\{u > t\}$ is at most $Ct^{-p}\|u\|^p_{N^{1,p}(X)}$. A stronger, integrated, version of this in the Euclidean setting can be found in the comprehensive book of Maz'ya [202, p. 109, Theorem 2.3.1 and p. 110, Remark]. This version due to Maz'ya appeared in [202] for the first time for general $1 \le p < \infty$. For the case $p = 2$ the result, due again to Maz'ya, was proved in [199].

Lemma 9.2.3 is standard. For instance, one can obtain this as a two-sided estimate by letting $R \to \infty$ in [202, p. 106, Section 2.2.4]. This lemma indicates, for example, that if the measure μ is Ahlfors Q-regular then, for $E \subset X$, $\mathrm{Cap}_p(E) = 0$ whenever $\mathcal{H}_{Q-p}(E) = 0$. This is, however, not a characterization of zero p-capacity sets; if a set is of p-capacity zero then its Hausdorff dimension is at most $Q - p$ but its $(Q - p)$-dimensional Hausdorff measure may be positive.

Theorem 9.2.8 does hold true for $p = 1$ as well, but the proof uses the concept of relative isoperimetric inequality, which is beyond the scope of this book. We refer interested readers to the paper [160].

The issues on measurability were dealt with in the paper [139], where it was also shown that if a metric space supports a p-Poincaré inequality then *any* real-valued function that has an upper gradient in $L^p_{\mathrm{loc}}(X)$ must necessarily be measurable. The concept of the MEC_p property was first studied by

Ohtsuka in the Euclidean setting [220]. The study of Borel, analytic, co-analytic, projective, and universally measurable subsets of a complete separable metric space falls into the category of descriptive set theory; see for example [149]. All Borel subsets of a complete separable metric space are both analytic sets and co-analytic sets (complements of analytic sets), but not all analytic sets are Borel. However, Suslin proved that analytic sets that are also co-analytic are necessarily Borel. Analytic sets are universally measurable, and hence μ-measurable for Borel-regular measures μ. For a proof of this (due originally to Suslin and Lusin) and for a comprehensive treatment of analytic sets see [46, Chapter 11], where an example of an analytic set that is not a Borel set can also be found ([46, Section 11.5]).

Theorem 9.4.1 is due to Korte [165]. Annular quasiconvexity is a useful tool in the study of metric spaces. For instance, it was used by Mackay [191], who identified new criteria for nontrivial lower bounds on the conformal dimension of metric spaces, and by Herron [132], who characterized uniform domains in complete, doubling, annularly quasiconvex spaces via qualitative assumptions on Gromov–Hausdorff tangent cones. See Chapter 11 for further information on Gromov–Hausdorff convergence.

10

Other definitions of Sobolev-type spaces

In this chapter we discuss the relations between the Sobolev space $N^{1,p}$ and various other abstract Sobolev spaces defined on metric spaces. These include the Sobolev-type spaces defined by Cheeger, Hajłasz, Hajłasz and Koskela, and Korevaar and Schoen. Under the assumption of a suitable Poincaré inequality, various inclusions hold between these spaces. If we assume that $p > 1$ and a slightly better Poincaré inequality holds, then all the spaces in question are equal as sets and have comparable norms. In many situations the assumption of a slightly better Poincaré inequality is not overly restrictive; see the discussion in Chapter 12. However, in the case $p = 1$ no better Poincaré inequality is available, and hence in many situations this case is special.

Throughout this chapter, (X, d, μ) denotes a metric measure space as defined in Section 3.3, V denotes a Banach space, and $1 \leq p < \infty$, unless otherwise specified.

10.1 The Cheeger–Sobolev space

A measurable function $u : X \to V$ is said to belong to the *Cheeger–Sobolev space* $Ch^{1,p}(X : V)$ if and only if $u \in L^p(X : V)$ and there exists a sequence (u_n) of functions in $N^{1,p}(X : V)$ converging to u in $L^p(X : V)$ and a sequence (ρ_n) such that ρ_n is an upper gradient for u_n for each n and $\liminf_{n\to\infty} \|\rho_n\|_{L^p(X)}$ is finite.

The space $Ch^{1,p}(X : V)$ is endowed with the norm

$$\|u\|_{Ch^{1,p}(X:V)} = \|u\|_{L^p(X:V)} + \inf_{(\rho_n)} \liminf_{n\to\infty} \|\rho_n\|_{L^p(X)},$$

where the infimum is taken over all sequences (ρ_n) as above. The functions u as above are sometimes said to belong to the Sobolev space in the *relaxed sense*. As usual we write $Ch^{1,p}(X) := Ch^{1,p}(X : \mathbb{R})$. The Sobolev space $Ch^{1,p}(X)$ was introduced by Cheeger in [53].

For $1 < p < \infty$, the space $Ch^{1,p}(X : V)$ coincides with the Sobolev space $N^{1,p}(X : V)$ without any additional assumptions on the underlying space X.

Theorem 10.1.1 *The L^p-equivalence class of a function in $N^{1,p}(X : V)$ belongs to $Ch^{1,p}(X : V)$. When $p > 1$, a function $u \in Ch^{1,p}(X : V)$ has a μ-representative in $N^{1,p}(X : V)$, and $\|u\|_{Ch^{1,p}(X:V)} = \|u\|_{N^{1,p}(X:V)}$.*

Proof The 1-Lipschitz embedding $N^{1,p}(X : V) \subset Ch^{1,p}(X : V)$ is clear: choose $u_n = u$. By Lemma 6.2.2 there is a sequence (ρ_n) of p-integrable upper gradients of u that approximates ρ_u in $L^p(X)$.

Suppose now that $u \in Ch^{1,p}(X : V)$ and let (u_n) and (ρ_n) be as in the statement of the theorem. By Theorem 7.3.9, u has a representative in $N^{1,p}(X : V)$ and, by (7.3.10),

$$\begin{aligned}\|u\|_{N^{1,p}(X:V)} &= \|u\|_{L^p(X:V)} + \|\rho_u\|_{L^p(X)} \\ &\le \liminf_{n\to\infty} \|u_n\|_{L^p(X:V)} + \liminf_{n\to\infty} \|\rho_{u_n}\|_{L^p(X)} \\ &= \|u\|_{L^p(X:V)} + \liminf_{n\to\infty} \|\rho_{u_n}\|_{L^p(X)} .\end{aligned}$$

Taking the infimum over all such sequences (ρ_n) completes the proof. □

In Chapter 13 we will present Cheeger's differentiation theorem for Lipschitz functions on doubling metric measure spaces supporting a Poincaré inequality. In keeping with the overall aims of this book, we will use the framework of the Sobolev space $N^{1,p}(X)$ in that chapter. Cheeger's original proof of his Rademacher theorem used the framework of the Sobolev space $Ch^{1,p}(X)$.

10.2 The Hajłasz–Sobolev space

A measurable function $u : X \to V$ belongs to the *Hajłasz–Sobolev space* $M^{1,p}(X : V)$ if and only if $u \in L^p(X : V)$ and there exists a nonnegative function $g \in L^p(X)$ such that the inequality

$$|u(x) - u(y)| \le d(x, y)(g(x) + g(y)) \tag{10.2.1}$$

holds for all $x, y \in X \setminus E$, for some $E \subset X$ with $\mu(E) = 0$. The space $M^{1,p}(X : V)$ is endowed with the norm

$$\|u\|_{M^{1,p}(X:V)} = \|u\|_{L^p(X:V)} + \inf_g \|g\|_{L^p(X)},$$

where the infimum is taken over all g for which (10.2.1) holds. As usual, we write $M^{1,p}(X) := M^{1,p}(X : \mathbb{R})$. Functions g that satisfy (10.2.1) are called the *Hajłasz gradients* of u. Note that (in contrast with $N^{1,p}$) $M^{1,p}(X : V)$ is naturally defined as a collection of μ-equivalence classes of functions agreeing almost everywhere with respect to the measure in X.

Recall that we have encountered an inequality similar to (10.2.1) before, in (8.1.10) of Theorem 8.1.7, with g above replaced by $(Mh^p)^{1/p}$ for some p-integrable function h. However, in general $(Mh^p)^{1/p}$ does not belong to $L^p(X)$, and so the requirement that g belongs to $M^{1,p}(X : V)$ is a priori more stringent. This observation motivates the notation $M^{1,p}$ for this Sobolev space. The Sobolev space $M^{1,p}(X)$ was introduced by Hajłasz in [108].

For continuous functions in $M^{1,p}(X : V)$, the Hajłasz gradient g can always be redefined on a set of measure zero so that the defining inequality (10.2.1) holds everywhere.

Lemma 10.2.2 *Suppose that μ is doubling. Let u be a continuous function in $M^{1,p}(X : V)$ with Hajłasz gradient $g \in L^p(X)$. Set*

$$\tilde{g}(x) := \limsup_{r\to 0} ⨍_{B(x,r)} g(z)\, d\mu(z). \tag{10.2.3}$$

Then $\tilde{g} = g$ almost everywhere and

$$|u(x) - u(y)| \le d(x, y)(\tilde{g}(x) + \tilde{g}(y)) \tag{10.2.4}$$

for all $x, y \in X$.

Proof The modified function $\tilde{g}$ agrees with g almost everywhere by the Lebesgue differentiation theorem (see Section 3.4). Let $x, y \in X$ be arbitrary and let $\epsilon > 0$. Choose $0 < \delta < \epsilon$ small enough that

$$⨍_{B(x,r)} g(z)\, d\mu(z) < \tilde{g}(x) + \epsilon$$

and

$$⨍_{B(y,r)} g(w)\, d\mu(w) < \tilde{g}(y) + \epsilon$$

whenever $0 < r < \delta$. Fix such an r. By assumption,

$$|u(z) - u(w)| \le d(z, w)(g(z) + g(w))$$

for almost every $z \in B(x, r)$ and almost every $w \in B(y, r)$. Consequently (recall that $\delta < \epsilon$),

$$\begin{aligned}
&\left| \fint_{B(x,r)} u(z)\, d\mu(z) - \fint_{B(y,r)} u(w)\, d\mu(w) \right| \\
&\quad \le \fint_{B(x,r)} \fint_{B(y,r)} |u(z) - u(w)|\, d\mu(w)\, d\mu(z) \\
&\quad \le (d(x, y) + 2\epsilon) \left(\fint_{B(x,r)} g\, d\mu + \fint_{B(y,r)} g\, d\mu \right) \\
&\quad < (d(x, y) + 2\epsilon)(\tilde{g}(x) + \tilde{g}(y) + 2\epsilon).
\end{aligned}$$

Since u is continuous, the left-hand side converges to $|u(x) - u(y)|$ as $r \to 0$. Letting $\epsilon \to 0$ yields (10.2.4) and the proof is complete. □

Lemma 10.2.5 *Let μ be a doubling measure. Each continuous function u in $M^{1,p}(X : V)$ is in $N^{1,p}(X : V)$, with*

$$\|u\|_{N^{1,p}(X:V)} \le 3\|u\|_{M^{1,p}(X:V)}. \tag{10.2.6}$$

Observe that no conditions on X are needed for this lemma, save that μ be doubling. Moreover, the result holds for all $p \ge 1$.

Proof Let $u \in M^{1,p}(X : V)$ be continuous. By the previous lemma, each Hajłasz gradient g of u can be modified on a set of measure zero in such a way that (10.2.1) holds for all $x, y \in X$. Moreover, the modified function from (10.2.3) is Borel by the proof of Lemma 6.2.5; notice that in the definition (10.2.3) we may replace the limit $r \to 0$ by the limit to zero through rational numbers and that the function $u_q(x) := \fint_{B(x,q)} g(z)\, d\mu(z)$ is lower semicontinous and hence Borel for each $q > 0$.

We claim that $3g$ is an upper gradient for u. Let $x, y \in X$ and let $\gamma : [0, L] \to X$ be a rectifiable curve joining x to y. As always, we assume that γ is parametrized by the arc length.

Fix $n \in \mathbb{N}$. For each $i = 0, \dots, n-1$, let γ_i denote the restriction of γ to $[iL/n, (i+1)L/n]$ and let x_i be a point in γ_i with

$$g(x_i) \le \frac{1}{\text{length}(\gamma_i)} \int_{\gamma_i} g\, ds = \frac{n}{L} \int_{\gamma_i} g\, ds.$$

Observe that $d(x_{i-1}, x_i) \le 2L/n$ and $d(x_{i-1}, x_i) + d(x_i, x_{i+1}) \le 3L/n$ for each i. By a telescoping argument,

$$\begin{aligned}
|u(x_0) - u(x_{n-1})| &\le \sum_{i=1}^{n-1} |u(x_{i-1}) - u(x_i)| \\
&\le \sum_{i=1}^{n-1} d(x_{i-1}, x_i)(g(x_{i-1}) + g(x_i)) \\
&= d(x_0, x_1)g(x_0) + \sum_{i=1}^{n-2}(d(x_{i-1}, x_i) + d(x_i, x_{i+1}))g(x_i) \\
&\quad + d(x_{n-2}, x_{n-1})g(x_{n-1}) \\
&\le \frac{2L}{n}\frac{n}{L}\int_{\gamma_0} g\, ds + \frac{3L}{n}\frac{n}{L}\int_{\gamma_1\cup\gamma_2\cup\cdots\cup\gamma_{n-2}} g\, ds + \frac{2L}{n}\frac{n}{L}\int_{\gamma_{n-1}} g\, ds \\
&\le 3\int_\gamma g\, ds.
\end{aligned}$$

Since u is continuous, we can let $n \to \infty$ to obtain the desired inequality $|u(x) - u(y)| \le 3\int_\gamma g\, ds$. Thus $u \in N^{1,p}(X:V)$ with $\|u\|_{N^{1,p}(X:V)} \le 3\|u\|_{M^{1,p}(X:V)}$ as desired. □

It is an interesting question whether the embedding norm factor 3 in (10.2.6) can be improved.

Lemma 10.2.7 *Suppose that the measure μ is doubling. The class of Lipschitz functions in $M^{1,p}(X:V)$ is dense in $M^{1,p}(X:V)$.*

Proof To prove this lemma, let $u \in M^{1,p}(X:V)$ and let $g \ge 0$ be a Hajłasz gradient of u. For $\lambda > 0$ let $E_\lambda := \{x \in X : g(x) > \lambda\}$. Then, since $\lambda^{-p}g^p \ge 1$ on E_λ, we obtain the estimate

$$\mu(E_\lambda) \le \frac{1}{\lambda^p}\left(\int_{E_\lambda} g^p\, d\mu\right)^{1/p}.$$

Because $\mu(E_\lambda) \to 0$ as $\lambda \to \infty$ and $p \ge 1$, we see from the above estimate and the absolute continuity of the integral $\int g^p\, d\mu$ that we also have $\lambda^p\mu(E_\lambda) \to 0$ as $\lambda \to \infty$.

Furthermore, $u|_{X\setminus E_\lambda}$ is 2λ-Lipschitz continuous. Let $u_\lambda : X \to V$ be a Lipschitz extension of u to X, as given for example by Theorem 4.1.21; then u_λ is $2C\lambda$-Lipschitz continuous on X. Let $g_\lambda := (g + 3C\lambda)\chi_{E_\lambda}$; we will now show that g_λ is a Hajłasz gradient of $u - u_\lambda$. This immediately yields that u_λ approximates u in $M^{1,p}(X:V)$. Observe that if $x, y \in X \setminus E_\lambda$ then $(u - u_\lambda)(x) = (u - u_\lambda)(y) = 0$, and hence the desired version of (10.2.1) is trivial on $X \setminus E_\lambda$. If $x, y \in E_\lambda$ then

$$\begin{aligned}|(u - u_\lambda)(x) - (u - u_\lambda)(y)| &\le |u(x) - u(y)| + |u_\lambda(x) - u_\lambda(y)| \\ &\le d(x, y)(g(x) + g(y) + 2C\lambda) \\ &= d(x, y)((g(x) + C\lambda) + (g(y) + C\lambda)) \\ &\le d(x, y)(g_\lambda(x) + g_\lambda(y)).\end{aligned}$$

Finally, if $x \in E_\lambda$ and $y \in X \setminus E_\lambda$ then $g_\lambda(x) = 3C\lambda + g(x)$ and $g_\lambda(y) = 0$ with $g(y) \le \lambda \le C\lambda$, and so

$$\begin{aligned}|(u - u_\lambda)(x) - (u - u_\lambda)(y)| &\le |u(x) - u(y)| + |u_\lambda(x) - u_\lambda(y)| \\ &\le d(x, y)(g(x) + g(y) + 2C\lambda) \\ &\le d(x, y)(g(x) + 3C\lambda) \\ &= d(x, y)(g_\lambda(x) + g_\lambda(y)).\end{aligned}$$

Combining the above cases, we see that g_λ is indeed a Hajłasz gradient of $u - u_\lambda$ as desired. □

Since two functions in $\tilde{N}^{1,p}$ which agree almost everywhere belong to the same equivalence class in $N^{1,p}$ (see Proposition 7.1.31), it follows that the set of equivalence classes of continuous functions in $M^{1,p}(X : V)$ embeds into $N^{1,p}(X : V)$. Thus the closure (in the $M^{1,p}$-norm) of this set is a subspace of $N^{1,p}$. By Lemma 10.2.7, Lipschitz-continuous functions are dense in $M^{1,p}(X : V)$. Thus this closure is the full Hajłasz–Sobolev space $M^{1,p}(X : V)$. We have proved the following.

Theorem 10.2.8 *For any $p \ge 1$, any metric space X equipped with a doubling measure μ, and any Banach space V, $M^{1,p}(X : V)$ embeds continuously into $N^{1,p}(X : V)$ with embedding norm at most* 3.

Combining the above theorem with (8.1.10) of Theorem 8.1.7 yields the following corollary.

Corollary 10.2.9 *Suppose that μ is doubling and X supports a q-Poincaré inequality for some $1 \le q < p$. Then $M^{1,p}(X : V) = N^{1,p}(X : V)$.*

As will be seen in Chapter 12, the requirement that X support a q-Poincaré inequality for some $q < p$ is not restrictive at all whenever X is complete.

10.3 Sobolev spaces defined via Poincaré inequalities

Another notion of Sobolev spaces can be defined directly in terms of Poincaré inequalities.

Fix a constant $\lambda \geq 1$. The *Poincaré–Sobolev space* $P^{1,p}(X : V)$ consists of all functions $u \in L^p(X : V)$ for which there exists a function $g \in \mathrm{L}^p(X)$ such that

$$\fint_B |u - u_B| \, d\mu \leq \operatorname{diam}(B) \left(\fint_{\lambda B} g^p \, d\mu \right)^{1/p} \tag{10.3.1}$$

for all balls B in X.

The definition of $P^{1,p}(X : V)$ clearly depends on the parameter λ, but we will suppress this dependence in our notation. We emphasize that we do *not* assume in the above definition that the function g is either an upper gradient or a Hajłasz gradient of the function u; the only connection between these two functions is that indicated by the Poincaré inequality (10.3.1). A function g such that (10.3.1) holds for all balls B in X will be called a *Poincaré gradient* of u. The Sobolev space $P^{1,p}(X : V)$ was introduced by Hajłasz and Koskela in [114]. As with $M^{1,p}$, the elements of $P^{1,p}$ are μ-equivalence classes of functions. Thus some care should be taken in comparing $P^{1,p}$ with $N^{1,p}$.

In this section we consider the relationships between the Poincaré–Sobolev space $P^{1,p}(X : V)$ and the Sobolev spaces $N^{1,p}(X : V)$ and $M^{1,p}(X : V)$. We begin with the simplest inclusion. Observe that no assumptions on X are needed in the following result.

Proposition 10.3.2 *If $u \in M^{1,p}(X : V)$ with Hajłasz gradient g, then u belongs to $P^{1,p}(X : V)$ with Poincaré gradient $2g$.*

Proof Integrating the inequality $|u(x) - u(y)| \leq d(x, y)(g(x) + g(y))$ (valid for x, y outside some set $E \subset X$ with $\mu(E) = 0$) twice over a ball B yields

$$\begin{aligned} &\fint_B \fint_B |u(x) - u(y)| \, d\mu(x) \, d\mu(y) \\ &\qquad \leq \operatorname{diam}(B) \left(\fint_B g(x) \, d\mu(x) + \fint_B g(y) \, d\mu(y) \right). \end{aligned}$$

The proof is completed by an application of Hölder's inequality. □

The following result is an immediate consequence of the definitions.

Proposition 10.3.3 *Assume that X supports a p-Poincaré inequality. If u is in $N^{1,p}(X : V)$ with upper gradient g then the L^p-equivalence class of u belongs to $P^{1,p}(X : V)$ with Poincaré gradient Cg, where C depends only on the data of the Poincaré inequality on X.*

To establish the reverse inclusion, we take advantage of the discrete-convolution approximations defined in Section 9.2. In the case $1 < p < \infty$ the result follows from the weak compactness of $N^{1,p}$ (Theorem 7.3.9). The case $p = 1$ is more challenging; we will comment on this case in the notes to this chapter. For simplicity we consider only the case $p > 1$ in the following theorem.

Theorem 10.3.4 *Assume that the measure on X is doubling, and that $1 < p < \infty$. If $u \in P^{1,p}(X : V)$ with Poincaré gradient g then some μ-representative of u belongs to $N^{1,p}(X : V)$ with upper gradient Cg, where C depends only on the doubling constant of μ.*

Proof We consider the discrete-convolution approximations (u_r) as in Section 9.2. That is, $u_r(x) = \sum_i \varphi_{r,i}(x) u_{B(x_i,r)}$, where $\{B_i = B(x_i, r)\}$ is a cover of X with $\sum_i \chi_{B(x_i,6\lambda r)} \le C$ and where $(\varphi_{r,i})$ denotes the corresponding Lipschitz partition of unity. Given the assumptions on the cover, we know that each u_r is locally Lipschitz continuous and hence is also measurable.

We claim that there exists a constant C, depending only on the doubling constant of μ, such that for each $r > 0$ the function

$$g_r(x) := C \sup_{i:B_i \ni x} \left(\fint_{3\lambda B_i} g(y)^p \, d\mu(y) \right)^{1/p},$$

is an upper gradient of u_r. To this end, we show that the inequality

$$\operatorname{Lip} u_r(x) \le \frac{C}{r} \sup_{i:3B_i \ni x} \fint_{3B_i} |u - u_{3B_i}| \, d\mu \tag{10.3.5}$$

holds true for every $x \in X$ and appeal to Lemma 6.2.6; the desired result then holds since $u \in P^{1,p}(X : V)$. Here

$$\operatorname{Lip} u_r(x) = \limsup_{\rho \to 0} \sup_{y \in B(x,\rho)} \frac{|u_r(x) - u_r(y)|}{\rho}$$

denotes the pointwise upper Lipschitz-constant function of u_r, as in (6.2.4).

To prove (10.3.5), fix a ball $B_j \ni x$ from the cover in the first paragraph of the proof. For $0 < \rho < r$ and $y \in B(x, \rho)$ we compute, using $\sum_i \varphi_{r,i} \equiv 1$ and the doubling property of μ, that

$$\begin{aligned} |u_r(y) - u_r(x)| &= \left| \sum_{i:3B_i \ni x} (\varphi_{r,i}(y) - \varphi_{r,i}(x))(u_{B_i} - u_{B_j}) \right| \\ &\le \frac{C}{r} d(x, y) \sum_{i:3B_i \ni x} |u_{B_i} - u_{B_j}| \\ &\le \frac{C}{r} d(x, y) \sum_{i:3B_i \ni x} (|u_{B_i} - u_{3B_i}| + |u_{3B_i} - u_{B_j}|) \end{aligned}$$

$$\leq \frac{C}{r}\rho \sup_{i:3B_i \ni x} \fint_{3B_i} |u - u_{3B_i}| \, d\mu.$$

The desired estimate (10.3.5) follows upon taking the supremum over all $y \in B(x, \rho)$, dividing by ρ and taking the limit superior as $\rho \to 0$.

We argue similarly to prove that $u_r \to u$ in $L^p(X : V)$. For $x \in B_j$,

$$\begin{aligned} |u_r(x) - u(x)| &= \left| \sum_i \varphi_{r,i}(x)[u_{B_i} - u(x)] \right| \\ &\leq \sum_i |u_{B_i} - u(x)| \, \varphi_{r,i}(x) \leq \sum_{i:2B_i \cap B_j \neq \emptyset} |u_{B_i} - u(x)| \\ &\leq \sum_{i:2B_i \cap B_j \neq \emptyset} \fint_{B_i} |u(y) - u(x)| \, d\mu(y). \end{aligned}$$

Using the fact that $X = \bigcup_i B_i$, the bounded overlap property $\sum_i \chi_{6\lambda B_i} \leq C$, the doubling property of μ, the fact that $5B_j \supset B_i$ whenever $2B_i \cap B_j$ is nonempty, and Hölder's inequality, we obtain

$$\begin{aligned} &\int_X |u_r(x) - u(x)|^p \, d\mu(x) \leq \sum_j \int_{B_j} |u_r(x) - u(x)|^p \, d\mu(x) \\ &\leq C \sum_j \int_{B_j} \left(\fint_{5B_j} |u(y) - u(x)| \, d\mu(y) \right)^p \, d\mu(x) \\ &\leq C \sum_j \int_{B_j} \fint_{5B_j} |u(y) - u(x)|^p \, d\mu(y) \, d\mu(x) \\ &\leq C \sum_j \int_{B_j} \left(\fint_{5B_j} |u(y) - u_{5B_j}|^p \, d\mu(y) + |u(x) - u_{5B_j}|^p \right) \, d\mu(x) \\ &\leq Cr^p \sum_j \left(\int_{B_j} \fint_{6\lambda B_j} g(y)^p \, d\mu(y) \, d\mu(x) + \int_{5B_j} |u(x) - u_{5B_j}|^p \, d\mu(x) \right) \\ &\leq Cr^p \sum_j \int_{6\lambda B_j} g^p \, d\mu \leq Cr^p \int_X g^p \, d\mu. \end{aligned}$$

Here we have used the fact that the left-hand side of (10.3.1) improves as follows:

$$\left(\fint_B |u - u_B|^p \, d\mu \right)^{1/p} \leq C \operatorname{diam}(B) \left(\fint_{\lambda B} g^p \, d\mu \right)^{1/p};$$

this can be verified by examining the proof of Theorem 9.1.32(i). Given that $g \in L^p(X)$ it is now clear that $\lim_{r \to 0} \|u_r - u\|_{L^p(X:V)} = 0$.

We have shown that $u_r \in N^{1,p}(X : V)$ for each $r > 0$ and that it converges in $L^p(X : V)$ to u. To conclude the proof we appeal to Theorem 7.3.9. To this end we must show that (u_r) is bounded in $N^{1,p}(X : V)$; it suffices to verify that (g_r) is bounded in $L^p(X)$. We have that

$$\begin{aligned}\int_X g_r^p \, d\mu &= C \sum_i \int_{3B_i} \fint_{3\lambda B_i} g(y)^p \, d\mu(y) \, d\mu(x) \\ &\leq C \sum_i \int_{3\lambda B_i} g(y)^p \, d\mu(y) \leq C \int_X g^p \, d\mu;\end{aligned}$$

since $g \in L^p(X)$ the proof is complete. □

10.4 The Korevaar–Schoen–Sobolev space

In connection with their study of harmonic mappings into metric spaces in [164], Korevaar and Schoen introduced a type of Sobolev space that we have not yet considered in this book. Historically, this was the first Sobolev space, among those considered in this chapter, to be defined. In the original definition of Korevaar and Schoen, the source space was a domain with smooth boundary in a Riemannian manifold and the target was a metric space. We will consider a variation on their definition in which the source is a metric measure space and the target is a Banach space. At the conclusion of this section we briefly discuss the relationship between our approach and the original theory of Korevaar and Schoen. This variation appeared first in [168].

Let $u : X \to V$. For $x \in X$ and $\epsilon > 0$, define

$$e_\epsilon^p(x; u) := \epsilon^{-p} \fint_{B(x,\epsilon)} |u(y) - u(x)|^p \, d\mu(y).$$

A function $u : X \to V$ is said to be in the *Korevaar–Schoen–Sobolev space* $KS^{1,p}(X : V)$ if $u \in L^p(X : V)$ and if

$$E^p(u) := \sup_B \limsup_{\epsilon \to 0} \int_B e_\epsilon^p(x; u) \, d\mu(x) < \infty, \tag{10.4.1}$$

where the supremum is taken over all metric balls B in X. The quantity $E^p(u)$ in (10.4.1) is called the *Korevaar–Schoen p-energy* of the function u. Note that the elements of $KS^{1,p}$ are L^p-equivalence classes. We equip $KS^{1,p}(X : V)$ with the norm

$$\|u\|_{KS^{1,p}(X:V)} := \|u\|_{L^p(X:V)} + (E^p(u))^{1/p}. \tag{10.4.2}$$

The principal results of this section are the following two theorems.

Theorem 10.4.3 *Assume that the measure on X is doubling. For $p > 1$, each element u in $KS^{1,p}(X : V)$ has a Lebesgue representative $\tilde{u}$ in $N^{1,p}(X : V)$ satisfying*

$$\int_X \rho_{\tilde{u}}^p \, d\mu \le CE^p(u). \tag{10.4.4}$$

Here C denotes a constant depending only on the doubling constant of μ and on p.

For the case $p = 1$, see the notes to this chapter.

Theorem 10.4.5 *If μ is a doubling measure then $P^{1,p}(X : V) \subset KS^{1,p}(X : V)$.*

Combining Proposition 10.3.3 and Theorems 10.4.3 and 10.4.5 yields the following corollary.

Corollary 10.4.6 *Assume that the measure on X is doubling, $1 < p < \infty$, and that X satisfies a p-Poincaré inequality. Then*

$$P^{1,p}(X : V) = N^{1,p}(X : V) = KS^{1,p}(X : V).$$

Observe that the definitions of neither the Korevaar–Schoen–Sobolev space nor the Poincaré–Sobolev space make any reference to upper gradients or to the Poincaré inequality defined using such gradients. The assumption of the Poincaré inequality in the above corollary is nevertheless necessary; see the example below.

Example 10.4.7 The example of the planar slit disk $X = \overline{B}(0, 1) \setminus ([0, 1] \times \{0\})$, equipped with the Lebesgue measure m_2 and the Euclidean metric, demonstrates that without a Poincaré inequality we do *not* in general have $N^{1,p}(X : V) \subset P^{1,p}(X : V)$, nor do we have $N^{1,p}(X : V) \subset KS^{1,p}(X : V)$. For this space X, with $V = \mathbb{R}$, we have $P^{1,p}(X) = KS^{1,p}(X) = W^{1,p}(\overline{B}(0, 1))$ but $N^{1,p}(X) = W^{1,p}(X) \neq W^{1,p}(\overline{B}(0, 1))$.

To see that without a Poincaré inequality we may have $KS^{1,p}(X) \not\subset P^{1,p}(X)$, we consider the set

$$X_\alpha := \mathbb{R}^2 \setminus \{(x, y) \in \mathbb{R}^2 : x > 1 \text{ and } |y| < x^{-\alpha}\}$$

for $\alpha > 1$. We equip $X = X_\alpha$ with the Lebesgue measure m_2 and the Euclidean metric. On the one hand, since X_α is a proper space and the setting is Euclidean, it is easy to verify that $KS^{1,p}(X) = W^{1,p}(X)$. On the other hand, note that with $1 < \theta < \beta < \alpha$, the function $u = \varphi \hat{u}$, where

$$\hat{u}(x, y) = \begin{cases} 1 & \text{if } x \geq 1 \text{ and } x^{-\alpha} \leq y \leq x^{-\beta}, \\ 0 & \text{if } x \geq 1 \text{ and } y \leq -x^{-\alpha}, \\ 0 & \text{if } x < 1 \text{ or else } x \geq 1 \text{ and } y \geq x^{-\theta}, \end{cases}$$

and where $\varphi \in C^\infty$ satisfies $\varphi(x, y) = 0$ for $(x, y) \in B((1, 0), 1)$ and $\varphi(x, y) = 1$ for $(x, y) \notin B((1, 0), 2)$, has an extension to a function in $W^{1,p}(X)$ when $p \geq 2$. Suppose that this extension, also denoted by u, is in $P^{1,p}(X)$. Let $g \in L^p(X)$, $g \geq 0$, be a Poincaré gradient of u. For $j \geq 1$, let $z_j = (j, j^{-\alpha})$ and consider the balls $B_j = B(z_j, j^{-\beta}/2)$. For sufficiently large j (for example, $j > 16^{1/(\alpha-\beta)}$), we know that

$$\frac{1}{2} \leq C \fint_{B_j} |u - u_{B_j}| \, dm_2 \leq C j^{-\beta} \left(\fint_{\lambda B_j} g^p \, dm_2 \right)^{1/p},$$

that is,

$$\int_{\lambda B_j} g^p \, dm_2 \geq \frac{1}{C} \frac{1}{j^{-\beta(p-2)}} = \frac{1}{C} j^{\beta(p-2)}.$$

However, for sufficiently large j the balls λB_j are pairwise disjoint. It follows that, if $g \in L^p(X)$, we must have $\sum_{j=j_0}^\infty j^{\beta(p-2)} < \infty$, which is not possible for $p \geq 2$. It follows that when $p \geq 2$ no μ-representative of the function u can be in $P^{1,p}(X)$.

Proof of Theorem 10.4.3 The argument is similar to that in the proof of Theorem 10.3.4, so we provide only a sketch. Again, as in Section 9.2, fix $\epsilon > 0$ and let $\{B_i = B(x_i, \epsilon)\}_i$ be a cover of X with $\sum_i \chi_{16B_i} \leq C$ and $(\varphi_{\epsilon,i})$ the corresponding partition of unity.

Fix a ball $B_0 = B(x_0, R)$ and $\delta > 0$. By the definition of $KS^{1,p}(X : V)$ there is a positive number ϵ_0 such that, for $\epsilon < \epsilon_0$,

$$\int_{B_0} \fint_{B(x,8\epsilon)} \frac{|u(x) - u(y)|^p}{\epsilon^p} \, d\mu(y) \, d\mu(x) \leq E^p(u) + \delta.$$

Set $u_\epsilon := \sum_i u_{B_i} \varphi_{\epsilon,i}$. Because $\sum_i \varphi_{\epsilon,i} \equiv 1$,

$$u_\epsilon(x) - u(x) = \sum_i (u_{B_i} - u(x)) \varphi_{\epsilon,i}(x).$$

For $x \in B_j$, the bounded-overlap property of the cover and the doubling property of μ imply that

$$
\begin{aligned}
|u_\epsilon(x) - u(x)|^p &\le \left(\sum_{i:2B_i \cap B_j \neq \emptyset} |u_{B_i} - u(x)| \right)^p \\
&\le C \fint_{B(x,8\epsilon)} |u(y) - u(x)|^p \, d\mu(y).
\end{aligned}
$$

Hence, assuming that ϵ is sufficiently small (relative to R),

$$
\begin{aligned}
\int_{\frac{1}{2}B_0} |u_\epsilon(x) - u(x)|^p \, d\mu(x) &\le \sum_{j:B_j \cap \frac{1}{2}B_0 \neq \emptyset} \int_{B_j} |u_\epsilon(x) - u(x)|^p \, d\mu(x) \\
&\le C \sum_{j:B_j \cap \frac{1}{2}B_0 \neq \emptyset} \int_{B_j} \fint_{B(x,8\epsilon)} |u(y) - u(x)|^p \, d\mu(y) \, d\mu(x) \\
&\le C\epsilon^p \int_{B_0} \fint_{B(x,8\epsilon)} \frac{|u(y) - u(x)|^p}{\epsilon^p} \, d\mu(y) \, d\mu(x) \\
&\le C\epsilon^p \big(E^p(u) + \delta \big).
\end{aligned}
$$

Since the upper bound in the preceding expression tends to zero as $\epsilon \to 0$, we conclude that $u_\epsilon \to u$ in $L^p(B_0 : V)$.

An argument similar to that in the proof of Theorem 10.3.4 shows that

$$
g_\epsilon(x) := \frac{C}{\epsilon} \sup_{j:B_j \ni x} \fint_{4B_j} |u - u_{4B_j}| \, d\mu
$$

is an upper gradient of u_ϵ. Note that

$$
g_\epsilon(x)^p \le C \sup_{j:B_j \ni x} \fint_{4B_j} \fint_{B(y,8\epsilon)} \frac{|u(y) - u(z)|^p}{\epsilon^p} \, d\mu(z) \, d\mu(y)
$$

and hence

$$
\begin{aligned}
&\int_{\frac{1}{2}B_0} g_\epsilon^p(x) \, d\mu(x) \\
&\quad \le C \sum_{j:B_j \cap \frac{1}{2}B_0 \neq \emptyset} \int_{B_j} \fint_{4B_j} \fint_{B(y,8\epsilon)} \frac{|u(y) - u(z)|^p}{\epsilon^p} \, d\mu(z) \, d\mu(y) \, d\mu(x) \\
&\quad \le C \sum_{j:B_j \cap \frac{1}{2}B_0 \neq \emptyset} \int_{4B_j} \fint_{B(y,8\epsilon)} \frac{|u(y) - u(z)|^p}{\epsilon^p} \, d\mu(z) \, d\mu(y) \\
&\quad \le C \int_{B_0} \fint_{B(y,8\epsilon)} \frac{|u(y) - u(z)|^p}{\epsilon^p} \, d\mu(z) \, d\mu(y) \le C\big(E^p(u) + \delta\big).
\end{aligned}
$$

In conclusion, we have constructed an increasing exhaustion of X by open balls $B_n = B(x_0, n)$, $n \in \mathbb{N}$, and functions $u_n \in N^{1,p}(B_n : V)$ with upper gradients $g_n \in L^p(B_n)$, so that, for each n_0, $u_n \to u$ in $L^p(B_n : V)$ and the values $\int_{B_n} g_n^p \, d\mu$ remain uniformly bounded in n. The claim now follows by an application of Lemma 7.3.22. □

The proof of Theorem 10.4.5 makes use of the Riesz potentials $J_{p,r}$, $p \geq 1$, $r > 0$, defined for nonnegative functions $g \in L^p_{\text{loc}}(X)$ by

$$J_{p,r}g(x) = \sum_{k=0}^{\infty} 2^{-k} r \left(⨍_{B(x,2^{-k}r)} g^p \, d\mu \right)^{1/p}.$$

The following integral estimates for the Riesz potentials will be proven after the proof of Theorem 10.4.5.

Lemma 10.4.8 *Assume that μ is a doubling measure on X and that $p \geq 1$. Then*

(i) *there exists a constant $C > 0$ such that*

$$\int_{B(x,\epsilon)} (J_{p,\epsilon} g)^p \, d\mu \leq C \epsilon^p \int_{B(x,2\epsilon)} g^p \, d\mu$$

for any $x \in X$ and any $\epsilon > 0$;

(ii) *there exists a constant $C > 0$ such that*

$$\int_X (J_{p,\epsilon} g)^p \, d\mu \leq C \epsilon^p \int_X g^p \, d\mu$$

for any $\epsilon > 0$.

Proof of Theorem 10.4.5 Let $u \in P^{1,p}(X : V)$ and choose $g \in L^p(X)$ such that the Poincaré inequality (10.3.1) holds. For Lebesgue points $x, y \in X$ of u, and for integers i, we set $B_i = B(x, 2^{-i} d(x, y))$ if $i \geq 0$ and $B_i = B(y, 2^{i} d(x, y))$ if $i < 0$. Then

$$\begin{aligned} |u(x) - u(y)| \leq \sum_{i \in \mathbb{Z}} |u_{B_i} - u_{B_{i+1}}| &\leq \sum_{i \in \mathbb{Z}} 2^{-|i|} d(x, y) \left(⨍_{2\lambda B_i} g^p \, d\mu \right)^{1/p} \\ &\leq C \left(J_{p,2\lambda d(x,y)} g(x) + J_{p,2\lambda d(x,y)} g(y) \right), \end{aligned}$$

by (10.3.1).

Because of the doubling property of μ, if $r_1 < r_2$ then $J_{p,r_1} g \le C J_{p,r_2} g$. For almost every $x \in X$ it follows that

$$\begin{aligned} e_\epsilon^p(x; u) &= \epsilon^{-p} \fint_{B(x,\epsilon)} |u(x) - u(y)|^p \, d\mu(y) \\ &\le C\epsilon^{-p} \left(J_{p,2\lambda\epsilon} g(x)^p + \fint_{B(x,\epsilon)} J_{p,2\lambda\epsilon} g(y)^p \, d\mu(y) \right) \\ &\le C \left(\epsilon^{-p} J_{p,4\lambda\epsilon} g(x)^p + \fint_{B(x,4\lambda\epsilon)} g^p \, d\mu \right), \end{aligned}$$

by Lemma 10.4.8(i). Integrating over a ball $B \subset X$ yields

$$\begin{aligned} \int_B e_\epsilon^p(x; u) \, d\mu(x) &\le C\epsilon^{-p} \int_B (J_{p,4\lambda\epsilon} g)^p \, d\mu + C \int_B \fint_{B(x,4\lambda\epsilon)} g^p \, d\mu \, d\mu(x) \\ &\le C \int_X g^p \, d\mu + C \int_X \fint_{B(x,4\lambda\epsilon)} g^p \, d\mu \, d\mu(x), \end{aligned}$$

by Lemma 10.4.8(ii). By Fubini's theorem and the doubling property of μ,

$$\begin{aligned} &\int_X \fint_{B(x,4\lambda\epsilon)} g(y)^p \, d\mu(y) \, d\mu(x) = \int_X \int_X \frac{\chi_{B(x,4\lambda\epsilon)}(y)}{\mu(B(x, 4\lambda\epsilon))} g(y)^p \, d\mu(y) \, d\mu(x) \\ &\le C \int_X \int_X \frac{\chi_{B(y,4\lambda\epsilon)}(x)}{\mu(B(y, 4\lambda\epsilon))} g(y)^p \, d\mu(x) \, d\mu(y) = C \int_X g^p \, d\mu \, . \end{aligned}$$

Hence $\limsup_{\epsilon\to 0} \int_B e_\epsilon^p(x; u) \, d\mu(x) \le C \int_X g^p \, d\mu$. By taking the supremum over all balls, we conclude from the definition (10.4.1) that $E^p(u) \le C \int_X g^p \, d\mu < \infty$. Thus $u \in KS^{1,p}(X : V)$ and the proof is complete. □

We will now prove Lemma 10.4.8.

Proof of Lemma 10.4.8 We break the proof into three cases, according to whether $p > Q$, $p = Q$, or $p < Q$. Here Q is the exponent in (9.1.14). The proof is similar in technique to that of Theorem 9.1.15.

Let $x \in X$ and $\epsilon > 0$. For $z \in B(x, \epsilon)$ and a nonnegative integer k_0,

$$\begin{aligned} J_{p,\epsilon} g(z) &= \sum_{k=0}^{k_0} 2^{-k}\epsilon \left(\fint_{B(z,2^{-k}\epsilon)} g^p \, d\mu \right)^{1/p} \\ &\quad + \sum_{k=k_0+1}^{\infty} 2^{-k}\epsilon \left(\fint_{B(z,2^{-k}\epsilon)} g^p \, d\mu \right)^{1/p} \\ &\le J_1 + 2^{-k_0}\epsilon M \chi_{B(x,2\epsilon)} g^p(z)^{1/p}, \end{aligned}$$

where $M\chi_{B(x,2\epsilon)}g^p$ is the Hardy–Littlewood maximal function of the zero extension of g outside $B(x,2\epsilon)$. By (9.1.14),

$$\begin{aligned} J_1 &:= \sum_{k=0}^{k_0} 2^{-k}\epsilon \left(\fint_{B(z,2^{-k}\epsilon)} g^p \, d\mu \right)^{1/p} \\ &\leq \sum_{k=0}^{k_0} \frac{2^{-k}\epsilon}{\mu(B(z,2^{-k}\epsilon))^{1/p}} \left(\int_{B(z,2\epsilon)} g^p \, d\mu \right)^{1/p} \\ &\leq C\epsilon \sum_{k=0}^{k_0} 2^{-k(1-Q/p)} \left(\fint_{B(x,2\epsilon)} g^p \, d\mu \right)^{1/p}. \end{aligned}$$

Now we split the argument into three cases.

Case $p > Q$. Then $\sum_{k=0}^{\infty} 2^{-k(1-Q/p)} < \infty$, and so

$$J_1 \leq C\epsilon \left(\fint_{B(x,2\epsilon)} g^p \, d\mu \right)^{1/p};$$

letting $k_0 \to \infty$, we obtain

$$J_{p,\epsilon} g(z) \leq C\epsilon \left(\fint_{B(x,2\epsilon)} g^p \, d\mu \right)^{1/p},$$

from which the desired inequality (i) follows.

Case $p < Q$. We may clearly assume that $\int_{B(x,2\epsilon)} g^p \, d\mu > 0$. Note that

$$J_1 \leq C\epsilon \, 2^{k_0(Q/p-1)} \left(\fint_{B(x,2\epsilon)} g^p \, d\mu \right)^{1/p}.$$

If we choose k_0 such that

$$\epsilon \, 2^{-k_0} M\chi_{B(x,2\epsilon)} g^p(z)^{1/p} \leq C\epsilon \, 2^{k_0(Q/p-1)} \left(\fint_{B(x,2\epsilon)} g^p \, d\mu \right)^{1/p} \tag{10.4.9}$$

then

$$J_{p,\epsilon} g(z) \leq 2C\epsilon \, 2^{k_0(Q/p-1)} \left(\fint_{B(x,2\epsilon)} g^p \, d\mu \right)^{1/p}.$$

Inequality (10.4.9) holds if and only if

$$2^{-k_0 Q/p} \leq \frac{C}{M\chi_{B(x,2\epsilon)} g^p(z)^{1/p}} \left(\fint_{B(x,2\epsilon)} g^p \, d\mu \right)^{1/p}; \tag{10.4.10}$$

it follows from $\int_{B(x,2\epsilon)} g^p \, d\mu > 0$ that $M\chi_{B(x,2\epsilon)} g^p(z) > 0$. If the term on the right-hand side of (10.4.10) is not larger than 1, then we can choose k_0 such that

$$2^{-k_0 Q/p} \approx \frac{C}{M\chi_{B(x,2\epsilon)} g^p(z)^{1/p}} \left(\fint_{B(x,2\epsilon)} g^p \, d\mu \right)^{1/p},$$

from which we get

$$J_{p,\epsilon} g(z) \le C\epsilon M\chi_{B(x,2\epsilon)} g^p(z)^{1/p-1/Q} \left(\fint_{B(x,2\epsilon)} g^p \, d\mu \right)^{1/Q}. \quad (10.4.11)$$

However, if the term on the right-hand side of (10.4.10) is greater than 1, i.e.

$$\frac{C}{M\chi_{B(x,2\epsilon)} g^p(z)^{1/p}} \left(\fint_{B(x,2\epsilon)} g^p \, d\mu \right)^{1/p} > 1,$$

then we can choose $k_0 = 0$, in which case

$$J_{p,\epsilon} g(z) \le C\epsilon \left(\fint_{B(x,2\epsilon)} g^p \, d\mu \right)^{1/p}. \quad (10.4.12)$$

Each point in $B(x, \epsilon)$ satisfies at least one of (10.4.11) and (10.4.12). Let A_1 denote the collection of all points for which (10.4.11) holds, and A_2 denote the collection of the remaining points. For $t > 0$, we have by the Hardy–Littlewood maximal theorem 3.5.6,

$$\begin{aligned}
&\mu(\{z \in B(x, \epsilon) \cap A_1 : J_{p,\epsilon} g(z) > t\}) \\
&\quad \le \mu\left(\left\{ z \in B(x, \epsilon) : M\chi_{B(x,2\epsilon)} g^p(z) \right.\right. \\
&\qquad \left.\left. > \frac{t^{pQ/(Q-p)}}{C\epsilon^{pQ/(Q-p)} \left(\fint_{B(x,2\epsilon)} g^p \, d\mu \right)^{p/(Q-p)}} \right\}\right) \\
&\quad \le C \frac{\epsilon^{pQ/(Q-p)}}{t^{pQ/(Q-p)}} \left(\fint_{B(x,2\epsilon)} g^p \, d\mu \right)^{p/(Q-p)} \int_{B(x,2\epsilon)} g^p \, d\mu \\
&\quad \le C \frac{\epsilon^{pQ/(Q-p)}}{t^{pQ/(Q-p)}} \mu(B(x, 2\epsilon)) \left(\fint_{B(x,2\epsilon)} g^p \, d\mu \right)^{Q/(Q-p)}.
\end{aligned}$$

By (10.4.12),

$$
\begin{aligned}
&\mu(\{z \in B(x,\epsilon)\cap A_2 : J_{p,\epsilon}g(z) > t\})\\
&\qquad \le \mu\left(\left\{z\in B(x,\epsilon) : \left(\fint_{B(x,2\epsilon)} g^p\,d\mu\right)^{1/p} > C^{-1}\frac{t}{\epsilon}\right\}\right)\\
&\qquad \le C\,\frac{\epsilon^p}{t^p}\int_{B(x,2\epsilon)} g^p\,d\mu \le C\,\frac{\epsilon^p}{t^p}\,\mu(B(x,2\epsilon))\fint_{B(x,2\epsilon)} g^p\,d\mu\\
&\qquad \le C\,\frac{\epsilon^{pQ/(Q-p)}}{t^{pQ/(Q-p)}}\,\mu(B(x,2\epsilon))\left(\fint_{B(x,2\epsilon)} g^p\,d\mu\right)^{Q/(Q-p)},
\end{aligned}
$$

where, in deriving the last inequality, we used the fact that, in order for the set

$$
\left\{z\in B(x,\epsilon) : M\chi_{B(x,2\epsilon)}g^p(z) > \frac{t^{pQ/(Q-p)}}{C\epsilon^{pQ/(Q-p)}\left(\fint_{B(x,2\epsilon)} g^p\,d\mu\right)^{p/(Q-p)}}\right\}
$$

to be nonempty, we need to have $t < 2C\epsilon\left(\fint_{B(x,2\epsilon)} g^p\,d\mu\right)^{1/p}$ and $pQ/(Q-p) > 1$. Combining the above two estimates, we have

$$
\begin{aligned}
&\mu(\{z\in B(x,\epsilon) : J_{p,\epsilon}g(x) > t\})\\
&\qquad \le 2C\left(\frac{\epsilon}{t}\right)^{Qp/(Q-p)}\mu(B(x,2\epsilon))\left(\fint_{B(x,2\epsilon)} g^p\,d\mu\right)^{Q/(Q-p)}.
\end{aligned}
$$

A careful tracking of the constant C_p in relation to the constant C_1 in the proof of Theorem 3.5.6 yields

$$
\begin{aligned}
&\left(\int_{B(x,\epsilon)} J_{p,\epsilon}g(z)^p\,d\mu(z)\right)^{1/p}\\
&\quad \le C\mu(B(x,2\epsilon))^{1/p-1/p^*}\epsilon\mu(B(x,2\epsilon))^{1/p^*}\left(\fint_{B(x,2\epsilon)} g^p\,d\mu\right)^{Q/(p^*(Q-p))},
\end{aligned}
$$

from which the desired inequality (i) follows.

Case $p = Q$. The method in this case is similar to that for the previous case, but with a slight variation: here, we have,

$$
J_1 \le C\epsilon\sum_{k=0}^{k_0}\left(\fint_{B(x,2\epsilon)} g^p\,d\mu\right)^{1/p} \le C\epsilon\,2^{k_0}\left(\fint_{B(x,2\epsilon)} g^p\,d\mu\right)^{1/p}.
$$

Now we proceed as before, choosing a suitable k_0. If

$$
C\left(\fint_{B(x,2\epsilon)} g^p\,d\mu\right)^{1/p} \ge M\chi_{B(x,2\epsilon)}g^p(z)^{1/p}
$$

then we choose $k_0 = 0$, and if

$$C \left(\fint_{B(x,2\epsilon)} g^p \, d\mu \right)^{1/p} < M\chi_{B(x,2\epsilon)} g^p(z)^{1/p},$$

then we choose k_0 such that $2^{-2k_0} \approx C \left(\fint_{B(x,2\epsilon)} g^p \, d\mu \right)^{1/p} / M\chi_{B(x,2\epsilon)} g^p(z)^{1/p}$. Proceeding as in the case $p < Q$, we again obtain inequality (i).

Inequality (ii) follows from inequality (i) via a covering argument. □

Korevaar–Schoen spaces on domains

We now give an alternate definition of the Korevaar–Schoen–Sobolev space $KS^{1,p}(X : V)$. In the case when X is proper, the two definitions coincide. Our new definition is more appropriate if we wish to consider such spaces as defined on domains $\Omega \subset X$. In this way we will eventually relate our definition to the original Sobolev spaces considered by Korevaar and Schoen.

For a map $u : X \to V$ from a metric measure space $X = (X, d, \mu)$ into a Banach space V, we define $e^p_\epsilon(x; u)$, $x \in X$, $\epsilon > 0$, $p \geq 1$, as before and we set

$$\tilde{E}^p(u) := \sup_\varphi \limsup_{\epsilon \to 0} \int_X \varphi(x) e^p_\epsilon(x; u) \, d\mu(x) , \tag{10.4.13}$$

where the supremum is taken over all continuous compactly supported functions $\varphi : X \to [0, 1]$. Then u is said to be in the *(modified) Korevaar–Schoen–Sobolev space* $\widetilde{KS}^{1,p}(X : V)$ if $\tilde{E}^p(u)$ is finite. It is clear that this definition agrees with the prior definition in the case when X is proper, i.e., when closed balls in X are compact.

The revised definition extends naturally to domains $\Omega \subset X$. For a map $u : \Omega \to V$ we consider

$$\tilde{E}^p(u) := \sup_\varphi \limsup_{\epsilon \to 0} \int_\Omega \varphi(x) e^p_\epsilon(x; u) \, d\mu(x) , \tag{10.4.14}$$

the supremum being taken over all continuous compactly supported functions $\varphi : \Omega \to [0, 1]$. Observe that $e^p_\epsilon(x; u)$ is defined provided x lies in

$$\Omega_\epsilon := \{z \in \Omega : \operatorname{dist}(z, X \setminus \Omega) > \epsilon\},$$

and the integral in (10.4.14) makes sense when $\epsilon < \operatorname{dist}(\operatorname{spt}(\varphi), X \setminus \Omega)$.

Theorems 10.4.5 and 10.4.3 remain true for the spaces $P^{1,p}(\Omega : V)$, $KS^{1,p}(\Omega : V)$, and $N^{1,p}(\Omega : V)$ under appropriate hypotheses.

Remark 10.4.15 The original definition of Korevaar and Schoen, based on (10.4.14), with even more general metric space targets, was in the setting of smoothly bounded domains in a Riemannian manifold M, with compact completion $\overline{\Omega}$.

10.5 Summary

We will summarize the results of this chapter in the following theorems.

A combination of Proposition 10.3.2 with Theorems 10.3.4, 10.1.1, and 10.4.5 gives the following theorem.

Theorem 10.5.1 *Let μ be a doubling measure and assume that $p > 1$. Then*

$$M^{1,p} \subset P^{1,p} \subset KS^{1,p} \subset N^{1,p} = Ch^{1,p}.$$

Combining Theorem 10.5.1, Corollary 10.4.6, (8.1.10), and Lemma 3.5.10 gives the following theorem.

Theorem 10.5.2 *Let μ be a doubling measure and assume that X satisfies the p-Poincaré inequality for some $p > 1$. Then*

$$M^{1,p} \subset P^{1,p} = KS^{1,p} = N^{1,p} = Ch^{1,p} \subset \bigcup_{q<p} M^{1,q}.$$

Furthermore, the norms $\|\cdot\|_{N^{1,p}} = \|\cdot\|_{Ch^{1,p}}$ and $\|\cdot\|_{KS^{1,p}}$ are comparable.

From Theorem 10.5.2 and Corollary 10.2.9 we obtain the following.

Theorem 10.5.3 *Let μ be a doubling measure, let $p > 1$ and assume that X satisfies the q-Poincaré inequality for some $1 \leq q < p$. Then*

$$M^{1,p} = P^{1,p} = KS^{1,p} = N^{1,p} = Ch^{1,p}.$$

Furthermore, the norms $\|\cdot\|_{M^{1,p}}$, $\|\cdot\|_{N^{1,p}} = \|\cdot\|_{Ch^{1,p}}$, and $\|\cdot\|_{KS^{1,p}}$ are all comparable.

In these theorems, all the Sobolev spaces consist of V-valued functions on a metric measure space X for a given Banach space V.

The assumption of a better Poincaré inequality in Theorem 10.5.3 is not restrictive, at least if the metric space X is complete. We will discuss this further in Chapter 12.

10.6 Notes to Chapter 10

The claim of Theorem 10.5.3 fails to hold for $p = 1$ even in the classical Euclidean setting. In this setting, one still has

$$W^{1,1} = N^{1,1} = KS^{1,1} = P^{1,1}$$

but $M^{1,1} \subsetneq N^{1,1}$ and $N^{1,1} \subsetneq Ch^{1,1}$. In fact, $M^{1,1}$ consists of those functions in L^1 whose first-order partial derivatives belong to the Hardy space H^1, and $Ch^{1,1}$ coincides with the space of functions of bounded variation. For these results see [164], [87], [169], and [81]. For results in the metric setting see [87], [6], [13], [209].

In the current literature there are further theories of Sobolev-type spaces of functions in a metric setting that we have not covered in this book. An axiomatic approach to the theory of Sobolev spaces was considered by Gol'dstein and Troyanov in [97], [98], and [96], where they identified axioms that drive the theory of Sobolev spaces in the metric setting.

Observe that in metric spaces such as the Sierpiński gasket, where there are not enough nonconstant rectifiable curves to support a Poincaré inequality, the theory of Sobolev spaces as considered in this book may not be suitable. In a subclass of such fractal metric spaces, including the so-called post-critically finite fractals, an alternative notion of Sobolev spaces (for $p = 2$) is based on the theory of Dirichlet forms developed by Beurling and Deny [28], [72]. For further information see Section 14.3.

11

Gromov–Hausdorff convergence and Poincaré inequalities

In this chapter we first review in detail the fundamental notion of the Gromov–Hausdorff convergence of metric spaces, including its measured version. In particular, we state and prove Gromov's compactness theorem, which, roughly speaking, states that each family of uniformly doubling metric spaces is precompact with respect to Gromov–Hausdorff convergence. Then we study the persistence of doubling measures and Poincaré inequalities under this convergence. Taken together, these results ensure that a central class of metric spaces considered in this book, i.e., doubling metric measure spaces supporting a Poincaré inequality, is complete when considered with respect to the Gromov–Hausdorff distance.

11.1 The Gromov–Hausdorff distance

Let (Z, d) be a metric space. For $\epsilon > 0$ and $A \subset Z$ nonempty, the *ϵ-neighborhood of A* is defined as the set

$$N_\epsilon(A) := \{z \in Z : \operatorname{dist}(z, A) < \epsilon\} = \bigcup_{a \in A} B(a, \epsilon). \tag{11.1.1}$$

The *Hausdorff distance* in Z between nonempty subsets $A, B \subset Z$ is

$$d_H^Z(A, B) := \inf\{\epsilon > 0 : A \subset N_\epsilon(B) \text{ and } B \subset N_\epsilon(A)\}. \tag{11.1.2}$$

It is clear that $d_H^Z(A, B) = d_H^Z(B, A)$, and an elementary calculation reveals that the triangle inequality is satisfied:

$$d_H^Z(A, B) \le d_H^Z(A, C) + d_H^Z(C, B)\,.$$

Thus d_H^Z behaves like a metric on the collection of all nonempty subsets of Z, bearing in mind two issues: it can take on the value ∞ and $d_H^Z(A, B) = 0$

does not necessarily mean that $A = B$. For example, the distance between any set $A \subset Z$ and its closure $\overline{A}$ is zero. However, we have the following result (whose proof is left to the reader).

Proposition 11.1.3 *We have $d_H^Z(A, B) > 0$ for every pair of distinct closed subsets A and B in Z. In particular, d_H^Z defines a metric on the collection of all nonempty closed and bounded subsets of Z.*

A set $A \subset Z$ is called an *ϵ-net*, where $\epsilon > 0$, if $N_\epsilon(A) = Z$, i.e., if every point in Z is within distance ϵ from some point in A. If A is an ϵ-net in Z then $d_H^Z(A, Z) \leq \epsilon$. Extending the notion of ϵ-nets, we say that $A \subset Z$ is a *0-net* if A is dense in Z.

We denote by $\mathcal{K}_Z$ the collection of all nonempty closed and bounded subsets of Z. Recall that a metric space (Z, d) is said to be *totally bounded* if it contains a finite ϵ-net for each $\epsilon > 0$.

Proposition 11.1.4

(i) *If Z is complete then $(\mathcal{K}_Z, d_H^Z)$ is complete.*
(ii) *If Z is totally bounded then $(\mathcal{K}_Z, d_H^Z)$ is totally bounded.*
(iii) *If Z is compact then $(\mathcal{K}_Z, d_H^Z)$ is compact.*

Proof To prove (i), let $A_1, A_2, \ldots$ be a Cauchy sequence in $(\mathcal{K}_Z, d_H^Z)$. We may assume that

$$d_H^Z(A_i, A_{i+1}) < 2^{-i} \tag{11.1.5}$$

for every i. Let

$$A = \bigcap_i \overline{\bigcup_{j \geq i} A_j}.$$

Then A is closed and bounded, and we claim that $d_H^Z(A_i, A) \to 0$ as $i \to \infty$. To this end, fix $\epsilon > 0$. Suppose that there are infinitely many sets $A_{i_1}, A_{i_2}, \ldots$ that meet $Z \setminus N_\epsilon(A)$; we assume that $i_k < i_{k+1}$ and that $2^{-i_1+1} < \epsilon/10$. Fix a point $a_{i_1} \in A_{i_1} \setminus N_\epsilon(A)$. By (11.1.5),

$$d_H^Z(A_{i_1}, A_{i_2}) < 2^{-i_1+1},$$

which yields a point $a_{i_2} \in A_{i_2}$ such that $d(a_{i_2}, a_{i_1}) < 2^{-i_1+1}$ and

$$\operatorname{dist}(a_{i_2}, A) \geq \operatorname{dist}(a_{i_1}, A) - 2^{-i_1+1} \geq 9\epsilon/10.$$

Suppose now that we have found points $a_{i_1} \in A_{i_1}, \ldots, a_{i_k} \in A_{i_k}$ such that

$$\operatorname{dist}(a_{i_k}, A) \geq (8 + 2^{-k+2})\epsilon/10.$$

Because

$$d_H^Z(A_{i_k}, A_{i_{k+1}}) < 2^{-i_k+1},$$

there is a point $a_{i_{k+1}} \in A_{i_{k+1}}$ such that $d(a_{i_k}, a_{i_{k+1}}) \leq 2^{-i_k+1}$ and hence

$$\begin{aligned} \operatorname{dist}(a_{i_{k+1}}, A) &\geq (8 + 2^{-k+2})\epsilon/10 - 2^{-i_k+1} \\ &\geq (8 + 2^{-k+2})\epsilon/10 - 2^{-k+1}\epsilon/10 \\ &= (8 + 2^{-(k+1)+2})\epsilon/10 . \end{aligned}$$

By construction, (a_{i_k}) is a Cauchy sequence in Z. It follows from the preceding discussion and from the completeness of Z that there is a point $a \in Z$ satisfying $\lim_{k\to\infty} a_{i_k} = a$ while $\operatorname{dist}(a, A) \geq 8\epsilon/10$. This contradicts the definition of A, and we conclude that $A_i \subset N_\epsilon(A)$ for all sufficiently large i. Similar reasoning shows that $A \subset N_\epsilon(A_i)$ for all sufficiently large i. Indeed, suppose that there are infinitely many indices $i_1 < i_2 < \cdots$ and points $a_{i_k} \in A \setminus N_\epsilon(A_{i_k})$. By the definition of A, there also are points $a_{j_k} \in B(a_{i_k}, \epsilon/10) \cap A_{j_k}$ for all k and for some sequence $j_1 < j_2 < \cdots$. Then

$$d_H^Z(A_{j_k}, A_{i_k}) \geq \operatorname{dist}(a_{i_k}, A_{i_k}) - d(a_{i_k}, a_{j_k}) \geq \epsilon - \epsilon/10 .$$

Because the left-hand side of the preceding inequality tends to zero as $k \to \infty$ we have a contradiction, as desired. This concludes the proof of part (i).

For part (ii), let $\epsilon > 0$ and let S be a finite ϵ-net in Z. Then the set of all subsets of S is a finite ϵ-net in $\mathcal{K}_Z$.

Part (iii) follows from parts (i) and (ii) and from the fact that a complete metric space is compact if and only if it is totally bounded.

The proof of the proposition is complete. □

The Hausdorff metric as introduced in the preceding discussion measures the distance between subsets of a fixed metric space. To measure the distance between two abstract metric spaces, we consider isometric realizations of the two spaces in a larger space and take the infimum of the resulting Hausdorff distances. For simplicity, we will consider only separable metric spaces, and employ the Fréchet embedding theorem (see Section 4.1) to this end.

The *Gromov–Hausdorff distance* between two separable metric spaces X and Y is

$$d_{GH}(X, Y) := \inf d_H^\infty(i(X), j(Y)), \tag{11.1.6}$$

where $d_H^\infty := d_H^{l^\infty}$ is the Hausdorff distance in l^∞ and the infimum is taken over all isometric embeddings $i : X \to l^\infty$ and $j : Y \to l^\infty$. Recall that the aforementioned Fréchet embedding theorem ensures that such embeddings always exist.

Lemma 11.1.7 *The distance function d_{GH} satisfies the triangle inequality on the class of all separable metric spaces.*

Proof Given three separable metric spaces X, Y, Z we are free to choose isometric embeddings i_X, i_Y, i_Z of X, Y, Z, respectively, into l^∞. Now the fact that the Hausdorff metric d_H^∞ on the subsets of l^∞ validates the triangle inequality implies that d_{GH} also satisfies the triangle inequality, upon taking the infimum over all embeddings i_X, i_Y, i_Z. □

We do not call d_{GH} a metric because it might well be that $d_{GH}(X, Y) = \infty$ for some separable metric spaces X, Y.

We write

$$X_i \overset{GH}{\to} X \tag{11.1.8}$$

if $X, X_1, X_2, \ldots$ are metric spaces such that $\lim_{i\to\infty} d_{GH}(X_i, X) = 0$. We also say that the sequence (X_i) *Gromov–Hausdorff-converges* to X in this case.

Given two metric spaces X, Y, a simpler formulation of the Gromov–Hausdorff distance is given in terms of metrics on the disjoint union $X \coprod Y := X \times \{0\} \cup Y \times \{1\}$. From now on, the copy of X within $X \coprod Y$, namely $X \times \{0\}$, will also be denoted by X, and similarly $Y \times \{1\}$ is identified with Y.

Proposition 11.1.9 *The value of $d_{GH}(X, Y)$ is the same as either of the following two quantities:*

(i) *the infimum of the values $d_H^Z(\iota(X), \iota'(Y))$ over all metric spaces (Z, d_Z) and isometric embeddings $\iota : X \to Z$ and $\iota' : Y \to Z$;*
(ii) *the infimum of the values $d_H^{Z_0}(X, Y)$ over all metrics on $Z_0 := X \coprod Y$ that agree with the given metrics on X and Y.*

Proof Denote by d_1 the value of the infimum in (i) and by d_2 the value of the infimum in (ii). It is clear that $d_1 \leq d_{GH}(X, Y)$. To complete the proof we show first that $d_{GH}(X, Y) \leq d_2$ and then that $d_2 \leq d_1$.

Let d be a metric on $Z_0 = X \coprod Y$ that agrees with the given metrics on X and Y. Choose an isometric embedding $\kappa : Z_0 \to l^\infty$. Then

$$d_{GH}(X, Y) \leq d_H^\infty(\kappa(X), \kappa(Y)) = d_H^{Z_0}(X, Y),$$

which proves the first inequality. Next, let (Z, d_Z), $\iota : X \to Z$, and $\iota' : Y \to Z$ be a triple as in (i) (observe that such a triple always exists, for example, when $Z = l^\infty$). For each $\delta > 0$ we can extend the metrics d_X on X and d_Y on Y to a metric d_δ on $Z_0 = X \coprod Y$ by setting

$$d_\delta(x, y) := d_Z(\iota(x), \iota'(y)) + \delta,$$

for $x \in X$ and $y \in Y$. Relative to the metric d_δ we have

$$d_H^{Z_0}(X, Y) \leq d_H^Z(\iota(X), \iota'(Y)) + \delta.$$

Taking the infimum over all δ and all triples Z, ι, ι' as above finishes the proof. □

For future purposes it will be useful to reformulate the notion of the Gromov–Hausdorff distance in terms of approximate isometries. Let (X, d_X) and (Y, d_Y) be two metric spaces, let $A \subset X$, and let $\delta, \epsilon_X, \epsilon_Y$ be nonnegative parameters. We say that a (possibly noncontinuous) map $f : A \to Y$ is a $(\delta, \epsilon_X, \epsilon_Y)$-*approximate isometry* if A is an ϵ_X-net in X, $f(A)$ is an ϵ_Y-net in Y, and

$$|d_Y(f(x_1), f(x_2)) - d_X(x_1, x_2)| \leq \delta \tag{11.1.10}$$

for all $x_1, x_2 \in A$. We say that X and Y are $(\delta, \epsilon_X, \epsilon_Y)$-*approximately isometric* if there exists a map $f : A \to Y$ as above for some ϵ_X-net $A \subset X$. If the condition holds for some triple of parameters $\delta, \epsilon_X, \epsilon_Y$, we say merely that X and Y are *approximately isometric*.

It is easy to see that the relation of approximate isometry is symmetric in the sense that if X and Y are $(\delta, \epsilon_X, \epsilon_Y)$-approximately isometric then Y and X are $(\delta, \epsilon_Y, \epsilon_X + \delta)$-approximately isometric. The following proposition ensures that the relation is transitive as well. Consequently, the relation of approximate isometry is an equivalence relation among separable metric spaces.

Proposition 11.1.11 *Let X and Y be two separable metric spaces. If $d_{GH}(X, Y) < \eta$ for some $\eta > 0$ then X and Y are $(2\eta, 0, 2\eta)$-approximately isometric. Conversely, if X and Y are $(\delta, \epsilon_X, \epsilon_Y)$-approximately isometric then*

$$d_{GH}(X, Y) \leq \max\{\epsilon_X, \epsilon_Y\} + \delta/2.$$

Proof Suppose first that $d_{GH}(X, Y) < \eta$. We may assume that $X, Y \subset l^\infty$ with $d_H^\infty(X, Y) < \eta$. For each $a \in X$, we may choose a point $a' \in Y$ with $\|a - a'\|_\infty < \eta$. Define $f : X \to Y$ by setting $a' = f(a)$. Then $f(X)$ is a 2η-net in Y because for every $y \in Y$ there is some $x_y \in X$ with $\|x_y - y\|_\infty < \eta$, and so $\|y - f(x_y)\|_\infty < 2\eta$. Moreover, for $x_1, x_2 \in X$,

$$\begin{aligned} d_X(x_1, x_2) - 2\eta = \|x_1 - x_2\|_\infty - 2\eta &< \|f(x_1) - f(x_2)\|_\infty \\ &< \|x_1 - x_2\|_\infty + 2\eta \end{aligned}$$

for $x_1, x_2 \in X$ as desired.

For the converse, assume that $f : A \to Y$ satisfies (11.1.10) for some $\delta > 0$, where A is an ϵ_X-net in X and $f(A)$ is an ϵ_Y-net in Y. Define a distance function d on $Z_0 = X \coprod Y$ that extends the metrics on X and Y by setting

$$d(x, y) = d(y, x) := \inf_{a \in A} \{d_X(x, a) + d_Y(f(a), y)\} + \delta/2 \qquad (11.1.12)$$

for $x \in X$ and $y \in Y$. Because of the additive term $\delta/2$ in the definition of $d(x, y)$ above, it is clear that $d(z, w) = 0$ if and only if both $z, w \in X$ with $z = w$, or both $z, w \in Y$ with $z = w$. It is easy to verify that d satisfies the triangle inequality, and hence defines a metric on Z_0. Another elementary calculation reveals that $d_H^{Z_0}(X, Y) \leq \max\{\epsilon_X, \epsilon_Y\} + \delta/2$. Part (ii) of Proposition 11.1.9 gives $d_{GH}(X, Y) \leq d_H^{Z_0}(X, Y)$, which completes the proof. □

In (11.1.12) we could replace $\delta/2$ by any positive number smaller than $\delta/2$.

Proposition 11.1.11 allows us to reformulate further the notion of Gromov–Hausdorff convergence in terms of ϵ-nets. The following proposition is used in the proof of Gromov's compactness theorem, see Section 11.2.

Proposition 11.1.13 *Let $X, X_1, X_2, \ldots$ be compact metric spaces. Then $X_i \stackrel{GH}{\to} X$ if and only if for each $\epsilon > 0$ there exist finite ϵ-nets $S_i \subset X_i$ and $S \subset X$ such that $S_i \stackrel{GH}{\to} S$.*

Proof Suppose first that the statement involving ϵ-nets is true. Fix $\epsilon > 0$. Then $d_{GH}(S, S_i) < \epsilon/2$ for all sufficiently large i. We also know that $d_{GH}(S, X)$ and $d_{GH}(S_i, X_i)$ are no more than ϵ. Thus, by Lemma 11.1.7, we see that $d_{GH}(X, X_i) < 3\epsilon$ if i is sufficiently large. Since ϵ is arbitrary, the proof of this half of the proposition is complete.

Next, assume that $X_i \stackrel{GH}{\to} X$ and let $\epsilon > 0$. Fix $0 < \delta \leq \epsilon/8$; for all sufficiently large i we have $d_{GH}(X_i, X) \leq \delta$. Let S be a finite $\epsilon/2$-net in X, and choose a $(2\delta, 0, 2\delta)$-approximate isometry $f_i : X \to X_i$, as guaranteed by Proposition 11.1.11. Writing $S_i := f_i(S)$ and using the restriction on δ, we find that S_i is an ϵ-net in X_i. Clearly f_i defines a $(2\delta, 0, 0)$-approximate isometry from S to S_i, whence $d_{GH}(S_i, S) \leq \delta$ by Proposition 11.1.11. Since δ is arbitrary, this inequality suffices to complete the proof. □

As a further consequence of Proposition 11.1.11 we deduce the following.

Theorem 11.1.14 *The Gromov–Hausdorff distance defines a metric on the collection of all isometry classes of compact metric spaces.*

By the *isometry class* of a metric space we mean the collection of all metric spaces that are isometric with the space (see Section 4.1).

Proof Let X and Y be compact metric spaces with $d_{GH}(X, Y) = 0$. We want to show that X and Y are isometric. By Proposition 11.1.11, for each $i = 1, 2, \ldots$ there exists a $(1/i, 0, 1/i)$-approximate isometry $f_i : X \to Y$. Choose a countable dense subset $S \subset X$. By a Cantor diagonal-type argument we may arrange, after passing to a subsequence (i_k), that the limit $f(x) := \lim_{k\to\infty} f_{i_k}(x)$ exists for every $x \in S$. Since f_{i_k} satisfies the approximate isometry criterion

$$|d_Y(f_{i_k}(x_1), f_{i_k}(x_2)) - d_X(x_1, x_2)| < 1/i_k$$

for $x_1, x_2 \in S$, it follows that f is an isometric embedding of S into Y. Moreover, $f(S)$ is dense in Y since $f_{i_k}(X)$ is a $1/i_k$-net in Y. Then f may be extended to an isometry from X onto Y.

The triangle inequality follows from Lemma 11.1.7. Since X, Y are compact metric spaces, $d_{GH}(X, Y)$ is finite. The theorem follows. □

Denote by $\mathcal{M}_\mathcal{C}$ the collection of all isometry classes of compact metric spaces. Then we have the following fundamental fact.

Theorem 11.1.15 *The metric space $(\mathcal{M}_\mathcal{C}, d_{GH})$ is complete, separable, and contractible.*

For the separability claim in the theorem, we observe that the collection of all finite metric spaces is dense in $\mathcal{M}_\mathcal{C}$. However, given a metric space X with n points, the isometry classes of metrics on X are described by a subclass of the class of symmetric $n \times n$ matrices, which is a subset of the separable space $\mathbb{R}^{n^2}$. Contractibility is established by considering, for each $X = (X, d) \in \mathcal{M}_\mathcal{C}$, the family of metric spaces $\lambda X = (X, \lambda d)$, $0 < \lambda \leq 1$. The hard part is to prove completeness. We defer this proof to the next section, until after the proof of Gromov's compactness theorem.

11.2 Gromov's compactness theorem

A family $\mathcal{X}$ of compact metric spaces is said to be *uniformly compact* if there exist a constant $D < \infty$ and a function $N : (0, \infty) \to (0, \infty)$ such that, for each $X \in \mathcal{X}$, the following two conditions hold true: (i) $\operatorname{diam}(X) \leq D$, and (ii) X contains, for every $\epsilon > 0$, an ϵ-net of at most $N(\epsilon)$ number of points.

Note that the family $\mathcal{K}_Z$ of compact subsets of a fixed compact metric space Z is uniformly compact. Proposition 11.1.4 asserts that this family forms a compact metric space when endowed with the Hausdorff metric d_H^Z. The following theorem, which is fundamental to the theory of Gromov–Hausdorff convergence, states an analogous result for abstract families of uniformly compact metric spaces endowed with the Gromov–Hausdorff distance.

Gromov compactness theorem *Every uniformly compact family of metric spaces is precompact in the Gromov–Hausdorff distance. More precisely, every* *sequence of metric spaces in a uniformly compact family of compact metric spaces contains a subsequence that converges in the Gromov–Hausdorff distance to a compact metric space.*

The proof of this theorem makes use of the following lemma, which we leave as an exercise for the interested reader.

Lemma 11.2.1 *Let (X_i) be a sequence of separable metric spaces and let $X = \{x_1, \dots, x_m\}$ be a finite metric space. Then the following are equivalent:*

(i) *(X_i) converges in the Gromov–Hausdorff metric to X;*
(ii) *each X_i may be expressed as a union of m nonempty subsets $X_{i,1}, \dots, X_{i,m}$ such that*

$$\max_k \operatorname{diam}(X_{i,k}) \to 0$$

and

$$\max_{k,l} \left| \operatorname{dist}(X_{i,k}, X_{i,l}) - d(x_k, x_l) \right| \to 0$$

as $i \to \infty$.

Proof of the Gromov compactness theorem Let $\mathcal{X}$ be a uniformly compact family with data D and N, and let $(X_1, d_1), (X_2, d_2), \dots$ be a sequence of spaces in $\mathcal{X}$. For each $i, k \geq 1$, choose a $1/k$-net $S_{i,k}$ in X_i of cardinality at most $N(1/k)$. By first choosing a subsequence, if necessary, we may assume that $\lim_i \#\bigcup_k S_{i,k} =: N$ exists. Here $\#K$ denotes the cardinality of the finite set K. If N is finite then we let $S = \{1, \dots, N\}$, and if $N = \infty$ we set $S = \mathbb{N}$. We need to equip S with a metric e. To do so we proceed as follows. Let $S_i := \bigcup_k S_{i,k}$. Since S_i is countable, we have an enumeration $S_i = \{x_{i,1}, x_{i,2}, \dots\}$ such that $S_{i,1} = \{x_{i,1}, \dots, x_{i,m_1}\}$, $S_{i,2} = \{x_{i,m_1+1}, \dots, x_{i,m_2}\}$, etc. By passing to a subsequence of the sequence (S_i) if necessary, we may assume that $\lim_i d_i(x_{i,1}, x_{i,2}) =: d_{1,2}$ exists. By passing to a further subsequence, we may assume also that $\lim_i d_i(x_{i,1}, x_{i,3}) =: d_{1,3}$ exists and that $\lim_i d_i(x_{i,2}, x_{i,3}) =: d_{2,3}$ exists. Proceeding inductively over $S \times S$, and then using a Cantor diagonalization process, we have a subsequence, also denoted (S_i), such that for $n, m \in S$ we have the existence of the limit

$$\lim_i d_i(x_{i,m}, x_{i,n}) =: d_{m,n}.$$

We set $e(n, m) := d_{n,m}$. Note that

$$e(m, n) := \lim_{i \to \infty} d_i(x_{i,m}, x_{i,n}). \tag{11.2.2}$$

It is easy to see that e is nonnegative and symmetric, but there might be distinct $n, m \in S$ for which $e(n, m) = 0$. A direct computation using the triangle inequality for d_i tells us that e satisfies a triangle inequality. Hence the space $S_\infty = S/\sim$, with $n \sim m$ if and only if $e(n, m) = 0$, is a metric space equipped with the metric e. Let X be the completion of (S_∞, e).

We claim that X is compact. For this it suffices to prove that X is totally bounded. For each $k \geq 1$, the set $S_i^k := S_{i,1} \cup \cdots \cup S_{i,k} \subset S_i$ is a $1/k$-net in X_i and has cardinality $N_{i,k} \leq N(1) + \cdots + N(1/k)$. Set $N_k = \lim_i N_{i,k}$. We now check that $S^k := \{1, \ldots, N_k\}$ is a $1/k$-net in (S_∞, e), which suffices.

Given $m \in S$, there is, for every i, a point $x_{i,n_i} \in S_i^k$ for some n_i in $\{1, \ldots, N_k\}$ such that $d_i(x_{i,m}, x_{i,n_i}) \leq 1/k$. Because N_k is independent of i and because the limit in (11.2.2) exists, we infer that $e(m, n) \leq 1/k$ for some $n = \{1, \ldots, N_k\}$. It follows that X is compact.

Finally, we claim that $X_i \overset{GH}{\to} X$. Since the spaces S_i^k and S^k are all finite metric spaces and $d_i(x_{i,m}, x_{i,n}) \to e(m, n)$ for all relevant $x_{i,m}, x_{i,n} \in S_i^k$ and $m, n \in S^k$, Lemma 11.2.1 implies that $S_i^k \overset{GH}{\to} S^k$ for each k. Therefore, by Proposition 11.1.13, $X_i \overset{GH}{\to} X$. The proof is complete. □

We are now ready to prove Theorem 11.1.15.

Proof of Theorem 11.1.15 The separability and contractibility of the space $(\mathcal{M}_\mathcal{C}, d_{GH})$ were established immediately after the statement of the theorem. As for completeness, it follows easily from the definitions that each Cauchy sequence (X_i) in $\mathcal{M}_\mathcal{C}$ is uniformly compact. Indeed, given $\epsilon > 0$ there is a positive integer i_ϵ such that $d_{GH}(X_i, X_j) < \epsilon/3$ for $i, j \geq i_\epsilon$. Let A_ϵ be an $\epsilon/3$-net in X_{i_ϵ}. The set A_ϵ is finite because X_{i_ϵ} is compact. For $j \geq i_\epsilon$, by the fact that $d_{GH}(X_{i_\epsilon}, X_j) < \epsilon/3$ we may find isometries of X_{i_ϵ}, X_j into l^∞ such that the Hausdorff distance between the embedded images of X_{i_ϵ} and X_j is smaller than $\epsilon/3$. Thus, for each $a \in A_\epsilon$ we can find a point $x_{a,j} \in X_j$ such that the distance between the embedded images of a and $x_{a,j}$ is smaller than $\epsilon/3$. It now follows from $d_{GH}(X_{i_\epsilon}, X_j) < \epsilon/3$ that the set $A_{\epsilon,j} = \{x_{a,j} : a \in A_\epsilon\}$ is an ϵ-net in X_j with cardinality at most $N_\epsilon = \#A_\epsilon$. For each $i < i_\epsilon$, the compact space X_i contains an ϵ-net of cardinality $N_{\epsilon,i}$. Now the choice $N(\epsilon) := \max\{N_{\epsilon,1}, \ldots, N_{\epsilon,i_\epsilon - 1}, N_\epsilon\}$ satisfies the uniform compactness condition for the sequence. By the Gromov compactness theorem, each such sequence has a Gromov–Hausdorff limit which is in turn compact. The isometry class of the limit is unique by Theorem 11.1.14. This completes the proof. □

11.3 Pointed Gromov–Hausdorff convergence

The definition of Gromov-Hausdorff convergence presented in the first section of this chapter makes sense for arbitrary (separable) metric spaces. However, it is unduly restrictive in the noncompact case. For example, consider the sequence of circles $C_i := \{x \in \mathbb{R}^2 : |x - ie_2| = i\}$, $i = 1, 2, \ldots$, as subsets of $\mathbb{R}^2$. For fixed $r > 0$, the (closed) balls of radius r centered at the origin $0 \in C_i$ look increasingly similar to the corresponding ball in $\mathbb{R}$, i.e., the interval $[-r, r]$. We would like to be able to say that C_i converges to $\mathbb{R}$ in the limit as $i \to \infty$. However, $d_{GH}(C_i, \mathbb{R}) = \infty$ for all i.

In what follows, we will only define Gromov-Hausdorff convergence (and not an actual distance) for noncompact spaces. We consider *pointed spaces*, i.e., the triple (X, d, a) of a metric space (X, d) together with a point $a \in X$, or simply pairs (X, a) with $a \in X$, and formulate the definition in terms of a version of the approximate isometry criterion of Proposition 11.1.11.

Definition 11.3.1 A sequence of pointed separable metric spaces

$$(X_1, d_1, a_1), (X_2, d_2, a_2), \ldots$$

is said to *pointed Gromov–Hausdorff-converge* to a pointed separable metric space (X, d, a) if, for each $r > 0$ and $0 < \epsilon < r$, there exists an i_0 such that for each $i \geq i_0$ there is a map $f_i = f_i^\epsilon : B(a_i, r) \to X$ satisfying:

(i) $f_i(a_i) = a$;
(ii) $|d(f_i(x), f_i(y)) - d_i(x, y)| < \epsilon$ for all $x, y \in B(a_i, r)$;
(iii) $B(a, r - \epsilon) \subset N_\epsilon(f_i(B(a_i, r)))$, where N_ϵ is defined in (11.1.1).

We denote this mode of convergence by $(X_i, d_i, a_i) \overset{GH}{\to} (X, d, a)$ or simply $(X_i, a_i) \overset{GH}{\to} (X, a)$.

Note that conditions (i) and (ii) imply that

$$f_i(B(a_i, r)) \subset B(a, r + \epsilon). \tag{11.3.2}$$

This and (iii) together give in turn that

$$B(a, r - \epsilon) \subset N_\epsilon(f_i(B(a_i, r))) \subset B(a, r + 2\epsilon). \tag{11.3.3}$$

It can be directly verified that if X is a length space (as in Section 5.1)) then (ii) and (iii) mean, in our terminology, that $f_i : B(a_i, r) \to B(a, r + 2\epsilon)$ is an $(\epsilon, 0, 3\epsilon)$-approximate isometry; see (11.1.10).

Remark 11.3.4 The preceding definition of pointed Gromov–Hausdorff convergence would make perfect sense even without the requirement of separability. Similarly, the separability requirement could be dropped from the definition of the Gromov–Hausdorff distance, by replacing it with either of the conditions (i) and (ii) of Proposition 11.1.9. To avoid unnecessary generalities we retain our standing separability assumption, which is essential for certain other aspects of this book. When dealing with Gromov–Hausdorff convergence we may not always mention the assumption that the spaces in question are separable; it will be tacitly assumed.

What is more, in our main applications we will be dealing with a yet more restrictive type of metric space, namely, length spaces. For such spaces, pointed Gromov–Hausdorff convergence takes an even simpler form. See Proposition 11.3.12 below and [49, Section 7.5].

We now show that this new notion of convergence coincides with the previous notion for bounded spaces.

Proposition 11.3.5 *Let $X, X_1, X_2, \dots$ be bounded and separable metric spaces.*

(i) *If $\sup_i \operatorname{diam}(X_i) < \infty$ and $(X_i, d_i, a_i) \overset{GH}{\to} (X, d, a)$ for some $a_i \in X_i$ and $a \in X$ then $X_i \overset{GH}{\to} X$.*
(ii) *If $X_i \overset{GH}{\to} X$ and $a \in X$ then there exist points $a_i \in X_i$ such that $(X_i, d_i, a_i) \overset{GH}{\to} (X, d, a)$.*

Proof To prove (i), set $R := \sup_i \operatorname{diam}(X_i)$. Fix $r > \max\{R, \operatorname{diam}(X)\}$ and let $0 < \epsilon < r - \max\{R, \operatorname{diam}(X)\}$. By assumption, from Definition 11.3.1(iii) and by the choice of ϵ and r (which guarantees that $X = B(a, r - \epsilon)$), we know that for all $i \geq i_\epsilon$ there exists an $(\epsilon, 0, \epsilon)$-approximate isometry f_i from $B(a_i, r) = X_i$ to $B(a, r - \epsilon) = X$. This and Proposition 11.1.11 then give that $d_{GH}(X, X_i) \leq 2\epsilon$ for all $i \geq i_\epsilon$. Thus $X_i \overset{GH}{\to} X$.

Next we turn to part (ii). Let $r > \epsilon > 0$ be arbitrary. Choose i_0 sufficiently large that $d_{GH}(X_i, X) \leq \epsilon/8$ for $i \geq i_0$. By Proposition 11.1.11 there exists an $(\epsilon/4, 0, \epsilon/4)$-approximate isometry $g_i : X_i \to X$. Choose points $a_i \in X_i$ satisfying $d(g_i(a_i), a) < \epsilon/4$ and define $f_i : X_i \to X$ by

$$f_i(x) = \begin{cases} g_i(x) & \text{if } x \neq a_i, \\ a & \text{if } x = a_i. \end{cases}$$

We claim that the restriction of f_i to the ball $B(a_i, r)$ satisfies the conditions in Definition 11.3.1. It suffices to verify condition (ii) of Definition 11.3.1 in the case $y = a_i$, $x \neq a_i$. In this case,

$$|d(f_i(x), f_i(a_i)) - d_i(x, a_i)| = |d(g_i(x), a) - d_i(x, a_i)|$$

$$
\begin{aligned}
&\le |d(g_i(x), a) - d(g_i(x), g_i(a_i))| \\
&\quad + |d(g_i(x), g_i(a_i)) - d_i(x, a_i)| \\
&\le d(a, g_i(a_i)) + \epsilon/4 \\
&\le \epsilon/4 + \epsilon/4 = \epsilon/2.
\end{aligned}
$$

For condition (iii) of Definition 11.3.1, let $x \in B(a, r-\epsilon)$ and choose $y \in X_i$ with $d(g_i(y), x) < \epsilon/4$. Since

$$d(g_i(y), g_i(a_i)) \le d(g_i(y), x) + d(x, a) + d(a, g_i(a_i)) < r - \epsilon/2,$$

we find that $y \in B(a_i, r)$. Because we also have $d(x, f_i(y)) \le d(x, g_i(y)) + d(g_i(y), f_i(y)) < \epsilon$, (iii) follows. The proposition is proved. □

The requirement that X be bounded is not needed a priori in part (i) of the above proposition, since if the sequence (X_i) of metric spaces is uniformly bounded (that is, $\sup_i \operatorname{diam}(X_i) < \infty$) then it automatically follows that X is also bounded whenever $X_i \overset{GH}{\to} X$ in the sense of Definition 11.3.1.

Remarks 11.3.6 (a) For the circles C_i given at the beginning of this section, we have $(C_i, 0) \overset{GH}{\to} (\mathbb{R}, 0)$.

(b) Let $X_i = \{0, i\}$, $a_i = 0$, and $X = \{0\}$, $a = 0$, where we use the Euclidean metric in X_i and X. Then (X_i, a_i) converges to (X, a) in the sense of Definition 11.3.1, but $d_{GH}(X_i, X) \to \infty$. Thus the condition $\sup_i \operatorname{diam}(X_i) < \infty$ is needed in the previous proposition.

(c) Let $X_i = \{0, 1 + 1/i\}$, $a_i = 0$, and $X = \{0, 1\}$, $a = 0$, where again we use the Euclidean metric in X_i and X. Then (X_i) Gromov–Hausdorff converges to X, but the closed balls $\overline{B}(a_i, 1) \subset X_i$ do not Gromov–Hausdorff converge to the closed ball $\overline{B}(0, 1) \subset X$. Thus one should not define pointed Hausdorff convergence by requiring that $\overline{B}(a_i, r)$ should Gromov–Hausdorff-converge to $\overline{B}(a, r)$ for each $r > 0$.

Recall that a metric space (X, d) constitutes a length space if $d(x, y) = \inf_\gamma \operatorname{length}(\gamma)$ for every pair $x, y \in X$, where the infimum is taken over all rectifiable curves γ joining x to y in X.

We require the following characterization of complete length spaces.

Lemma 11.3.7 *A complete metric space (X, d) is a length space if and only if to every pair of points $x, y \in X$ and every $\epsilon > 0$ there corresponds $z \in X$ such that*

$$\max\{d(x, z), d(z, y)\} \le \frac{1}{2} d(x, y) + \epsilon. \tag{11.3.8}$$

The points z as in the above lemma are called *approximate midpoints* in [45, Section I.2, page 30].

Proof The necessity part of the assertion is immediate; given $x, y \in X$ and $\epsilon > 0$, choose for z the midpoint of a curve from x to y with length at most $d(x, y) + \epsilon/2$.

To prove the sufficiency, fix $x, y \in X$. Without loss of generality we may assume that $d(x, y) = 1$. Then fix $\epsilon > 0$. We need to show that there is a curve from x to y with length at most $1 + \epsilon$. To this end, denote by $D_n = \{k2^{-n} : k = 0, 1, \ldots, 2^n\}$ the set of dyadic rationals of level $n = 0, 1, 2, \ldots$ in the unit interval $[0, 1]$. We define inductively maps $f_n : D_n \to X$ as follows. Let $f_0(0) = x$ and $f_0(1) = y$, and observe that

$$d(f_0(0), f_0(1)) = 1 .$$

Suppose now that f_n has been defined and that it satisfies

$$d(f_n(\kappa), f_n(\tau)) \le 2^{-n} + 2^{-n}\epsilon \sum_{k=0}^{n} 2^{-k} \tag{11.3.9}$$

for every pair of consecutive points $\kappa, \tau \in D_n$, i.e., we require that $(\kappa, \tau) \cap D_n = \emptyset$. Next, let $\kappa \in D_{n+1}$. If $\kappa \in D_n$, so that $f_n(\kappa)$ is defined, we let $f_{n+1}(\kappa) = f_n(\kappa)$. Otherwise, κ is the mid point of an interval I_n of length 2^{-n} such that f_n is defined at the end points κ_n and τ_n of I_n. Choose a point $z_\kappa \in X$ such that

$$\max\left\{d(f_n(\kappa_n), z_\kappa), d(z_\kappa, f_n(\tau_n))\right\} \le \frac{1}{2} d(f_n(\kappa_n), f_n(\tau_n)) + 2^{-2n-2}\epsilon$$

and put $f_{n+1}(\kappa) = z_\kappa$. Then the induction hypothesis (11.3.9) gives that (11.3.9) can be assumed to hold for every n. There is an obvious mapping $f : D \to X$, extending each f_n, defined in the dense set $D = \bigcup_n D_n \subset [0, 1]$. It is easy to check using (11.3.9) that f is $(1 + 2\epsilon)$-Lipschitz; since X is complete, f extends to a $(1 + 2\epsilon)$-Lipschitz map $F : [0, 1] \to X$. Because $F(0) = x$ and $F(1) = y$, the assertion follows from (5.1.2). □

We record two additional facts about length spaces; the proofs, which are simple, are left to the reader. For the first fact, recall the notation from (11.1.1).

Lemma 11.3.10 *Let X be a length space. Then*

$$N_r(N_s(A)) = N_{r+s}(A) \tag{11.3.11}$$

whenever $A \subset X$ and $r, s > 0$. Moreover, the closure $\overline{B(q, r)}$ of an open ball coincides with the closed ball $\overline{B}(q, r)$ whenever $q \in X$ and $r > 0$.

Proposition 11.3.12 *Let (X_i, a_i) be a sequence of pointed length spaces. If (X, a) is a complete pointed Gromov–Hausdorff limit of (X_i, a_i) then X is a length space. Moreover, in this case we have both that $B(a_i, r) \overset{GH}{\to} B(a, r)$ and that $\overline{B}(a_i, r) \overset{GH}{\to} \overline{B}(a, r)$ for every $r > 0$.*

Proof The first statement follows from the definitions and Lemma 11.3.7. Fix $r > 0$, $0 < \epsilon < r$, and functions f_i as in the definition of pointed Gromov–Hausdorff convergence. From (11.3.3), (11.3.2), and (11.3.11) we deduce that $B(a, r) = N_\epsilon(B(a, r - \epsilon)) \subset N_{2\epsilon}(f_i(B(a_i, r)))$ and that $f_i(B(a_i, r)) \subset N_\epsilon(B(a, r))$, which in turn implies that $d_H^X(f_i(B(a_i, r)), B(a, r)) \leq 2\epsilon$, for all sufficiently large i. Furthermore the map $f_i : B(a_i, r) \to f_i(B(a_i, r))$ is an $(\epsilon, 0, 0)$-approximate isometry, and we can deduce from the preceding and Proposition 11.1.11 that $d_{GH}(B(a_i, r), B(a, r)) \leq 3\epsilon$ for all sufficiently large i that depend only on ϵ. The conclusion $B(a_i, r) \overset{GH}{\to} B(a, r)$ follows.

Finally, the statement about closed balls follows from the corresponding statement about open balls, from Lemma 11.1.7 and from the fact that the Gromov–Hausdorff distance between a set and its closure is zero. The proposition follows. □

Remark 11.3.13 The example in Remark 11.3.6(c) shows that the claim of Proposition 11.3.12 may fail if one drops the length space assumption.

Recall that a metric space is proper if each closed ball in it is compact. Proper spaces are always separable. A sequence $(X_1, a_1), (X_2, a_2), \ldots$ of pointed spaces is said to be *eventually proper* if for every $r > 0$ there is an i_r such that the ball $\overline{B}(a_i, r) \subset X_i$ is compact for every $i \geq i_r$.

For example, if (X, d) is a locally compact metric space and $a \in X$ then the sequence $(X, d, a), (X, 2d, a), (X, 3d, a), \ldots$ is eventually proper.

Proposition 11.3.14 *Let (X_i, d_i, a_i) be an eventually proper sequence of pointed metric spaces. If (X, d_X, a) is a complete pointed Gromov–Hausdorff limit of the sequence (X_i, d_i, a_i) then X is proper. If (X, d_X, a) and (Y, d_Y, q) are proper pointed Gromov–Hausdorff limits of (X_i, d_i, a_i) then there is an isometry $f : X \to Y$ with $f(a) = q$.*

Proof Given that X is complete, to show that X is proper it suffices to show that the closed balls $\overline{B}(a, r)$ are totally bounded for each $r > 0$; that is, for each $\epsilon > 0$ the ball $\overline{B}(a, r)$ has a finite ϵ-net. We argue as follows.

Let (X_i, a_i) be an eventually proper sequence of pointed metric spaces that pointed Gromov–Hausdorff-converges to (X, a). Then there exists i_r such that $\overline{B}(a_i, r + 3\epsilon)$ is compact, and hence also totally bounded, whenever $i > i_r$.

We can choose a positive integer $i > i_r$ large enough that there is a map $f_i : B(a_i, r + 3\epsilon) \to X$ satisfying conditions (i)–(iii) of Definition 11.3.1 for $\epsilon/10$. In particular,

$$\overline{B}(a, r) \subset B(a, r + 2\epsilon) \subset N_{\epsilon/10}(f_i(B(a_i, r + 3\epsilon))).$$

Let $\{z_1, \ldots, z_k\}$ be a finite $\epsilon/20$-net of $\overline{B}(a_i, r + 3\epsilon)$. By perturbing these points slightly if necessary, we obtain a finite $\epsilon/10$-net of $B(a_i, r + 3\epsilon)$, also denoted by $\{z_1, \ldots, z_k\}$. We modify $\{f_i(z_1), \ldots, f_i(z_k)\}$ to obtain a finite ϵ-net of $\overline{B}(a, r)$. To do so, note that whenever $z \in \overline{B}(a, r)$ there is a point $y_z \in B(a_i, r + 3\epsilon)$ such that $d(z, f_i(y_z)) < \epsilon/10$. We then find z_j from the $\epsilon/10$-net such that $d_i(y_z, z_j) < \epsilon/10$. It follows from condition (ii) of Definition 11.3.1 that

$$\left| d(f_i(y_z), f_i(z_j)) - d_i(y_z, z_j) \right| < \epsilon/10,$$

and so we have $d(f_i(z_j), z) < 3\epsilon/10$. Consequently, the $3\epsilon/10$-balls centered at the points $f_i(z_j)$, $j = 1, \ldots, k$, cover $\overline{B}(a, r)$. We replace $f_i(z_j)$ with a point in $\overline{B}(a, r) \cap B(f_i(z_j), 3\epsilon/10)$ if this set is nonempty and $f_i(z_j)$ is not in $\overline{B}(a, r)$. We discard $f_i(z_j)$ if $\overline{B}(a, r) \cap B(f_i(z_j), 3\epsilon/10)$ is empty. Thus we obtain a finite ϵ-net in $\overline{B}(a, r)$. This concludes the verification that X is proper.

Let (X, d_X, a) and (Y, d_Y, q) be two such pointed metric spaces. To prove the second claim, we fix $r > 0$ and show that there is an isometry between $B(a, r)$ and $B(q, r)$. To do so, we choose $0 < \epsilon < r/10$ and then choose i large enough that there are maps $f_i^{\epsilon/2} : B(a_i, r + \epsilon) \to B(a, r + 2\epsilon)$ and $F_i^{\epsilon/2} : B(a_i, r + \epsilon) \to B(q, r + 2\epsilon)$ satisfying conditions (i)–(iii) of Definition 11.3.1. We construct a version of an inverse map $g_i : B(a, r) \to B(a_i, r + \epsilon)$ for $f_i^{\epsilon/2}$ as follows. Given $z \in B(a, r)$ we choose a point $y_z \in B(a_i, r + \epsilon)$ such that $d_X(z, f_i^{\epsilon/2}(y_z)) < \epsilon/2$ and set $g_i(z) = y_z$. It can be directly verified that, for $z, w \in B(a, r)$,

$$\left| d_X(z, w) - d_i(g_i(z), g_i(w)) \right| < 3\epsilon.$$

Let $H_i = f_i^{\epsilon/2} \circ g_i : B(a, r) \to B(q, r + 2\epsilon)$; then, for $z, w \in B(a, r)$,

$$\left| d_X(z, w) - d_Y(H_i(z), H_i(w)) \right| < 4\epsilon.$$

Furthermore, it can be verified that $N_\epsilon(H_i(B(a, r)))$ contains $B(q, r)$. Now, an Arzelà–Ascoli type argument, together with the fact proved above that X and Y are proper, yields an isometry between $B(a, r)$ and $B(q, r)$. □

Remark 11.3.15 Finite subsets of $\mathbb{R}$ can Gromov–Hausdorff converge to $\mathbb{Q} \subset \mathbb{R}$, showing that the assumption in Proposition 11.3.14 that X be complete cannot be dropped.

Next we present a version of the Gromov compactness theorem for pointed proper spaces.

A family $\{(X_\alpha, a_\alpha) : \alpha \in \mathcal{A}\}$ of pointed metric spaces is said to be *pointed totally bounded* if there is a function $N : (0, \infty) \times (0, \infty) \to (0, \infty)$ such that, for each $0 < \epsilon < R$ and $\alpha \in \mathcal{A}$, the closed ball $\overline{B}(a_\alpha, R)$ in X_α contains an ϵ-net of cardinality at most $N(\epsilon, R)$.

We have the following pointed version of Theorem 11.2.

Theorem 11.3.16 *Every eventually proper sequence in a pointed totally bounded family of pointed metric spaces contains a subsequence that pointed Gromov–Hausdorff converges to a proper pointed metric space.*

The proof of Theorem 11.3.16 is a straightforward variation on that of the Gromov compactness theorem; see Section 11.2.

Proof We proceed as in the proof of the Gromov compactness theorem, but with a modification to take into account the fact that we do not have a uniform bound on the diameters of our metric spaces.

For each pair of positive integers $k, i \in \mathbb{N}$ we can choose a $1/k$-net $S_{i,k}$ in $B(a_i, k)$, with cardinality at most $N(1/k, k) + 1$, such that $a_i \in S_{i,k}$. We can arrange that, for each i, $S_{i,1} \subset S_{i,2} \subset S_{i,3} \subset \cdots$. Set

$$S_i = \bigcup_{k \in \mathbb{N}} S_{i,k}.$$

It is clear that S_i is countable and dense in X_i. We enumerate points in S_i as follows. Let

$$\begin{aligned} S_{i,1} &= \{x_{i,1} = a_i, x_{i,2}, \ldots, x_{i,m_{i,1}}\} && \text{with } m_{i,1} \le N(1,1) + 1, \\ S_{i,2} \setminus S_{i,1} &= \{x_{i,m_{i,1}+1}, \ldots, x_{i,m_{i,2}}\} && \text{with } m_{i,2} \le N(1/2, 2) + 1, \\ S_{i,j+1} \setminus S_{i,j} &= \{x_{i,m_{i,j}+1}, \ldots, x_{i,m_{i,j+1}}\} && \text{with } m_{i,j+1} \\ & && \le N(1/(j+1), j+1) + 1. \end{aligned}$$

Then S_i inherits this enumeration.

The case $\lim_{i\to\infty} \#S_i < \infty$ having already been covered in the Gromov compactness theorem, we may assume that $\lim_{i\to\infty} \#S_i = \infty$. We will follow the proof of that theorem by constructing a metric on $S = \mathbb{N}$. To do so, we enumerate $S \times S \setminus \Delta = \{(n_1, m_1), (n_2, m_2), \ldots\}$, with $\Delta = \{(z, z) : z \in S\}$ such that $(n_1, m_1) = (1, 2)$, and we choose a subsequence $(X_{i_{1,j}}, d_{i_{1,j}}, a_{i_{1,j}})$ such that

$$\lim_{j\to\infty} d_{i_{1,j}}(x_{i_{1,j},1}, x_{i_{1,j},2}) = d_{1,2}$$

exists and for $j \geq 1$ we have

$$|d_{i_{1,j}}(x_{i_{1,j},1}, x_{i_{1,j},2}) - d_{1,2}| < 2^{-j}.$$

Inductively, we can choose subsequences $(X_{i_{m,j}}, d_{i_{m,j}}, p_{i_{m,j}})$ for each positive integer $m \geq 2$ such that:

1. $(X_{i_{m,j}}, d_{i_{m,j}}, a_{i_{m,j}})$ is a subsequence of $(X_{i_{m-1,j}}, d_{i_{m-1,j}}, a_{i_{m-1,j}})$;
2. $\lim_{j\to\infty} d_{i_{m,j}}(x_{i_{m,j},\alpha_m}, x_{i_{m,j},\beta_m}) = d_{\alpha_m,\beta_m}$ exists, where $(\alpha_m, \beta_m) \in S \times S \setminus \Delta$ is the mth term in the enumeration of this set;
3. $|d_{i_{m,j}}(x_{i_{m,j},\alpha_m}, x_{i_{m,j},\beta_m}) - d_{\alpha_m,\beta_m}| < 2^{-j}$ whenever $j \geq m$.

Note that $d_{\alpha_m,\beta_m} = d_{\beta_m,\alpha_m}$. Now, as in the proof of the Gromov compactness theorem, we obtain a metric e on $S_\infty = S/\sim$, where for $\alpha, \beta \in S$ we set $\alpha \sim \beta$ if and only if $d_{\alpha,\beta} = 0$, and $e([i],[j]) = d_{i,j} = d_{j,i}$ (for elements $i \in S$, we denote the equivalence class in S_∞ that contains i by $[i]$). Let X be the completion of S_∞ in this metric, and let $a := [1]$.

It remains to show that the diagonal sequence $(X_{i_{m,m}}, d_{i_{m,m}}, a_{i_{m,m}})$, henceforth denoted simply as (X_i, d_i, a_i), converges to (X, e, a). That is, for each positive real number ϵ and for every $r > \epsilon$ we want to show that there is a positive integer i_0 such that, whenever $i > i_0$, there is a mapping $f_i : B(a_i, r) \to X$ satisfying the three conditions of Definition 11.3.1.

To do so, we fix $0 < \epsilon < r < \infty$ and pick a positive integer k such that $1/k < \epsilon/10$ and $k > r$. By the pointed total boundedness property, we can find a $1/k$-net T_k in $B(a, r - \epsilon/2) \subset S_\infty$ such that the cardinality of T_k is at most $N(1/k, k) + 1$ and $p \in T_k$. Let $\{[j_1], \ldots, [j_m]\}$ be this set, with $j_1 = 1$; we then have $m \leq N(1/k, k) + 1$. By the choice of the subsequence of pointed metric spaces, we know that there is a positive integer i_1 such that whenever $i \geq i_1$ we have

$$|d_i(x_{i,j_l}, x_{i,j_n}) - e([j_l], [j_n])| < \epsilon/100$$

for each pair of points $[j_l]$, $[j_n]$ in this set.

Note that for each positive integer i the cardinality of $S_{i,k}$ is at most $N(1/k, k) < \infty$. So, in considering $d_i(\overline{x}, \overline{y})$, $\overline{x}, \overline{y} \in S_{i,k}$, we are considering at most $2^{N(1/k,k)+1}$ real numbers. Therefore we can find a positive real number $i_0 \geq i_1$ such that, for each $i \geq i_0$ and $j_1, j_2 \in \{1, \ldots, m_i\}$, $m_i \leq N(1/k, k) + 1$, we have

$$|d_i(x_{i,j_1}, x_{i,j_2}) - e([j_1], [j_2])| < \epsilon/10.$$

We define $f_i : B(a_i, r) \to X$ as follows.

First, let $f(a_i) = [1] = a$. For $x \in B(a_i, r) \setminus \{a_i\}$ we can find $\overline{x} \in S_{i,k}$ such that $d_i(x, \overline{x}) < \epsilon/10$. We choose one such $\overline{x}$ and, with the labeling $\overline{x} = x_{i,j} \in S_{i,k}$, set $f_i(x) = [j]$.

By construction, f_i satisfies the first condition of Definition 11.3.1. To show that it satisfies the second condition, let $x, y \in B(p_i, r)$ and $\overline{x}, \overline{y}$ be the correspondingly chosen points in $S_{i,k}$ with $f_i(x) = [j_x], f_i(y) = [j_y]$. Then

$$\begin{aligned}|e(f_i(x), f_i(y)) - d_i(x, y)| &= |e([j_x], [j_y]) - d_i(x, y)| \\ &\le |e([j_x], [j_y]) - d_i(\overline{x}, \overline{y})| + |d_i(\overline{x}, \overline{y}) - d_i(x, y)| \\ &\le \epsilon/10 + d_i(x, \overline{x}) + d_i(\overline{y}, y) < 3\epsilon/10 < \epsilon;\end{aligned}$$

that is, the second condition is satisfied.

Finally, to verify the third condition of Definition 11.3.1, note that, by the choice of i_0, given $i_1 \le i_0$, we have $T_k \subset f_i(S_{i,k} \cap B(a_i, r))$, and this condition on T_k tells us that $f_i(S_{i,k} \cap B(a_i, r))$ forms a $1/k$-net in $B(a, r - \epsilon/2)$; hence

$$B(a, r - \epsilon) \subset B(a, r - \epsilon/2) \subset N_\epsilon(f_i(B(a_i, r))).$$

This completes the proof. □

The doubling condition introduced in Section 4.1 gives rise to families of metric spaces to which Theorem 11.3.16 can be applied. We next consider a local variant of this condition.

A family $\{(X_\alpha, a_\alpha) : \alpha \in \mathcal{A}\}$ of pointed metric spaces is said to be *pointed boundedly doubling* if there is a function $M : (0, \infty) \to (0, \infty)$ such that for every index $\alpha \in \mathcal{A}$, the ball $B(a_\alpha, r)$ is doubling with constant $M(r)$.

Observe that every pointed boundedly doubling family is pointed totally bounded.

The proof of the following proposition is left to the reader.

Proposition 11.3.17 *Let (X_i, a_i) be a sequence of pointed boundedly doubling metric spaces. If (X, a) is a pointed Gromov–Hausdorff limit of (X_i, a_i) then X is boundedly doubling.*

If (X, a) is a pointed boundedly doubling space then it follows from Lemma 4.1.12 that every ball $B(a, r) \subset X$ contains an ϵ-net of cardinality at most $M(r)(r/\epsilon)^s$, where $s = \log M(r)/\log 2$. We thus have the following corollary to Theorem 11.3.16 and Proposition 11.3.17.

Theorem 11.3.18 *Every eventually proper sequence in a pointed boundedly doubling family of pointed metric spaces contains a subsequence that pointed Gromov–Hausdorff converges to a pointed boundedly doubling proper pointed metric space.*

11.4 Pointed measured Gromov–Hausdorff convergence

In this section we discuss Gromov–Hausdorff convergence in the presence of measures.

Let (μ_i) be a sequence of Borel measures on a metric space Z such that $\mu_i(B) < \infty$ for every ball $B \subset Z$ and for every i. The measures μ_i are said to *converge weakly* to a Borel measure μ on Z if $\int_Z \varphi\, d\mu_i \to \int_Z \varphi\, d\mu$ as $i \to \infty$ for every *boundedly supported* continuous function φ on Z. (A real-valued function on metric space is said to be boundedly supported if the function vanishes outside a ball.) We denote this convergence by

$$\mu_i \stackrel{*}{\rightharpoonup} \mu .$$

Strictly speaking, the above convergence occurs via the operation of these measures on the normed vector space $C_b(Z)$ of all boundedly supported continuous functions on Z, and hence it should be termed weak* convergence; however, for ease of terminology we merely refer to it as weak convergence. Although it is more traditional to consider φ to be compactly supported in the above definition, here we need to consider as well measures on l^∞, in which case one does not have nontrivial compactly supported continuous functions. While the requirement of compact support for φ is standard for proper metric spaces, in applications to Gromov–Hausdorff convergence it is more profitable to consider the more general class of boundedly supported functions φ.

Remark 11.4.1 If a sequence of measures μ_i converges weakly to μ, U is a bounded open set, and K is a compact set then

$$\mu(U) \le \liminf_{i\to\infty} \mu_i(U) \qquad \text{and} \qquad \mu(K) \ge \limsup_{i\to\infty} \mu_i(K). \tag{11.4.2}$$

The above claims are justified by the fact that if U is an open set then

$$\mu(U) = \sup\left\{ \int_Z \varphi\, d\mu : \varphi \in C_b(Z),\ \mathrm{spt}(\varphi) \subset U,\ \text{and } |\varphi| \le 1 \right\},$$

together with the definition of weak convergence of measures given above.

We also have the following useful inequality: for a compact set $K \subset Z$, a nonnegative continuous function u, and a bounded set W with $K \subset W$ and $\mathrm{dist}(K, Z \setminus W) > 0$, we have

$$\int_K u\, d\mu \le \liminf_{i\to\infty} \int_W u\, d\mu_i. \tag{11.4.3}$$

To see this, consider the function $\varphi \in C_b(Z)$ defined as

$$\varphi(x) := \left(1 - \frac{\mathrm{dist}(x, K)}{\mathrm{dist}(K, Z \setminus W)}\right)_+ .$$

Then $u\varphi \in C_b(Z)$ and

$$\int_K u\,d\mu \le \int_Z u\varphi\,d\mu = \lim_{i\to\infty}\int_Z u\varphi\,d\mu_i \le \liminf_{i\to\infty}\int_W u\,d\mu_i .$$

In the above, given a real number t we write $t_+ := \max\{t, 0\}$.

Recall the definition $f_\#\mu(A) = \mu(f^{-1}(A))$ of push-forward measures from Section 3.3.

Definition 11.4.4 Let $(X_1, d_1, \mu_1), (X_2, d_2, \mu_2), \ldots$ be a sequence of compact metric measure spaces. We say that a compact metric measure space (X, d, μ) is a *measured Gromov–Hausdorff limit* of (X_i, d_i, μ_i) if there exist isometric embeddings $\iota_i : X_i \to l^\infty$, $\iota : X \to l^\infty$, such that $d_H^\infty(\iota_i(X_i), \iota(X)) \to 0$ and $(\iota_i)_\#\mu_i \overset{*}{\rightharpoonup} \iota_\#\mu$ as measures on l^∞. (Recall the weak convergence from above.) We denote this convergence by

$$(X_i, d_i, \mu_i) \overset{GH}{\to} (X, d, \mu)\,.$$

We define *pointed measured Gromov–Hausdorff convergence* for proper length spaces only; that is the setting required in this book.

Definition 11.4.5 Let

$$(X_1, d_1, a_1, \mu_1), (X_2, d_2, a_2, \mu_2), \ldots$$

be a sequence of proper pointed metric measure spaces, where each metric space (X_i, d_i) is also a length space. We say that the sequence (X_i, d_i, a_i, μ_i) *pointed measured Gromov–Hausdorff-converges* to a proper pointed metric measure space (X, d, a, μ) if $(X_i, d_i, a_i) \overset{GH}{\to} (X, d, a)$ in the sense of Definition 11.3.1 and if

$$(\overline{B}(a_i, r), d_i, \mu_i \lfloor \overline{B}(a_i, r)) \overset{GH}{\to} (\overline{B}(a, r), d, \mu \lfloor \overline{B}(a, r)) \tag{11.4.6}$$

in the sense of Definition 11.4.4 for every $r > 0$. We denote this convergence by

$$(X_i, d_i, a_i, \mu_i) \overset{GH}{\to} (X, d, a, \mu)\,.$$

Note that the assumption $(X_i, d_i, a_i) \overset{GH}{\to} (X, d, a)$ implies, by Proposition 11.3.12, that $\overline{B}(a_i, r) \overset{GH}{\to} \overline{B}(a, r)$, so that requirement (11.4.6) makes sense for measures restricted to appropriate balls. Also note that the limit space is necessarily a length space, by the same proposition, because we also require the length spaces (X_i, a_i) to converge in the pointed Gromov–Hausdorff sense to the proper, and hence complete, (X, a).

In view of Proposition 11.3.14, the limit ball in (11.4.6), up to isometry, coincides with the corresponding pointed ball from the previous paragraph.

Gromov's compactness theorem for measured limits takes the following form:

Theorem 11.4.7 *Let (X_i, d_i, a_i, μ_i) be a pointed totally bounded sequence of pointed proper length-metric measure spaces satisfying*

$$\sup_i \mu_i(B(a_i, r)) < \infty \tag{11.4.8}$$

for each $r > 0$. Then (X_i, d_i, a_i, μ_i) contains a subsequence that converges in the pointed measured Gromov–Hausdorff sense to a pointed proper length metric measure space (X, d, a, μ) in such a way that $\mu(B(a, r)) \leq \sup_i \mu_i(B(a_i, r))$ for all $r > 0$.

Proof For each i let $\iota_i : X_i \to l^\infty$ be an isometric embedding as in Definition 11.4.4. We have a corresponding sequence of push-forward measures $(\iota_i)_\# \mu_i$ on l^∞, and the goal is to obtain a limit measure μ on l^∞; the support of such a limit measure is a viable candidate for the limit metric space. We will apply the weak version of the Banach–Steinhaus theorem 2.3.3 to an appropriate Banach space. Observe that each $\overline{B}(a_i, r)$ is separable, and hence we can choose a set $S_i \subset l^\infty$ which is countable and dense in $\iota_i(\overline{B}(a_i, r))$. Then $\bigcup_{i\in\mathbb{N}} S_i$ is dense in the closure of $\bigcup_{i\in\mathbb{N}} \iota_i(\overline{B}(a_i, r))$. Without loss of generality we can assume that $\iota_i(a_i) = 0$.

Let U be the collection of all continuous functions with support in $B(0, r) \subset l^\infty$. Observe that, in the supremum norm, U is a nonseparable Banach space because l^∞ is nonseparable. However, in considering the actions of $(\iota_i)_\# \mu_i$ on U, only the values of $f \in U$ taken at points in the closure K of $\bigcup_{i\in\mathbb{N}} \iota_i(\overline{B}(a_i, r))$ matter. Thus the supremum norm is not the correct norm to impose on U. Instead, given $f \in U$ we set $\|f\| := \sup_{z\in K} |f(z)|$. It is easily verified that $\|\cdot\|$ is a seminorm on U. Using the equivalence relation $\sim$ on U given by $f \sim g$ if and only if $\|f - g\| = 0$, we obtain a complete separable Banach space $V := U/\sim$. The separability of V follows, by a short argument, from the separability of K. We are now at liberty to apply Theorem 2.3.3 to the operators T_i on V given by $T_i(f) = \int_{l^\infty} f \, d(\iota_i)_\# \mu_i$, to obtain a limit map T for a subsequence; an application of the Riesz representation theorem now yields the weak limit measure on l^∞. □

A measure μ on a metric space X is called a *local doubling measure* if for each $r_0 < \infty$ there exists a constant $C_D(r_0) < \infty$ such that

$$\mu(B(x, r)) \leq C_D(r_0) \mu(B(x, r/2))$$

for each $x \in X$ and $0 < r \leq r_0$. Recall that if C_D can be chosen independent of r_0 then μ is a *doubling measure*. Every metric space which carries a (locally) doubling measure is boundedly doubling in the sense of the previous section. A family $(X_\alpha, d_\alpha, a_\alpha, \mu_\alpha)$ of pointed metric measure spaces is said to be *uniformly locally doubling in measure*, or just uniformly locally doubling, if μ_α is a locally doubling measure on X_α for each α, with the constant $C_D(r_0)$ independent of α for each $r_0 > 0$. Such a family of pointed metric measure spaces is uniformly boundedly doubling in the sense of the previous section provided that it also satisfies (11.4.8).

Corollary 11.4.9 *Let (X_i, d_i, a_i, μ_i) be a family of pointed proper uniformly locally doubling length metric measure spaces satisfying* (11.4.8). *Then (X_i, d_i, a_i, μ_i) has a subsequence that pointed measured Gromov–Hausdorff-converges to a pointed proper length-metric measure space.*

Example 11.4.10 Fix $\kappa \in \mathbb{R}$ and $V < \infty$ and consider the collection $\mathcal{M}(\kappa, V)$ consisting of all pointed Riemannian n-manifolds (M^n, a) (endowed with the Riemannian distance and volume) which have Ricci curvature bounded below by κ and volume bounded above by V. Then $\mathcal{M}(\kappa, V)$ is uniformly locally doubling and hence precompact with respect to pointed and measured Gromov–Hausdorff convergence, by Corollary 11.4.9. This is a consequence of the *Bishop–Gromov volume comparison inequality*; see for instance [49, Theorem 10.6.6].

11.5 Persistence of doubling measures under Gromov–Hausdorff convergence

This section is devoted to the proof of the following theorem.

Theorem 11.5.1 *Let (X_i, d_i, a_i, μ_i) be a sequence of complete length spaces which converges in the sense of pointed measured Gromov–Hausdorff convergence to a complete space (X, d, a, μ). If each measure μ_i is doubling with constant C_D then μ is also doubling with constant C_D^2.*

Proof Let $(X_i, d_i, a_i, \mu_i) \overset{GH}{\to} (X, d, a, \mu)$ as in the statement of the theorem. Since complete doubling spaces are proper, it follows from Proposition 11.3.12 and Definition 11.4.4 that X is a length space and that there exist isometric embeddings $\iota_i : X_i \to l^\infty$ and $\iota : X \to l^\infty$ with $d_H^\infty(\iota_i(B(a_i, R)), \iota(B(a, R))) \to 0$ for each $R > 0$ such that $(\iota_i)_\# \mu_i$ converges weakly to $(\iota)_\# \mu$ (as measures on l^∞).

For $z \in l^\infty$ and $\rho > 0$ we denote by $B_\infty(z, \rho)$ the ball in l^∞ with center z and radius ρ.

Fix $x \in X$ and $r > 0$ and set $R = r + d(a, x)$. Choose $x_i \in B(a_i, R)$ such that $\delta_i := \|\iota_i(x_i) - \iota(x)\|_\infty \to 0$ as $i \to \infty$.

Since μ (resp. μ_i) is supported on X (resp. X_i) and $(\iota_i)_\# \mu_i$ converges weakly to $(\iota)_\# \mu$, we deduce from (11.4.2) that

$$\begin{aligned}\mu(B(x, r)) = (\iota)_\#\mu(B_\infty(\iota(x), r)) &\le \liminf_{i\to\infty} (\iota_i)_\#\mu_i(B_\infty(\iota(x), r))\\ &\le \liminf_{i\to\infty} (\iota_i)_\#\mu_i(B_\infty(\iota_i(x_i), r + \delta_i)) = \liminf_{i\to\infty} \mu_i(B(x_i, r + \delta_i)).\end{aligned}$$

Similarly, applying (11.4.2) to the compact set $\overline{B}_\infty(\iota(x), r/3)$ we obtain that

$$\mu(B(x, r/2)) \ge \mu(\overline{B}(x, r/3)) \ge \limsup_{i\to\infty} \mu_i(\overline{B}(x_i, r/3 - \delta_i)). \qquad (11.5.2)$$

The proof is complete. □

The length space assumption in the previous theorem was used only to guarantee the Hausdorff convergence in l^∞ for our sequence of balls. With a little more work, this assumption can be removed.

Moreover, we can improve the conclusion of the theorem to obtain the same doubling constant C_D for the limit space. Since this improvement relies on tools which we will use in subsequent chapters, we provide the details.

We first state a volume decay property for doubling measures in length spaces.

Proposition 11.5.3 *Let (X, d) be a length space and assume that μ is a doubling measure on X with doubling constant C_D. Then there exist constants $C < \infty$ and $0 < \beta \le 1$ depending only on C_D such that*

$$\mu(B(x, r) \setminus \overline{B}(x, (1 - \epsilon)r)) \le C\epsilon^\beta \mu(B(x, r)) \qquad (11.5.4)$$

for every $x \in X$, $r > 0$, and $0 < \epsilon \le 1$.

Remark 11.5.5 Proposition 11.5.3 fails in the absence of the length space assumption. Let $X = \mathbb{R} \cup \mathbb{S}^1 = \{z \in \mathbb{C} : |z| = 1 \text{ or } \operatorname{Im} z = 0\}$ with the metric inherited from $\mathbb{C} = \mathbb{R}^2$. Let $\mu = \mathcal{H}_1$, the one-dimensional Hausdorff measure on X. It is easy to see that (11.5.4) cannot hold for any choice of constants C and β for balls centered at the origin.

Proof of Proposition 11.5.3 To simplify the notation, we introduce the abbreviation $A(x, s, t) := B(x, t) \setminus \overline{B}(x, s)$ for the open annulus with radii $s < t$ centered at $x \in X$. First, we prove the estimate

$$\mu(A(x, r-t, r)) \leq C_D^4\, \mu(A(x, r-3t, r-t)) \tag{11.5.6}$$

for all $x \in X$, $r > 0$, and $0 < t < r/3$.

For each $y \in A(x, r-t, r)$ choose a curve γ_y, of length at most $d(x, y)+t$, joining x to y. Let $z_y \in \gamma_y$ satisfy $d(x, z_y) = r - 2t$. Then $B_y := B(z_y, t)$ is a subset of $A(x, r-3t, r-t)$ and $y \in 3B_y$. By the $5B$-covering lemma (see Section 3.3), we may choose a countable collection of points $y_1, y_2, \ldots$ from $A(x, r-t, r)$ such that the balls $3B_{y_j}$ are pairwise disjoint and that $A(x, r-t, r) \subset \bigcup_j 15B_{y_j}$. Then

$$\begin{aligned}\mu(A(x, r-t, r)) &\leq \sum_j \mu(15B_{y_j}) \leq C_D^4 \sum_j \mu(B_{y_j}) \\ &\leq C_D^4\, \mu(A(x, r-3t, r-t)),\end{aligned}$$

as required.

We now turn to the proof of (11.5.4). Fix $x \in X$ and $r > 0$. Choosing $t = r/6$ in (11.5.6), we find that

$$\begin{aligned}\mu(A(x, 5r/6, r)) &\leq C_D^4\, \mu(A(x, r/2, 5r/6)) \\ &\leq C_D^4 \left(\mu(B(x, r)) - \mu(A(x, 5r/6, r))\right),\end{aligned}$$

and hence that

$$\mu(A(x, 5r/6, r)) \leq \frac{C_D^4}{C_D^4 + 1} \mu(B(x, r)). \tag{11.5.7}$$

Choosing $t_m = 1/(2 \times 3^m)$ for $m = 0, 1, 2, \ldots$ and applying (11.5.6) gives

$$\begin{aligned}\mu(A(x, (1-t_m)r, r)) &\leq C_D^4\, \mu(A(x, (1-t_{m-1})r, (1-t_m)r)) \\ &\leq C_D^4 \big(\mu(A(x, (1-t_{m-1})r, r)) \\ &\quad - \mu(A(x, (1-t_m)r, r))\big),\end{aligned}$$

whence by (11.5.7) and by induction we arrive at

$$\mu(A(x, (1-t_m)r, r)) \leq \left(\frac{C_D^4}{C_D^4 + 1}\right)^m \mu(B(x, r)) \tag{11.5.8}$$

for every $m = 1, 2, \ldots$ Now let $0 < \epsilon \leq 1$. If $\epsilon \leq 1/2$, we choose a nonnegative integer m such that

$$\frac{1}{2 \times 3^{m+1}} < \epsilon \leq \frac{1}{2 \times 3^m}. \tag{11.5.9}$$

Using (11.5.8) and (11.5.9), we deduce the annular decay estimate (11.5.4) with $C = 6^\beta$ and $\beta = \log(1 + C_D^{-4})/\log 3$. If, $\epsilon \geq 1/2$ then $1 \leq (2\epsilon)^\beta$ and so again (11.5.4) holds. $\square$

Using Proposition 11.5.3 we will show that the constant C_D^2 in Theorem 11.5.1 can be improved to C_D. We resume the proof of Theorem 11.5.1 after (11.5.2). From the annular decay property (Proposition 11.5.3), it follows that $\mu(\partial B(x,\rho)) = 0$ for every $\rho > 0$. Consequently, by Proposition 11.5.3,

$$\begin{aligned}\frac{\mu(B(x,r))}{\mu(B(x,r/2))} &\le \limsup_{i\to\infty} \frac{\mu_i(B(x_i, r+\delta_i))}{\mu_i(B(x_i, r/2-\delta_i))} \\ &\le C_D \limsup_{i\to\infty} \frac{\mu_i(B(x_i, r+\delta_i))}{\mu_i(B(x_i, r-2\delta_i))} = C_D.\end{aligned}$$

This completes the proof.

11.6 Persistence of Poincaré inequalities under Gromov–Hausdorff convergence

We now come to the principal aim of this chapter: the persistence of Poincaré inequalities under the Gromov–Hausdorff convergence of uniformly doubling metric measure spaces.

Theorem 11.6.1 *Let (X_i, d_i, a_i, μ_i) be a sequence of complete length spaces which converges in the sense of pointed and measured Gromov–Hausdorff convergence to a complete space (X, d, a, μ). Let $1 \le p < \infty$, $C_D, C_P < \infty$, and $\lambda \ge 1$ be fixed. If each measure μ_i is doubling with constant C_D and each space (X_i, d_i, μ_i) satisfies the p-Poincaré inequality with constants C_P and λ then (X, d, μ) also satisfies the p-Poincaré inequality, with constants C'_P and λ' depending only on p, C_P, λ, and C_D.*

To prove Theorem 11.6.1 we embed our sequence of spaces into l^∞. The following proposition allows us to assume that the function–upper-gradient pair, a priori only defined in the limit space, is defined on the whole of l^∞ as a function–upper-gradient pair. It is important here that both the function and the upper gradient are assumed to be Lipschitz; in our setting this is guaranteed by Theorem 8.4.1.

Proposition 11.6.2 *Let X be a length space which is a subset of a geodesic metric space Z. Let u and ρ be bounded Lipschitz functions on X such that $\inf_X \rho > 0$ and ρ is an upper gradient of u. Fix $\delta > 0$. Then there exist Lipschitz functions $\overline{u}$ and $\overline{\rho}$ on Z which extend u and ρ respectively. Moreover, $\overline{\rho}$ is bounded and $(1+\delta)\overline{\rho}$ is an upper gradient of $\overline{u}$ (on Z).*

Assuming for the moment the validity of Proposition 11.6.2, we can give a proof of Theorem 11.6.1.

Proof of Theorem 11.6.1 Let $(X_i, d_i, a_i, \mu_i) \overset{GH}{\to} (X, d, a, \mu)$ as in the statement of the theorem. As in the proof of Theorem 11.5.1 there exist isometric embeddings $\iota_i : X_i \to l^\infty$ and $\iota : X \to l^\infty$ with $d_H^\infty(\iota_i(B(a_i, R)), \iota(B(a, R))) \to 0$ for each $R > 0$ such that $(\iota_i)_\#\mu_i$ converges weakly to $(\iota)_\#\mu$ as measures on l^∞.

Fix a ball $B = B(x, r)$ in X. By Theorem 8.4.1 it suffices to verify the Poincaré inequality for each pair consisting of a bounded Lipschitz function u together with its bounded Lipschitz continuous upper gradient ρ. Fixing $0 < \delta \leq 1$, the hypotheses of Proposition 11.6.2 are satisfied for the pair u, $\rho + \delta/2$. Thus there exist Lipschitz extensions $\overline{u}$ and $\overline{\rho}$ of these functions to the whole of l^∞ such that $(1+\delta)\overline{\rho}$ is an upper gradient of $\overline{u}$.

As in the proof of Theorem 11.5.1, choose $x_i \in X_i$ such that $\iota_i(x_i) \to \iota(x)$ in l^∞. For ease of notation, denote by $B^\infty = B_\infty(\iota(x), r)$ and $B_i^\infty = B_\infty(\iota_i(x_i), r)$ the balls of radius r centered at the points $\iota(x), \iota_i(x_i)$ in l^∞. From the definition of Gromov–Hausdorff convergence, there is an N such that the inclusions $2B^\infty \subset 4B_i^\infty \subset 6B^\infty$ and $2\lambda B^\infty \subset 4\lambda B_i^\infty \subset 6\lambda B^\infty$ hold for each $n \geq N$. By the Poincaré inequality in X_i,

$$\begin{aligned}\fint_{2B^\infty} |\overline{u} - \overline{u}_{4B_i^\infty}|\, d((\iota_i)_\#\mu_i) &\leq C_P\, C_D\, r \left(\fint_{4\lambda B_i^\infty} (1+\delta)^p\, \overline{\rho}^p\, d((\iota_i)_\#\mu_i)\right)^{1/p} \\ &\leq C_P\, C_D^2\, r \left(\fint_{6\lambda B^\infty} (1+\delta)^p\, \overline{\rho}^p\, d((\iota_i)_\#\mu_i)\right)^{1/p}.\end{aligned}$$

Since $\overline{u}$ is continuous and X_i is proper, the quantities $(\overline{u}_{4B_i^\infty})$ are uniformly bounded. By passing to a subsequence if necessary, we can ensure that $\overline{u}_{4B_i^\infty} \to \alpha$ for some $\alpha \in \mathbb{R}$. From (11.4.3) and Theorem 11.5.1 we have

$$\begin{aligned}\fint_{B^\infty} |\overline{u} - \alpha|\, d((\iota)_\#\mu) &\leq \frac{1}{(\iota)_\#\mu(B^\infty)} \int_{2B_i^\infty} |\overline{u} - \alpha|\, d((\iota)_\#\mu) \\ &\leq C \liminf_i \fint_{2B_i^\infty} |\overline{u} - \overline{u}_{4B_i^\infty}|\, d((\iota_i)_\#\mu) \\ &\leq Cr \limsup_i \left(\fint_{6\lambda \overline{B}^\infty} (1+\delta)^p \overline{\rho}^p\, d((\iota_i)_\#\mu)\right)^{1/p} \\ &\leq Cr \left(\fint_{7\lambda B^\infty} (1+\delta)^p \overline{\rho}^p\, d((\iota)_\#\mu)\right)^{1/p}.\end{aligned}$$

Letting $\delta \to 0$ we obtain

$$\fint_{B^\infty} |\overline{u} - \alpha|\, d((\iota)_\#\mu) \leq Cr \left(\fint_{7\lambda B^\infty} \overline{\rho}^p\, d((\iota)_\#\mu)\right)^{1/p}.$$

Since μ is supported on X and $\fint_B |u - u_B|\,d\mu \le 2\fint_B |u - \alpha|\,d\mu$, the inequality

$$\fint_B |u - u_B|\,d\mu \le Cr \left(\fint_{7\lambda B} \rho^p\,d\mu \right)^{1/p}$$

follows. The proof is complete. □

Proof of Proposition 11.6.2 Let u and ρ be bounded Lipschitz functions on X as in the statement. Let L be a Lipschitz constant for ρ on X. Finally, let $\delta > 0$; since the conclusion is stronger for smaller values of δ we will assume that $\delta \le 1$.

We now construct the desired extensions $\overline{u}$ and $\overline{\rho}$ to $Z \supset X$ in several steps.

First, we extend ρ to a bounded Lipschitz function ρ_1 on Z such that

$$|u(x) - u(y)| \le (1 + \delta) \int_\gamma \rho_1\,ds \tag{11.6.3}$$

for all $x, y \in X$ and all rectifiable curves $\gamma \subset N_\epsilon(X) \subset Z$ joining x to y. Here $\epsilon > 0$ is a suitably chosen small constant whose value will be determined in the proof.

We define ρ_1 by the McShane extension of ρ to Z, truncated in such a way that $\inf_X \rho \le \rho_1(z) \le \sup_X \rho$ for all $z \in Z$. Explicitly, set

$$\hat{\rho}_1(z) = \inf_{x \in X} \rho(x) + L\,d(x, z)$$

and

$$\rho_1(z) = \max \left\{ \inf_X \rho, \min \left\{ \hat{\rho}_1(z), \sup_X \rho \right\} \right\}.$$

Thus ρ_1 is an L-Lipschitz function on Z with $\rho_1(x) = \rho(x)$ for $x \in X$. For each $r > 0$ and $z_0 \in Z$ we record the estimate

$$\frac{\sup_{B(z_0,r)} \rho_1}{\inf_{B(z_0,r)} \rho_1} \le 1 + \frac{2L}{\inf_X \rho}\, r, \tag{11.6.4}$$

which follows from the Lipschitz property of ρ_1.

Set $\eta = (1/6) \min\{1, (6L)^{-1} \inf_X \rho\}\delta$ and $\epsilon = \eta^2/2$. Let $x, y \in X$ and let γ be a rectifiable curve in $N_\epsilon(X)$ joining x to y. We distinguish two cases.

Case (i), length$(\gamma) < \eta$. Choose a curve $\beta \subset X$ joining x to y with length$(\beta) < (1 + \delta/3)d(x, y)$. Then $\beta \subset B(x, 2\eta)$, whence

$$\int_\beta \rho\,ds = \int_\beta \rho_1\,ds \le \sup_{B(x,2\eta)} \rho_1\,(1 + \delta/3)\,\text{length}(\gamma). \tag{11.6.5}$$

However, $\gamma \subset B(x, \eta)$ and hence

$$\int_\gamma \rho_1 \, ds \geq \inf_{B(x,\eta)} \rho_1 \, \text{length}(\gamma). \tag{11.6.6}$$

Combining (11.6.5), (11.6.6), and (11.6.4) yields

$$\int_\beta \rho \, ds \leq \left(1 + \frac{4L\eta}{\inf_X \rho}\right)\left(1 + \frac{\delta}{3}\right) \int_\gamma \rho_1 \, ds \leq (1+\delta) \int_\gamma \rho_1 \, ds,$$

from which (11.6.3) follows since ρ is an upper gradient of u on X.

Case (ii), length$(\gamma) \geq \eta$. We can reduce to the previous case by a decomposition argument. Write γ as the union of consecutively connected subcurves γ_i, $i = 1, \ldots, N$ such that $\eta/2 \leq \text{length}(\gamma_i) < \eta$. For $i = 1, \ldots, N$, let z_i^1 and z_i^2 be the end points of γ_i, ordered so that $z_i^2 = z_{i+1}^1$. Since $\gamma_i \subset N_\epsilon(X)$ we may choose points $x_i^1, x_i^2 \in X$ with $x_1^1 = x$, $x_N^2 = y$, $x_i^2 = x_{i+1}^1$, and $d(x_i^1, z_i^1) < \epsilon$. As in the previous case, choose curves β_i joining x_i^1 to x_i^2 with length$(\beta_i) \leq (1 + \delta/3) d(x_i^1, x_i^2)$. Let β be the curve obtained as the union of the consecutively intersecting curves β_i, $i = 1, \ldots, N$.

As before, it suffices to prove the estimate

$$\int_\beta \rho \, ds \leq (1+\delta) \int_\gamma \rho_1 \, ds$$

and, since $\int_\beta \rho \, ds = \sum_i \int_{\beta_i} \rho \, ds$ and $\int_\gamma \rho \, ds = \sum_i \int_{\gamma_i} \rho \, ds$, it suffices for us to prove that

$$\int_{\beta_i} \rho \, ds \leq (1+\delta) \int_{\gamma_i} \rho_1 \, ds$$

for each i. To this end, we observe that

$$\begin{aligned} d(x_i^1, x_i^2) &\leq d(z_i^1, z_i^2) + 2\epsilon \leq \text{length}(\gamma_i) + \eta^2 \\ &\leq (1 + 2\eta) \, \text{length}(\gamma_i) \leq (1 + \delta/3)\eta. \end{aligned}$$

Thus

$$\beta_i \subset B(x_i^1, (1+\delta/3) d(x_i^1, x_1^2)) \subset B(x_i, 2\eta)$$

and so

$$\int_{\beta_i} \rho \, ds \leq \left(\sup_{B(x_i^1, 2\eta)} \rho_1 \right) (1+\delta/3)(1+2\eta) \, \text{length}(\gamma_i). \tag{11.6.7}$$

Next, since $\gamma_i \subset B(z_i^1, \eta) \subset B(x_i^1, 2\eta)$, we have

$$\int_{\gamma_i} \rho_1 \, ds \geq \left(\inf_{B(x_i^1, 2\eta)} \rho_1 \right) \text{length}(\gamma_i). \tag{11.6.8}$$

Combining (11.6.7), (11.6.8), and (11.6.4) yields

$$\int_{\beta_i} \rho\, ds \le \left(1 + \frac{4L\eta}{\inf_X \rho}\right)\left(1 + \frac{\delta}{3}\right)(1+2\eta)\int_{\gamma_i} \rho_1\, ds \le (1+\delta)\int_{\gamma_i} \rho_1\, ds.$$

This completes the proof of (11.6.3) for case (ii).

Next, we extend ρ_1 to a bounded Lipschitz function $\overline{\rho}$ on the whole of Z in such a way that (11.6.3) holds (with ρ_1 replaced by $\overline{\rho}$) for all $x, y \in X$ and all rectifiable curves in Z joining x to y.

The construction uses Lipschitz partitions of unity. Since u is bounded, it has finite oscillation $\operatorname{osc}_X u := \sup\{|u(x) - u(y)| : x, y \in X\}$. Choose a nonnegative Lipschitz function h on Z which vanishes on $N_{\epsilon/4}(X)$ and achieves its maximum $2\operatorname{osc}_X u/\epsilon$ on the whole of $Z \setminus N_{\epsilon/2}(X)$. For example, one may take

$$h(z) = \frac{2\operatorname{osc}_X u}{\epsilon} \min\left\{\frac{\operatorname{dist}(z, N_{\epsilon/4}(X))}{\operatorname{dist}(Z \setminus N_{\epsilon/2}(X), N_{\epsilon/4}(X))}, 1\right\}.$$

Set $\overline{\rho} = \rho_1 + h$ and let γ be a rectifiable curve in Z joining $x, y \in X$. If $\gamma \subset N_\epsilon(X)$ then (11.6.3) already holds, so assume that γ meets $Z \setminus N_\epsilon(X)$. Then the length of that portion of γ which lies in $N_\epsilon(X) \setminus N_{\epsilon/2}(X)$ is at least $\epsilon/2$ and so

$$\begin{aligned}\int_\gamma \overline{\rho}\, ds &\ge \frac{2\operatorname{osc}_X u}{\epsilon} \operatorname{length}\big(\gamma \cap (N_\epsilon(X) \setminus N_{\epsilon/2}(X))\big) \\ &\ge \operatorname{osc}_X u \ge |u(x) - u(y)|,\end{aligned}$$

which proves the desired inequality.

Finally, we extend u to a Lipschitz function $\overline{u}$ on the whole of Z in such a way that (11.6.3) holds (with ρ_1 replaced by $\overline{\rho}$ and u replaced by $\overline{u}$) for all $x, y \in Z$ and all rectifiable curves Z joining x to y.

Define $\overline{u} : Z \to \mathbb{R}$ by

$$\overline{u}(z) = \inf\left\{(1+\delta)\int_\gamma \overline{\rho}\, ds + u(x)\right\},$$

where the infimum is taken over all $x \in X$ and all rectifiable curves γ joining x to z. It is clear that $\overline{u}$ is an extension of u. Furthermore $-\infty < \overline{u}(z) < \infty$ for each z because $\overline{\rho}$ is bounded on Z and u is bounded on X. Hence the fact that $(1+\delta)\overline{\rho}$ is an upper gradient for $\overline{u}$ on Z follows from the definition; see for example the proof of Lemma 7.2.13. Finally, $\overline{u}$ is Lipschitz since $\overline{\rho}$ is bounded and Z is geodesic.

The proof of Proposition 11.6.2 is complete. □

Throughout this and the previous sections we have assumed that the metric spaces X_i are complete. We now point out why the assumption of completeness (which is equivalent to properness, in the presence of the doubling property) is not very restrictive.

By Lemma 8.2.3, if any space (X_i, d_i, μ_i) considered in this section or the previous section is not complete, then we can complete the space to obtain a sequence of spaces that converge to the same Gromov–Hausdorff limit as the original sequence. Note that the Gromov–Hausdorff distance between X_i and $\overline{X}_i$ is zero. Furthermore, the requirement that the X_i be length spaces is also not essential. Indeed, if (X_i, d_i, μ_i) is doubling, supports a p-Poincaré inequality, and is complete then by Theorem 8.3.2 it is quasiconvex with quasiconvexity constant dependent solely on the doubling and Poincaré constants. The corresponding inner metric $\hat{d}_i$, as considered in Proposition 8.3.12, is a length metric on X_i and is bilipschitz equivalent to the original metric d_i. By Lemma 8.3.18 the space $(X_i, \hat{d}_i, \mu_i)$ is also doubling, complete, and supports a p-Poincaré inequality, with constants dependent solely on the original data relating to (X_i, d_i, μ_i).

Combining Theorem 11.6.1 with Corollary 11.4.9, we obtain the following result. Note that the uniform doubling condition with constant C_D implies that the sequence of pointed metric measure spaces satisfies the hypothesis (11.4.8) of Corollary 11.4.9.

Theorem 11.6.9 *Let (X_i, d_i, a_i, μ_i) be a sequence of pointed metric measure spaces, each of which is a length space. Let $1 \le p < \infty$, $C_D, C_P < \infty$, and $\lambda \ge 1$ be fixed. If each measure μ_i is doubling with constant C_D and each space (X_i, d_i, μ_i) satisfies the p-Poincaré inequality with constants C_P and λ then a subsequence of (X_i, d_i, a_i, μ_i) pointed measured Gromov–Hausdorff-converges to a pointed complete metric measure space (X, d, a, μ) such that (X, d, μ) also satisfies the p-Poincaré inequality with constants C'_P and λ' depending only on p, C_P, λ, and C_D.*

11.7 Notes to Chapter 11

The space $\mathcal{K}_Z$ is called the *(compacta) hyperspace* of Z. For a comprehensive account of the modern theory of hyperspaces, see [215]. The existence of invariant sets for iterated function systems (i.e., fractals) relies on the completeness of the compacta hyperspace of $\mathbb{R}^n$.

In discussing pointed metric spaces, the topology we considered was not defined via a metric. However, this topology is indeed metrizable; see [131].

The notion of Gromov–Hausdorff topology on classes of manifolds was first considered by Gromov [107]. Since then this notion has been found useful in the study of geometry; for example, Perelman *et al.* used it to prove the Poincaré conjecture [50], [225]. The paper [147] gives further details on a result of Perelman that two compact (same-dimensional) Alexandrov spaces are homeomorphic if the Gromov–Hausdorff distance between them is sufficiently small. More on Gromov–Hausdorff convergence can be found in the books [49] and [45].

Given a complete doubling metric measure space (X, d, μ) supporting a p-Poincaré inequality, and a point $x_0 \in X$, we can "zoom into" X in the vicinity of x_0 by considering a sequence of pointed metric measure spaces (X_n, x_0, d_n, μ_n), where $X_n = X$, $d_n(x, y) = nd(x, y)$, and $\mu_n = \mu(B(x_0, 1/n))^{-1}\mu$. It is straightforward to see that (X_n, d_n, μ_n) also is doubling and supports a p-Poincaré inequality, with the same constants as for X. Pointed measured Gromov–Hausdorff limits of suitable subsequences of such a sequence also exist and are doubling and support a p-Poincaré inequality, by the results of the previous section; see Theorem 11.6.9. Such limit spaces are called *tangent spaces* to X at x_0; they have a wide variety of uses. In the present context, they were used by Cheeger in [53] to study the infinitesimal behavior of Lipschitz functions on metric measure spaces. For further information see Section 13.6.

Theorem 11.5.1 is due to Cheeger [53, Theorem 9.1]. Proposition 11.5.3 was proved independently by Colding and Minicozzi [65] and Buckley [47, Corollary 2.2].

A result weaker than Theorem 11.6.1 was proved by Cheeger in [53, Theorem 9.6], namely, that the limit space satisfies the q-Poincaré inequality for each $q > p$ (with constants $C'_P = C'_P(q)$ and $\lambda' = \lambda'(q)$ which a priori may blow up as $q \to p$). As given in this chapter, Theorem 11.6.1 was proved (independently) by Cheeger, Koskela (both of whose proofs were unpublished), and Keith [150]. The proof given here is modeled on that from [150].

The proof of the density of Lipschitz functions in $N^{1,p}(X)$ (for $p > 1$) given in [16] also shows that the minimal p-weak upper gradient of a function $f \in N^{1,p}(X)$ can be approximated by a stronger notion of local Lipschitz-constant functions for some sequence of Lipschitz approximations of f. In the verification of Poincaré inequalities, the use of such a strong notion of local Lipschitz-constant functions, called *asymptotic Lipschitz functions* in [16], would enable us to replace the constants C'_P and λ' in Theorem 11.6.1 by C_P and λ respectively. However, the aforementioned approximation requires the technology of optimal mass transportation. Since this technology is outside the scope of this book, we omit this improved result from our exposition. See [16, Remark 8.3] for further details.

12

Self-improvement of Poincaré inequalities

The focus of this chapter is the Keith–Zhong theorem on the self-improvement of p-Poincaré inequalities for $1 < p < \infty$. In [153], Keith and Zhong proved that whenever X is a complete metric space equipped with a doubling measure and supporting a p-Poincaré inequality for some $1 < p < \infty$ then X also supports a q-Poincaré inequality for some $q \geq 1$ with $q < p$. Stated another way, for complete and doubling metric measure spaces the Poincaré inequality is an open-ended condition, that is, the collection of p for which X supports a p-Poincaré inequality is a relatively open subset of $[1, \infty)$. This result has numerous applications and corollaries; for a sample of these see Theorems 12.3.13 and 12.3.14.

Throughout this chapter our standing assumptions are that $X = (X, d)$ is a complete metric space, that μ is a doubling measure on X, and that the metric measure space (X, d, μ) supports a p-Poincaré inequality for some $1 < p < \infty$.

As discussed in Corollary 8.3.16 and Lemma 8.3.18 we may, and will, also assume without loss of generality that X *is a geodesic space*. Finally, in view of Theorem 9.1.15(i) and Hölder's inequality, we may assume that the integrals on both sides of the Poincaré inequality are taken over the same ball, i.e., that the parameter λ in (8.1.1) is equal to 1. We occasionally repeat these assumptions for emphasis.

For a positive real number x, we write $\lceil x \rceil$ for the smallest integer greater than or equal to x.

12.1 Geometric properties of geodesic doubling metric measure spaces

We begin with a few miscellaneous facts about metric measure spaces.

In many arguments in this chapter we will need to consider inclusions between dilations of balls. Note that if B and B' are balls in X with $B \subset B'$ and $\lambda > 0$, it is *not* necessarily the case that $\lambda B \subset \lambda B'$. For instance, consider $X = [0, \infty)$ equipped with the Euclidean metric and let $B = B(0, 2) = [0, 2)$ and $B' = B(1, 1 + \epsilon) = [0, 2 + \epsilon)$ for any $\epsilon > 0$. Then $\lambda B \not\subset \lambda B'$ for any $\lambda > 1/(1 - \epsilon)$, even though $B \subset B'$.

It is easy to see that if X is connected then $\lambda B \subset (2\lambda + 1)B'$ whenever $B \subset B'$ and $\lambda > 1$. In geodesic spaces we can obtain a slight improvement.

Lemma 12.1.1 *Let X be a geodesic space, let $B \subset B'$ be balls in X, and let $\lambda > 1$. Then $\lambda B \subset (2\lambda - 1)B'$.*

Note that the conclusion of the lemma is asymptotically sharp as $\lambda \to 1$.

Proof Let $B = B(y, r)$ and $B' = B(x, R)$. Suppose that $z \in B(y, \lambda r)$. Let γ be a geodesic joining y to z, and choose $w \in \gamma$ such that $d(w, y) = (1/\lambda)d(z, y)$. Then $w \in B(y, r)$, so $w \in B(x, R)$. Moreover, $d(z, w) = (\lambda - 1)d(w, y)$. Using the triangle inequality we obtain an estimate as follows:

$$\begin{aligned} d(z, x) &\le d(z, w) + d(w, x) \\ &< (\lambda - 1)d(w, y) + R \\ &\le (\lambda - 1)(d(w, x) + d(x, y)) + R \\ &< (\lambda - 1)(R + R) + R = (2\lambda - 1)R. \end{aligned}$$

The proof is complete. □

We next show that if X is a geodesic space and $E \subset X$ is measurable then the relative measure density function $r \mapsto \mu(B(x, r) \cap E)/\mu(B(x, r))$ is continuous. The proof uses the geodesic property via Proposition 11.5.3.

Lemma 12.1.2 *Let E be a measurable set and let $x \in X$. Then the function*

$$r \mapsto \frac{\mu(B(x, r) \cap E)}{\mu(B(x, r))}$$

is continuous.

Proof Fix $r > 0$ and $\delta > 0$, and consider r' with $r < r' < r + \delta$. We will show that

$$\left| \frac{\mu(B(x, r) \cap E)}{\mu(B(x, r))} - \frac{\mu(B(x, r') \cap E}{\mu(B(x, r'))} \right| \le \omega_r(\delta),$$

where $\omega_r(\delta) \to 0$ as $\delta \to 0$. An application of the triangle inequality shows that the left-hand side of the above inequality is less than or equal to

$$\left|\frac{\mu(B(x,r)\cap E)-\mu(B(x,r')\cap E)}{\mu(B(x,r'))}\right|+\left|\frac{\mu(B(x,r)\cap E)}{\mu(B(x,r))}-\frac{\mu(B(x,r)\cap E)}{\mu(B(x,r'))}\right|$$
$$\leq \frac{\mu(A(x,r,r')\cap E)}{\mu(B(x,r'))}+\frac{\mu(A(x,r,r'))\,\mu(B(x,r)\cap E)}{\mu(B(x,r'))\,\mu(B(x,r))}$$
$$\leq 2\,\frac{\mu(A(x,r,r'))}{\mu(B(x,r'))},$$

where, as in the proof of Proposition 11.5.3, we write $A(x,s,t) = B(x,t)\setminus \overline{B}(x,s)$. Applying Proposition 11.5.3 with $\epsilon = 1 - r/r'$ yields

$$\left|\frac{\mu(B(x,r)\cap E)}{\mu(B(x,r))}-\frac{\mu(B(x,r')\cap E)}{\mu(B(x,r'))}\right| \leq 2C\left(1-\frac{r}{r'}\right)^{\beta} < 2C\left(\frac{\delta}{r+\delta}\right)^{\beta},$$

where $\beta \in (0,1]$ and C both depend only on the doubling constant C_μ.

A similar argument, with $A(x,r,r')$ replaced by $A(x,r',r)$, gives a similar estimate when $r-\delta < r' < r$. The proof is complete. □

We complete this section by giving estimates on the overlap of the covering of a ball by balls of equal radii.

We fix $0 < s < r < \operatorname{diam}(X)/2$, and $x_0 \in X$. If $z_1,\ldots,z_N$ are points in $B(x_0,r)$ that are maximally s-separated (and so $d(z_i,z_j) \geq s$ if $i \neq j$ and $B(x_0,r) \subset \bigcup_{j=1}^N B(z_j,2s)$) then, from the pairwise disjointness property of the balls $B(z_j,s/2) \subset B(x_0,2r)$, we obtain

$$\sum_{j=1}^{N} \mu(B(z_j,s)) \leq C_\mu^2\,\mu(B(x_0,r)).$$

Now, from the above and the relative lower decay property of μ,

$$\frac{1}{C}\sum_{j=1}^{N}\left(\frac{s}{r}\right)^{Q} \leq C_\mu^2,$$

that is,

$$N \leq C\left(\frac{r}{s}\right)^{Q} =: N_2 = N_2\left(C_\mu, \frac{r}{s}\right), \tag{12.1.3}$$

where the constant C depends solely on the doubling constant C_μ.

Now fix $\alpha \geq 1$ and suppose that $x \in B(x_i,\alpha s)$ for $i \in I \subset \{1,\ldots,N\}$. Then for these i we have that $x_i \in B(x,\alpha s)$, and so a repetition of the above argument with $r = \alpha s$ gives via (12.1.3) that

$$\#I \leq C \left(\frac{\alpha s}{s}\right)^Q = C\alpha^Q =: N_1 = N_1(C_\mu, \alpha), \tag{12.1.4}$$

where $\#I$ refers to the cardinality of I.

12.2 Preliminary local arguments

The self-improvement of Poincaré inequalities follows from an application of Cavalieri's principle (3.5.5) once we obtain the self-improvement of certain estimates of level sets. In this section we first consider local versions of the level set estimates.

Given a Lipschitz function $u : X \to \mathbb{R}$ and $\tau > 0$, we set

$$M_\tau^\# u(x) := \sup_{x \in B \subset \tau B_0} \frac{1}{\operatorname{rad}(B)} \fint_B |u - u_B| \, d\mu.$$

Here B_0 is a fixed ball in X. The quantity $M_\tau^\# u$ denotes a type of maximal function which measures the maximum local deviation (in L^1) of u from its average value on balls, divided by the radius of the ball, computed over balls contained in τB_0.

For $\lambda > 0$ we define the level set

$$U_\lambda := \{x \in 128 B_0 : M_{128}^\# u(x) > \lambda\}.$$

In what follows we fix $\alpha > 3$; it will be chosen more specifically later.

Proposition 12.2.1 *For sufficiently large positive integers k whenever $u : X \to \mathbb{R}$ is Lipschitz and*

$$\lambda < \frac{1}{\operatorname{rad}(B_0)} \fint_{B_0} |u - u_{B_0}| \, d\mu,$$

we have

$$\begin{aligned} \mu(B_0) \leq\ & 2^{kp-\alpha} \mu(U_{2^k\lambda}) + 8^{kp-\alpha} \mu(U_{8^k\lambda}) \\ & + 8^{k(p+1)} \mu\left(\{x \in 2B_0 : \operatorname{Lip} u(x) > 8^{-k}\lambda\}\right). \end{aligned} \tag{12.2.2}$$

The proof of Proposition 12.2.1 is the goal of this section and is accomplished via a series of lemmas.

By rescaling the metric and measure, we can assume without loss of generality that $\mu(B_0) = 1 = \operatorname{rad}(B_0)$. By replacing u with $\lambda^{-1} u$, we may assume that $\lambda = 1$. (Note that a choice of the integer k that works for the value $\lambda = 1$

in the scaled function $\lambda^{-1}u$ will also work for the original value of λ and the original function u.)

The above reductions understood, we observe that it suffices to prove, for sufficiently large k and for all Lipschitz functions $u : X \to \mathbb{R}$ such that

$$\int_{B_0} |u - u_{B_0}|\, d\mu > 1, \tag{12.2.3}$$

that

$$\begin{aligned} 1 \le\ & 2^{kp-\alpha}\mu(U_{2^k}) + 8^{kp-\alpha}\mu(U_{8^k}) \\ & + 8^{k(p+1)}\,\mu\left(\{x \in 2B_0 : \operatorname{Lip} u(x) > 8^{-k}\}\right). \end{aligned} \tag{12.2.4}$$

Suppose that $k \ge 10$ is a positive integer such that (12.2.4) does not hold for some u that satisfies (12.2.3). Then

$$\mu(U_{2^k}) < 2^{\alpha-kp}, \tag{12.2.5}$$

$$\mu(U_{8^k}) < 8^{\alpha-kp}, \tag{12.2.6}$$

and

$$\mu\left(\{x \in 2B_0 : \operatorname{Lip} u(x) > 8^{-k}\}\right) < 8^{-k(p+1)}. \tag{12.2.7}$$

As emphasized above, we seek an upper bound on k that is independent of u. Towards this end, we wish to keep a careful track of the constants in order to obtain an upper bound on the values of k for which (12.2.4) does not hold. Throughout this chapter, C_μ will denote the doubling constant of μ, C_{HL} will denote the constant associated with the L^1 to $L^{1,\infty}$ estimate (3.5.7) in the Hardy–Littlewood maximal function theorem 3.5.6, and C_P will denote the constant in the p-Poincaré inequality (8.1.1). Recall, however, that the constant C_{HL} itself depends only on the doubling constant C_μ. Later on we will encounter other natural constants depending only on C_μ; to simplify the coming formulas we will indicate these constants with individualized notation.

For the remainder of this section we assume that u is Lipschitz and satisfies (12.2.3).

Lemma 12.2.8 *There is a positive integer k_* such that if $k \ge k_*$ and (12.2.5) holds then*

$$\int_{2B_0\setminus U_{2^k}} |u - u_{2B_0\setminus U_{2^k}}|\, d\mu \ge \frac{1}{4(1+2C_\mu^3)}. \tag{12.2.9}$$

Proof First note that if U_{2^k} is empty then

$$\int_{2B_0\setminus U_{2^k}} |u - u_{2B_0\setminus U_{2^k}}|\, d\mu = \int_{2B_0} |u - u_{2B_0}|\, d\mu \ge \frac{1}{2}\int_{B_0} |u - u_{B_0}|\, d\mu > \frac{1}{2},$$

and (12.2.9) holds. Hence, without loss of generality we may assume that U_{2^k} is nonempty. Furthermore, by subtracting a constant from u if necessary, we may assume that $u_{2B_0\setminus U_{2^k}} = 0$ (note that the conditions (12.2.3), (12.2.5), (12.2.6), and (12.2.7) are each stable under the subtraction of a constant).

If $x \in U_{2^k}$ then there is some ball $B \subset 128B_0$ with $x \in B$ such that $\mathrm{rad}(B)^{-1} \fint_B |u - u_B|\, d\mu > 2^k$. It follows that $B \subset U_{2^k}$ and hence that U_{2^k} is open. Hence, for $x \in B_0 \cap U_{2^k}$ and sufficiently small $r > 0$, we have $B(x, r) \subset U_{2^k}$, so that on the one hand

$$\frac{\mu(B(x, r) \cap U_{2^k})}{\mu(B(x, r))} = 1 \qquad \text{for } x \in B_0 \cap U_{2^k} \text{ and } r > 0 \text{ small.} \qquad (12.2.10)$$

On the other hand, $\mu(B(x, 1/5) \cap U_{2^k}) \le \mu(U_{2^k}) < 2^{\alpha - kp}$, and so

$$\frac{\mu(B(x, 1/5) \cap U_{2^k})}{\mu(B(x, 1/5))} = \mu(B(x, 1/5) \cap U_{2^k}) \frac{\mu(B_0)}{\mu(B(x, 1/5))} \le 2^{\alpha - kp}\, C_\mu^4 .$$

(Recall our normalizing assumption that $\mu(B_0) = 1 = \mathrm{rad}(B_0)$.) Choose k_0 large enough that $2^{\alpha - k_0 p}\, C_\mu^4 \le 1/5$, for example,

$$k_0 := \left\lceil \frac{\log(5 \times 2^\alpha\, C_\mu^4)}{p \log 2} \right\rceil . \qquad (12.2.11)$$

Thus

$$\frac{\mu(B(x, 1/5) \cap U_{2^k})}{\mu(B(x, 1/5))} \le \frac{1}{5} \qquad \text{for } x \in B_0 \cap U_{2^k}, \qquad (12.2.12)$$

whenever $k \ge k_0$.

We will proceed under the assumption that $k \ge k_0$. In view of (12.2.10), (12.2.12), and Lemma 12.1.2, for each $x \in B_0 \cap U_{2^k}$ there is some $0 < r_x < 1/5$ such that

$$\frac{\mu(B(x, r_x) \cap U_{2^k})}{\mu(B(x, r_x))} = \frac{1}{2} \qquad \text{and} \qquad \frac{\mu(B(x, r_x) \setminus U_{2^k})}{\mu(B(x, r_x))} = \frac{1}{2}. \qquad (12.2.13)$$

The collection $\{B(x, r_x) : x \in B_0 \cap U_{2^k}\}$ covers $B_0 \cap U_{2^k}$; an application of the $5B$-covering lemma (see Section 3.3) provides a countable pairwise disjoint subcollection $\{B_i\}$ such that $\{5B_i\}$ covers $B_0 \cap U_{2^k}$. Observe that for each i we have $5B_i \subset 2B_0$. Now

$$1 < \int_{B_0} |u - u_{B_0}|\, d\mu \le 2 \int_{B_0} |u|\, d\mu,$$

and so by (12.2.13) we get

$$
\begin{aligned}
\frac{1}{2} &\le \int_{B_0\setminus U_{2^k}} |u|\,d\mu + \int_{B_0\cap U_{2^k}} |u|\,d\mu \\
&\le \int_{2B_0\setminus U_{2^k}} |u|\,d\mu + \sum_i \int_{5B_i} |u|\,d\mu \\
&\le \int_{2B_0\setminus U_{2^k}} |u|\,d\mu + \sum_i \int_{5B_i} |u - u_{B_i\setminus U_{2^k}}|\,d\mu + \sum_i \mu(5B_i) \fint_{B_i\setminus U_{2^k}} |u|\,d\mu \\
&\le \int_{2B_0\setminus U_{2^k}} |u|\,d\mu + \sum_i \int_{5B_i} |u - u_{B_i\setminus U_{2^k}}|\,d\mu + 2C_\mu^3 \sum_i \int_{B_i\setminus U_{2^k}} |u|\,d\mu \\
&\le (1+2C_\mu^3) \int_{2B_0\setminus U_{2^k}} |u|\,d\mu + \sum_i \int_{5B_i} |u - u_{B_i\setminus U_{2^k}}|\,d\mu.
\end{aligned}
$$

Since $5B_i \setminus U_{2^k}$ is nonempty for each i, and $\mathrm{rad}(B_i) \le 1/5$, by (12.2.13) we have

$$
\begin{aligned}
\fint_{5B_i} |u - u_{B_i\setminus U_{2^k}}|\,d\mu &\le \fint_{5B_i} |u - u_{5B_i}|\,d\mu + |u_{5B_i} - u_{B_i\setminus U_{2^k}}| \\
&\le \fint_{5B_i} |u - u_{5B_i}|\,d\mu + \frac{\mu(5B_i)}{\mu(B_i \setminus U_{2^k})} \fint_{5B_i} |u - u_{5B_i}|\,d\mu \\
&\le (1+2C_\mu^3) \fint_{5B_i} |u - u_{5B_i}|\,d\mu \le 5(1+2C_\mu^3)\, 2^k\, \mathrm{rad}(B_i) \\
&\le (1+2C_\mu^3)\, 2^k.
\end{aligned}
$$

Therefore, by the choice of the cover and by (12.2.13),

$$
\begin{aligned}
\frac{1}{2} &\le (1+2C_\mu^3) \int_{2B_0\setminus U_{2^k}} |u|\,d\mu + (1+2C_\mu^3)\, 2^k \sum_i \mu(5B_i) \\
&\le (1+2C_\mu^3) \int_{2B_0\setminus U_{2^k}} |u|\,d\mu + (1+2C_\mu^3)\, 2^k\, C_\mu^3 \sum_i \mu(B_i) \\
&\le (1+2C_\mu^3) \int_{2B_0\setminus U_{2^k}} |u|\,d\mu + (1+2C_\mu^3)\, 2^{k+1}\, C_\mu^3 \sum_i \mu(B_i \cap U_{2^k}) \\
&\le (1+2C_\mu^3) \int_{2B_0\setminus U_{2^k}} |u|\,d\mu + (1+2C_\mu^3)\, 2^{k+1}\, C_\mu^3\, \mu(U_{2^k}).
\end{aligned}
$$

Applying (12.2.5) yields

$$\frac{1}{2} \leq (1+2C_\mu^3) \int_{2B_0 \setminus U_{2^k}} |u| \, d\mu + (1+2C_\mu^3)\, 2^{k+1}\, C_\mu^3\, 2^{\alpha-kp}$$
$$= (1+2C_\mu^3) \int_{2B_0 \setminus U_{2^k}} |u| \, d\mu + (1+2C_\mu^3)\, 2^{1+\alpha}\, C_\mu^3\, 2^{-k(p-1)}.$$

Recalling that $p > 1$, choose k_1 large enough that

$$2^{1+\alpha}\,(1+2C_\mu^3)\, C_\mu^3\, 2^{-k_1(p-1)} \leq \frac{1}{4};$$

for example, choose

$$k_1 := \left\lceil \frac{\log(2^{3+\alpha}\, C_\mu^3\,(1+2C_\mu^3))}{(p-1)\log 2} \right\rceil. \tag{12.2.14}$$

If $k \geq k_* := \max\{k_0, k_1\}$ then

$$\frac{1}{4} \leq (1+2C_\mu^3) \int_{2B_0 \setminus U_{2^k}} |u| \, d\mu,$$

and so, by the renormalization assumption that $u_{2B_0 \setminus U_{2^k}} = 0$,

$$\int_{2B_0 \setminus U_{2^k}} |u - u_{2B_0 \setminus U_{2^k}}| \, d\mu = \int_{2B_0 \setminus U_{2^k}} |u| \, d\mu \geq \frac{1}{4(1+2C_\mu^3)},$$

as desired. □

For future needs we will apply a telescoping argument towards obtaining a Lipschitz estimate. Fix $y \in \tau B_0$, $r > 0$, and $x \in B(y, r)$. We will consider the telescoping family of balls given by $B_0' = B(y, r)$ and, for integers $i > 0$, $B_i' := B(x, 2^{1-i} r)$. Since u is Lipschitz, x is a Lebesgue point of u and so

$$\begin{aligned}
|u(x) - u_{B(y,r)}| &\leq \sum_{i=0}^{\infty} |u_{B_i'} - u_{B_{i+1}'}| \\
&\leq C_\mu^2 \sum_{i=1}^{\infty} \fint_{2B_i'} |u - u_{2B_i'}| \, d\mu + |u_{B_0'} - u_{B_1'}| \\
&\leq C_\mu^2 \sum_{i=1}^{\infty} \fint_{2B_i'} |u - u_{2B_i'}| \, d\mu \\
&\quad + C_\mu^2 \left(\fint_{2B_1'} |u - u_{2B_1'}| \, d\mu + |u_{2B_1'} - u_{B_1'}| \right)
\end{aligned}$$

$$\leq C_\mu^2 \sum_{i=1}^{\infty} \fint_{2B_i'} |u - u_{2B_i'}| \, d\mu$$

$$+ C_\mu^2 (1 + C_\mu) \fint_{2B_1'} |u - u_{2B_1'}| \, d\mu$$

$$\leq C_\mu^2 (2 + C_\mu) \sum_{i=1}^{\infty} \fint_{2B_i'} |u - u_{2B_i'}| \, d\mu$$

$$\leq C_\mu^2 (2 + C_\mu) \, M_{\tau+3r}^{\#} u(x) \left(\sum_{i=1}^{\infty} 2^{2-i} r \right).$$

Hence

$$|u(x) - u_{B(y,r)}| \leq 4 C_\mu^2 (2 + C_\mu) r M_{\tau+3r}^{\#} u(x). \tag{12.2.15}$$

Here we have made use of the fact that $\mathrm{rad}(B_0) = 1$ and that if $y \in \tau B_0$, $x \in B(y, r)$, and $i \geq 1$ then

$$2B_i' = B(x, 2^{2-i} r) \subset (\tau + r + 2^{2-i} r) B_0 \subset (\tau + 3r) B_0.$$

If $x, y \in 2^i B_0$ for some i then we may apply (12.2.15) twice, with $\tau = 2^i$ and $r = 1.01 \, d(x, y) \leq 2^{i+2}$, to obtain

$$\begin{aligned} |u(x) - u(y)| &\leq |u(x) - u_{B(y,r)}| + |u(y) - u_{B(y,r)}| \\ &\leq 4 C_\mu^2 (2 + C_\mu) r (M_{2^i+3r}^{\#} u(x) + M_{2^i+3r}^{\#} u(y)) \\ &\leq 8 C_\mu^2 (2 + C_\mu) d(x, y) (M_{2^{i+3}}^{\#} u(x) + M_{2^{i+3}}^{\#} u(y)) \, . \end{aligned}$$

Consequently, if $\lambda > 0$ and $x, y \in 16 B_0 \setminus U_\lambda$ then we see from the above, with $i = 4$, that

$$|u(x) - u(y)| \leq 16 C_\mu^2 (2 + C_\mu) \lambda d(x, y), \tag{12.2.16}$$

i.e., $u|_{16 B_0 \setminus U_{8^k}}$ is $16 C_\mu^2 (2 + C_\mu) 8^k$-Lipschitz. We will apply the preceding with $\lambda = 8^k$.

We next extend $u|_{16 B_0 \setminus U_{8^k}}$ as a Lipschitz function on the whole of $2B_0$, while preserving good bounds on a suitable modified maximal function $M_\tau^{\#}$ in the complement of U_{8^k}.

Lemma 12.2.17 *There is an extension v of $u|_{16 B_0 \setminus U_{8^k}}$ to $16 B_0$ such that:*

(i) *the extension v is $C_1 8^k$-Lipschitz continuous on $2B_0$;*
(ii) $\mathrm{Lip}\, v \leq C_1 8^k$ *on $8 B_0$;*
(iii) *we have*

$$M_8^{\#} v \leq 16 C_\mu^3 M_{128}^{\#} u \qquad \text{on } 2B_0 \setminus U_{8^k}. \tag{12.2.18}$$

Here the constant C_1 depends solely on the doubling constant C_μ.

Proof The argument is similar to that in the proof of Theorem 4.1.21. We use a Whitney-type decomposition of U_{8^k} as in Proposition 4.1.15. For x in the open set U_{8^k}, we define

$$r(x) := \frac{1}{8}\operatorname{dist}(x, X \setminus U_{8^k})$$

and consider the collection $\mathcal{G} = \{B(x, r(x)) : x \in U_{8^k}\}$ which covers U_{8^k}. An application of the Whitney decomposition proposition 4.1.15 provides a countable subcollection $\mathcal{G}_0 = \{B_i = B(x_i, r_i)\} \subset \mathcal{G}$ with $r_i = r(x_i)$, which continues to cover U_{8^k}, such that $\{B_i/5\}$ is pairwise disjoint and $\{2B_i\}$ has bounded overlap. Let $\{\varphi_i\}$ be a Lipschitz partition of unity as in Section 4.1, that is, $0 \le \varphi_i \le 1$, φ_i is C_0/r_i-Lipschitz continuous, φ_i is supported on $2B_i$, and $\sum_i \varphi_i \equiv 1$ on U_{8^k}. Here $C_0 \ge 1$ depends solely on C_μ and is independent of u, k, and B_0.

We define the extension v as follows:

$$v(x) = \begin{cases} u(x) & \text{if } x \in X \setminus U_{8^k}, \\ \sum_i u_{B_i}\varphi_i(x) & \text{if } x \in U_{8^k}. \end{cases}$$

Proof of (i) and (ii). First, we establish conclusion (i), the Lipschitz continuity of v. The desired estimate is clear for points $x, y \in 16B_0 \setminus U_{8^k}$ since, by (12.2.16), u is $16C_\mu^2(2 + C_\mu)8^k$-Lipschitz on $16B_0 \setminus U_{8^k}$. To establish the Lipschitz continuity in the remaining cases, we appeal to the fact that X is geodesic and reduce the desired claim to the pointwise estimate

$$\operatorname{Lip} v(x) \le C_1\, 8^k \qquad \text{for } x \in 8B_0. \tag{12.2.19}$$

Now we note that any two points x and y in $2B_0$ are joined by a geodesic which is contained in $4B_0$; the desired Lipschitz estimate follows on integrating $\operatorname{Lip} v$ along this geodesic and using (12.2.19). The value of C_1 depends only on the doubling constant C_μ and will be given later in the proof; see (12.2.20).

We turn to the proof of (12.2.19), which then also proves (ii) of the lemma. First, assume that $x \in 8B_0 \setminus U_{8^k}$, that $y \in 8B_0 \cap U_{8^k}$, and that $d(x, y) < \frac{11}{10}\operatorname{dist}(y, X \setminus U_{8^k})$. By the definition of v we have

$$v(y) - v(x) = \sum_i \left(u_{B_i} - u(x)\right)\varphi_i(y).$$

Since $\operatorname{dist}(y, X \setminus U_{8^k}) \le d(x, y) < \frac{11}{10}\operatorname{dist}(y, X \setminus U_{8^k})$, we see that for each i for which $\varphi_i(y) \ne 0$ we have

$$6r_i \le d(x, y) \le 11r_i \qquad \text{and} \qquad 4r_i \le d(x, x_i) \le 13r_i\,.$$

Moreover, $r_i \le \frac{1}{6} d(x, y) \le \frac{1}{6} \operatorname{diam}(8B_0) \le \frac{8}{3}$.

Setting $\tilde{B}_i = B(x_i, 14r_i)$ and using (12.2.15) together with the fact that $x \notin U_{8^k}$, we obtain

$$\begin{aligned}
|u_{B_i} - u(x)| &\le |u(x) - u_{\tilde{B}_i}| + |u_{\tilde{B}_i} - u_{B_i}| \\
&\le 4C_\mu^2(2 + C_\mu)(14r_i)M_{128}^{\#}u(x) + C_\mu^6 \fint_{\tilde{B}_i} |u - u_{\tilde{B}_i}|\, d\mu \\
&\le 16C_\mu^2(2 + C_\mu)d(x, y)M_{128}^{\#}u(x) + C_\mu^6(14r_i)M_{128}^{\#}u(x) \\
&\le \big(16C_\mu^2(2 + C_\mu) + 4C_\mu^6\big)d(x, y)M_{128}^{\#}u(x) \\
&\le \big(16C_\mu^2(2 + C_\mu) + 4C_\mu^6\big)8^k\, d(x, y).
\end{aligned}$$

It follows that if $d(x, y) < \frac{11}{10} \operatorname{dist}(y, X \setminus U_{8^k})$ then

$$\begin{aligned}
|v(y) - v(x)| &\le \sum_i |u_{B_i} - u(x)|\varphi_i(y) \\
&\le \sum_i \big(16C_\mu^2(2 + C_\mu) + 4C_\mu^6\big)\, 8^k\, d(x, y)\varphi_i(y) \\
&= \big(16C_\mu^2(2 + C_\mu) + 4C_\mu^6\big)8^k\, d(x, y).
\end{aligned}$$

For $y \in 8B_0 \cap U_{8^k}$, for which $d(x, y) \ge \frac{11}{10} \operatorname{dist}(y, X \setminus U_{8^k})$, we can find $x' \in 8B_0 \setminus U_{8^k}$ such that $d(x', y) < \frac{11}{10} \operatorname{dist}(y, X \setminus U_{8^k})$, and the above argument gives

$$|v(y) - v(x')| \le \big(16C_\mu^2(2 + C_\mu) + 4C_\mu^6\big)\, 8^k\, d(x', y).$$

Note that $d(x', y) + d(x, x') \le 3d(x, y)$. By (12.2.16) we also know that

$$|v(x) - v(x')| \le \big(16C_\mu^2(2 + C_\mu) + 4C_\mu^6\big)\, 8^k\, d(x, x').$$

Combining the above two inequalities and the previous argument, we obtain

$$|v(x) - v(y)| \le 3\big(16C_\mu^2(2 + C_\mu) + 4C_\mu^6\big)\, 8^k\, d(x, y)$$

whenever $y \in 8B_0 \cap U_{8^k}$. Since (as observed above) the same estimate holds if $x, y \in 8B_0 \setminus U_{8^k}$, we conclude that

$$|v(y) - v(x)| \le 3\big(16C_\mu^2(2 + C_\mu) + 4C_\mu^6\big)\, 8^k\, d(x, y)$$

whenever $x \in 8B_0 \setminus U_{8^k}$. Consequently, we obtain

$$\operatorname{Lip} v(x) \le 3\big(16C_\mu^2(2 + C_\mu) + 4C_\mu^6\big)\, 8^k \qquad \text{for all } x \in 16B_0 \setminus U_{8^k}.$$

This proves (12.2.19) for $x \in 8B_0 \setminus U_{8^k}$.

Now suppose that $x \in 8B_0 \cap U_{8^k}$. Pick j such that $x \in B_j$. For points $y \in B_j$ we have

$$\begin{aligned}|v(x) - v(y)| &= \left|\sum_i \big(u_{B_i} - u_{B_j}\big)(\varphi_i(x) - \varphi_i(y))\right| \\ &\leq \sum_i |u_{B_i} - u_{B_j}|\,|\varphi_i(x) - \varphi_i(y)|.\end{aligned}$$

The only terms which contribute to the sum are those corresponding to the indices i for which $x \in 2B_i$ or $y \in 2B_i$. For such i we have $B_i \subset 6B_j$; see (4.1.19). Note that $9B_j \subset 128B_0$ and that $9B_j \setminus U_{8^k}$ is nonempty. By the doubling property of μ, for $z \in 9B_j \setminus U_{8^k}$ we have

$$\begin{aligned}|u_{B_i} - u_{B_j}| &\leq |u_{B_i} - u_{9B_j}| + |u_{B_j} - u_{9B_j}| \\ &\leq \left(\frac{\mu(9B_j)}{\mu(B_i)} + \frac{\mu(9B_j)}{\mu(B_j)}\right) \fint_{9B_j} |u - u_{9B_j}|\, d\mu \\ &\leq 2C_\mu^5 \fint_{9B_j} |u - u_{9B_j}|\, d\mu \\ &\leq 18\, C_\mu^5 \operatorname{rad}(B_j)\, M_{128}^{\#} u(z) \\ &\leq 18\, C_\mu^5 \operatorname{rad}(B_j)\, 8^k.\end{aligned}$$

Hence

$$\begin{aligned}|v(x) - v(y)| &\leq 18\, C_\mu^5 \operatorname{rad}(B_j)\, 8^k \sum_i |\varphi_i(x) - \varphi_i(y)| \\ &\leq 18\, C_\mu^5 \operatorname{rad}(B_j)\, 8^k \frac{C_0}{\operatorname{rad}(B_i)}\, d(x, y)\, C_2,\end{aligned}$$

where $C_2 \geq 1$ is a bound for the maximum number of balls B_i such that $2B_i \cap B_j \neq \emptyset$. Such a C_2 depends solely on C_μ (see Proposition 4.1.15).

The construction of $\mathcal{G}_0$ ensures that $\operatorname{rad}(B_j) \leq 2\operatorname{rad}(B_i)$ whenever $2B_i \cap B_j$ is nonempty. Hence

$$|v(x) - v(y)| \leq 36 C_0 C_2 C_\mu^5 8^k\, d(x, y).$$

Consequently,

$$\operatorname{Lip} v(x) \leq 36 C_0 C_2 C_\mu^5 8^k \qquad \text{for all } x \in 8B_0 \cap U_{8^k}.$$

This proves (12.2.19) for $x \in 16B_0 \cap U_{8^k}$. Set

$$C_1 := 3 \max\left\{16 C_\mu^2 (2 + C_\mu) + 4C_\mu^6,\ 12 C_0 C_2 C_\mu^5\right\} \tag{12.2.20}$$

and observe that C_1 depends only on C_μ. Combining the above estimates, we have that

$$\operatorname{Lip} v(x) \le C_1\, 8^k \qquad \text{for all } x \in 16B_0.$$

As noted above, the geodesic property of X now ensures that u is $C_1 8^k$-Lipschitz on $2B_0$. This completes the proof of the first part of the lemma.

Proof of (iii) Now we verify the estimate $M_8^\# v \le 16 C_\mu^3 M_{128}^\# u$ on $2B_0 \setminus U_{8^k}$. Fix $x \in 2B_0 \setminus U_{8^k}$ and let $B \subset 8B_0$ be a ball such that $x \in B$ and

$$\frac{1}{2} M_8^\# v(x) \le \frac{1}{\operatorname{rad}(B)} \fint_B |v - v_B|\, d\mu. \tag{12.2.21}$$

If B does not intersect U_{8^k} then $v|_B = u|_B$ and the result is immediate; so, we can assume that $B \cap U_{8^k}$ is nonempty.

If $u_{4B} \neq 0$, we may replace u with $u - u_{4B}$; the extension of $u - u_{4B}$ from the first part of the lemma will be $v - u_{4B}$ and the estimate in (12.2.21) is unchanged. Hence, without loss of generality we may, for the remainder of this proof, assume that $u_{4B} = 0$.

From (12.2.21) we immediately deduce that

$$M_8^\# v(x) \le \frac{4}{\operatorname{rad}(B)} \fint_B |v|\, d\mu.$$

Then

$$\begin{aligned}
\frac{1}{\operatorname{rad}(4B)} \fint_{4B} |u - u_{4B}|\, d\mu &= \frac{1}{4\operatorname{rad}(B)} \fint_{4B} |u|\, d\mu \\
&= \frac{1}{4\mu(4B)\operatorname{rad}(B)} \left(\int_{4B \setminus U_{8^k}} |u|\, d\mu + \int_{4B \cap U_{8^k}} |u|\, d\mu \right) \\
&\ge \frac{1}{4\mu(4B)\operatorname{rad}(B)} \left(\int_{B \setminus U_{8^k}} |v|\, d\mu + \sum_{i:2B_i \cap B \neq \emptyset} \int_{2B_i \cap 4B} |u| \varphi_i\, d\mu \right)
\end{aligned}$$

where we have used the facts that $\sum_i \varphi_i(x) = 1$ when $x \in U_{8^k}$ and that the support of φ_i is contained in $2B_i$. Since $\operatorname{rad}(B_i) = 8^{-1} \operatorname{dist}(x_i, X \setminus U_{8^k})$ (where x_i is the center of B_i) and $B \setminus U_{8^k} \neq \emptyset$, it follows that $2B_i \subset 4B$ whenever $2B_i \cap B$ is nonempty. Hence, on the one hand,

$$\begin{aligned}
&\frac{1}{\operatorname{rad}(4B)} \fint_{4B} |u - u_{4B}|\, d\mu \\
&\qquad \ge \frac{1}{4\mu(4B)\operatorname{rad}(B)} \left(\int_{B \setminus U_{8^k}} |v|\, d\mu + \sum_{i:2B_i \cap B \neq \emptyset} \int_{2B_i} |u| \varphi_i\, d\mu \right).
\end{aligned}$$

On the other hand,

$$\begin{aligned}\int_{B\cap U_{8^k}} |v|\, d\mu &= \int_{B\cap U_{8^k}} \left|\sum_i u_{B_i}\varphi_i\right| d\mu \le \int_{B\cap U_{8^k}} \sum_i |u_{B_i}|\varphi_i\, d\mu \\ &\le \sum_{i:2B_i\cap B\neq\emptyset} \int_{2B_i} |u_{B_i}|\varphi_i\, d\mu \\ &\le \sum_{i:2B_i\cap B\neq\emptyset} \int_{2B_i} |u|_{B_i}\varphi_i\, d\mu \\ &\le \sum_{i:2B_i\cap B\neq\emptyset} \mu(2B_i)\, |u|_{B_i} \\ &\le C_\mu \sum_{i:2B_i\cap B\neq\emptyset} \int_{B_i} |u|\, d\mu \le \frac{C_\mu}{c_0} \sum_{i:2B_i\cap B\neq\emptyset} \int_{2B_i} |u|\varphi_i\, d\mu.\end{aligned}$$

In the last line we used the fact that $\varphi_i \ge c_0$ on B_i, where c_0 depends only on the bounded overlap of the balls B_i and hence only on C_μ. It follows that

$$\begin{aligned}&\frac{1}{\mathrm{rad}(4B)} \fint_{4B} |u - u_{4B}|\, d\mu \\ &\qquad \ge \frac{c_0}{4C_\mu\, \mu(4B)\, \mathrm{rad}(B)} \left(\int_{B\setminus U_{8^k}} |v|\, d\mu + \int_{B\cap U_{8^k}} |v|\, d\mu \right).\end{aligned}$$

Since $B \subset 8B_0$ we deduce from Lemma 12.1.1 that $4B \subset (7\times 8)B_0 \subset 128B_0$ and so

$$\begin{aligned}M^{\#}_{128}u(x) &\ge \frac{1}{\mathrm{rad}(4B)} \fint_{4B} |u - u_{4B}|\, d\mu \\ &\ge \frac{1}{4C_\mu\, \mu(4B)\, \mathrm{rad}(B)} \int_B |v|\, d\mu \ge \frac{1}{16C_\mu^3} M^{\#}_8 v(x).\end{aligned}$$

This completes the proof of Lemma 12.2.17. □

We continue to denote the extension of u from the previous lemma by v. For $s > 0$ we define

$$F_s := \{x \in 4B_0 \,:\, M^{\#}_8 v(x) > s\}.$$

Lemma 12.2.22 *Assume that* (12.2.6) *and* (12.2.7) *hold and that* $k > k_*$, *where* $k_* = \max\{k_0, k_1\}$ *and* k_0 *and* k_1 *are given in* (12.2.11) *and* (12.2.14) *respectively. Then, with* C_1 *as in* (12.2.20),

$$\mu(F_s) \le \frac{2\, C_{HL}(C_1\, C_P)^p 8^\alpha}{s^p} \tag{12.2.23}$$

and

$$\int_{8B_0\setminus U_{8^k}} (\operatorname{Lip} v)^p \, d\mu \le 2C_1^p 8^{-k}. \tag{12.2.24}$$

Recall (as pointed out in the introduction of this chapter) that we are assuming the Poincaré inequality (8.1.1) with $\lambda = 1$, i.e., the integrals on both sides of the Poincaré inequality are taken over the same ball.

Proof By construction, $v = u$ on $16B_0 \setminus U_{8^k}$, and so $\operatorname{Lip} v = \operatorname{Lip} u$ μ-almost everywhere on this set. However, by Lemma 12.2.17(ii) we know that $\operatorname{Lip} v$ is bounded above by $C_1 8^k$ on $8B_0$, and hence, by (12.2.7) and by the normalization $\mu(B_0) = 1$,

$$\begin{aligned}
\int_{8B_0\setminus U_{8^k}} (\operatorname{Lip} v)^p \, d\mu &= \int_{8B_0\setminus U_{8^k}} (\operatorname{Lip} u)^p \, d\mu \\
&\le C_1^p 8^{kp} \mu\left(\{x \in 8B_0 : \operatorname{Lip} u(x) > 8^{-k}\}\right) + 8^{-kp}\mu(8B_0) \\
&\le C_1^p 8^{-k} + C_\mu^2 8^{-kp} \\
&\le 2C_1^p 8^{-k},
\end{aligned}$$

which proves the second estimate. To prove the first estimate we use (for the first time) the fact that X supports a p-Poincaré inequality. For $B \subset 8B_0$, we have

$$\frac{1}{\operatorname{rad}(B)} \fint_B |v - v_B| \, d\mu \le C_P \left(\fint_B (\operatorname{Lip} v)^p \, d\mu \right)^{1/p}.$$

Hence

$$M_8^{\#} v(x) \le C_P M_*(\chi_{8B_0} \operatorname{Lip} v^p)(x)^{1/p},$$

where $M_*(\chi_{8B_0} h)$ denotes the noncentered Hardy–Littlewood maximal function as in (3.5.12) of the zero-extension of h outside $8B_0$. By the weak estimate (3.5.7) on the maximal function (which also holds for noncentered maximal functions), we see that

$$\begin{aligned}
\mu(F_s) &\le \mu\left(\{x \in 4B_0 : (s/C_P)^p \le M_*(\chi_{8B_0} \operatorname{Lip} v)^p(x)\}\right) \\
&\le C_{HL} C_P^p s^{-p} \int_{8B_0} (\operatorname{Lip} v)^p \, d\mu.
\end{aligned}$$

Now recall again that, from Lemma 12.2.17(ii), $\operatorname{Lip} v$ is bounded above by $C_1\, 8^k$ on $8B_0$. We combine Lemma 12.2.17(ii) (proved above) with (12.2.6) to obtain

$$\begin{aligned}\int_{8B_0} (\mathrm{Lip}\, v)^p \, d\mu &= \int_{8B_0 \setminus U_{8^k}} (\mathrm{Lip}\, v)^p \, d\mu + \int_{8B_0 \cap U_{8^k}} (\mathrm{Lip}\, v)^p \, d\mu \\ &\le 2C_1^p 8^{-k} + (C_1 8^k)^p \mu(U_{8^k}) \\ &\le 2C_1^p 8^{-k} + C_1^p 8^\alpha \\ &\le 2C_1^p 8^\alpha .\end{aligned}$$

Combining this with the above, we obtain

$$\mu(F_s) \le \frac{2C_{HL}(C_1 C_P)^p 8^\alpha}{s^p},$$

thus completing the proof of Lemma 12.2.22. □

By repeating the proof of inequality (12.2.16) (applied to the modified function v rather than to u, with $\tau = 1$, $B = 2B_0$, and $r = 2$ and noting that $M_7^\# v \le M_8^\# v$) we see that if $x, y \in 2B_0 \setminus F_s$ for some $s > 0$ then

$$|v(x) - v(y)| \le 16C_\mu^2(2 + C_\mu)sd(x, y) . \tag{12.2.25}$$

In F_s we have control over Lip v by $C_1 8^k$. This is not sufficient. We therefore modify v on F_s via the McShane extension lemma (see Section 4.1) to obtain a $16C_\mu^2(2 + C_\mu)s$-Lipschitz function v_s that agrees with v on $2B_0 \setminus F_s$. We consider the function

$$h := \frac{1}{k} \sum_{j=2k}^{3k-1} v_{2^j} .$$

We now suppose that $k > k_2$, where

$$k_2 := \left\lceil \frac{\log(16C_\mu^3)}{\log 2} \right\rceil . \tag{12.2.26}$$

If $x \in F_{2^j}$ for some $2k \le j \le 3k - 1$ then $M_8^\# v(x) > 2^j$. If, additionally, $x \notin U_{8^k}$ then, by (12.2.18),

$$M_{128}^\# u(x) > \frac{2^j}{16C_\mu^3} \ge \frac{2^{2k}}{16C_\mu^3} > 2^k$$

and so $x \in U_{2^k}$. Since $U_{8^k} \subset U_{2^k}$ we conclude that $2B_0 \setminus U_{2^k} \subset 2B_0 \setminus F_{2^j}$ for each $2k \le j \le 3k - 1$. Thus $v_{2^j}(x) = v(x)$ for all $x \in 2B_0 \setminus U_{2^k}$ and all j as above, and so $h = v = u$ on $2B_0 \setminus U_{2^k}$. From Lemma 12.2.8, we conclude that

$$\int_{2B_0 \setminus U_{2^k}} |h - h_{2B_0 \setminus U_{2^k}}| \, d\mu \ge \frac{1}{4(1 + 2C_\mu^3)}.$$

Lemma 12.2.27 *Suppose that* (12.2.5), (12.2.6), *and* (12.2.7) *hold, and that* k *is a positive integer with* $k > \max\{k_0, k_1, k_2\}$. *Then*

$$\int_{2B_0} (\mathrm{Lip}\, h)^p \, d\mu \geq \frac{1}{(16 C_P C_\mu (1 + 2C_\mu^3))^p}, \tag{12.2.28}$$

and μ*-almost everywhere on* $2B_0$ *we also have*

$$\mathrm{Lip}\, h \leq (\mathrm{Lip}\, u)\, \chi_{2B_0 \setminus U_{8k}} + \frac{16 C_\mu^2 (2 + C_\mu)}{k} \sum_{j=2k}^{3k-1} 2^j \chi_{U_{8k} \cup F_{2^j}}. \tag{12.2.29}$$

Proof From the discussion before the statement of the lemma,

$$\frac{1}{4(1+2C_\mu^3)} \leq \int_{2B_0 \setminus U_{2^k}} |h - h_{2B_0 \setminus U_{2^k}}| \, d\mu \leq 2 \int_{2B_0} |h - h_{2B_0}| \, d\mu.$$

An application of the p-Poincaré inequality and the fact that $\mu(B_0) = 1 = \mathrm{rad}(B_0)$ now gives

$$\begin{aligned}\frac{1}{8(1+2C_\mu^3)} &\leq C_\mu \fint_{2B_0} |h - h_{2B_0}| \, d\mu \leq 2 C_P C_\mu \left(\fint_{2B_0} (\mathrm{Lip}\, h)^p \, d\mu \right)^{1/p} \\ &\leq 2 C_P C_\mu \left(\int_{2B_0} (\mathrm{Lip}\, h)^p \, d\mu \right)^{1/p},\end{aligned}$$

from which the first claim of the lemma follows.

To prove the second claim, notice that, as $h = u$ on $2B_0 \setminus U_{2^k}$, the above inequality holds on $2B_0 \setminus U_{2^k}$. For $2k \leq j \leq 3k-1$, $\mathrm{Lip}\, v_{2^j}(x) = \mathrm{Lip}\, v(x)$ for μ-a.e. $x \in 2B_0 \setminus F_{2^j}$, and v_{2^j} is $16C_\mu^2(2+C_\mu)2^j$-Lipschitz on $2B_0$. Hence

$$\mathrm{Lip}\, v_{2^j} \leq (\mathrm{Lip}\, u)\, \chi_{2B_0 \setminus U_{8k}} + 16 C_\mu^2 (2 + C_\mu) 2^j \chi_{U_{8k} \cup F_{2^j}}.$$

Here we have used the fact that on $2B_0 \setminus U_{8k}$ we have $v = u$. The second claim of the lemma follows from the above. □

Now we are ready to prove (12.2.2). We formulate the required bounds on k in the following proposition. In addition to the previous lower bounds k_0, k_1, k_2, a further lower bound will be needed. We set

$$k_3 := \left\lceil \left[8^{1+\alpha} C_{HL} \left(1024\, C_1 C_P^2 C_\mu^3 (2 + C_\mu)(1 + 2C_\mu^3) \right]^p \right)^{1/(p-1)} \right\rceil. \tag{12.2.30}$$

Proposition 12.2.31 *Assume that* $k > \max\{k_0, k_1, k_2, k_3\}$, *where* k_0, k_1, k_2, k_3 *are given by* (12.2.11), (12.2.14), (12.2.26), *and* (12.2.30), *and also that* k *satisfies*

$$k^p < 8^k. \tag{12.2.32}$$

Then at least one of the estimates (12.2.5), (12.2.6), *and* (12.2.7) *must fail; consequently* (12.2.2) *is necessarily true.*

Proof Note that if $0 < s_1 \le s_2$ then $F_{s_2} \subset F_{s_1}$. Hence for $2k \le j \le 3k-1$, we see that $F_{2^{j+1}} \subset F_{2^j}$. By Lemma 12.2.22,

$$\mu(F_{2^j}) \le \frac{2\, C_{HL}(C_1 C_P)^p 8^\alpha}{2^{jp}}.$$

Suppose that (12.2.5), (12.2.6), and (12.2.7) hold. Observe that by choice, $k > \max\{k_0, k_1, k_2\}$. For $2k \le j \le 3k-1$, we know from the first line of this proof that $F_{2^j} = F_{2^{3k-1}} \cup (F_{2^{3k-2}} \setminus F_{2^{3k-1}}) \cup \cdots \cup (F_{2^j} \setminus F_{2^{j+1}})$. Therefore

$$\begin{aligned}
\int_{2B_0} & \left(\frac{1}{k}\sum_{j=2k}^{3k-1} 2^j\, \chi_{U_{8^k}\cup F_{2^j}}\right)^p d\mu \\
&= \frac{1}{k^p}\left(\sum_{i=2k}^{3k-1} 2^i\right)^p \mu(2B_0 \cap (U_{8^k} \cup F_{2^{3k-1}})) \\
&\quad + \sum_{j=2k}^{3k-2} \frac{1}{k^p}\left(\sum_{i=2k}^{j} 2^i\right)^p \mu\left(2B_0 \cap (U_{8^k} \cup (F_{2^j} \setminus F_{2^{j+1}}))\right).
\end{aligned}$$

Summing the geometric series gives

$$\begin{aligned}
\int_{2B_0} \left(\frac{1}{k}\sum_{j=2k}^{3k-1} 2^j\, \chi_{U_{8^k}\cup F_{2^j}}\right)^p d\mu &\le \frac{1}{k^p} 2^{3kp}\left(\mu(U_{8^k}) + \mu(F_{2^{3k-1}})\right) \\
&\quad + \sum_{j=2k}^{3k-2} \frac{1}{k^p} 2^{(j+1)p}\left(\mu(U_{8^k}) + \mu(F_{2^j})\right) \\
&= \sum_{j=2k}^{3k-1} \frac{1}{k^p} 2^{(j+1)p}\left(\mu(U_{8^k}) + \mu(F_{2^j})\right).
\end{aligned}$$

We now use (12.2.6) and (12.2.23) to deduce that

$$\begin{aligned}
\int_{2B_0} \left(\frac{1}{k}\sum_{j=2k}^{3k-1} 2^j\, \chi_{U_{8^k}\cup F_{2^j}}\right)^p d\mu &\le \frac{2^p}{k^p}\sum_{j=2k}^{3k-1} 2^{jp}\left(8^{\alpha-kp} + \frac{2\,C_{HL}(C_1 C_P)^p 8^\alpha}{2^{jp}}\right) \\
\le \frac{2^p 8^\alpha}{(2^p-1)k^p} + \frac{2^{p+1} C_{HL}(C_1 C_P)^p 8^\alpha}{k^{p-1}} &\le 2^{p+2} C_{HL}(C_1 C_P)^p 8^\alpha k^{1-p}.
\end{aligned}$$

Recall that $v = u$ on $2B_0 \setminus U_{8^k}$. By (12.2.24), (12.2.29), and (12.2.32),

$$\begin{aligned}
\int_{2B_0} (\mathrm{Lip}\, h)^p \, d\mu &= \int_{2B_0 \setminus U_{8^k}} (\mathrm{Lip}\, h)^p \, d\mu + \int_{2B_0 \cap U_{8^k}} (\mathrm{Lip}\, h)^p \, d\mu \\
&\le 2^p \int_{2B_0 \setminus U_{8^k}} (\mathrm{Lip}\, u)^p \, d\mu \\
&\quad + 2^p (16 C_\mu^2 (2 + C_\mu))^p \int_{2B_0} \left(\frac{1}{k} \sum_{j=2k}^{3k-1} 2^j \, \chi_{U_{8^k} \cup F_{2^j}} \right)^p d\mu \\
&\le 2^{p+1} C_1^p 8^{-k} + (32 C_\mu^2 (2 + C_\mu))^p \, 2^{p+2} C_{HL} (C_1 C_P)^p 8^\alpha k^{1-p} \\
&\le 8 C_{HL} (64\, C_1 C_P C_\mu^2 (2 + C_\mu))^p 8^\alpha k^{1-p} .
\end{aligned}$$

Combining this estimate with (12.2.28) gives

$$k^{p-1} \le 8^{1+\alpha} C_{HL} \left(1024\, C_1 C_P^2 C_\mu^3 (2 + C_\mu)(1 + 2 C_\mu^3) \right)^p .$$

Since $k > k_3$ we obtain a contradiction. We conclude that at least one of the estimates (12.2.5), (12.2.6), (12.2.7) must fail. This completes the proof of the proposition. □

12.3 Self-improvement of the Poincaré inequality

In this section we prove the main result of this chapter, Theorem 12.3.9: the self-improving character of p-Poincaré inequalities.

We again fix a ball $B_1 \subset X$, and for $t \ge 1$ and $x \in B_1$ we set

$$M_\tau^* u(x) := \sup_{\substack{x \in B \\ \tau B \subset B_1}} \frac{1}{\mathrm{rad}(B)} \fint_B |u - u_B| \, d\mu .$$

In contrast with the definition of $M_\tau^\# u$, where the supremum was over all balls containing x that are subsets of the τ-fold enlargement of the fixed ball B_0, here we take the supremum over all balls containing x whose τ-fold enlargements are subsets of the fixed ball B_1. Note that if $\tau_1 \ge \tau_2$ then $M_{\tau_1}^* u(x) \le M_{\tau_2}^* u(x)$, whereas $M_{\tau_1}^\# u(x) \ge M_{\tau_2}^\# u(x)$.

For $\lambda > 0$ let

$$U_\lambda^* := \{ x \in B_1 \; : \; M_{256}^* u(x) > \lambda \}$$

and

$$U_\lambda^{**} := \{ x \in B_1 \; : \; M_2^* u(x) > \lambda \} .$$

From the above discussion, $U_\lambda^* \subset U_\lambda^{**}$.

Lemma 12.3.1 *For* $\lambda > 0$ *we have* $\mu(U_\lambda^{**}) \le C_\mu^{13}\,\mu(U_{\lambda/C_A}^*)$*, where*

$$C_A := 11\,C_\mu^5\,256^{\log(C_\mu)}. \tag{12.3.2}$$

Proof For $x \in U_\lambda^{**}$ there is a ball B_x such that $x \in B_x$ and $2B_x \subset B_1$, with

$$\frac{1}{\mathrm{rad}(B_x)} \fint_{B_x} |u - u_{B_x}|\, d\mu > \lambda.$$

The family $\{B_x : x \in U_\lambda^{**}\}$ covers U_λ^{**}; by the $5B$-covering lemma, there is a countable subcollection $\{B_i\}_i$, with the balls $2B_i$ pairwise disjoint, such that $\{10B_i\}_i$ covers U_λ^{**}. It follows that

$$\mu(U_\lambda^{**}) \le \sum_i \mu(10B_i) \le C_\mu^4 \sum_i \mu(B_i).$$

Fixing i, let $B_i = B(x_i, r_i)$, and let $\mathcal{F}_i$ be the collection of all balls with their center in B_i and with radius $r_i/512$. If there is a ball $B_i' \in \mathcal{F}_i$ such that

$$\fint_{2B_i'} |u - u_{2B_i'}|\, d\mu > \frac{\lambda}{256\,C_A}\, r_i,$$

where C_A is as in (12.3.2), then (since $256B_i' \subset B_1$) it follows that $2B_i' \subset 2B_i \cap U_{\lambda/C_A}^*$ and hence

$$\mu(B_i) \le C_\mu^9\,\mu(2B_i') \le C_\mu^9\,\mu(2B_i \cap U_{\lambda/C_A}^*). \tag{12.3.3}$$

The desired estimate $\mu(U_\lambda^{**}) \le C_\mu^{13}\mu(U_{\lambda/C_A}^*)$ will follow if the above inequality holds for each i, since the balls $2B_i$ are pairwise disjoint.

Towards this end, let $C_A > 0$ be as in (12.3.2) and suppose that there is some index i such that

$$\fint_{2\widehat{B}} |u - u_{2\widehat{B}}|\, d\mu \le \frac{\lambda}{512\,C_A}\, r_i \tag{12.3.4}$$

for each ball $\widehat{B} = B(y, r_i/512)$ with $y \in B(x_i, r_i)$. We may assume for the remainder of this proof that $u_{B(x_i, r_i/256)} = 0$. Fix a point $y \in B(x_i, r_i) \setminus B(x_i, r_i/512)$ and let $B_i' = B(y, r_i/512)$. Let γ be a geodesic connecting x_i to y, and note that the length of γ is $d(x_i, y) < r_i$. Because $d(x_i, y) \ge r_i/512$, we can find points $z_1, \dots, z_{n-1}$ on γ such that $r_i/700 \le d(z_j, z_{j+1}) \le r_i/600$ for $j = 0, \dots, n-1$, where $z_0 = x_i$ and $z_n = y$. Let w_j be the mid point between z_j and z_{j+1} on γ. Set $B_j = B(z_j, r_i/512)$. Then $B_j^+ := B(w_j, r_i/1200) \subset B_j \cap B_{j+1}$ and $B_j \cup B_{j+1} \subset B(w_j, r_i/256)$, and it follows that $\max\{\mu(B_j), \mu(B_{j+1})\} \le C_\mu^3\, \mu(B_j^+)$. Now, we have

$$\begin{aligned}|u_{2B_j} - u_{2B_{j+1}}| &\le |u_{2B_j} - u_{B_j^+}| + |u_{B_j^+} - u_{2B_{j+1}}| \\ &\le \fint_{B_j^+} |u - u_{2B_j}|\, d\mu + \fint_{B_j^+} |u - u_{2B_{j+1}}|\, d\mu \\ &\le C_\mu^4 \left(\fint_{2B_j} |u - u_{2B_j}|\, d\mu + \fint_{2B_{j+1}} |u - u_{2B_{j+1}}|\, d\mu \right) \\ &\le 2C_\mu^4 \frac{\lambda}{512\, C_A}\, r_i.\end{aligned}$$

Since $u_{B(x_i, r_i/256)} = 0$, it follows that

$$|u_{2B_i'}| \le \sum_{j=0}^{n-1} |u_{2B_j} - u_{2B_{j+1}}| \le 2nC_\mu^4 \frac{\lambda}{512\, C_A}\, r_i.$$

Because $n \le r_i/(r_i/700) = 700$, it follows that

$$|u_{2B_i'}| \le \frac{175}{64}\, C_\mu^4 \frac{\lambda}{C_A}\, r_i.$$

Combining this with the assumption that (12.3.4) holds, we see that

$$\fint_{2B_i'} |u|\, d\mu \le \frac{\lambda}{512\, C_A} r_i + \frac{175}{64} C_\mu^4 \frac{\lambda}{C_A} r_i \le \frac{175}{32}\, C_\mu^4\, \frac{\lambda}{C_A}\, r_i$$

when $B_i' = B(y, r_i/512)$ and $y \in B(x_i, r_i) \setminus B(x_i, r_i/512)$. Since

$$u_{B(x_i, r_i/256)} = 0,$$

the above inequality holds also for $y \in B(x_i, r_i/512)$ by (12.3.4).

By the doubling property of μ, we can cover $B(x_i, r_i)$ by balls $\widehat{B}_k$, $k = 1, \dots, m$, with $m \le C_\mu^4\, 256^{\log(C_\mu)}$, centered at points in $B(x_i, r_i)$ and having radii $r_i/256$ (see (8.1.14) and (12.1.3)). We obtain

$$\begin{aligned}\frac{\lambda r_i}{2} \le \frac{1}{2} \fint_{B(x_i, r_i)} |u - u_{B(x_i, r_i)}|\, d\mu &\le \fint_{B(x_i, r_i)} |u|\, d\mu \\ &\le \sum_k \frac{\mu(\widehat{B}_k)}{\mu(B(x_i, r_i))} \fint_{\widehat{B}_k} |u|\, d\mu \\ &\le \sum_k C_\mu \frac{175}{32}\, C_\mu^4\, \frac{\lambda}{C_A}\, r_i \\ &\le \frac{175}{32} \times 256^{\log(C_\mu)}\, C_\mu^5\, \frac{\lambda}{C_A}\, r_i,\end{aligned}$$

that is,

$$C_A \le \frac{175}{16}\, C_\mu^5\, 256^{\log(C_\mu)}.$$

This is a contradiction of the choice of C_A from (12.3.2). Hence we have that, for each i, some ball of radius $r_i/256$ with center in $B_i = B(x_i, r_i)$ violates (12.3.4). This completes the proof. □

We now prove a global analog of (12.2.2).

Lemma 12.3.5 *There is a positive integer k_4 which depends solely on C_μ and C_P such that if $k > k_4$ and $k^p < 8^k$ then, for all $\lambda > 0$,*

$$\mu(U_\lambda^*) \le 2^{kp-3}\mu(U_{2^k\lambda}^*) + 8^{kp-3}\mu(U_{8^k\lambda}^*) + 8^{(k+1)(p+1)}\, C_\mu^{24}\, C_A^{3(p+1)}\, \mu\left(\{x \in B_1 : \operatorname{Lip} u(x) > 8^{-k}\lambda/C_A^3\}\right). \tag{12.3.6}$$

Here C_A is given by (12.3.2).

Proof Let $\mathcal{G}$ denote the collection of all balls B for which $256B \subset B_1$ and

$$\frac{1}{\operatorname{rad}(B)} \fint_B |u - u_B|\, d\mu > \lambda.$$

Thus $\mathcal{G}$ covers U_λ^* and so, by the $5B$-covering lemma, we can extract a countable subcollection $\mathcal{G}_0 = \{B_i\}_i$ such that $\{256B_i\}_i$ is a pairwise disjoint collection and $\{1280B_i\}_i$ covers U_λ^*. We have $\mu(U_\lambda^*) \le C_\mu^{11} \sum_i \mu(B_i)$.

We set

$$k_4 := \max\{k_0, k_1, k_2, k_3\} - \left\lceil \frac{\log(C_A)}{\log 2} \right\rceil, \tag{12.3.7}$$

with k_0, k_1, k_2, k_3 given by (12.2.11), (12.2.14), (12.2.26) and (12.2.30). Assume that $k > k_4$ and $k^p < 8^k$, and let $\hat{k} = k + \lceil \log(C_A)/\log 2 \rceil$. Then we are in a position to invoke Proposition 12.2.31 with $B_0 := B_i$ for each i. Here we should keep in mind that the sets U_λ referred to in that proposition depend on B_i as well; $U_\lambda = U_\lambda(B_i) = \{x \in 128B_i : M_{128}^{\#}u(x) > \lambda\}$. By the choice of B_i we know that $256B_i \subset B_1$, and so $U_\lambda = U_\lambda \cap 128B_i \subset U_\lambda^{**}$. Hence, by (12.2.2),

$$\mu(B_i) \le 2^{\hat{k}p-\alpha}\mu\left(U_{2^{\hat{k}}\lambda}^{**} \cap 128B_i\right) + 8^{\hat{k}p-\alpha}\mu\left(U_{8^{\hat{k}}\lambda}^{**} \cap 128B_i\right) + 8^{\hat{k}(p+1)}\mu\left(\{x \in 2B_i : \operatorname{Lip} u(x) > 8^{-\hat{k}}\lambda\}\right).$$

Summing over the indices i and noting that the family $\{128B_i\}_i$ is pairwise disjoint, we obtain

$$C_\mu^{-11}\mu(U_\lambda^*) \le 2^{\hat{k}p-\alpha}\mu\left(U^{**}_{2^{\hat{k}}\lambda}\right) + 8^{\hat{k}p-\alpha}\mu\left(U^{**}_{8^{\hat{k}}\lambda}\right) + 8^{\hat{k}(p+1)}\mu\left(\{x \in B_1 : \operatorname{Lip} u(x) > 8^{-\hat{k}}\lambda\}\right).$$

By Lemma 12.3.1,

$$C_\mu^{-24}\mu(U_\lambda^*) \le 2^{\hat{k}p-\alpha}\mu\left(U^{*}_{2^{\hat{k}}\lambda/C_A}\right) + 8^{\hat{k}p-\alpha}\mu\left(U^{*}_{8^{\hat{k}}\lambda/C_A}\right) + 8^{\hat{k}(p+1)}\mu\left(\{x \in B_1 : \operatorname{Lip} u(x) > 8^{-\hat{k}}\lambda\}\right).$$

Since $C_A > 1$, we see that

$$C_\mu^{-24}\mu(U_\lambda^*) \le 2^{\hat{k}p-\alpha}\mu\left(U^{*}_{2^{\hat{k}}\lambda/C_A}\right) + 8^{\hat{k}p-\alpha}\mu\left(U^{*}_{8^{\hat{k}}\lambda/C_A^3}\right) + 8^{\hat{k}(p+1)}\mu\left(\{x \in B_1 : \operatorname{Lip} u(x) > 8^{-\hat{k}}\lambda\}\right).$$

Recalling the relation of k to $\hat{k}$, we see that

$$C_\mu^{-24}\mu(U_\lambda^*) \le 2^{(k+1)p-\alpha}\, C_A^p \mu(U^*_{2^k\lambda}) + 8^{(k+1)p-\alpha}\, C_A^{3p}\, \mu(U^*_{8^k\lambda}) + 8^{(k+1)(p+1)}\, C_A^{3(p+1)}\, \mu\left(\{x \in B_1 : \operatorname{Lip} u(x) > 8^{-k}\lambda/C_A^3\}\right).$$

Observe that C_A and C_μ are independent of the choices of α and k. Choose $\alpha > 3$ such that $2^{p-\alpha}C_A^p C_\mu^{24} < 2^{-3}$. For example, choose

$$\alpha = p + 4 + \frac{\log(C_A^p C_\mu^{24})}{\log 2}. \tag{12.3.8}$$

Then we also have $8^{p-\alpha}\, C_A^{3p}\, C_\mu^{22} < 8^{-3}$, and we obtain the desired result. □

Now we are ready to prove the main result of this chapter, the Keith–Zhong self-improvement theorem for Poincaré inequalities on complete doubling metric measure spaces.

Theorem 12.3.9 *Suppose that X is complete, μ is doubling, and X supports a p-Poincaré inequality for some $1 < p < \infty$. Then there exists $q \ge 1$ with $1 \le q < p$ such that X supports a q-Poincaré inequality. The improvement $p - q$ in the exponent depends solely on the data C_μ, C_P, and p.*

Proof Recall the simplifying assumption, adopted in this chapter, that X is geodesic. Fix a Lipschitz function u on X and a ball $B \subset X$.

In the preceding computations, we took $B_1 = 256B$. As in the statement of Lemma 12.3.5, we now assume that k is an integer satisfying $k > k_4$ and $k^p < 8^k$, where k_4 is defined as in (12.3.7). We choose $\epsilon > 0$ such that

$$8^{k\epsilon} < 2,$$

that is, $0 < \epsilon < 1/(3k)$. We now show that $q = p - \epsilon$ satisfies the conclusion of the theorem.

Integrating (12.3.6) against $\lambda^{p-\epsilon}$, we obtain

$$\begin{aligned}\int_0^\infty &\mu(U_\lambda^*)\, d(\lambda^{p-\epsilon}) \\ &\le 2^{kp-3} \int_0^\infty \mu(U_{2^k\lambda}^*)\, d(\lambda^{p-\epsilon}) + 8^{kp-3} \int_0^\infty \mu(U_{8^k\lambda}^*)\, d(\lambda^{p-\epsilon}) \\ &\quad + 8^{(k+1)(p+1)}\, C_\mu^{22}\, C_A^{3(p+1)} \int_0^\infty \mu\big(\{x \in B_1 : \operatorname{Lip} u(x) \\ &\qquad\qquad > 8^{-k}\lambda / C_A^3\}\big)\, d(\lambda^{p-\epsilon}).\end{aligned}$$

Changing variables in each integral on the right-hand side, we obtain

$$\begin{aligned}\int_0^\infty &\mu(U_\lambda^*)\, d(\lambda^{p-\epsilon}) \\ &\le 2^{k\epsilon-3} \int_0^\infty \mu(U_\lambda^*)\, d(\lambda^{p-\epsilon}) + 8^{k\epsilon-3} \int_0^\infty \mu(U_\lambda^*)\, d(\lambda^{p-\epsilon}) \\ &\quad + 8^{k(2p+1-\epsilon)+(p+1)}\, C_\mu^{22}\, C_A^{3(2p+1-\epsilon)} \int_0^\infty \mu\big(\{x \in B_1 : \operatorname{Lip} u(x) \\ &\qquad\qquad > \lambda\}\big)\, d(\lambda^{p-\epsilon}).\end{aligned}$$

By the choice of ϵ, we deduce that $2^{k\epsilon-3} < 1/4$ and $8^{k\epsilon-3} < 1/4$, and so

$$\begin{aligned}\frac{1}{2} \int_0^\infty &\mu(U_\lambda^*)\, d(\lambda^{p-\epsilon}) \\ &\le 8^{k(2p+1-\epsilon)+(p+1)}\, C_\mu^{22}\, C_A^{3(2p+1-\epsilon)} \int_0^\infty \mu\big(\{x \in B_1 : \operatorname{Lip} u(x) \\ &\qquad\qquad > \lambda\}\big)\, d(\lambda^{p-\epsilon}),\end{aligned}$$

whence we obtain from Cavalieri's principle (3.5.5) that

$$\begin{aligned}\int_B (M_{256}^* u)^{p-\epsilon}\, d\mu &\le \int_{B_1} (M_{256}^* u)^{p-\epsilon}\, d\mu \\ &\le 2 \times 8^{k(2p+1-\epsilon)+(p+1)}\, C_\mu^{22}\, C_A^{3(2p+1-\epsilon)} \int_{B_1} (\operatorname{Lip} u)^{p-\epsilon}\, d\mu.\end{aligned}$$

Observe that if $x \in B_1/256 = B$ then

$$M_{256}^* u(x) \geq \frac{1}{\mathrm{rad}(B)} \fint_B |u - u_B| \, d\mu;$$

thus we obtain

$$\left(\frac{1}{\mathrm{rad}(B)} \fint_B |u - u_B| \, d\mu\right)^{p-\epsilon} \mu(B) \leq C \int_{B_1} (\mathrm{Lip}\, u)^{p-\epsilon} \, d\mu,$$

that is (recall that we set $B_1 = 256B$ at the beginning of the proof), if B is a ball in X then, by the doubling property of μ, we get

$$\fint_B |u - u_B| \, d\mu \leq C \,\mathrm{rad}(B) \left(\fint_{256B} (\mathrm{Lip}\, u)^{p-\epsilon} \, d\mu\right)^{1/(p-\epsilon)}.$$

We have thus established a $(p - \epsilon)$-Poincaré inequality for Lipschitz functions u and the corresponding pointwise Lipschitz-constant functions $\mathrm{Lip}\, u$. Finally, an appeal to Theorem 8.4.2 completes the proof. □

The self-improving property of the p-Poincaré inequality given above in Theorem 12.3.9 required X to be complete (and hence also proper, since the measure on X is doubling). The requirement of completeness can be relaxed to local completeness.

Proposition 12.3.10 *Suppose that X is locally complete and μ is doubling. If X supports a p-Poincaré inequality for some $1 < p < \infty$ then there is some $q \geq 1$ with $1 \leq q < p$ such that every function in $N^{1,p}(X)$, together with any of its upper gradients, satisfies a q-Poincaré inequality.*

Proof Since X supports a p-Poincaré inequality, Theorem 8.2.1 implies that Lipschitz functions are dense in $N^{1,p}(X)$. As in Lemma 8.2.3 we set $\widehat{X}$ to be the completion of X, equipped with the zero-extension $\widehat{\mu}$ of μ to $\widehat{X} \setminus X$. By Lemma 8.2.3 we see that functions in $N^{1,p}(X)$ extend to functions in $N^{1,p}(\widehat{X})$ and that $\widehat{X}$, together with $\widehat{\mu}$, supports a p-Poincaré inequality. Now, from Theorem 12.3.9 we know that $\widehat{X}$ supports a q-Poincaré inequality for some $1 \leq q < p$. The desired conclusion follows from the facts that X is an open subset of $\widehat{X}$ and that $\widehat{\mu}(\widehat{X} \setminus X) = 0$. □

It is important to be careful in analyzing the self-improving properties of the Poincaré inequality on noncomplete spaces. Proposition 12.3.10 does *not* imply that X supports a q-Poincaré inequality.

Remark 12.3.11 Fix $n \geq 2$ and $1 < p \leq n$. Then there exists a locally compact Ahlfors n-regular metric measure space which supports a p-Poincaré inequality but does not support a q-Poincaré inequality for any $1 \le q < p$. The space X can be chosen to be a subset of $\mathbb{R}^n$, equipped with the Euclidean metric and the Lebesgue measure. More specifically, we choose a sufficiently large Cantor-type set $E \subset \mathbb{R}^{n-1}$ such that $\mathbb{R}^n \setminus (E \times \{0\})$ is the example space; with a correct choice of E, every function in the $N^{1,p}$-class of this metric space extends to $N^{1,p}(\mathbb{R}^n)$ but not every function in the $N^{1,q}$-class extends to $N^{1,q}(\mathbb{R}^n)$. Since E has measure zero, self-improvement of the Poincaré inequality fails for such functions. See [167] for details.

It is natural to ask whether, in a complete doubling metric measure space, a p-Poincaré inequality always improves to the best possible one, namely a 1-Poincaré inequality. This is not the case. See Section 14.2.

Theorem 12.3.9 has numerous important consequences. Several results in the previous chapters relied on the validity of a better Poincaré inequality; by Theorem 12.3.9 this hypothesis can be relaxed to the p-Poincaré inequality for complete spaces. To fill in the picture, we now record some results of this type.

For the following result, compare Lemma 9.2.6.

Lemma 12.3.12 *Let X be complete and let μ be a doubling measure on X. Assume that X supports a p-Poincaré inequality for some $p > 1$. For $u \in N^{1,p}(X : V)$ and $\lambda > 0$, $\mathrm{Cap}_p(\{Mu > \lambda\}) \le C\lambda^{-p}\|u\|^p_{N^{1,p}(X:V)}$.*

The refined estimates on the size of the set of Lebesgue points of an $N^{1,p}$-function (Theorem 9.2.8) admit a similar improvement.

Theorem 12.3.13 *Let X be complete, let μ be a doubling measure on X, and assume that X supports a p-Poincaré inequality for some $p > 1$. Let u be a function in $N^{1,p}(X : V)$. Then p-a.e. point in X is a Lebesgue point for u.*

Finally, all the standard notions regarding Sobolev spaces coincide in the case $p > 1$, provided that the underlying space is a complete doubling metric measure space supporting a p-Poincaré inequality. Compare the following theorem with Theorems 10.5.2 and 10.5.3.

Theorem 12.3.14 *Let X be complete, let μ be a doubling measure on X, and assume that X supports a p-Poincaré inequality for some $p > 1$. Then*

$$M^{1,p} = P^{1,p} = KS^{1,p} = N^{1,p} = Ch^{1,p}.$$

Moreover, the norms $\|\cdot\|_{M^{1,p}}$, $\|\cdot\|_{N^{1,p}} = \|\cdot\|_{Ch^{1,p}}$ and $\|\cdot\|_{KS^{1,p}}$ are all comparable.

12.4 Notes to Chapter 12

The results on the self-improvement of Poincaré inequalities in this chapter were established by Keith and Zhong [153]. Our exposition follows their original reasoning but we have kept careful track of the constants in order to obtain estimates for the size ϵ of the self-improvement.

Examples of complete Ahlfors-regular metric measure spaces that support a p-Poincaré but not a q-Poincaré inequality for some $1 \leq q < p$ appear in [125]; see also Section 14.2. These examples show that the degree of self-improvement necessarily depends on the given data.

The counterexamples to self-improvement referred to in Remark 12.3.11 are from [167]. These are based on a careful analysis of removable sets of a certain type for Sobolev spaces. The setting is that of an n-dimensional Euclidean space, and only exponents $1 < p \leq n$ are covered. We do not know of any relevant examples in the super-critical case $p > n$.

For a study of Orlicz–Poincaré inequalities and their self-improvement, we refer the reader to [270], [71], and [148].

13
An introduction to Cheeger's differentiation theory

Euclidean Lipschitz functions are differentiable almost everywhere with respect to the Lebesgue measure; this is a fundamental result by Rademacher [226]. The focus of this chapter is to establish a Rademacher-type differentiability result via a linear differential structure for Sobolev spaces on doubling metric measure spaces supporting a p-Poincaré inequality. As a consequence we will show that $N^{1,p}(X)$ is reflexive if $p > 1$.

The results of this chapter are due to Cheeger [53].

In this chapter we will assume that X is complete, that the measure is doubling, and that a p-Poincaré inequality holds true for some $p > 1$. By Lemma 8.2.3, the assumption that X is complete is not overly restrictive. However, we assume completeness so that, by Theorem 8.3.2, X is known to be a quasiconvex space. Now a bilipschitz change in the metric on X produces a complete metric measure space with doubling measure supporting a p-Poincaré inequality and, in addition, X is a geodesic space. It then follows that, with this new metric, we have access to the annular decay property described in Proposition 11.5.3.

13.1 Asymptotic generalized linearity

To construct a linear differential structure we need coordinate functions. A substitute for coordinate functions in the metric setting will be constructed in Section 13.4, using the tool of asymptotic generalized linearity developed in the present section. To do this, we first need the notion of p-harmonicity. In the Euclidean setting, the standard coordinate functions are p-harmonic for each $p > 1$.

A function $f \in N^{1,p}_{\mathrm{loc}}(\Omega)$, for an open set $\Omega \subset X$, is *p-harmonic* if

$$\int_{\mathrm{spt}(u)} \rho_f^p \, d\mu \le \int_{\mathrm{spt}(u)} \rho_{f+u}^p \, d\mu$$

whenever $u \in N^{1,p}(X)$ has compact support in Ω. Here ρ_f refers to the minimal p-weak upper gradient of f guaranteed by Theorem 6.3.20. In the Euclidean setting, such minimizers are precisely those functions that are p-harmonic. Using this concept as a model, we consider the following asymptotic version of p-harmonicity.

Given an open set $U \subset X$, let $N^{1,p}_c(U)$ be the collection of all functions $u \in N^{1,p}(X)$ with compact support contained in U.

Definition 13.1.1 A function $f \in N^{1,p}_{\mathrm{loc}}(X)$ is *asymptotically p-harmonic* at a point $x_0 \in X$ if

$$\lim_{r \to 0} \left(\fint_{B(x_0,r)} \rho_f^p \, d\mu - \inf_{u \in N^{1,p}_c(B(x_0,r))} \fint_{B(x_0,r)} \rho_{f+u}^p \, d\mu \right) = 0.$$

Note that, for each $r > 0$,

$$\fint_{B(x_0,r)} \rho_f^p \, d\mu \ge \inf_{u \in N^{1,p}_c(B(x_0,r))} \fint_{B(x_0,r)} \rho_{f+u}^p \, d\mu.$$

For the remainder of this chapter we will assume that $p > 1$.

Definition 13.1.2 We say that f is *asymptotically generalized linear* at $x_0 \in X$ if f is asymptotically p-harmonic at x_0 and

$$\lim_{r \to 0} \fint_{B(x_0,r)} \rho_f^p \, d\mu = \rho_f(x_0)^p.$$

By the Lebesgue differentiation theorem (see Section 3.4), we know that μ-a.e. point in X is a Lebesgue point of ρ_f^p because $\rho_f \in L^p_{\mathrm{loc}}(X)$.

Theorem 13.1.3 *Suppose that μ is doubling. Then every Lipschitz function on X is asymptotically generalized linear at μ-a.e. point in X.*

Proof Let f be an L-Lipschitz continuous function on X. Since the analysis is local, we may assume without loss of generality that f has compact support in X.

As pointed out above, μ-a.e. point in X is a Lebesgue point of ρ_f^p, where $\rho_f \le \operatorname{Lip} f \in L^p_{\mathrm{loc}}(X)$. Hence it suffices to show that f is asymptotically p-harmonic at μ-a.e. point in X. Suppose that this is not the case. Then there is a set $A \subset X$ with $\mu(A) > 0$ such that at no point of A is f asymptotically p-harmonic. So, for each $x_0 \in A$, we have

$$\limsup_{r \to 0} \left(\fint_{B(x_0,r)} \rho_f^p \, d\mu - \inf_{u \in N_c^{1,p}(B(x_0,r))} \fint_{B(x_0,r)} \rho_{f+u}^p \, d\mu \right) > 0.$$

Hence we can find $\epsilon > 0$ and a set $A_0 \subset A$ with $\mu(A_0) > 0$ such that, for each $x_0 \in A$, there is a sequence of radii, $r_i(x_0) \le 1/i$, satisfying

$$\fint_{B(x_0,r_i(x_0))} \rho_f^p \, d\mu - \inf_{u \in N_c^{1,p}(B(x_0,r_i(x_0)))} \fint_{B(x_0,r_i(x_0))} \rho_{f+u}^p \, d\mu > \epsilon,$$

and so for each i we can find $u_{x_0,i} \in N_c^{1,p}(B(x_0, r_i(x_0)))$ such that

$$\fint_{B(x_0,r_i(x_0))} \rho_f^p \, d\mu \ge \fint_{B(x_0,r_i(x_0))} \rho_{f+u_{x_0,i}}^p \, d\mu + \epsilon. \tag{13.1.4}$$

Since truncation does not increase the p-weak upper gradient (see Proposition 6.3.23), we can assume without loss of generality that

$$\max_{\overline{B}(x_0,r_i(x_0))} (f + u_{x_0,i}) \le \max_{\overline{B}(x_0,r_i(x_0))} f$$

and that

$$\min_{\overline{B}(x_0,r_i(x_0))} (f + u_{x_0,i}) \ge \min_{\overline{B}(x_0,r_i(x_0))} f.$$

This is done by replacing $u_{x_0,i}$ with

$$\max \left\{ \min \left\{ f + u_{x_0,i}, \max_{\overline{B}(x_0,r_i(x_0))} f \right\}, \min_{\overline{B}(x_0,r_i(x_0))} f \right\} - f$$

if necessary. It follows from the L-Lipschitz continuity of f that

$$\|(f + u_{x_0,i}) - f\|_{L^\infty(\overline{B}(x_0,r_i(x_0)))} \le 2Lr_{i(x_0)}. \tag{13.1.5}$$

We now fix $k \in \mathbb{N}$. The balls $B(x_0, r_i(x_0))$, $x_0 \in A_0$ and $i \in \mathbb{N}$ with $i \ge k$, form a fine cover of A_0 (see Section 3.4 on Vitali measures) and, because μ is a doubling measure, we can appeal to the Vitali covering theorem (see Section 4.2) to obtain a pairwise disjoint countable subfamily, $B(x_j, r_j)$, $j \in \mathbb{N}$ and $r_j \le 1/k$, such that

$$\mu\left(A_0 \Big\backslash \bigcup_{j \in \mathbb{N}} B(x_j, r_j) \right) = 0.$$

Corresponding to each $j \in \mathbb{N}$, by the above discussion we have a function $u_j := u_{x_j,k}$ in $N^{1,p}_c(B(x_j, r_j))$ that satisfies (13.1.4). We now define $f_k : X \to \mathbb{R}$ by setting

$$f_k(x) = \begin{cases} f(x) + u_j(x) & \text{if } x \in B(x_j, r_j) \text{ for some } j, \\ f(x) & \text{otherwise.} \end{cases}$$

By inequality (13.1.5) we know that $\|f_k - f\|_{L^\infty(X)} \le 2L/k$, and it follows that $f_k \to f$ in $L^p_{\text{loc}}(X)$. Furthermore, by Lemma 6.3.14 and by the pairwise disjointness property of the subfamily, the function

$$\rho_f \chi_{X \setminus \bigcup_j B(x_j,r_j)} + \sum_j \rho_{f+u_j} \chi_{B(x_j,r_j)}$$

is a p-weak upper gradient of f_k. Note that, by (13.1.4) and by the pairwise disjointness property of the subfamily,

$$\begin{aligned} \int_X \rho_{f_k}^p \, d\mu &= \int_{X \setminus \bigcup_j B(x_j,r_j)} \rho_f^p \, d\mu + \sum_j \int_{B(x_j,r_j)} \rho_{f+u_j}^p \, d\mu \\ &\le \int_{X \setminus \bigcup_j B(x_j,r_j)} \rho_f^p \, d\mu + \sum_j \left(\int_{B(x_j,r_j)} \rho_f^p \, d\mu - \epsilon \mu(B(x_j, r_j)) \right) \\ &\le \int_X \rho_f^p \, d\mu \; - \; \epsilon \, \mu(A_0). \end{aligned}$$

Because f has compact support, A_0 is bounded and so $\mu(A_0) < \infty$. It follows that (f_k) is a bounded sequence in $N^{1,p}(X)$. Since $p > 1$, an application of Theorem 7.3.8 together with Proposition 7.3.7 shows that $\int_X \rho_f^p \, d\mu \le \liminf_k \int_X \rho_{f_k}^p \, d\mu$; recall from (13.1.5) that $f_k \to f$ in $L^p(X)$. It follows that

$$\int_X \rho_f^p \, d\mu \le \liminf_k \int_X \rho_{f_k}^p \, d\mu \le \int_X \rho_f^p \, d\mu \; - \; \epsilon \, \mu(A_0),$$

which is not possible because $\epsilon \mu(A_0) > 0$. Therefore the basic premise that f fails to be asymptotically p-harmonic on a positive measure subset of X is false, and the theorem is proved. □

The following proposition extends the above result. Given functions $f_1, \ldots, f_k$ on X, we define $\vec{f} : X \to \mathbb{R}^k$ by setting $\vec{f}(x) = (f_1(x), \ldots, f_k(x))$.

Proposition 13.1.6 *Suppose that μ is doubling. Let $f_1, \ldots, f_k$ be L-Lipschitz functions. Let Z denote the collection of all $x \in X$ such that, for each $\vec{a} = (a_1, \ldots, a_k)$ in $\mathbb{R}^k$, the function $\vec{a} \cdot \vec{f} = \sum_{i=1}^k a_i f_i$ is asymptotically* *generalized linear at x. Then $\mu(X \setminus Z) = 0$. Furthermore, for all $\vec{a}, \vec{b} \in \mathbb{R}^k$,*

$$|\rho_{\vec{a}\cdot\vec{f}+u} - \rho_{\vec{b}\cdot\vec{f}+u}| \leq L \sum_{i=1}^{k} |a_i - b_i| \tag{13.1.7}$$

whenever $u \in N^{1,p}(X)$.

Proof The inequality (13.1.7) is directly verified using (6.3.18); the inequality holds at Lebesgue points of the two functions.

For each $\vec{a} \in \mathbb{R}^k$, let $Z(\vec{a})$ be the collection of points $x \in X$ at which the Lipschitz function $\vec{a} \cdot \vec{f}$ is asymptotically generalized linear. By Theorem 13.1.3, $\mu(X \setminus Z(\vec{a})) = 0$. It follows that $Z_0 := \bigcap_{\vec{a}\in\mathbb{Q}^k} Z(\vec{a})$ also satisfies $\mu(X \setminus Z_0) = 0$. By replacing $X \setminus Z_0$ with a Borel set of zero measure containing $X \setminus Z_0$, and then replacing Z_0 with the complement of this zero-measure Borel set if necessary, we may assume that Z_0 is Borel. Furthermore, for each $x \in Z_0$, inequality (13.1.7) holds for each $\vec{a}, \vec{b} \in \mathbb{Q}^k$. It suffices to prove that $Z_0 \subset Z$. Towards this end, for $\vec{a} \in \mathbb{R}^k$ choose a sequence of points $\vec{a}^j \in \mathbb{Q}^k$ such that $\lim_j \vec{a}^j = \vec{a}$. Then, by (13.1.7), we know that $\rho_{\vec{a}^j\cdot\vec{f}} \to \rho_{\vec{a}\cdot\vec{f}}$ uniformly in Z_0. Thus every point of Z_0, being a Lebesgue point of each $\rho_{\vec{a}^j\cdot\vec{f}}$, is also a Lebesgue point of $\rho_{\vec{a}\cdot\vec{f}}$.

Let $x_0 \in Z_0$, and suppose that $\vec{a} \cdot \vec{f}$ is not asymptotically p-harmonic at x_0. Then there is a positive real number ϵ, a sequence of radii $r_m \to 0$, and a corresponding choice of functions $u_m \in N_c^{1,p}(B(x_0, r_m))$, such that

$$\fint_{B(x_0,r_m)} \rho^p_{\vec{a}\cdot\vec{f}}\, d\mu \geq \epsilon + \fint_{B(x_0,r_m)} \rho^p_{\vec{a}\cdot\vec{f}+u_m}\, d\mu.$$

We may choose k large enough that

$$\left| \fint_{B(x_0,r_m)} \rho^p_{\vec{a}^k\cdot\vec{f}}\, d\mu - \fint_{B(x_0,r_m)} \rho^p_{\vec{a}\cdot\vec{f}}\, d\mu \right| < \frac{\epsilon}{10}$$

and that

$$\left| \fint_{B(x_0,r_m)} \rho^p_{\vec{a}^k\cdot\vec{f}+u_m}\, d\mu - \fint_{B(x_0,r_m)} \rho^p_{\vec{a}\cdot\vec{f}+u_m}\, d\mu \right| < \frac{\epsilon}{10}.$$

Such a choice of k is independent of m (by the uniform convergence in Z_0 discussed above), and it follows that

$$\fint_{B(x_0,r_m)} \rho^p_{\vec{a}^k\cdot\vec{f}}\, d\mu \geq \frac{4\epsilon}{5} + \fint_{B(x_0,r_m)} \rho^p_{\vec{a}^k\cdot\vec{f}+u_m}\, d\mu.$$

Because this holds for each m, we see that x_0 is not a point of asymptotic p-harmonicity of $\vec{a}^k \cdot \vec{f}$, which would violate the fact that $x_0 \in Z_0$. Thus x_0

must be a point of asymptotic p-harmonicity of $\vec{a} \cdot \vec{f}$ as well, from which the claim of the proposition follows. □

13.2 Caccioppoli-type estimates

A fundamental property of p-harmonic functions is a Caccioppoli-type estimate that allows one to locally control integrals of the gradient of the function in terms of integrals of the function itself. In this section we show that functions that are asymptotically generalized linear in the sense of Definition 13.1.2 satisfy such an inequality at small scales, at points of asymptotic generalized linearity.

Given an open set $U \subset X$ with $X \setminus U$ nonempty, and a positive real number η, we set $U_\eta := \{x \in U : \operatorname{dist}(x, X \setminus U) > \eta\}$.

Lemma 13.2.1 *Let $0 \le \delta < \nu \le 1$, $\eta > 0$, and $f \in N^{1,p}(U)$. Suppose that f satisfies*

$$\int_{U \setminus U_{2\eta}} \rho_f^p \, d\mu \le \delta^p \int_U \rho_f^p \, d\mu \tag{13.2.2}$$

and that, for each $u \in N_c^{1,p}(U)$,

$$\nu^p \int_U \rho_f^p \, d\mu \le \int_U \rho_{f+u}^p \, d\mu. \tag{13.2.3}$$

Then

$$\int_U \rho_f^p \, d\mu \le \frac{1}{\eta^p(\nu - \delta)^p} \int_{U_\eta \setminus U_{2\eta}} |f|^p \, d\mu.$$

Proof We can choose a $1/\eta$-Lipschitz function φ on X such that $\varphi = 1$ on $U_{2\eta}$, $\varphi = 0$ on $X \setminus U_\eta$, and $0 \le \varphi \le 1$ on X. By Proposition 6.3.28,

$$\frac{1}{\eta}|f|\chi_{U_\eta \setminus U_{2\eta}} + (1-\varphi)\rho_f$$

is a p-weak upper gradient of $(1-\varphi)f$. Because $\varphi f \in N_c^{1,p}(U)$, it follows from (13.2.3) and (13.2.2) that

$$\begin{aligned}
\nu \left(\int_U \rho_f^p \, d\mu \right)^{1/p} &\le \left(\int_U \rho_{(1-\varphi)f}^p \, d\mu \right)^{1/p} \\
&\le \left(\int_{U_\eta \setminus U_{2\eta}} \frac{1}{\eta^p} |f|^p \, d\mu \right)^{1/p} + \left(\int_U (1-\varphi)^p \rho_f^p \, d\mu \right)^{1/p}
\end{aligned}$$

$$\le \frac{1}{\eta}\left(\int_{U_\eta\setminus U_{2\eta}} |f|^p\,d\mu\right)^{1/p} + \left(\int_{U\setminus U_{2\eta}} \rho_f^p\,d\mu\right)^{1/p}$$
$$\le \frac{1}{\eta}\left(\int_{U_\eta\setminus U_{2\eta}} |f|^p\,d\mu\right)^{1/p} + \delta\left(\int_{U} \rho_f^p\,d\mu\right)^{1/p}.$$

Thus

$$(\nu-\delta)\left(\int_U \rho_f^p\,d\mu\right)^{1/p} \le \frac{1}{\eta}\left(\int_{U_\eta\setminus U_{2\eta}} |f|^p\,d\mu\right)^{1/p},$$

from which the lemma follows. □

Theorem 13.2.4 *Suppose that μ is doubling and that X is a length space. Then there exist $0<\tau<1$ and $C\ge 1$, both depending only on the doubling constant C_μ of the measure μ, such that, whenever $f\in N^{1,p}_{\mathrm{loc}}(X)$ is asymptotically generalized linear at $x_0\in X$ with $\rho_f(x_0)>0$, there exists a positive real number r_0 such that*

$$\left(\int_{B(x_0,r)} \rho_f^p\,d\mu\right)^{1/p} \le \frac{C}{r}\left(\int_{B(x_0,(1+\tau)r/2)\setminus B(x_0,\tau r)} |f|^p\,d\mu\right)^{1/p}$$

for $0<r<r_0$.

Proof The conclusion of the theorem follows from Lemma 13.2.1 if we can show that f satisfies the hypotheses of the lemma with $U=B(x_0,r)$, for some choice of $0\le\delta<\nu\le 1$ and $\eta=(1-\tau)r/2$ that is independent of f, x_0, and r.

Since x_0 is a Lebesgue point of ρ_f^p and $\rho_f(x_0)>0$, there is a positive real number $r_1\le 1$ such that, whenever $0<r<r_1$,

$$0<\rho_f(x_0)^p \le 2^p \fint_{B(x_0,r)} \rho_f^p\,d\mu. \tag{13.2.5}$$

By the asymptotic p-harmonicity of f at x_0, we have the existence of $r_2>0$ such that, whenever $0<r<\min\{r_2,r_1\}$ and $u\in N^{1,p}_c(B(x_0,r))$,

$$0<\frac{\rho_f(x_0)^p}{2^p} \le \fint_{B(x_0,r)} \rho_f^p\,d\mu \le 2^p\fint_{B(x_0,r)} \rho_{f+u}^p\,d\mu;$$

thus f satisfies (13.2.3) for $U=B(x_0,r)$ with $\nu=1/2$ whenever $0<r<\min\{r_1,r_2\}\le 1$.

Again, because x_0 is a Lebesgue point of ρ_f^p and $\rho_f(x_0)>0$, it follows that for each $\delta\in(0,1/8)$ there is a positive real number $r_3<1$ such that, whenever $0 < \rho < r < r_3$, we have

$$\fint_{B(x_0,r)} \rho_f^p \, d\mu - \fint_{B(x_0,\rho)} \rho_f^p \, d\mu < \delta^p \rho_f(x_0)^p,$$

and hence, setting $\tau := \rho/r$, $B = B(x_0, r)$, and $\tau B = B(x_0, \tau r)$, we see that

$$\frac{1}{\mu(B)} \int_{B \setminus \tau B} \rho_f^p \, d\mu - \left(\frac{1}{\mu(\tau B)} - \frac{1}{\mu(B)} \right) \int_{\tau B} \rho_f^p \, d\mu < \delta^p \rho_f(x_0)^p.$$

By the hypothesis of the theorem, X is a length space and μ is doubling. It follows from Proposition 11.5.3 that

$$\frac{1}{\mu(B)} \int_{B \setminus \tau B} \rho_f^p \, d\mu \le \delta^p \rho_f(x_0)^p + \frac{C_\mu (1-\tau)^\beta}{\mu(B)} \int_B \rho_f^p \, d\mu.$$

Thus, for $0 < r < \min\{r_1, r_2, r_3\}$, we obtain

$$\int_{B \setminus \tau B} \rho_f^p \, d\mu < \left(2^p \delta^p + C_\mu (1-\tau)^\beta \right) \int_B \rho_f^p \, d\mu.$$

Let $\delta = 1/(3^{1/p} \times 8)$ and choose τ close to 1 and such that $C_\mu(1-\tau)^\beta < \delta^p$. We choose $r_0 = \min\{r_1, r_2, r_3\} \le 1$, with r_3 corresponding to the above choice of δ as well as to x_0. Then, by the above,

$$\int_{B \setminus \tau B} \rho_f^p \, d\mu < 3^p \delta^p \int_B \rho_f^p \, d\mu.$$

Thus (13.2.2) is satisfied with $\eta = (1-\tau) r/2$ and with 3δ playing the role of δ there. An application of Lemma 13.2.1 now completes the proof. □

13.3 Minimal weak upper gradients of distance functions are nontrivial

In this section we show that, for each $x_0 \in X$, the minimal p-weak upper gradient of the Lipschitz function $x \mapsto d(x, x_0)$ is positive on a large set.

Given a set $A \subset X$ and a point $x_0 \in X$, we say that x_0 is a *point of density* of A if

$$\liminf_{r \to 0} \frac{\mu(B(x_0, r) \cap A)}{\mu(B(x_0, r))} = 1.$$

Note that necessarily $\mu(A) > 0$ if A has a point of density.

Lemma 13.3.1 *Suppose that μ is doubling and supports a p-Poincaré inequality. Let $x_0 \in X$, and consider the function f given by $f(x) = d(x_0, x)$. If x_0 is a point of density of a set $A \subset X$ then there is a set $A_1 \subset A$ with*

$\mu(A_1) > 0$ such that $\rho_f > 0$ on A_1. Furthermore, we can choose A_1 in such a way that x_0 is also a point of density of A_1.

Proof Suppose that $\rho_f = 0$ almost everywhere in A. For each $0 < \epsilon < 1$, we can find an $r_0 > 0$ such that, whenever $0 < r < r_0$,

$$1 - \epsilon^p \leq \frac{\mu(B(x_0, r) \cap A)}{\mu(B(x_0, r))} \leq 1.$$

We will show that there is some $\epsilon_0 > 0$ such that when $\epsilon < \epsilon_0$ we will have a contradiction with the fact that $\rho_f = 0$ almost everywhere on A.

An application of the p-Poincaré inequality to the balls $B(x_0, r)$ with $0 < r < r_0/\lambda$, together with the fact that $\rho_f \leq 1$, gives on the one hand

$$\begin{aligned} \fint_{B(x_0,r)} |f - f_{B(x_0,r)}| \, d\mu &\leq Cr \left(\fint_{B(x_0,\lambda r)} \rho_f^p \, d\mu \right)^{1/p} \\ &\leq Cr \left(\frac{\mu(B(x_0, \lambda r) \setminus A)}{\mu(B(x_0, \lambda r))} \right)^{1/p} \leq Cr\epsilon. \end{aligned}$$

Hence

$$\fint_{B(x_0,r)} |f - f_{B(x_0,r)}| \, d\mu \leq Cr\epsilon. \tag{13.3.2}$$

On the other hand, if $f_{B(x_0,r)} \geq r/4$ then

$$\begin{aligned} Cr\epsilon &\geq \fint_{B(x_0,r)} |f - f_{B(x_0,r)}| \, d\mu \\ &\geq \frac{1}{\mu(B(x_0, r))} \int_{B(x_0,r/8)} (f_{B(x_0,r)} - d(x, x_0)) \, d\mu(x) \\ &\geq \frac{\mu(B(x_0, r/8))}{\mu(B(x_0, r))} \left(\frac{r}{4} - \frac{r}{8} \right) \geq \frac{r}{8C_\mu^3}, \end{aligned}$$

and so in this case $\epsilon \geq 1/(8CC_\mu^3) > 0$. If $f_{B(x_0,r)} < r/4$ then

$$\begin{aligned} Cr\epsilon &\geq \frac{1}{\mu(B(x_0, r))} \int_{B(x_0,r) \setminus B(x_0,r/2)} (d(x, x_0) - f_{B(x_0,r)}) \, d\mu(x) \\ &\geq \frac{\mu(B(x_0, r) \setminus B(x_0, r/2))}{\mu(B(x_0, r))} \left(\frac{r}{2} - \frac{r}{4} \right) \\ &\geq \frac{r}{C}, \end{aligned}$$

where we have used (8.1.16).

In both cases we have $\epsilon \geq \epsilon_0 > 0$ (where ϵ_0 depends on the doubling constant C_μ). As mentioned above, this leads to a contradiction. Thus it is not possible to have $\rho_f = 0$ almost everywhere in A.

The above argument shows that given *any* measurable $A_2 \subset A$ that contains x_0 as a point of density, ρ_f cannot vanish on A_2. Thus, if we set A_1' to be the collection of all $x \in A$ that are Lebesgue points of ρ_f with $\rho_f(x) > 0$ then x_0 is a point of density of A_1'. Setting $A_1 = A_1' \cup \{x_0\}$ completes the proof of the lemma. □

13.4 The differential structure

Henceforth in this chapter we will assume that μ is doubling and that X is a geodesic space supporting a p-Poincaré inequality.

Let $0 < r < \operatorname{diam}(X)/2$, $0 < \eta < r/2$, and $0 < s < \eta$. Given $x_0 \in X$ and an open set $U \subset B(x_0, r)$, we can find points $x_1, \dots, x_N$ in $B(x_0, r)$, as in Section 12.1, such that the collection $\{B(x_i, s)\}_i$ covers U_η, with $d(x_i, x_j) \geq s/2$ if $i \neq j$ and $B(x_i, s) \subset B(x_0, r)$. Correspondingly, for a function $f \in L^p(B(x_0, r))$, set

$$\phi_{s,r}(f) := \left(\mu(B(x_1, s))^{1/p} ⨍_{B(x_1,s)} f \, d\mu, \dots, \mu(B(x_N, s))^{1/p} ⨍_{B(x_N,s)} f \, d\mu \right). \tag{13.4.1}$$

Then $\phi_{s,r} : L^p(B(x_0, r)) \to \mathbb{R}^N$, where $N \leq N_2 = N_2(C_\mu, r/s)$ is as in (12.1.3). Now equip $\mathbb{R}^N$ with the ℓ_p-norm

$$|(a_1, \dots, a_N)| := \left(\sum_{j=1}^{N} |a_j|^p \right)^{1/p}.$$

Proposition 13.4.2 *Let $f \in N^{1,p}_{\mathrm{loc}}(X)$. Suppose that η, K are positive real numbers and $U \subset B(x_0, r)$ is an open set such that*

$$\eta^p \int_U \rho_f^p \, d\mu \leq K^p \int_{U_\eta} |f|^p \, d\mu. \tag{13.4.3}$$

Let $0 < s < \beta\eta$, where $\beta = \min\{1, [4CKN_1^{1/p}]^{-1}\}$ with C the constant from Remark 9.1.19. Then

$$\int_{U_\eta} |f|^p \, d\mu \leq 2^{p+1} |\phi_{s,r}(f)|^p.$$

Consequently, if $\mathcal{F}$ is a vector space of functions in $N^{1,p}_{\mathrm{loc}}(X)$ that satisfy (13.4.3) then the vector space dimension of $\mathcal{F}$ is at most N_2.

Proof Since the balls $B(x_i, s)$ cover U_η,

$$\begin{aligned}\int_{U_\eta} |f|^p \, d\mu &\le \sum_{i=1}^{N} \int_{B(x_i,s)} |f|^p \, d\mu \\ &\le 2^p \sum_{i=1}^{N} \int_{B(x_i,s)} |f - f_{B(x_i,s)}|^p \, d\mu \\ &\quad + 2^p \sum_{i=1}^{N} |f_{B(x_i,s)}|^p \mu(B(x_i, s)).\end{aligned}$$

By Remark 9.1.19 together with Hölder's inequality, followed by the bounded overlap property from (12.1.4) with $\alpha = 1$, we now have

$$\begin{aligned}\int_{U_\eta} |f|^p \, d\mu &\le 2^p C^p s^p \sum_{i=1}^{N} \int_{B(x_i,s)} \rho_f^p \, d\mu \\ &\quad + 2^p \sum_{i=1}^{N} |f_{B(x_i,s)}|^p \mu(B(x_i, s)) \\ &\le (2Cs)^p N_1 \int_U \rho_f^p + 2^p \sum_{i=1}^{N} |f_{B(x_i,s)}|^p \mu(B(x_i, s)) \\ &\le (2Cs)^p N_1 \int_U \rho_f^p + 2^p |\phi_{s,r}(f)|^p.\end{aligned}$$

An application of (13.4.3) now yields

$$\begin{aligned}\int_{U_\eta} |f|^p \, d\mu &\le \frac{(2KCs)^p}{\eta^p} N_1 \int_{U_\eta} |f|^p \, d\mu + 2^p |\phi_{s,r}(f)|^p \\ &\le 2^{-p} \int_{U_\eta} |f|^p \, d\mu + 2^p |\phi_{s,r}(f)|^p \\ &\le \frac{1}{2} \int_{U_\eta} |f|^p \, d\mu + 2^p |\phi_{s,r}(f)|^p.\end{aligned}$$

Thus we obtain the first part of the claim.

Towards the second part of the claim, note that $\phi_{s,r} : \mathcal{F} \to \mathbb{R}^N$ is a linear map. By the above discussion this map is injective. It follows that the vector space dimension of $\mathcal{F}$ cannot be larger than the dimension of $\mathbb{R}^N$, which is $N \le N_2$. This completes the proof of the proposition. □

Theorem 13.4.4 *There is a countable collection of measurable sets $\{U_\alpha\}_\alpha$ with $\mu(U_\alpha) > 0$ satisfying $\mu(X \setminus \bigcup_\alpha U_\alpha) = 0$, and, for each U_α, there is* *a collection of* 1*-Lipschitz functions* $f_1^\alpha, \ldots, f_{N(\alpha)}^\alpha$ *on* X*, where* $N(\alpha) \le N_2$*, such that*

(i) *each* f_j^α, $j = 1, \ldots, N(\alpha)$*, is asymptotically generalized linear at each point of* U_α*;*
(ii) *for each* $\vec{a} \in \mathbb{R}^{N(\alpha)} \setminus \{\vec{0}\}$ *and each* $x_0 \in U_\alpha$ *we have* $\rho_{\vec{a}\cdot\vec{f^\alpha}}(x_0) > 0$*;*
(iii) *whenever* $u : X \to \mathbb{R}$ *is an L-Lipschitz function, there is a set* $V_\alpha(u) \subset U_\alpha$*, with* $\mu(U_\alpha \setminus V_\alpha(u)) = 0$*, and Borel functions* $b_j^\alpha(u) : V_\alpha(u) \to \mathbb{R}$*,* $j = 1, \ldots, N(\alpha)$*, such that for* $x_0 \in V_\alpha(u)$ *we have* $\rho_{\vec{a}\cdot\vec{f}^\alpha - u}(x_0) = 0$ *if and only if*

$$\vec{a} = \vec{b}^\alpha(u)(x_0) = (b_1^\alpha(u)(x_0), \ldots, b_{N(\alpha)}^\alpha(u)(x_0)).$$

Observe that condition (ii) in the above theorem guarantees that the collection $f_1^\alpha, \ldots, f_{N(\alpha)}^\alpha$ is linearly independent.

Proof By first decomposing X into a countable number of measurable sets, each of positive and finite measure, we note that it suffices to show that, given a measurable set A of positive measure, there is a measurable set $U \subset A$ satisfying the above conditions. Hence we fix such a set A. Then by Lemma 13.3.1 we know that there is at least one Lipschitz function f and a measurable set $A_1 \subset A$ with $\mu(A_1) > 0$ such that $\rho_f > 0$ on A_1. By throwing away a set of measure zero if necessary, by Theorem 13.1.3 we can also assume that f is asymptotically generalized linear at each point in A_1.

Let F be a maximal collection of Lipschitz functions on X such that each $f \in F$ is asymptotically generalized linear at each point $x_0 \in A_1$ with $\rho_f(x_0) > 0$ and such that whenever $\vec{a} = (a_1, \ldots, a_k) \in \mathbb{R}^k \setminus \{\vec{0}\}$ we have, for each choice of distinct $f_1, \ldots, f_k \in F$,

$$\rho_{\sum_{j=1}^k a_j f_j}(x_0) > 0. \tag{13.4.5}$$

We know from the previous paragraph that F is nonempty.

By Proposition 13.1.6 we know that the vector space $\mathcal{F}$ constructed using the functions in F as a basis consists of Lipschitz functions, each of which is asymptotically generalized linear at each point in $A_2 \subset A_1$ for some measurable set A_2, independent of the choice of function, such that $\mu(A_1 \setminus A_2) = 0$. By the above constraint on F, we know that for each nonzero function $h \in \mathcal{F}$ we have $\rho_h > 0$ on $A_2 \subset A_1$. Therefore, by Theorem 13.2.4, we know that each function in $\mathcal{F}$ satisfies the Caccioppoli-type inequality (13.4.3). Now an application of Proposition 13.4.2 yields that $\mathcal{F}$ has vector space dimension at most $N \le N_2$, with N_2 given by (12.1.3) with $s = \tau r/4$. Recall that N_2 depends solely on C_μ and $\tau/4$, which in turn, via Proposition 11.5.3, depends solely on C_μ. Thus the basis F consists of at most N Lipschitz functions $f_1, \dots, f_N$.

It now remains to verify the last condition on A_2 for the above choice of $f_1, \dots, f_N$. To do so, we consider a Lipschitz function $u : X \to \mathbb{R}$, and let $V(u)$ denote the collection of all $x_0 \in A_2$ for which $\vec{a} \cdot \vec{f} - u$ is asymptotically generalized linear at x_0 whenever $\vec{a} \in \mathbb{Q}^N$. Again by Theorem 13.1.3, we know that $\mu(A_2 \setminus V(u)) = 0$ and a repetition of the proof of Proposition 13.1.6 (with the aid of (13.1.7)) demonstrates that $\vec{a} \cdot \vec{f} - u$ is asymptotically generalized linear at x_0 whenever $\vec{a} \in \mathbb{R}^N$.

Suppose that there is a set of positive measure, $W(u)$, with $W(u) \subset V(u)$, such that at each point $x_0 \in W(u)$ we have that $\rho_{\vec{a}\cdot\vec{f}-u}(x_0) > 0$ for each choice of $\vec{a} \in \mathbb{R}^N$. Then $F \cup \{u\}$ would be a larger collection that satisfies the requirements of F given in the second paragraph of this proof, with $W(u)$ replacing A_1. In this case we can add u to the collection F and replace A_1 with $W(u)$.

The above procedure can be repeated at most $N_2 - \#F$ times. After that, we can no longer find a Lipschitz function u that satisfies the supposition given in the previous paragraph. Thus now we have a collection F and a measurable set A_2 of positive measure such that whenever u is a Lipschitz function on X, for μ-a.e. $x_0 \in V(u)$ there is a choice of $\vec{a} \in \mathbb{R}^N$ such that $\rho_{\vec{a}\cdot\vec{f}-u}(x_0) = 0$. Let $\vec{b}(x_0)$ be this choice of $\vec{a}$; by replacing $V(u)$ with this full measure subset, we can assume that such a $\vec{b} = \vec{b}(x_0)$ exists for each $x_0 \in V(u)$.

To complete the proof, suppose that $\vec{a} \in \mathbb{R}^N$ is such that $\rho_{\vec{a}\cdot\vec{f}-u}(x_0) = 0$. By (6.3.18),

$$\rho_{[\vec{a}-\vec{b}]\cdot\vec{f}}(x_0) = \rho_{\vec{a}\cdot\vec{f}-\vec{b}\cdot\vec{f}}(x_0) \le \rho_{\vec{a}\cdot\vec{f}-u}(x_0) + \rho_{\vec{b}\cdot\vec{f}-u}(x_0) = 0,$$

because x_0 is a Lebesgue point for each of the three functions above, whence we have a violation of (13.4.5) unless $\vec{a} = \vec{b}(x_0)$. This completes the proof of the theorem. □

The function $\vec{b}^\alpha(u) : V_\alpha(u) \to \mathbb{R}^{N(\alpha)}$ is called a derivative of u and is denoted $D^\alpha u$. If the sets U_α are pairwise disjoint then we can write

$$D(u) = \sum_\alpha D^\alpha u \chi_{U_\alpha}. \tag{13.4.6}$$

The above proof shows that we can always arrange for the sets U_α to be pairwise disjoint. It is easily verified that

$$D(u+v) = D(u) + D(v) \qquad \text{and} \qquad D(\beta u) = \beta D(u) \tag{13.4.7}$$

whenever u and v are Lipschitz functions on X and $\beta \in \mathbb{R}$. We will see in the next section that the operator D also satisfies a Leibniz rule; see

Remark 13.5.6(i). Note that the above discussion is local, and so the differential structure naturally extends to locally Lipschitz functions on X as well as to functions that are locally Lipschitz continuous on an open subset of X.

13.5 Comparisons between ρ_u and Lip u, Taylor's theorem, and the reflexivity of $N^{1,p}(X)$

A function u on an open subset of $\mathbb{R}^n$ is differentiable at a point x_0 in this open set if and only if a first-order Taylor approximation holds at that point:

$$\lim_{y\to x_0} \frac{|u(y) - u(x_0) - \nabla u(x_0)\cdot(y - x_0)|}{\|y - x_0\|} = 0.$$

In this section we demonstrate that for Lipschitz functions on X a similar result holds. Using this Taylor-approximation result, we will show the equality of ρ_u and $\operatorname{Lip} u$ and prove the reflexivity of $N^{1,p}(X)$. The proof of the Taylor approximation uses a comparison of the minimal p-weak upper gradient ρ_u of a Lipschitz function u on X with its pointwise Lipschitz-constant functions $\operatorname{Lip} u$ and $\operatorname{lip} u$ as defined in (6.2.4) and (6.2.3).

In addition to the doubling property of μ and the p-Poincaré inequality, recall that in this chapter we have the mild assumption that X is a geodesic space (after a bilipschitz change in the metric). The key result of this section is the following theorem relating $\operatorname{lip} u$ and $\operatorname{Lip} u$ to ρ_u.

Theorem 13.5.1 *If u is a Lipschitz function on X then for μ-a.e. $x \in X$ we have*

$$\operatorname{lip} u(x) = \rho_u(x) = \operatorname{Lip} u(x).$$

To prove this theorem, we need some auxiliary results first. We know from Lemma 6.2.6 that if u is Lipschitz then $\rho_u \le \operatorname{lip} u$.

Proposition 13.5.2 *There is a constant $C > 0$ such that if u is a Lipschitz function on X and $x_0 \in X$ is a Lebesgue point of ρ_u^p then $\operatorname{Lip} u(x_0) \le C\rho_u(x_0)$. In particular, $\operatorname{lip} u \le \operatorname{Lip} u \le C\rho_u \le C \operatorname{lip} u$ at μ-a.e. point in X.*

Proof Let $x_0 \in X$ be such that

$$\lim_{r\to 0} \fint_{B(x_0,r)} |\rho_u - \rho_u(x_0)|^p \, d\mu = 0.$$

Note that μ-a.e. point in X is such a point; see for example the Lebesgue differentiation theorem, in Section 3.4. Let $R > 0$ and $x, y \in B(x_0, R/4)$.

Because u is continuous and hence every point of X is a Lebesgue point of u, we obtain once again by a telescoping argument together with an application of the p-Poincaré inequality with $\lambda = 1$ (see Theorem 9.1.15) that

$$|u(x) - u(y)| \leq Cd(x, y) \left(M_p\rho_u(x) + M_p\rho_u(y)\right),$$

where

$$M_p\rho_u(w) := \sup_{0<r<2\operatorname{dist}(w, X\setminus B(x_0, R))} \left(\fint_{B(w,r)} \rho_u^p \, d\mu\right)^{1/p}.$$

By Minkowski's inequality, $M_p u + M_p v \geq M_p(u + v)$ whenever $u, v \in L^p(X)$. Then, with v_1 the constant function $v_1(w) = \rho_u(x_0)$ and v_2 the function $v_2(w) = \rho_u(w) - \rho_u(x_0)$, we see that

$$M_p\rho_u = M_p(v_1 + v_2) \leq \rho_u(x_0) + M_p(\rho_u - \rho_u(x_0)).$$

Consequently,

$$\begin{aligned} |u(x) - u(y)| \leq Cd(x, y) \big(2\rho_u(x_0) &+ M_p(\rho_u - \rho_u(x_0))(x) \\ &+ M_p(\rho_u - \rho_u(x_0))(y)\big). \end{aligned} \tag{13.5.3}$$

By the choice of x_0, we know that $\lim_{r\to 0} \tau_r = 0$ for

$$\tau_r := \fint_{B(x_0,r)} |\rho_u - \rho_u(x_0)|^p \, d\mu.$$

We now use a Calderón–Zygmund type decomposition to control $M_p(\rho_u - \rho_u(x_0))$.

For $K > 0$ let E_K be the set of all points $x \in B(x_0, R)$ for which there is a positive real number $r < \operatorname{dist}(x, X \setminus B(x_0, R))$ such that

$$\fint_{B(x,r)} |\rho_u - \rho_u(x_0)|^p \, d\mu > K\tau_R.$$

We can find a cover of E_K by balls $B(x, r)$, $x \in E_K$ and $0 < r < \operatorname{dist}(x, X \setminus B(x_0, R))$, such that

$$\fint_{B(x,r)} |\rho_u - \rho_u(x_0)|^p \, d\mu > K\tau_R.$$

By applying the $5B$-covering lemma in Section 3.3, we extract a countable pairwise disjoint subcollection $\{B(x_i, r_i)\}_i$ such that $E_K \subset \bigcup_i B(x_i, 5r_i)$.

Using the doubling property of μ, we obtain

$$\begin{aligned}\mu(E_K) \le C \sum_i \mu(B(x_i, r_i)) &\le \frac{C}{K\tau_R} \sum_i \int_{B(x_i,r_i)} |\rho_u - \rho_u(x_0)|^p \, d\mu \\ &\le \frac{C}{K\tau_R} \int_{B(x_0,R)} |\rho_u - \rho_u(x_0)|^p \, d\mu \\ &= \frac{C}{K} \mu(B(x_0, R)).\end{aligned}$$

Fix $0 < \epsilon < 1$. If $x \in B(x_0, R) \setminus E_K$ then $⨍_{B(x,r)} |\rho_u - \rho_u(x_0)|^p \, d\mu \le K\tau_R$ for every $0 < r < \operatorname{dist}(x, X \setminus B(x_0, R))$, that is, $M_p(\rho_u - \rho_u(x_0))(x)^p \le K\tau_R$. If $x \in E_K$ and $0 < r < R$ then, by Lemma 8.1.13 (see the discussion in Section 13.3), we know that $\mu(B(x, \epsilon r))/\mu(B(x_0, R)) \ge C^{-1}(\epsilon r/R)^Q$. Thus, when

$$K = 2C_0 \left(\frac{R}{\epsilon r}\right)^Q,$$

we know that $\mu(B(x, \epsilon r)) > \mu(E_K)$ and so there is a point in $B(x, \epsilon r) \setminus E_K$. So, in either case, whenever $x \in B(x_0, R/2)$ and $0 < r < \operatorname{dist}(x, X \setminus B(x_0, R))$, there is a point in $B(x, \epsilon r) \setminus E_K$.

Let $z \in B(x_0, R/2)$ be such that $d(x_0, z) = R/4$; such a point exists because X is a geodesic space. So, letting $r = R/4$ in the above discussion, we see that when $K = C_1/\epsilon^Q$ we can find an $x \in B(x_0, \epsilon R/4) \setminus E_K$ and $y \in B(z, \epsilon R/4) \setminus E_K$ and obtain from (13.5.3) and the Lipschitz continuity of u that

$$\begin{aligned}|u(x_0) - u(z)| &\le |u(x_0) - u(x)| + |u(x) - u(y)| + |u(y) - u(z)| \\ &\le 2L\epsilon \frac{R}{4} + |u(x) - u(y)| \\ &\le 2L\epsilon \frac{R}{4} + Cd(x, y) \left(2\rho_u(x_0) + 2(K\tau_R)^{1/p}\right) \\ &\le 2L\epsilon\, d(x_0, z) + C(2\epsilon + 1) d(x_0, z) \left(2\rho_u(x_0) + 2(\tau_R/\epsilon^Q)^{1/p}\right),\end{aligned}$$

where we have used the triangle inequality in the last line. Thus, whenever $z \in X$ satisfies $d(x_0, z) = R/4$,

$$\frac{|u(x_0) - u(z)|}{d(x_0, z)} \le 2L\epsilon + C\left(2\rho_u(x_0) + 2(K\tau_R)^{1/p}\right).$$

Because $\lim_{R \to 0} \tau_R = 0$, we have

$$\operatorname{Lip} u(x_0) = \limsup_{R \to 0} \sup_{d(z,x_0)=R/4} \frac{|u(x_0) - u(z)|}{d(x_0, z)} \le 2L\epsilon + 2C\rho_u(x_0).$$

Letting $\epsilon \to 0$ completes the proof. □

An immediate corollary of Proposition 13.5.2 is the following result.

Corollary 13.5.4 *If $u \in N^{1,p}(X)$ and (u_i) is a sequence of Lipschitz functions such that $u_i \to u$ in $N^{1,p}(X)$ then $\|\operatorname{Lip}(u_i - u_j)\|_{L^p(X)} \to 0$ as $i, j \to \infty$. Furthermore, if u is also Lipschitz continuous then* $\operatorname{Lip}(u - u_i) \to 0$ *in $L^p(X)$ and* $\operatorname{Lip} u_i \to \operatorname{Lip} u$ *in $L^p(X)$.*

The following corollary is obtained by combining Proposition 13.5.2 with Theorem 13.4.4 and Proposition 13.1.6.

Corollary 13.5.5 *Given a Lipschitz function u on X, for μ-a.e. $x_0 \in X$, when $x_0 \in U_\alpha$ we have*

$$\lim_{y \to x_0} \frac{\left| u(y) - u(x_0) - \vec{b}^\alpha(u)(x_0) \cdot \left(\vec{f}^\alpha(y) - \vec{f}^\alpha(x_0)\right) \right|}{d(x_0, y)} = 0.$$

Remarks 13.5.6 (a) As a consequence of the above corollary, $\vec{b}^\alpha(u) : U_\alpha \to \mathbb{R}^{N(\alpha)}$ is measurable, with $|\vec{b}^\alpha(u)| \le L$ when u is L-Lipschitz. Furthermore, a Leibniz-type result holds: if u, v are two Lipschitz functions on X then $D(uv) = vD(u) + uD(v)$. Also, if $\phi : \mathbb{R} \to \mathbb{R}$ is a C^1-function that is also Lipschitz, then $D(\phi \circ u) = \phi' \circ u D(u)$. In general, the requirement that ϕ is C^1 cannot be removed, since if all we know is that ϕ is Lipschitz (and hence by the classical Rademacher theorem that ϕ is differentiable only almost everywhere in $\mathbb{R}$), ϕ' may not exist anywhere on $u(U_\alpha)$ and so the above chain rule may not make sense.
(b) Note that, for each α,

$$\rho_{\vec{a}\cdot\vec{f}^\alpha} - \rho_{\vec{a}\cdot\vec{f}^\alpha - u} \le \rho_u \le \rho_{\vec{a}\cdot\vec{f}^\alpha - u} + \rho_{\vec{a}\cdot\vec{f}^\alpha}$$

and so, for $x_0 \in V_\alpha(f)$, by choosing $\vec{a} = \vec{b}^\alpha(u)(x_0)$ we see that

$$\rho_{\vec{b}^\alpha(u)(x_0)\cdot\vec{f}}(x_0) = \rho_u(x_0).$$

The above identity, loosely interpreted, states that the minimal weak upper gradient of the first-order Taylor-approximation function $x \mapsto u(x_0) + \vec{b}^\alpha(u)(x_0) \cdot (\vec{f}(x) - \vec{f}(x_0))$ coincides with the minimal weak upper gradient of u.

Theorem 13.5.7 *Let μ be doubling and X a geodesic space supporting a p-Poincaré inequality. Then $N^{1,p}(X)$ is reflexive provided that $p > 1$.*

Proof The map $\|\cdot\| : \mathbb{R}^{N(\alpha)} \to [0, \infty)$ given by $\|\vec{a}\| := \rho_{\vec{a}\cdot\vec{f}^\alpha}(x_0)$ for any fixed $x_0 \in U_\alpha$ forms a norm, by (6.3.18) and the fact that whenever $\vec{a} \ne \vec{0}$

we have $\rho_{\vec{a}\cdot\vec{f}^\alpha}(x_0) > 0$. Now, an application of Theorem 2.4.25 tells us that there is an inner product $\langle\cdot,\cdot\rangle_{x_0}$ on $\mathbb{R}^{N(\alpha)}$ such that with $|\vec{a}|_{x_0} = \sqrt{\langle\vec{a},\vec{a}\rangle_{x_0}}$ we have $|\vec{a}|_{x_0} \le \|\vec{a}\| \le \sqrt{N(\alpha)}|\vec{a}|_{x_0}$. When u is a Lipschitz function on X, $V_\alpha(u) \ni x_0 \mapsto \|\vec{b}^\alpha(u)(x_0)\|$ is measurable (see Remark 13.5.6(i) above) and, by the proof of Theorem 2.4.25, we can ensure that, for the chosen inner product $\langle\cdot,\cdot\rangle_{x_0}$, for $x_0 \in U_\alpha$ the map $x_0 \mapsto |\vec{b}^\alpha(u)(x_0)|_{x_0}$ is measurable for each Lipschitz function u on X and that by Remark 13.5.6(ii) above

$$|\vec{b}^\alpha(u)(x_0)|_{x_0} \approx \rho_u(x_0)$$

with comparison constant $\sqrt{N(\alpha)}$. Hence we can talk about the integrability of this map, and the integral of the pth power of the map is comparable with $\int_{U_\alpha} \rho_u^p\, d\mu$; that is, $N^{1,p}(X)$ has an equivalent norm that is uniformly convex when $1 < p < \infty$. It follows that $N^{1,p}(X)$ is reflexive as a Banach space. □

The following useful result gives a way of extending the operator D given in (13.4.7) from Lipschitz functions to $N^{1,p}(X)$.

Proposition 13.5.8 *Suppose that μ is doubling and X is a geodesic space supporting a p-Poincaré inequality for some $p > 1$. Then there is a positive integer N and a bounded linear differential operator*

$$D : N^{1,p}(X) \to L^p(X : \mathbb{R}^N)$$

such that $D(uv) = uD(v) + vD(u)$ whenever $u, v \in N^{1,p}(X)$ are Lipschitz functions, $|D(u)(x_0)|_{x_0} \le g_u(x_0) \le \sqrt{N}\,|D(u)(x_0)|_{x_0}$ for almost every $x_0 \in X$, and $D(u)$ coincides, for Lipschitz functions u, with the operator defined in (13.4.7).

Proof We choose $N = \sup_\alpha N(\alpha) \le N_2$. For each α and Lipschitz function u on X, we have that $D(u) \in L^p(X_\alpha, R^{N(\alpha)})$. By embedding $R^{N(\alpha)}$ into $\mathbb{R}^N$, we can assume that $D(u)(x) \in \mathbb{R}^N$ for almost every $x \in X$.

For Lipschitz functions $u \in N^{1,p}(X)$ we have a candidate $D(u)$ from (13.4.6). In light of Remarks 13.5.6 it suffices to show that the operator D extends uniquely from the subclass of Lipschitz functions to $N^{1,p}(X)$.

To this end, let $u \in N^{1,p}(X)$. Then, by Theorem 8.2.1, we have a sequence (u_k) of Lipschitz functions converging in $N^{1,p}(X)$ to u. It follows that $\rho_{u_k-u} \to 0$ as $k \to \infty$ and that $\rho_{u_k-u_l} \to 0$ as $k, l \to \infty$. The argument in the proof of Theorem 13.5.7 then shows that

$$\int_X |D(u_k - u_l)(x)|_x^p\, d\mu \to 0 \qquad \text{as } k, l \to \infty.$$

It follows from the Banach-space property of $L^p(X : \mathbb{R}^N)$ that $(D(u_k))$ has a limit in $L^p(X : \mathbb{R}^N)$; we denote this limit by $\vec{v}$.

If (v_k) is another sequence of Lipschitz functions converging in $N^{1,p}(X)$ to u then the sequence (w_k) given by $w_{2k-1} = u_k$ and $w_{2k} = v_k$ is also a sequence of Lipschitz functions converging in $N^{1,p}(X)$ to u. Thus the limit of $(D(v_k))$ coincides with $\vec{v}$ almost everywhere in X. Therefore the limit function $\vec{v}$ is unique.

Since $\rho_{u_k-u} \to 0$, by an application of (6.3.18) we have that for almost every $x \in X$,

$$\begin{aligned}|\vec{v}(x)|_x &= \lim_k |D(u_k)(x)|_x \le \limsup_k \rho_{u_k}(x)\\ &= \limsup_k (\rho_{u_k}(x) - \rho_{u_k-u}(x)) \le \rho_u(x)\\ &\le \limsup_k (\rho_{u_k}(x) + \rho_{u-u_k}(x)) = \limsup_k \rho_{u_k}(x)\\ &\le \sqrt{N}\,|D(u_k)(x)|_x = \sqrt{N}\,|\vec{v}(x)|_x.\end{aligned}$$

Repeating the argument that led to the uniqueness of $\vec{v}$ shows that whenever (u_k) is a sequence of functions from $N^{1,p}(X)$ (that are not necessarily Lipschitz) converging in $N^{1,p}(X)$ to u then $\vec{v}_{u_k} \to \vec{v}_u$ in $L^p(X : \mathbb{R}^N)$, where $\vec{v}_{u_k}, \vec{v}_u$ denote the limiting vector-valued functions obtained from the above Lipschitz approximation argument for u_k and u respectively. Set $D(u) = \vec{v}$. The above series of inequalities shows that D is a bounded operator on $N^{1,p}(X)$. Linearity and the Leibniz properties follow from Remarks 13.5.6. This completes the proof of the proposition. □

Another auxiliary result we need is the following approximation lemma. We postpone its proof until after the proof of Theorem 13.5.1.

Lemma 13.5.9 *Let $u : X \to \mathbb{R}$ be a Lipschitz function and g be a lower semicontinuous countably valued upper gradient of u such that $g \ge \eta > 0$ for some positive η. Then there is a sequence of Lipschitz functions $u_k : X \to \mathbb{R}$ with continuous upper gradients $\rho_k \in L^p_{\mathrm{loc}}(X)$ such that $u_k \to u$ in $L^p_{\mathrm{loc}}(X)$ and* $\limsup_k \rho_k \le g$ *almost everywhere in X.*

Proof of Theorem 13.5.1 By the proof of the Vitali–Carathéodory theorem in Section 4.2, we can approximate the minimal p-weak upper gradient ρ_u of u from above by a monotone decreasing sequence (h_k) of countably valued lower semicontinuous functions, each of which is bounded away from zero and is an upper gradient of u. If we can show that Lip $u \le h_k$ almost everywhere in X for each k then we have that Lip $u \le \rho_u$ almost everywhere, from which the conclusion of the theorem follows. Thus it suffices to show that Lip $u \le g$ for any countably valued lower semicontinuous upper gradient g of f for which there is a positive real number η with $g \ge \eta$.

By Lemma 13.5.9 we can find a sequence (u_k) of Lipschitz functions, with a corresponding sequence (ρ_k) of continuous upper gradients, such that $u_k \to u$ in $L^p_{\text{loc}}(X)$ and $\limsup_k \rho_k \le g$. Then (u_k) is a bounded sequence in $N^{1,p}_{\text{loc}}(X)$ and so, by the reflexivity property from Theorem 13.5.7, we can extract a convex combination sequence (h_k) of (u_k) that converges in $N^{1,p}_{\text{loc}}(X)$ to u. Convex combinations (g'_k) of (ρ_k) corresponding to the convex combination of (u_k) are upper gradients thereof, and $\limsup_k g'_k \le g$ almost everywhere. Thus we may assume that the sequence from Lemma 13.5.9 in addition satisfies $u_k \to u$ in $N^{1,p}_{\text{loc}}(X)$. By Corollary 13.5.4 we know that $\text{Lip}(u_k - u) \to 0$ in $L^p_{\text{loc}}(X)$ and that $\text{Lip}\, u_k \to \text{Lip}\, u$ in $L^p_{\text{loc}}(X)$. Passing to a subsequence if necessary, we can also assume that these two convergences also take place pointwise almost everywhere in X.

Fix a positive integer k. Then, because X is a geodesic space and g_k is an upper gradient of u_k, whenever $x \in X$ we have, for $y \in X$,

$$\frac{|u_k(x) - u_k(y)|}{d(x, y)} \le \frac{1}{d(x, y)} \int_\gamma g_k \, ds,$$

where γ is any geodesic connecting x to y. Now, because g_k is continuous we obtain that

$$\text{Lip}\, u_k(x) \le g_k(x).$$

Combining this with the discussion in the previous paragraph, we see that, for almost every $x \in X$,

$$\text{Lip}\, u(x) = \lim_k \text{Lip}\, u_k(x) \le \limsup_k g_k(x) \le g(x),$$

from which the desired conclusion stated in the first paragraph of this proof follows. This completes the proof of the theorem. □

Proof of Lemma 13.5.9 Because u is a Lipschitz function, we can assume that g is bounded above. Since g is a countably valued lower semicontinuous function and $g \ge \eta$, there is a countable collection of positive real numbers $\{a_j\}_{j \in I \subset \mathbb{N}}$ and open sets U_j, $j \in I$, such that

$$g = \eta + \sum_{j \in I} a_j \chi_{U_j}.$$

For each $j \in I$ we can exhaust U_j from inside by compact sets; that is, we have a sequence of compact sets $\{K_{j,k}\}_k$ with $K_{j,k} \subset K_{j,k+1} \subset U_j$ and $U_j = \bigcup_{k \in \mathbb{N}} K_{j,k}$. We correspondingly have a sequence of nonnegative Lipschitz functions $\psi_{j,k}$, supported on U_j, with $\psi_{j,k} = 1$ on $K_{j,k}$ and $0 \le \psi_{j,k} \le 1$

on X. Let

$$g_k := \eta + \sum_{j\in I,\, j\le k} a_j \psi_{j,k}.$$

Then it is easily seen that (g_k) monotonically increases to g and so $\limsup_k g_k \le g$.

Next, we build functions u_k, whose upper gradients are g_k, such that $u_k \to u$ in $L^p_{\text{loc}}(X)$. We first fix $x_0 \in X$ and $R > 0$ and let $B = B(x_0, R)$. For $\epsilon > 0$ let G_ϵ be a maximal ϵ-net contained in $2B$; that is, if $x, y \in G_\epsilon$ with $x \ne y$ then $d(x, y) \ge \epsilon$ and $2B \subset \bigcup_{x\in G_\epsilon} B(x, 2\epsilon)$. Therefore, by (12.1.3) the number of elements in G_ϵ is bounded by a constant that depends only on R/ϵ and the doubling constant C_μ. Furthermore, we can ensure that $G_\epsilon \subset G_\nu$ if $\nu \le \epsilon$. For each $z \in G_\epsilon$ we know that the function $u_{z,\epsilon}$ given by

$$u_{z,\epsilon}(x) = \inf_\gamma \int_\gamma g\, ds$$

is Lipschitz continuous with upper gradient g and, for each k, the function $u_{z,\epsilon,k}$ given by

$$u_{z,\epsilon,k}(x) = \inf_\gamma \int_\gamma g_k\, ds$$

is also Lipschitz continuous with upper gradient g_k. Here the infimum is taken over rectifiable curves γ connecting x to the fixed point $z \in G_\epsilon$. Using these functions, in analogy with the McShane extension lemma (see Section 4.1), we construct functions $u_{\epsilon,k}$ and $u_{\epsilon,\infty}$:

$$\begin{aligned} u_{\epsilon,k}(x) &= \inf_{z\in G_\epsilon} \left(u(z) + u_{z,\epsilon,k}(x)\right), \\ u_{\epsilon,\infty}(x) &= \inf_{z\in G_\epsilon} \left(u(z) + u_{z,\epsilon}(x)\right). \end{aligned}$$

Because g is an upper gradient of u, it follows that $u_{\epsilon,\infty} = u$ on G_ϵ. So, for each $x \in 2B$, by the facts that g is an upper gradient of $u_{\epsilon,\infty}$ and G_ϵ is an ϵ-net of $2B$, we have for $L = 1 + \sup_X g$ that $|u_{\epsilon,\infty}(x) - u(x)| \le 2L\epsilon$.

Note that g_k is an upper gradient of $u_{\epsilon,k}$ on X. To see this, note that, whenever $x, y \in X$, for each $\nu > 0$ we can find $z \in G_\epsilon$ and a path $\gamma_{y,\nu}$ connecting z to y such that $u_{\epsilon,k}(y) \ge u(x) + \int_{\gamma_{y,\nu}} g_k\, ds - \nu$. Then, whenever γ is a rectifiable curve connecting x to y, the concatenated path $\gamma + \gamma_{y,\nu}$ connects x to z, and so

$$u_{\epsilon,k}(x) - u_{\epsilon,k}(y) \le u(z) + \int_{\gamma+\gamma_{y,\nu}} g_k\, ds - u(z) - \int_{\gamma_{y,\nu}} g_k\, ds + \nu.$$

Taking the limit as $\nu \to 0$, we obtain $u_{\epsilon,k}(x) - u_{\epsilon,k}(y) \le \int_\gamma g_k \, ds$. Reversing the roles of x and y in the above argument completes the proof that g_k is an upper gradient of $u_{\epsilon,k}$. It remains to control $u_{\epsilon,k}$ in terms of $u_{\epsilon,\infty}$; to do so, we need to look at $u_{z,\epsilon,k}$ for $z \in G_\epsilon$. First, observe that, for each $z \in G_\epsilon$, the functions $u_{z,\epsilon,k}$ monotonically increase with respect to k and converge to $u_{z,\epsilon,\infty}$. This can be seen by an argument similar to that found in the proof of Theorem 8.4.2, in the part following the inequality (8.4.11). Keeping in mind that G_ϵ is a finite set and hence that the infima in the definitions of $u_{\epsilon,k}$, $u_{\epsilon,\infty}$ are actually minima, we obtain also that $u_{\epsilon,k}$ increases monotonically to $u_{\epsilon,\infty}$ as $k \to \infty$ on G_ϵ, after we pass to a subsequence if necessary. However, this is not sufficient for us: we need boundedness in $L^p(B)$ (then we can conclude, by an application of Theorem 7.3.8 together with the uniform convexity of $L^p(B)$).

Note that, by construction, $u_{\epsilon,k}(z) \le u(z)$ for each $z \in G_\epsilon$. For $x \in 2B$ there is a point $z \in G_\epsilon$ such that $d(z, x) \le \epsilon$. Thus

$$\begin{aligned} |u_{\epsilon,k}(x)| &\le |u_{\epsilon,k}(x) - u_{\epsilon,k}(z)| + |u_{\epsilon,k}(z) - u(z)| + |u(z)| \\ &\le L\epsilon + 3|u(z)| \le 4L\epsilon + 3|u(x)|. \end{aligned}$$

This gives the desired L^p-bounds on $u_{\epsilon,k}$ in terms of $\int_{2B} |u|^p \, d\mu$, which now completes the proof upon the taking of convex combinations of the sequence $(u_{2^{-j},k})$ and corresponding functions (g_k). Observe that, as $j \to \infty$, the sets $G_{2^{-j}}$ become dense in B. For each j we know that $u_{2^{-j},k}$ converges pointwise in $G_{2^{-j}}$ as $k \to \infty$. Note that the limit (after taking convex combinations with respect to k, of course) for each j is also L-Lipschitz on B. A further convex combination over j produces a further subsequence, which converges in $\bigcup_j G_{2^{-j}}$ to u, and the limit function is again L-Lipschitz on B. Now the density implies that the limit function must be u. This completes the proof of Lemma 13.5.9. □

13.6 Notes to Chapter 13

The content of the present chapter covers only a small fraction of Cheeger's paper [53]. In the later sections of [53], Cheeger couples his differentiation theorem to the notion of Gromov–Hausdorff tangent cones (see Section 11.7). There, the persistence of the doubling condition and Poincaré inequality under such a convergence is used to prove that the induced tangent functions on such cones, stemming from Lipschitz functions on the original space, necessarily satisfy a property which he terms *generalized linearity*. This property asserts that the minimal weak upper gradient of the tangent function is constant. In the context of a Riemannian manifold, such a property reduces to the classical linearity of the (almost everywhere defined) differential of a Lipschitz function. An exposition on this perspective of the differentiation theorem can also be found in [163].

Keith's thesis, published in [151], [150] and [152], made additional contributions to the Cheeger differentiation machinery. In [152], Keith showed that the coordinate functions in the differentiation theorem 13.4.4 can always be selected to be distance functions to points. The work of Keith [151], Gong [102], [99], [101], [100], Bates and Speight [24], and others has contributed significantly to weakening the hypotheses of Theorem 13.4.4. A related approach to the differentiation of Lipschitz functions on metric measure spaces was proposed by Weaver [282].

The differentiability of real-valued Lipschitz functions on a space X can occasionally be used to rule out the existence of bilipschitz embeddings of X into nice spaces. An early version of this principle appeared in [53], where it was shown that certain doubling spaces with a Poincaré inequality (specifically, nonabelian Carnot groups, the spaces of Bourdon and Pajot, and the spaces of Laakso, see Chapter 14 for a more extensive list of examples of spaces supporting a Poincaré inequality) admit no bilipschitz embedding into any finite-dimensional Euclidean space.

The interplay between differentiability and (non)embeddability has turned out to have a rich and intricate storyline. Substantial subsequent work in this direction has been done by Cheeger, Kleiner, Naor, and others; see for example [54–56], [59], [60], [57], [58], [185]. Such nonembedding results have turned out to be of significant interest in algorithmic computer science. We refer the reader to the article [216] by Naor for an illustrative survey of these matters.

While the notion of minimal weak upper gradients is intrinsic to the metric measure space theory, its use in the associated study of potential theory is occasionally made difficult by the fact that such weak upper gradients are not linear in nature and hence might not be associated with a differential equation. Cheeger differential structure, however, is not completely intrinsic to the metric space and more than one possible differential structure can be obtained from the construction. However, as discussed above, the use of Cheeger differential structures in the study of potential theory can sometimes overcome the difficulties encountered in the use of minimal weak upper gradients.

The topic of identifying metric measure spaces in which the inner product norm induced by a Cheeger differential structure agrees with the minimal weak upper gradient is currently under active development: see for example [170], [173]. For a sample of results related to the potential theory of Cheeger differential structure, see [171], [140], [67], [172], [159], [36], [35], [195]. Readers who wish to learn more may also wish to consult the monograph [31].

14

Examples, applications, and further research directions

In this final chapter we briefly discuss the theory of quasiconformal mappings, the initial motivation for our study of Sobolev spaces in the metric setting. We also describe a variety of examples of spaces supporting a Poincaré inequality. Our aim is to demonstrate the diversity of contexts in which the theory developed in this book finds its application. We also review several geometric criteria, due to Semmes and Keith, sufficient for the validity of the Poincaré inequality. We conclude this chapter by indicating several applications and extensions of the theory of Sobolev spaces on and between metric spaces, especially spaces supporting a Poincaré inequality, as well as current directions of research. The wealth of subjects to which this theory pertains indicates its central role within contemporary analysis and geometry.

14.1 Quasiconformal and quasisymmetric mappings

The principal original motivation in [123–125] for the study of upper gradients and spaces supporting a Poincaré inequality was the theory of quasiconformal and quasisymmetric mappings between metric spaces. Various characterizations of quasiconformality hold in the Euclidean setting, some of which have a natural extension to the metric setting. Some properties such as absolute continuity in measure (Lusin's condition N, as it is known in the literature) require the use of Sobolev classes in the definition of quasiconformality as well as Sobolev embedding theorems. The simplest characterizations of quasiconformality in the Euclidean setting (for instance, the notion of quasisymmetry) require only metric concepts and, indeed, the study of quasi-isometric mappings between Gromov hyperbolic spaces is closely tied to the metric theory of quasisymmetric mappings between their boundaries at infinity. However, a complete understanding of quasiconformal mappings between metric spaces requires a comparison of these various competing notions, including analytic definitions, in the metric setting.

Definition 14.1.1 A homeomorphism $f : X \to Y$, when $X = (X, d_X)$ and $Y = (Y, d_Y)$ are metric spaces, is said to be *η-quasisymmetric* for some homeomorphism $\eta : [0, \infty) \to [0, \infty)$ if

$$\frac{d_Y(f(x), f(y))}{d_Y(f(x), f(z))} \leq \eta\left(\frac{d_X(x, y)}{d_X(x, z)}\right)$$

whenever $x, y, z \in X$ are three distinct points. A homeomorphism $f : X \to Y$ between metric spaces is a *local η-quasisymmetry* if each point in X has a neighborhood in which f is η-quasisymmetric.

Definition 14.1.2 A homeomorphism $f : X \to Y$ between metric spaces is an *H-quasiconformal mapping* (or *metrically H-quasiconformal mapping*) for some $H \geq 1$ if

$$\limsup_{r \to 0^+} \frac{\sup_{y \in \overline{B}(x,r)} d_Y(f(y), f(x))}{\inf_{z \in X \setminus B(x,r)} d_Y(f(x), f(z))} \leq H$$

for each $x \in X$.

Definition 14.1.3 A homeomorphism $f : X \to Y$ between locally Ahlfors Q-regular metric measure spaces $X = (X, d_X, \mu_X)$ and $Y = (Y, d_Y, \mu_Y)$ is an *(analytically) K-quasiconformal mapping* for some $K \geq 1$ if f is in $N^{1,Q}_{\mathrm{loc}}(X : Y)$ and

$$\operatorname{Lip} f(x)^Q \leq K J_f(x)$$

for μ_X-a.e. $x \in X$. Here J_f is the Radon–Nikodym derivative (Jacobian) of the pull-back measure under f and is given by

$$J_f(x) = \limsup_{r \to 0^+} \frac{\mu_Y(f(B(x, r)))}{\mu_X(B(x, r))}.$$

Definition 14.1.4 A homeomorphism $f : X \to Y$ between locally Ahlfors Q-regular metric measure spaces is a *(geometric) C-quasiconformal mapping* for some $C \geq 1$ if

$$C^{-1} \operatorname{Mod}_Q(f \circ \Gamma) \leq \operatorname{Mod}_Q(\Gamma) \leq C \operatorname{Mod}_Q(f \circ \Gamma)$$

whenever Γ is a family of curves in X. Here $f \circ \Gamma$ denotes the family of curves in Y obtained as images of curves from Γ under f.

Metrically quasiconformal mappings were simply called quasiconformal mappings in [125].

In the Euclidean setting (for dimension $n \geq 2$) the notions of local quasisymmetry, metric quasiconformality, analytic quasiconformality, and geometric quasiconformality coincide; see for example Väisälä [273]. The equivalence of the definitions of quasiconformality and quasisymmetry in more general metric settings was a primary impetus for the development of the theory of analysis on metric spaces presented in this monograph. Quasiconformal maps of Carnot groups and other sub-Riemannian manifolds featured prominently in Mostow's celebrated rigidity theorem for lattices in rank one symmetric spaces. See Section 14.2 for more information. For his purposes, Mostow required the equivalence of the metric and analytic notions of quasiconformality. Technical complications in the proof of the absolute continuity of metrically quasiconformal mappings along curves led various authors (including the authors of [123]) to pursue other approaches to the foundations of quasiconformal mapping theory in metric spaces.

In [123] a direct proof, avoiding analytic machinery, was provided for the equivalence of metric quasiconformality and (local) quasisymmetry for mappings between Carnot groups. The technology developed for this proof inspired the authors of [123] to introduce the class of metric measure spaces supporting a Poincaré inequality. In the subsequent paper [125], the methods from [123] were extended to show that metrically quasiconformal mappings, from an Ahlfors Q-regular metric measure space (X, d_X, μ_X) supporting a Q-Poincaré inequality, onto a linearly locally connected Ahlfors Q-regular metric measure space (Y, d_Y, μ_Y), $Q > 1$, are necessarily locally quasisymmetric. Simultaneously, it was shown in [271] that quasisymmetric maps between (locally compact) Ahlfors Q-regular metric measure spaces, $Q > 1$, are necessarily geometrically quasiconformal. It should be noted that, on Ahlfors Q-regular metric measure spaces, $Q > 1$, the validity of the Q-Poincaré inequality is quantitatively equivalent to the so-called *Loewner condition*, which asserts the existence of a positive decreasing function $\varphi : (0, \infty) \to (0, \infty)$ such that

$$\operatorname{Mod}_Q \Gamma(E, F) \geq \varphi(t)$$

for all disjoint nondegenerate continua E and F that satisfy the condition $\operatorname{dist}(E, F) \leq t \min\{\operatorname{diam} E, \operatorname{diam} F\}$, where $\Gamma(E, F)$ denotes the family of all curves connecting E to F in X. The quasi-invariance of the Q-modulus in Q-regular spaces (Definition 14.1.4) suggests that such a reformulation of the Poincaré inequality in terms of quantitative control on the Q-modulus of curve families may be productive. For a finer analysis of the relationship between Poincaré inequalities and lower bounds on the p-moduli of curve families, see Section 14.2.

Under the a priori stronger assumption that the source space X supports a p-Poincaré inequality for some $p < Q$, in [125] additional properties of quasisymmetric mappings were established, namely, absolute continuity along curves and absolute continuity in measure (Lusin's condition N). Moreover, a version of the celebrated higher integrability theorem of Gehring was also obtained. We remind the reader that if X is assumed to be complete then the results of Keith and Zhong [153] (see Chapter 12) indicate that any metric measure space X satisfying the assumptions of the previous paragraph must indeed support such a stronger Poincaré inequality.

The paper [129] employed the emerging theory of Banach-space-valued Sobolev spaces to show that if X is locally compact and Ahlfors Q-regular and supports a Q-Poincaré inequality (but not necessarily a stronger Poincaré inequality) then quasisymmetric mappings $f : X \to Y$ onto metric spaces Y with locally finite Hausdorff Q-measure lie in the local Sobolev class $N^{1,Q}_{\mathrm{loc}}(X : Y)$. Combined with earlier results, this implies that, for maps between locally compact Ahlfors Q-regular metric measure spaces supporting a Q-Poincaré inequality, the notions of local quasisymmetry, metric quasiconformality, analytic quasiconformality, and geometric quasiconformality coincide and such maps satisfy the Lusin condition N.

Taking advantage of the theory already developed in this book, we briefly sketch the proof of the Sobolev regularity of quasisymmetric maps, as in the first sentence of the previous paragraph. We embed the target Y isometrically into a Banach space V (see, e.g., the Kuratowski embedding theorem or the Fréchet embedding theorem in Section 4.1). In the same way as in Section 9.2 and Theorem 10.3.4, we consider locally Lipschitz discrete convolution approximations $f_r : X \to V$ given by $f_r(x) := \sum_i \varphi_{r,i}(x) f(x_i)$ associated with a cover $\{B(x_i, r)\}$ of X with $\sum_i \chi_{B(x_i,6\lambda r)} \leq C$, with a corresponding Lipschitz partition of unity $(\varphi_{r,i})$. Analogously to the proof of Theorems 10.3.4 or 10.4.3, one first establishes that f_r is locally Lipschitz and hence lies in $N^{1,Q}_{\mathrm{loc}}$ for small r and that $f_r \to f$ uniformly on bounded sets. (The proofs of these results are slightly different from those in previous chapters, as here one extensively uses the quasisymmetry of the original map f.) Next, one introduces the Borel function

$$\rho_r(x) := \sum_i \frac{\operatorname{diam} f(B(x_i, r))}{r} \chi_{B(x_i,r)}(x) \,.$$

It follows, again from quasisymmetry, that $\operatorname{Lip} f_r \leq C\rho_r$ and hence $C\rho_r$ is an upper gradient of f_r. Finally, one shows that the functions ρ_r lie in L^Q_{loc} uniformly; the membership of f in $N^{1,Q}_{\mathrm{loc}}$ then follows from Theorem 7.3.9.

The Sobolev regularity of f being understood, the analytic formulation of quasiconformality follows easily from the metric assumption and the Lebesgue–Radon–Nikodym theorem.

In [168, Theorem 2.3], Koskela and MacManus proved that spaces supporting a stronger Poincaré inequality (e.g., a p-Poincaré inequality for some $1 \leq p < Q$) are invariant under quasisymmetric maps. More precisely, they showed that if X is Ahlfors Q-regular, $Q > 1$, and supports a p-Poincaré inequality for some $p < Q$, and if $f : X \to Y$ is a quasisymmetric mapping onto another Ahlfors Q-regular space Y, then Y also supports a q-Poincaré inequality for some $q < Q$ (q may depend on all the given data, including the quasisymmetry function η, on Q, and on p). The borderline case $p = Q$ (in which case $q = Q$ also) follows from the main result of [271], noting again the standard caveat concerning the results of Keith–Zhong and their implication for the nonexistence of complete Ahlfors Q-regular spaces supporting a Q-Poincaré inequality but no stronger p-Poincaré inequality for any $p < Q$. The paper [168] also establishes, under suitable hypotheses, that quasisymmetric mappings between Ahlfors Q-regular spaces preserve the Hajłasz–Sobolev space $M^{1,Q}(X : Y)$.

A recent paper by Williams [284] showed that analytic quasiconformality of index Q (which need not be associated with an Ahlfors regularity exponent for the underlying measures) is equivalent to geometric quasiconformality with the same index Q. Thus it seems that a consideration of curves and upper gradients is a correct approach to the study of quasiconformal mappings between metric measure spaces. It also follows from the results of [284] that if the metric measure spaces are "uniformly locally" Ahlfors Q-regular then quasiconformal mappings preserve the Dirichlet class $D^{1,Q}$, that is, if $f : X \to Y$ is quasiconformal and f^{-1} is also quasiconformal then the composition morphism $f_\# : D^{1,Q}(Y) \to D^{1,Q}(X)$, $f_\#(u) = u \circ f$, is bounded. Furthermore, if X contains a compact set K such that the family of curves which start in K and escape every compact set has positive Q-modulus then Y has the same property. Stated differently, either both X and Y are Q-hyperbolic or both are Q-parabolic. In particular, there is no quasiconformal mapping between the Euclidean space $\mathbb{R}^n$ and the hyperbolic n-space, nor does there exist a quasiconformal mapping between $\mathbb{R}^n$ and an open ball in $\mathbb{R}^n$.

The preceding discussion indicates that the preservation of Poincaré inequalities is associated with quasisymmetry, whereas the preservation of the Dirichlet class $D^{1,Q}$ is more closely associated with (analytic and geometric) quasiconformality. In the absence of Q-Poincaré inequality, it is not true that the inverse of an analytically quasiconformal mapping is analytically quasiconformal, as the example in [284, Remark 4.2] shows.

For further information on quasiconformal maps we recommend [120] and the references given above.

14.2 Spaces supporting a Poincaré inequality

Euclidean space and Riemannian manifolds

For each $n \geq 1$, the Euclidean space $\mathbb{R}^n$ (with Lebesgue measure) supports the 1-Poincaré inequality. This result is classical and admits a variety of proofs. We have already given one such proof, relying on a polar coordinate integration and boundedness results for operators defined in terms of Riesz potentials, in Remark 7.1.4. See also the further discussion surrounding (8.1.2). It is also possible to obtain the relevant estimates for Riesz potentials by starting from the fundamental solution of the Laplacian.

The underlying ideas in the preceding approach naturally translate to an abstract metric context. We discuss this observation in more detail below.

An alternate version of the Sobolev–Poincaré inequality (9.1.16), on the Euclidean space $\mathbb{R}^n$ for $1 \leq p < n$, asserts the existence of a constant $C(n, p)$ such that the inequality

$$\left(\int_{\mathbb{R}^n} |u|^{p^*}\right)^{1/p^*} \leq C(n,p) \left(\int_{\mathbb{R}^n} |\nabla u|^p\right)^{1/p} \tag{14.2.1}$$

holds for all smooth compactly supported functions u. The inequality (14.2.1) is sometimes known as the *Gagliardo–Nirenberg–Sobolev inequality*. When $p = 1$, (14.2.1) is closely related to the *isoperimetric inequality* via the study of level sets [37], [81], [202], [198]. The truncation method that allows one to pass from weak-type estimates to strong-type estimates is due to the work of Maz'ya [198]; see the proof of Lemma 8.1.31.

The Poincaré inequality transfers easily from Euclidean space to compact Riemannian manifolds if one works in charts. The situation for noncompact manifolds is significantly more complex; the validity or non-validity of such inequalities is dependent on the large-scale geometry of the manifolds and, in particular, on the behavior of the various curvatures. For complete noncompact manifolds with nonnegative Ricci curvature, the doubling property for the volume measure follows from the volume comparison inequality of Bishop and Gromov [49, Theorem 10.6.6], while the (2-)Poincaré inequality was first established by Buser [51]. A detailed discussion of Sobolev–Poincaré inequalities on noncompact Riemannian manifolds can be found in the book by Hebey [118]. The analytic and stochastic properties of Poincaré inequalities were treated in [240].

Weighted Euclidean spaces

Yet another large class of doubling metric measure spaces supporting a p-Poincaré inequality is the collection of Euclidean spaces equipped with the standard Euclidean metric but with a weighted measure; cf. the discussion at the end of Section 6.4 of this book. The weights ω for which the weighted Euclidean space is doubling and satisfies a p-Poincaré inequality are called p-admissible weights. Such weighted Euclidean spaces were considered extensively in the monograph [128]. (Note that the definition of p-admissibility given in the first edition of [128] contained extraneous conditions that were shown in the second edition to be redundant. In particular, the doubling condition and the validity of the weighted p-Poincaré inequality are the only requirements for a weight ω to be p-admissible). It was shown by Fabes, Kenig, and Serapioni [82] (see also Chapter 15 of [128]) that the Jacobian J_f of a quasiconformal self-map f of Euclidean space has the property that the weight $\omega = J_f^{1-p/n}$ is p-admissible for each $1 < p < n$.

Another large class of p-admissible weights are those to be found in the Muckenhoupt $\mathcal{A}_p$-class.

Definition 14.2.2 A weight ω is said to be in the $\mathcal{A}_p$-class if

$$\sup_B \left(\fint_B \omega \, dm_n \right) \left(\fint_B \omega^{-1/(p-1)} \, dm_n \right)^{p-1} < \infty,$$

where the above supremum is over all balls $B \subset \mathbb{R}^n$.

For $1 < p < \infty$, the Muckenhoupt $\mathcal{A}_p$-weights were shown by Muckenhoupt in [213, Theorem 9] to be the only weights on $\mathbb{R}^n$ under which the classical Hardy–Littlewood maximal function operators are bounded on $L^p(\mathbb{R}^n, \omega \, dm_n)$. By a result of Fabes, Kenig, and Serapioni [82], $\mathcal{A}_p$-weights are p-admissible. The class $\mathcal{A}_\infty := \bigcup_{1<p<\infty} \mathcal{A}_p$ therefore consists of weights in $\mathbb{R}^n$ that generate doubling spaces supporting a p-Poincaré inequality for some $1 < p < \infty$. In studying the question which weights are comparable with the Jacobians of quasiconformal mappings, David and Semmes [68] introduced so-called strong $\mathcal{A}_\infty$-weights, that is, weights ω on $\mathbb{R}^n$ for which the associated measure $\mu_\omega := \omega \, dm_n$ is doubling and for which there is a metric d_ω and a constant C such that

$$C^{-1} d(x, y) \le \mu_\omega(B_{x,y})^{1/n} = \left(\int_{B_{x,y}} \omega \, dm_n \right)^{1/n} \le C d(x, y)$$

whenever $x, y \in \mathbb{R}^n$ are distinct. Here $B_{x,y}$ is the ball centered at $(x+y)/2$ with radius $|x-y|/2$.

It was shown in [68] that Euclidean spaces equipped with a strong $\mathcal{A}_\infty$-weight support a weighted 1-Poincaré inequality, where the norm of the weak derivative $|\nabla u|$ of a Sobolev function u is replaced with $\omega^{-1/n}|\nabla u|$. Note that, while this term acts as the "metric d_ω" version of the weak derivative of u, this is a far cry from the Poincaré inequality considered in this book for we need the inequality for all upper gradients, some of which could in principle be smaller than $\omega^{-1/n}|\nabla u|$. However, it was shown in [161, Proposition 6.10] that the modified space $(\mathbb{R}^n, d_\omega, \mu_\omega)$ does indeed support a 1-Poincaré inequality as considered in this book. (It is easy to see that the identity map from $\mathbb{R}^n$ to $(\mathbb{R}^n, d_\omega)$ is quasisymmetric and that the metric measure space $(\mathbb{R}^n, d_\omega, \mu_\omega)$ is Ahlfors n-regular; hence, the validity of a q-Poincaré inequality on $(\mathbb{R}^n, d_\omega, \mu_\omega)$ for some $1 \le q < n$ follows from the result of Koskela and MacManus discussed in Section 14.1. The fact that one can take $q = 1$, however, requires a finer analysis and reflects the geometric properties of the source Euclidean space.)

Thus strong $\mathcal{A}_\infty$-weights, together with their associated metric and measure, provide a further class of examples of doubling metric measure spaces supporting a Poincaré inequality.

Topological manifolds and geometric decomposition spaces

The Poincaré inequality holds for some non-Riemannian metrics on topological manifolds as well. In [244], Semmes provided a large class of topological manifolds equipped with (potentially nonsmooth) metrics and measures supporting Poincaré inequalities. In order to state his result precisely, we recall that a metric space (X, d) is said to be *linearly locally contractible* if there exists a constant $C \ge 1$ such that every ball $B(x, r)$ can be contracted to a point inside the concentric ball $B(x, Cr)$.

Theorem 14.2.3 (Semmes) *Let (X, d) be a metric space that is a topological n-manifold which is Ahlfors n-regular when equipped with the Hausdorff n-measure $\mathcal{H}^n$. Assume also that X is linearly locally contractible. Then $(X, d, \mathcal{H}^n)$ supports the 1-Poincaré inequality.*

The requirement that X be a topological n-manifold can be relaxed. It suffices to assume that suitable relative (singular) homology groups of X agree with those of $\mathbb{R}^n$, i.e., that X is a *homology n-manifold.*

Theorem 14.2.3 extends the context of this book well beyond the Riemannian setting. It follows from a theorem of Sullivan that every topological n-manifold ($n \neq 4$) can be metrized as a Lipschitz manifold (i.e., the coordinate chart maps are locally bilipschitz). Such manifolds, if compact, verify the hypotheses of Theorem 14.2.3 and consequently support the Poincaré inequality. The situation is rather complicated and subtle for 4-manifolds; we refer to [122, §11.4] for a more detailed discussion of that case.

We next describe an interesting application of the preceding result to *metrized decomposition spaces*.

Let $\mathcal{F} = \{F\}$ be a collection of closed subsets partitioning a topological space X. The *decomposition space* $X/\mathcal{F}$ is the collection of equivalence classes $[x]$, where two points x and y in X are equivalent if and only if there exists $F \in \mathcal{F}$ such that $\{x, y\} \subset F$. Equip $X/\mathcal{F}$ with the quotient topology. It is a classical problem of geometric topology to understand the topology of such decomposition spaces. For instance, if X *is* a topological manifold, one may ask for conditions on $\mathcal{F}$ which guarantee that $X/\mathcal{F}$ is also a manifold. Even when $X/\mathcal{F}$ is not a manifold, it is sometimes the case that the product space $(X/\mathcal{F}) \times \mathbb{R}^m$ is a manifold for some integer $m \geq 1$; in the latter case we say that $X/\mathcal{F}$ is a *manifold factor* of $X \times \mathbb{R}^m$. Classical examples were considered by Whitehead and Bing; in these examples $X = \mathbb{R}^3$ and the collection $\mathcal{F}$ consists of a single topologically wild compact set (typically, a wild knot or link or a Cantor-type set), together with all the remaining points of X as singleton equivalence classes.

Semmes [245], [246] gave answers to some open questions on the (non-)existence of bilipschitz parametrizations of metric spaces by Euclidean spaces in at least three dimensions by constructing explicit self-similar metrics on suitable decomposition spaces $\mathbb{R}^3/\mathcal{F}$, or on the product spaces $(\mathbb{R}^3/\mathcal{F}) \times \mathbb{R}^m$, in such a way that the resulting space is linearly locally contractible, Ahlfors n-regular and a topological n-manifold (where $n = 3 + m$). According to Theorem 14.2.3, these spaces support Poincaré inequalities. For additional examples, see [127] and [222].

Poincaré inequalities and well-distributed curve pencils

We describe a very flexible and intuitive approach to the derivation of Poincaré inequalities which has been strongly advocated by Stephen Semmes. Roughly speaking, the Poincaré inequality is a consequence of the existence of well-behaved curve families ("pencils of curves") joining arbitrary pairs of points of the space. The relevant geometric axiom (14.2.5) is closely related to the representation formulas used in Chapters 7 and 8 for the proof of the Poincaré inequality in the Euclidean setting.

Definition 14.2.4 A metric measure space (X, d, μ) supports *well-distributed curve pencils* if there exists a constant $C > 0$ and, for each pair of points

$x, y \in X$, a family $\Gamma = \Gamma_{x,y}$ of rectifiable curves in X equipped with a probability measure $d\sigma = d\sigma_{x,y}$ such that each $\gamma \in \Gamma_{x,y}$ is a C-quasiconvex curve joining x to y and, for each Borel set $A \subset X$, the map $\gamma \mapsto$ length $(\gamma \cap A)$ is σ-measurable and satisfies

$$\int_\Gamma \text{length}(\gamma \cap A)\, d\sigma(\gamma) \leq C \int_{CB_{x,y}\cap A} \left(\frac{d(x,z)}{\mu(B(x,d(x,z)))} + \frac{d(y,z)}{\mu(B(y,d(y,z)))} \right) d\mu(z). \tag{14.2.5}$$

Here, for $\tau > 0$, $\tau B_{x,y} := B(x, \tau d(x,y)) \cup B(y, \tau d(x,y))$.

By standard methods (i.e. approximating with simple functions) one easily sees that (14.2.5) is equivalent to the inequality

$$\int_\Gamma \int_\gamma g\, ds\, d\sigma(\gamma) \leq C \int_{CB_{x,y}} g(z) \left(\frac{d(x,z)}{\mu(B(x,d(x,z)))} + \frac{d(y,z)}{\mu(B(y,d(y,z)))} \right) d\mu(z), \tag{14.2.6}$$

for Borel functions $g : CB_{x,y} \to \mathbb{R}$.

It is relatively easy to show that doubling metric measure spaces supporting well-distributed curve pencils admit a 1-Poincaré inequality. Indeed, for a fixed pair of points x, y in a ball $B(x_0, r)$, one can integrate the defining inequality for the upper gradient condition over the curve family $\Gamma_{x,y}$ with respect to $\sigma_{x,y}$. Applying (14.2.6) and Fubini's theorem and using the doubling property of the measure, the weak 1-Poincaré inequality follows without much effort.

Well-distributed curve pencils are easy to construct in Euclidean spaces. Given a pair of points $x, y \in \mathbb{R}^n$, $n \geq 2$, consider the family $\Gamma_{x,y}$ consisting of piecewise linear curves each composed of two line segments joining x and y to a common point on the hyperplane bisecting the segment $[x, y]$. In order to maintain a uniform quasiconvexity constant we restrict to curves which lie in a set of the form $CB_{x,y}$. The desired probability measure $\sigma_{x,y}$ on $\Gamma_{x,y}$ is the normalized spherical measure parametrizing such piecewise linear curves in terms of their initial direction (in $\mathbb{S}^{n-1}$) from, say, x. It is straightforward to check that such curve pencils satisfy the Semmes condition (14.2.5).

The existence of well-distributed curve families suffices for one to deduce the 1-Poincaré inequality but does not distinguish spaces supporting p-Poincaré inequalities for other p (but not the 1-Poincaré inequality). Typically the probability measure $\sigma_{x,y}$ is related to the (average) perimeter measure of balls centered at x and balls centered at y, as indicated by the estimates of such perimeters given in [13].

Alexandrov spaces

For a real number κ, a geodesic metric space (X, d) lies in the Cartan–Alexandrov–Toponogov class CAT(κ) if sufficiently small geodesic triangles in X are thinner than comparison triangles in the simply connected two-dimensional Riemannian-model surface M_κ of curvature κ. More specifically, X is a *$CAT(\kappa)$ space* if every point $x_0 \in X$ has a neighborhood U for which the following condition is satisfied: *whenever $x, y, z \in U$ are distinct and α, β, γ are arc length parametrized geodesics connecting x to y, x to z, and y to z in U, respectively, then, denoting by $\alpha_0, \beta_0, \gamma_0$ arc length parametrized geodesics in M_κ connecting three points x_0, y_0, z_0 with equal corresponding distances, we have $d(\alpha(s), \beta(t)) \leq d(\alpha_0(s), \beta_0(t))$ for all relevant choices of s and t and similarly for the pairs α, γ and β, γ.* The model surface M_κ is a Euclidean sphere $\mathbb{S}^2(r)$ with suitably chosen radius $r = r(\kappa)$ if $\kappa > 0$, the Euclidean plane $\mathbb{R}^2$ if $\kappa = 0$, or the hyperbolic plane $H^2_{\mathbb{R}}$ equipped with a suitable dilation $r \times g_0$, $r = r(\kappa)$, of the standard hyperbolic metric g_0 if $\kappa < 0$.

Spaces in the class CAT(κ) are also known as metric spaces *with curvature at most κ*; similarly, one defines metric spaces *with curvature at least κ* by reversing the inequality in the previous definition. The term *Alexandrov space* is sometimes used as a catch-all for spaces satisfying a curvature bound (upper or lower, for some choice of κ).

The theory of Alexandrov spaces is vast and we can do no more here than refer the reader to the excellent references [45], [49], and [225] and the forthcoming book [5]. There are numerous examples of spaces satisfying such curvature bounds, and the rather flexible hypotheses are preserved under various gluing, product, subspace, and limiting constructions. This abstract metric notion corresponds precisely to the classical Riemannian notion of sectional curvature bounds: a Riemannian manifold has curvature at most (resp. at least) κ in the above sense if and only if all sectional curvatures are bounded above (resp. below) by κ.

Let $X = (X, d)$ be a complete geodesic space with curvature at least κ (for some $\kappa \in \mathbb{R}$) and finite Hausdorff dimension. Then the Hausdorff dimension of X is an integer n, the Hausdorff n-measure $\mathcal{H}^n$ is locally doubling, and the metric measure space $(X, d, \mathcal{H}^n)$ supports a local 1-Poincaré inequality. If $\kappa \geq 0$ then the conclusions hold without the local modifier. See, for instance, [176] and the recent developments in [94], as well as other references therein. It follows that the theory developed in this book applies in the setting of spaces with Alexandrov lower curvature bounds.

Sphericalization and flattening

The Euclidean plane $\mathbb{R}^2$ and the 2-sphere $\mathbb{S}^2$ are both Ahlfors 2-regular spaces supporting a 1-Poincaré inequality, with $\mathbb{R}^2$ complete and unbounded and $\mathbb{S}^2$ compact and bounded. Stereographic projection is a conformal mapping identifying $\mathbb{R}^2$ with the punctured sphere $\mathbb{S}^2 \setminus \{p\}$, $p \in \mathbb{S}^2$. It is natural to ask whether an operation similar to stereographic projection identifies complete unbounded doubling metric measure spaces supporting a Poincaré inequality with corresponding bounded metric measure spaces supporting a Poincaré inequality.

The group of conformal maps acting on the (Riemann) sphere $\mathbb{S}^2$ identifies with the group of Möbius transformations. In higher dimensions and in connection with the theory of quasiconformal maps, one is naturally led to the study of *quasimöbius maps* characterized by quantitative control on the distortion of cross ratios of quadruples of points. Stereographic projection is a conformal map between the punctured sphere $\mathbb{S}^2 \setminus \{p\}$ and the plane $\mathbb{R}^2$ that is also a Möbius map and thus satisfies the following definition with $\eta(t) = t$.

Definition 14.2.7 A homeomorphism $f : X \to Y$ between two metric spaces (X, d_X), (Y, d_Y) is *ϑ-quasimöbius* for some homeomorphism $\vartheta : [0, \infty) \to [0, \infty)$, if

$$\frac{d_Y(f(x), f(y))\, d_Y(f(z), f(w))}{d_Y(f(x), f(z))\, d_Y(f(y), f(w))} \le \vartheta\left(\frac{d_X(x, y)\, d_X(z, w)}{d_X(x, z)\, d_X(y, w)}\right)$$

whenever $x, y, z, w \in X$ are four distinct points.

Deformations of metric spaces using procedures called *sphericalization* and *flattening*, were introduced by Balogh and Buckley in [20] and studied further in [48] and [133]. Given a point $a \in X$, define the *sphericalization density* $\rho_{S,a}(x, y)$ on the one-point compactification $X_a := X \cup \{\infty\}$ to be

$$\frac{d(x, y)}{\big(1 + d(x, a)\big)\big(1 + d(y, a)\big)}$$

if $x, y \in X$; to be

$$\frac{1}{1 + d(x, a)}$$

if $x \in X$ and $y = \infty$; and to be equal to zero if $x = y = \infty$. Next, define the *flattening density*

$$\rho_{F,a}(x, y) = \frac{d(x, y)}{d(x, a)\, d(y, a)}$$

on the punctured space $X^a := X \setminus \{a\}$. There are metrics $d_{S,a}$ on X_a and $d_{F,a}$ on X^a such that $d_{S,a} \approx \rho_{S,a}$ and $d_{F,a} \approx \rho_{F,a}$ with comparison constant 4;

see [20]. It can be seen easily that the sphericalization $(X_a, d_{S,a})$ is a bounded metric space, while $(X^a, d_{F,a})$ is unbounded if a is not an isolated point of X.

It was shown in [48] that the natural identification between X and $(X_a)^\infty$ is bilipschitz and that the natural embeddings of X into X_a and of X^a into X are quasimöbius. Furthermore, if X is quasiconvex then so are X_a and X^a, while if X is annularly quasiconvex then so are X_a and X^a, [48]. Here, recall from Chapter 8 that a metric space (X, d) is annularly quasiconvex if there is a constant $A \geq 1$ such that, whenever $x \in X$ and $0 < r < \text{diam}(X)/2$, each pair of points $y, z \in B(x, r) \setminus B(x, r/2)$ can be connected by a rectifiable curve of length at most $Ad(y, x)$, in the annulus $B(x, Ar) \setminus B(x, r/A)$. A result of Korte (see Theorem 9.4.1) ensures that complete Ahlfors Q-regular metric measure spaces supporting a p-Poincaré inequality for some $1 \leq p < Q$ are necessarily annularly quasiconvex. Thus, from the perspective adopted in this book, the assumptions of quasiconvexity and annular quasiconvexity are natural for X and hence also for X_a and X^a. However, the identifications between X and X_a or X^a are known only to be quasimöbius, not quasiconformal. Thus the issue of whether X_a or X^a is equipped with a doubling measure supporting a Poincaré inequality if X is so equipped is not obvious from the result of [168].

However, with suitable assumptions, sphericalizations and flattening do preserve the doubling and Poincaré inequality properties. Given a measure μ on X, we can consider an induced measure μ_a on X_a and an induced measure μ^a on X^a as follows: when $A \subset X_a$ and $F \subset X^a$ are Borel sets,

$$\mu_a(A) := \int_{A\setminus\{\infty\}} \frac{1}{\mu(B(a, 1 + d(z, a)))^2}\, d\mu(z)$$

and

$$\mu^a(F) := \int_F \frac{1}{\mu(B(a, d(z, a)))^2}\, d\mu(z).$$

In [186] it was shown that if X is quasiconvex and annularly quasiconvex, and μ is a doubling measure on X such that the metric measure space (X, d, μ) supports a p-Poincaré inequality, then $(X_a, d_{S,a}, \mu_a)$ and $(X^a, d_{F,a}, \mu^a)$ are doubling metric measure spaces supporting a p-Poincaré inequality. In conclusion, the sphericalization and flattening procedures yield further examples of doubling metric measure spaces supporting Poincaré inequalities.

Sub-Riemannian spaces

Moving still further from the Riemannian setting, we next discuss the validity of Poincaré inequalities on Carnot groups and more general sub-Riemannian spaces.

Mostow's hyperbolic rigidity theorem asserts the isometric equivalence of homeomorphic closed, negatively curved, manifolds of finite volume and dimension at least 3. As mentioned above, the equivalence of the definitions of quasiconformality, as discussed in Section 14.1, features in the proof of the Mostow rigidity theorem with regard to analysis on the boundaries of the universal covers of the original manifolds.

The boundary of the usual real hyperbolic space can be identified with the standard Euclidean sphere. The validity of Poincaré inequalities and the fundamental properties of quasiconformal maps on that boundary has already been discussed, in Sections 14.1 and 14.2. In the case of complex hyperbolic space, the boundary at infinity corresponds to a sub-Riemannian manifold locally modeled on the Heisenberg group. Let us recall the definitions.

Definition 14.2.8 The *(first) Heisenberg group* $\mathbb{H}$ is the nilpotent Lie group whose underlying space is the Euclidean space $\mathbb{R}^3$ equipped with the group law

$$(x, y, t) * (x', y', t') = (x + x', y + y', t + t' + 2(xy' - x'y))$$

and the left-invariant *Heisenberg metric* $d_H(p, q) = \|p^{-1} * q\|_H$, where

$$\|(x, y, t)\|_H = \left((x^2 + y^2)^2 + t^2\right)^{1/4}.$$

It is an instructive exercise to verify that d_H is a metric. Observe that the topology induced by d_H coincides with the Euclidean topology. Next, equip $\mathbb{H}$ with the Lebesgue measure μ of $\mathbb{R}^3$ (which is also a Haar measure for $\mathbb{H}$).

Proposition 14.2.9 *$(\mathbb{H}, d_H, \mu)$ is a locally compact, quasiconvex, Ahlfors 4-regular metric measure space satisfying the 1-Poincaré inequality.*

Local compactness is a consequence of the topological equivalence of the Heisenberg and Euclidean metrics. The anisotropic dilations $(\delta_r)_{r>0}$ defined by $\delta_r(x, y, t) = (rx, ry, r^2 t)$ act as group automorphisms and also as similarities of the Heisenberg metric. The Jacobian determinant of δ_r is equal to r^4. Denoting by $B_H(p, r)$ the open ball in the metric d_H with center p and radius $r > 0$, it follows that $\mu(B_H(p, r)) = cr^4$ for all p and r, where $c = \mu(B_H(o, 1))$ and where $o = (0, 0, 0)$ denotes the neutral element. Hence $(\mathbb{H}^n, d_H)$ is Ahlfors 4-regular. Since bilipschitz maps preserve Hausdorff dimension, the metric d_H is not bilipschitz equivalent to the Euclidean metric nor to any Riemannian metric on $\mathbb{R}^3$. Instead, d_H is bilipschitz equivalent to a geodesic *sub-Riemannian* metric, the so-called *Carnot–Carathéodory metric* d_{cc}. The latter metric is defined by infimization of the length of the horizontal curves connecting two given points, where the horizontality of a piecewise smooth curve $\gamma : [a, b] \to \mathbb{H}$ means that $\gamma'(s) \in H_{\gamma(s)}\mathbb{H}$ for almost all s. Here the fiber of the *horizontal subbundle* $H\mathbb{H}$ of the tangent bundle $T\mathbb{H}$ at a point p is defined to be

$$H_p\mathbb{H} = \operatorname{span}\{X(p), Y(p)\}, \qquad X = \frac{\partial}{\partial x} + 2y\frac{\partial}{\partial t}, \quad Y = \frac{\partial}{\partial y} - 2x\frac{\partial}{\partial t}$$

and is equipped with a smoothly varying family of inner products with respect to which the left-invariant vector fields X, Y are an orthonormal frame. Since d_{cc} is a geodesic metric, d_H is quasiconvex.

The validity of the Poincaré inequality on the metric measure space $(\mathbb{H}, d_H, \mu)$ can be seen by various methods. The Riesz potential and representation formula approach previously discussed is applicable. Another more geometric approach is described below. An elegant proof, which is due to Varopoulos [274] and can also be found in [114, Proposition 11.17], uses only the group structure, the homogeneous metric structure, the translation invariance of the Haar measure, and the geodesic property.

More generally, a *Carnot group* is a connected and simply connected Lie group $\mathbb{G}$ whose Lie algebra $\mathfrak{g}$ admits a stratified vector space decomposition $\mathfrak{g} = \mathfrak{v}_1 \oplus \cdots \oplus \mathfrak{v}_\iota$ such that $[\mathfrak{v}_1, \mathfrak{v}_j] = \mathfrak{v}_{j+1}$ for all $1 \le j < \iota$ and $[\mathfrak{v}_1, \mathfrak{v}_\iota] = 0$. Identifying the Lie algebra $\mathfrak{g}$ with the space of left-invariant vector fields and fixing a basis $X_1, \ldots, X_m$ for $\mathfrak{v}_1$, one again introduces the *horizontal bundle* $H\mathbb{G}$, whose fiber at $p \in \mathbb{G}$ is $H_p\mathbb{G} = \operatorname{span}\{X_1(p), \ldots, X_m(p)\}$. The Carnot–Carathéodory metric d_{cc} is defined as before. It is a geodesic metric, and the metric measure space $(\mathbb{G}, d_{cc}, \mu)$, where μ denotes the Haar measure, is again Ahlfors Q-regular, with $Q = \sum_{j=1}^{\iota} j \dim \mathfrak{v}_j$, and supports the 1-Poincaré inequality. The Loewner property of Carnot groups was proved directly by Reimann [230], in the Heisenberg group $\mathbb{H}$, and later in [119] in the setting of general Carnot groups.

By a result of Mitchell [210], Carnot groups occur as the Gromov–Hausdorff tangent spaces of (equiregular) sub-Riemannian manifolds. A typical example of such a manifold is the sub-Riemannian unit sphere $\mathbb{S}^3 \subset \mathbb{R}^4$. At each point $p \in \mathbb{S}^3$ there is a unique direction $\pm W(p)$ in $T_p\mathbb{S}^3$ such that $JW(p) \notin T_p\mathbb{S}^3$. Here J denotes the standard complex structure in $\mathbb{R}^4$ arising from the canonical identification with $\mathbb{C}^2$. The horizontal bundle $H\mathbb{S}^3$ is defined to be the orthocomplement of the vector field W. Since $\mathbb{S}^3$ is parallelizable, we may select two other nonvanishing vector fields U and V such that the pair $\{U, V\}$ is a global orthonormal frame for the horizontal bundle. As in the Heisenberg case, any two points of $\mathbb{S}^3$ can be joined by a horizontal curve, and there is an induced Carnot–Carathéodory metric d_{cc}. A version of stereographic projection identifies the Heisenberg group $\mathbb{H}$ with the punctured sphere

$\mathbb{S}^3 \setminus \{p\}$, both equipped with the Carnot–Carathéodory metric. (In fact, up to bilipschitz equivalence of the relevant metrics this identification corresponds to the sphericalization–flattening procedure described above.) The metric measure space $(\mathbb{S}^3, d_{cc}, \mathcal{H}^4_{cc})$ is Ahlfors 4-regular and supports the 1-Poincaré inequality. Its Gromov–Hausdorff tangent spaces coincide with the Heisenberg group $\mathbb{H}$ equipped with its Carnot–Carathéodory metric d_{cc}.

In Mostow's rigidity theorem, the induced maps on the boundary of a complex hyperbolic space are quasiconformal maps of $(\mathbb{S}^3, d_{cc})$. The equivalence of definitions of quasiconformality, which in turn relies on the first-order analytic structure coming from the Poincaré inequality, plays an essential role in his analysis. As discussed in Section 14.1, the modern theory of analysis in metric spaces, as presented in this book, arose from attempts to clarify the foundations of quasiconformal mapping theory in Carnot groups. This setting remains an important testing ground for the techniques and methodology of abstract metric space analysis.

The Sobolev space $N^{1,p}(\mathbb{G}, d_{cc})$ is identified with the so-called *horizontal Sobolev space* $W^{1,p}_H(\mathbb{G})$, which is defined similarly to the Euclidean case as the space of L^p functions f whose distributional derivatives $X_j f$, $j = 1, \ldots, m$, are also in L^p. The usual caveat regarding the choice of representative remains in force; cf. Theorem 7.4.5. Moreover, as shown in [114, Proposition 11.6 and Theorem 11.7], the minimal p-weak upper gradient of a Lipschitz function $u : \mathbb{G} \to \mathbb{R}$ is independent of p and coincides μ-almost everywhere with the norm of the horizontal gradient $\nabla_H u = (X_1 u, \ldots, X_m u)$.

The Rademacher theorem for Lipschitz maps on Carnot groups was known prior to Cheeger's work. Pansu [224] showed that Lipschitz maps between Carnot groups are almost everywhere differentiable, for a suitable notion of differentiability adapted to the setting and nowadays known as *Pansu differentiability*. In the case of maps $u : \mathbb{G} \to \mathbb{R}$, such differentiability reduces to differentiability along horizontal directions (e.g., the existence of the horizontal first-order derivatives $X_j u$, $j = 1, \ldots, m$), and the coordinate functions associated with the first layer variables may serve as Cheeger coordinate functions. (In $\mathbb{H}$, this means that the two functions $x : \mathbb{H} \to \mathbb{R}$ and $y : \mathbb{H} \to \mathbb{R}$ given by $(x, y, t) \mapsto x$ and $(x, y, t) \mapsto y$ are a suitable choice of Cheeger coordinates.)

Additional information on sub-Riemannian geometry and analysis can be found, for example, in [106], [211], and [52].

Nonmanifold examples

It was recognized early in the theory of analysis in metric spaces that the Poincaré inequality is a robust criterion which survives under various gluing or amalgamation procedures. Via such procedures it is easy to construct examples of spaces with nonmanifold points supporting Poincaré inequalities. In [125, Section 6.14], metric conditions were given on a triple (X, Y, A) of spaces, where X and Y are assumed to be locally compact and Ahlfors Q-regular for a common exponent $Q > 1$ and the space A is assumed a priori to admit isometric embeddings $\iota_X : A \hookrightarrow X$ and $\iota_Y : A \hookrightarrow Y$, so that the metric gluing space $X \coprod_A Y$ supports suitable Poincaré inequalities. Here $X \coprod_A Y$ is the decomposition space obtained from the usual disjoint union $X \coprod Y$ by the identification of two-element subsets $\{\iota_X(a), \iota_Y(a)\}$, $a \in A$, and is equipped with the L^1 metric $d(x, y) = \inf\{d_X(x, \iota_X(a)) + d_Y(\iota_Y(a), y) \ : \ a \in A\}$. For instance, the metric gluing space $\mathbb{R}^4 \coprod_{\mathbb{R}} \mathbb{H}$, where $\iota_X : \mathbb{R} \to \mathbb{R}^4$ and $\iota_Y : \mathbb{R} \to \mathbb{H}$ are isometric embeddings, is an Ahlfors 4-regular space supporting a p-Poincaré inequality for each $p > 3$, and each point of $\iota_X(\mathbb{R}) = \iota_Y(\mathbb{R})$ is not a manifold point.

The paper [117] provides examples of compact geodesic Ahlfors regular metric measure spaces supporting a 1-Poincaré inequality for which every point is a nonmanifold point. These examples have integral dimension n (in fact, the Ahlfors regularity dimension agrees with the topological dimension) and have the further property that at almost every point there exists a unique Gromov–Hausdorff tangent space which coincides with the Euclidean space $\mathbb{R}^n$.

The first examples of metric measure spaces supporting Poincaré inequalities with nonintegral Hausdorff dimension were provided by Bourdon and Pajot [43]. The local geometry of these spaces is topologically complex and is modeled on a classical self-similar fractal (the *Menger sponge*). The Bourdon–Pajot examples are compact geodesic metric spaces arising as boundaries at infinity of suitable hyperbolic "buildings". They are Ahlfors Q-regular for a suitable real number $Q > 1$ and support the 1-Poincaré inequality. The range of allowed dimensions for the Bourdon–Pajot examples comprises a countable dense subset of the interval $(1, +\infty)$. Later, Laakso [177] exhibited similar examples for *every* real number $Q > 1$; his examples are iterated decomposition spaces obtained by successive identifications of "wormhole" points in different fibers $I \times \{p\}$ and $I \times \{q\}$, $p, q \in K$, in the product space $I \times K$, where K is a suitable self-similar Cantor set. In view of the discussion in Chapter 13, both the Bourdon–Pajot and Laakso examples admit measurable differentiable structures, as in Theorem 13.4.4. In both classes of examples, the dimension $N(\alpha)$ of the Cheeger tangent space is everywhere equal to 1, although the Hausdorff dimension is strictly greater than 1.

The paper [192] established the validity of Poincaré inequalities for fat Sierpiński carpets $S_{\mathbf{a}}$. (The nomenclature was suggested by Jun Kigami.) These spaces are compact Ahlfors 2-regular subsets of the plane, equipped with the Euclidean metric and Lebesgue measure, which have no interior and contain no manifold points. More precisely, let $\mathbf{a} = (a_1, a_2, a_3, \ldots)$, where each a_j is the reciprocal of an odd integer greater than or equal to 3. Define $S_\mathbf{a} = \bigcup_{m=0}^\infty S_{\mathbf{a},m}$, where $S_{\mathbf{a},0} = [0, 1]\times[0, 1]$ and, for each $m \geq 1$, $S_{\mathbf{a},m}$ is obtained from $S_{\mathbf{a},m-1}$ by removing from each constituent square Q its concentric square Q' with relative size a_m and subdividing $Q\setminus Q'$ into $a_m^{-2}-1$ essentially disjoint squares of the same size as Q'. If the relative scaling sequence $\mathbf{a}$ is in ℓ^2, the resulting compact set $S_\mathbf{a}$ has positive Lebesgue measure (but empty interior) and indeed is Ahlfors 2-regular when equipped with the Euclidean metric d_E and Lebesgue measure μ. In [192] the following results were established:

1. $(S_\mathbf{a}, d_E, \mu)$ supports a 1-Poincaré inequality if and only if $\mathbf{a} \in \ell^1$;
2. $(S_\mathbf{a}, d_E, \mu)$ supports a p-Poincaré inequality for some $p < \infty$ if and only if $\mathbf{a} \in \ell^2$.

A typical example is the sequence $\mathbf{a} = (1/3, 1/5, 1/7, \ldots)$, which is contained in $\ell^2 \setminus \ell^1$. According to the above result, the corresponding fat Sierpiński carpet $S_\mathbf{a}$ supports a p-Poincaré inequality for each $p > 1$ but does not support a 1-Poincaré inequality. (Compare Keith and Zhong's Theorem 12.3.9.) See Figures 14.1 and 14.2 for illustrations of the standard Sierpiński carpet $S_{(1/3,1/3,1/3,\ldots)}$ as well as the fat Sierpiński carpet $S_{(1/3,1/5,1/7,\ldots)}$.

Poincaré inequalities and Loewner-type conditions

Stephen Keith [150] introduced a Loewner-type condition for a suitable weighted p-modulus, which turns out in rather great generality to be equivalent to the validity of the p-Poincaré inequality. To state his definition precisely we define a family of weighted measures $d\nu_{x,y}^C$ on a metric measure space (X, d, μ) as in (14.2.5):

$$d\nu_{x,y}^C(z) = \left(\frac{d(x, z)}{\mu(B(x, d(x, z)))} + \frac{d(y, z)}{\mu(B(y, d(y, z)))}\right) \chi_{CB_{x,y}}\, d\mu(z).$$

Proposition 14.2.10 (Keith) *Let (X, d, μ) be a complete doubling metric measure space and let $p \in [1, \infty)$. Then X supports a p-Poincaré inequality if and only if there exists a constant $C > 0$ such that for all $x, y \in X$ we have*

$$\operatorname{Mod}_p(\Gamma_{x,y}, \nu_{x,y}^C) \geq C^{-1} d(x, y)^{1-p}. \tag{14.2.11}$$

Here $\Gamma_{x,y}$ denotes the family of all rectifiable curves joining x to y, while $\operatorname{Mod}_p(\cdot, \nu)$ denotes the p-modulus considered in the metric measure space (X, d, ν).

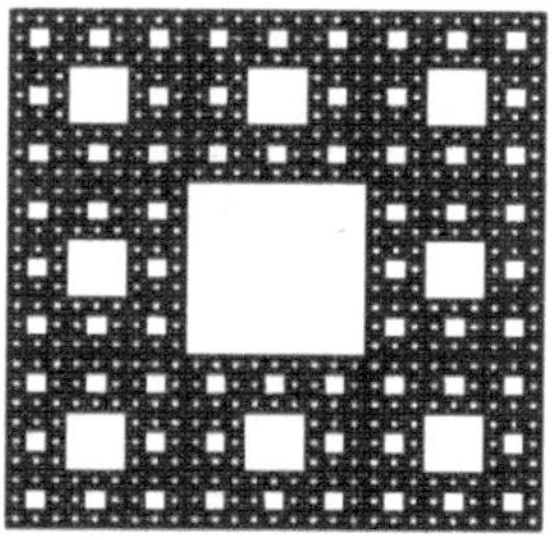

Figure 14.1 The standard Sierpiński carpet $S_{(1/3,1/3,1/3,\dots)}$.

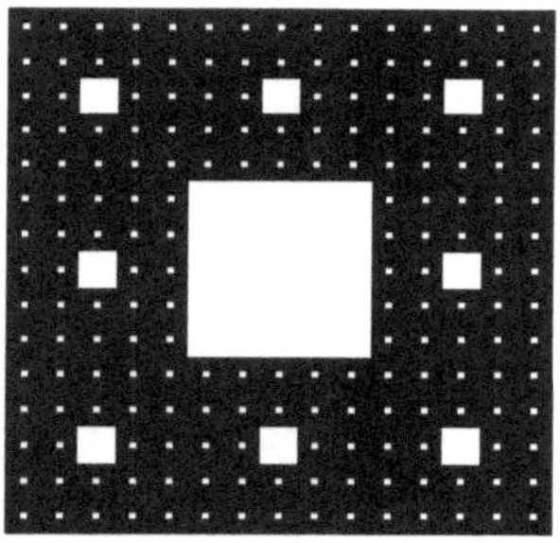

Figure 14.2 The fat Sierpiński carpet $S_{(1/3,1/5,1/7,\dots)}$.

Note that the existence of well-distributed curve pencils immediately implies the validity of (14.2.11). We give the proof for $p = 1$; the proof for $p > 1$ is similar. Let $(\Gamma_{x,y}, \sigma_{x,y})$ be as in Definition 14.2.4 and let g be admissible for $\Gamma_{x,y}$. Then

$$C \int_X g \, d\nu_{x,y}^C \geq \int_{\Gamma_{x,y}} \int_\gamma g \, ds \, d\sigma \geq 1.$$

Taking the infimum over all such g completes the proof.

Poincaré inequalities on the Bourdon–Pajot and Laakso spaces have been established by verifying the existence of well-distributed curve families. In the case of the Bourdon–Pajot examples, $(\Gamma_{x,y}, \sigma_{x,y})$ arise naturally from the construction of the associated hyperbolic space and its boundary. In the case of the Laakso examples, such curve families are constructed by hand.

The derivation of Poincaré inequalities on fat Sierpiński carpets in [192] takes advantage of the characterization of such inequalities in Proposition 14.2.10. Here, as in the Laakso examples, the desired curve families are constructed in a manner adapted to the natural structure of the underlying fractals.

It is more challenging to prove the Poincaré inequality on the Heisenberg group or more general Carnot groups by constructing Semmes curve pencils. On the Heisenberg group, such pencils were constructed by Korte, Lahti, and Shanmugalingam [166]. Further examples of metric measure spaces supporting such curve pencils are discussed in [166]. Note that the construction of such pencils on the Heisenberg group in [166] relies on some involved numerical calculations. The existence of such curve pencils on more complicated sub-Riemannian spaces remains a challenging open problem.

The ∞-Poincaré inequality

The existence of Semmes curve pencils is not necessary for a space to satisfy a p-Poincaré inequality. However, a larger family of curves is needed than is guaranteed by the quasiconvexity of the space. To this end, here we discuss the notion of ∞-Poincaré inequality and its relation to the existence of thick quasiconvex curve families.

Letting $p \to \infty$ in the p-Poincaré inequality (8.1.1) we obtain the following ∞-Poincaré inequality.

Definition 14.2.12 We say that (X, d, μ) supports an *∞-Poincaré inequality* if there are positive constants C, λ such that, whenever B is a ball in X and g is an upper gradient of a function u on X,

$$\fint_B |u - u_B| \, d\mu \le C \operatorname{diam}(B) \, \|g\|_{L^\infty(\lambda B)}.$$

By Hölder's inequality it is clear that a space supporting a p-Poincaré inequality for any finite p will necessarily support an ∞-Poincaré inequality. The converse assertion is false, even for complete doubling spaces; see the discussion at the end of this section.

In Ahlfors Q-regular spaces, for $p > Q$ we can reformulate Keith's condition with the usual (unweighted) modulus, take the pth root, and let $p \to \infty$. This suggests the heuristic that the ∞-Poincaré inequality should be characterized by quantitative lower bounds for the ∞-modulus of curve families joining pairs of points x, y. However, such an heuristic is not completely correct. In [79] it was shown that a complete doubling metric measure space supports an ∞-Poincaré inequality if and only if there is a constant $C > 0$ such that $\operatorname{Mod}_\infty(\Gamma^C_{x,y,E,F})$ is positive (without any estimates of the lower bound) whenever x, y are distinct points and $E \subset B(x, d(x, y)/C)$ and $F \subset B(y, d(x, y)/C)$ with E and F both of positive measure. The latter property is called *thick quasiconvexity* and the constant C is referred to as the *thick quasiconvexity constant*. Here $\Gamma^C_{x,y,E,F}$ denotes the family of all C-quasiconvex curves connecting E to F. This geometric characterization was recently improved in [78], where it was shown that a complete doubling metric measure space X supports an ∞-Poincaré inequality if and only if there is a constant $C > 0$ such that

$$\mathrm{Mod}_\infty(\Gamma^C_{x,y}) > 0$$

whenever x and y are distinct points in X and $\Gamma^C_{x,y}$ is the family of all C-quasiconvex curves in X with end points x and y. Thus, unlike the case of p-Poincaré inequalities for finite p, the quantitative control in the ∞-Poincaré inequality resides in the choice of the thick quasiconvexity constant, not in the lower bound for the ∞-modulus. Thick quasiconvexity is, in principle, easier to verify for a given metric measure space, and hence the ∞-Poincaré inequality provides a handy way of detecting metric measure spaces that do not support any p-Poincaré inequalities. Examples of such spaces include the standard Sierpiński carpet.

This is not the only difference between p-Poincaré inequalities for finite p and the ∞-Poincaré inequality. There exist complete doubling metric measure spaces supporting an ∞-Poincaré inequality but no p-Poincaré inequality for any finite p. One such example, given in [80], is the *Sierpiński strip*, obtained by pasting together in an infinite strip a sequence of finite stages in the usual construction of the standard Sierpiński carpet. This example also shows that the ∞-Poincaré inequality does not persist under pointed measured Gromov–Hausdorff limits. Thus the ∞-Poincaré inequality fails to have the self-improving and stability properties studied in Chapter 12 and Chapter 11 for the p-Poincaré inequality for finite p.

14.3 Applications and further research directions

Quasisymmetric uniformization

Let X be a metric space. The *quasisymmetric uniformization* problem for X asks for a list of metric conditions necessary and sufficient for another metric space Y, assumed a priori to be homeomorphic to X, to be quasisymmetrically homeomorphic to X. Motivated by Cannon's conjecture on the structure of finitely generated Gromov hyperbolic groups with 2-sphere boundary at infinity, Bonk and Kleiner used analysis on metric measure spaces supporting Poincaré inequalities to address quasisymmetric uniformization for the sphere $\mathbb{S}^2$. Notably, they showed in [40] that if Y is a metric space homeomorphic to $\mathbb{S}^2$ which is assumed to be Q-regular for some $Q \geq 2$ and also to satisfy the

Q-Poincaré inequality then in fact Y is quasisymmetrically equivalent to $\mathbb{S}^2$, $Q = 2$, and Y satisfies the 1-Poincaré inequality. (The final conclusion follows from Semmes' Theorem 14.2.3 upon noting that the linear local contractibility follows in this setting from the Poincaré inequality.)

The *Ahlfors regular conformal dimension* of a metric space (X, d) is the infimum of all exponents $Q > 0$ such that X is quasisymmetrically equivalent to an Ahlfors Q-regular metric space. Combining the above result with those in [41] yields the following strong conclusion in the direction of Cannon's conjecture.

Theorem 14.3.1 (Bonk–Kleiner) *Let Y be the Gromov boundary of a Gromov hyperbolic group equipped with a visual metric. Assume that Y is homeomorphic to $\mathbb{S}^2$ and that Y is minimal for the Ahlfors regular conformal dimension. Then Y is quasisymmetrically equivalent to $\mathbb{S}^2$.*

Gromov [107], [105] introduced a metric criterion which nowadays goes by the name of *Gromov hyperbolicity*. A geodesic metric space (X, d_X) is said to be *Gromov δ-hyperbolic* for some $\delta > 0$ if, whenever x, y, z is a distinct triple of points in X, and $\gamma_1, \gamma_2, \gamma_3$ are geodesics connecting x to y, y to z, and z to x respectively, each point in γ_3 is within a distance δ of the union $\gamma_1 \cup \gamma_2$. A group is Gromov hyperbolic if it is finitely generated and its Cayley graph, equipped with the path metric, is a Gromov hyperbolic space. There are numerous excellent resources for the theory of hyperbolic groups and their boundaries, including [146], [45], and [91]. Definitions of the terms in the theorem can be found there.

Analogous results for the 3-sphere $\mathbb{S}^3$ are false. The sub-Riemannian sphere $(\mathbb{S}^3, d_{cc})$ provides an example of a 4-regular space, satisfying the 1-Poincaré inequality, which is minimal for Ahlfors regular conformal dimension and arises as the Gromov boundary of the complex hyperbolic plane and which is not quasisymmetrically equivalent to the standard $\mathbb{S}^3$.

Quasisymmetric uniformization is of interest for other model spaces. A variant of Cannon's conjecture (the so-called *Kapovich–Kleiner conjecture*) concerns Gromov hyperbolic groups with Sierpiński carpet boundaries and leads to the study of the quasisymmetric uniformization problem for the Sierpiński carpet. Bourdon and Kleiner [42] introduced the *combinatorial Loewner property* and investigated its relationship to the classical Loewner property of the modulus of curve families and to the validity of Poincaré inequalities. They showed that the standard Sierpiński carpet $X = S_{(1/3,1/3,1/3,\dots)}$ satisfies the combinatorial Loewner property; whether X is quasisymmetrically equivalent to any space supporting the classical Loewner property or to any space supporting a Poincaré inequality remains a well-known open problem.

For further reading in this direction, we refer the reader to the excellent survey articles by Bonk [39] and by Kleiner [162] in the *Proceedings of the 2006 International Congress of Mathematicians*.

Analysis on fractals and the harmonic Sierpiński gasket

Fractals such as the Sierpiński gasket (Figure 14.3) and the Sierpiński carpet (Figure 14.1) contain an insufficient number of nonconstant rectifiable curves to support a Poincaré inequality. Indeed, for each $p \geq 1$ the p-modulus of the collection of all nonconstant rectifiable curves in such sets is equal to zero. Thus the theory of Sobolev spaces via upper gradients is not viable in this setting. In a certain class of fractals and in the case $p = 2$, an alternate theory of Sobolev spaces was developed using the formalism of Dirichlet forms (see Beurling and Deny [28]). This approach was described in detail in [155] in the special case of so-called *post-critically finite fractals*, the canonical example of which is the Sierpiński gasket. Further literature on this rapidly developing field includes [259], [258], [260], [261], [156], [262], [263], [130], [66], [155], [154], [22], [174], [85], [116]; see also the references therein. The Dirichlet form on such fractals is constructed as a limit of discrete energy forms on finite graph approximations. An excellent reference source for the theory of Dirichlet forms is the book by Fukushima, Oshima, and Takeda [90].

In general, Dirichlet forms on fractals are not strongly local. (A Dirichlet form $\mathcal{E}$ is strongly local if, whenever a function f in the domain of $\mathcal{E}$ is constant on a Borel set A, the Dirichlet energy density of f, in the sense of Beurling and Deny, vanishes on A.) This is in stark contrast with the theory of Sobolev functions considered in this book, where the energy density is given by the minimal p-weak upper gradient; such energy densities are always strongly local (see Lemma 6.3.8). The Dirichlet forms usually considered on

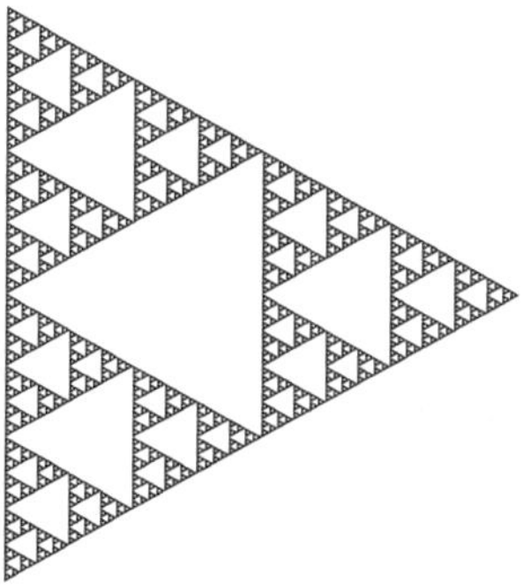

Figure 14.3 The standard Sierpiński gasket.

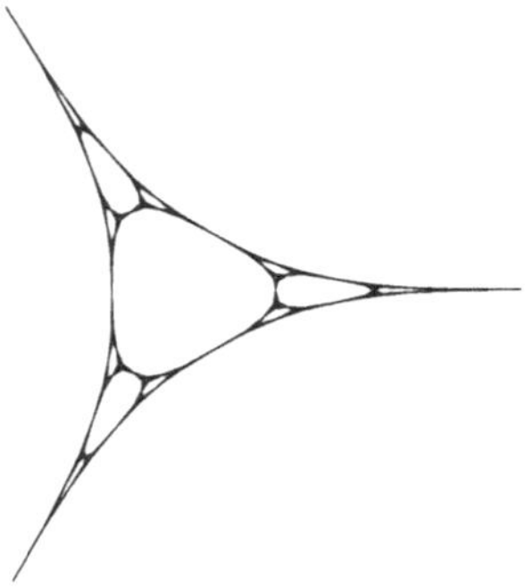

Figure 14.4 The harmonic gasket.

post-critically finite fractals are strongly local. If the Dirichlet form on a doubling metric measure space is strongly local and supports a Poincaré-type inequality (with the Dirichlet energy density playing the role of the upper gradients in the inequality) then the corresponding Sobolev-type space, called the Dirichlet domain, coincides with the Sobolev-type space $N^{1,2}(X)$; see [172] or [250]. For example, on the Sierpiński gasket there is a measure and a geodesic metric, induced by the aforementioned Dirichlet form, such that the modified metric measure space (known as the *harmonic Sierpiński gasket*) supports a 2-Poincaré inequality in the sense of the definition near the start of Section 8.1. Moreover, the resulting space admits an embedding into the plane such that the geodesic metric in question corresponds to the induced path metric on the image. See [145] and [144] for further details, and see Figure 14.4 for an image of the harmonic gasket.

Nonlinear potential theory on metric measure spaces

The study of harmonic functions in Euclidean domains led to the development of the potential theory associated with the Laplacian operator and, later on, to axiomatic potential theory. Good surveys of this theory can be found in the books [202], [44], [3], or [17]. The regularity theory for nonlinear subelliptic operators (see, for instance, [178], [92] and [93]) paved the way for an analogous theory of potentials for the p-Laplacian operator and its variants. For a sample of the latter topic we recommend [280], [281], and [128]. One is naturally led to extend nonlinear potential theory and the study of p-harmonic functions to the setting of metric measure spaces equipped with a doubling measure and supporting a Poincaré inequality. Early works in this direction include [158] and [249]. An informative survey of recent developments in this area can be found in [31], but this theory remains under active investigation. The metric measure space perspective has led unexpectedly to a number of new results even in the Euclidean theory. For instance, the resolutivity of p-quasicontinuous Sobolev functions in relation to the Perron method for the Dirichlet problem see [33] and [31, Section 10.4], and the self-improving property of Poincaré inequalities as in [153] (see also Chapter 12) were established first in the metric setting. The former was new even in the classical Euclidean setting while the latter was new in the weighted Euclidean setting.

Synthetic Ricci curvature bounds in metric measure spaces

One of the most exciting recent developments in analysis in metric spaces has been the emerging theory of synthetic Ricci curvature lower bounds [188], [189], [266], [265], [278], [264]. These investigations have identified remarkable connections to the study of optimal mass transport and measure contraction under gradient flows. The $CD(\kappa, N)$ condition (for curvature lower bound κ and dimension upper bound N) has been utilized by Ambrosio, Gigli, and Savaré. For instance, they related the $CD(\kappa, \infty)$ notion to probability measures on paths and to a Bakry–Émery curvature condition $BE(\kappa, \infty)$ associated with a bilinear energy form associated with the Sobolev space $N^{1,2}$ (which, in this context, is called the *Dirichlet domain* while the associated bilinear form is called the *Dirichlet form*). This is a rapidly growing field, which we will not attempt to survey in any detail. A small but representative sample of the literature includes [15], [14], [9], [10], and [241]. For related papers on the Bakry–Émery curvature condition, see also [171], [140], and [141].

In the process of developing the theory of such abstract curvature conditions, Ambrosio, Colombo, Di Marino, Gigli, and Savaré demonstrated that if X is a complete metric measure space equipped with a doubling measure then Lipschitz functions are dense in $N^{1,p}(X)$ and $N^{1,p}(X)$ is reflexive for all p with $1 < p < \infty$, even if X does not support a p-Poincaré inequality. See [16] and [12] for these remarkable results. A recent paper of Ambrosio, Di Marino, and Savaré [11] explains the connections between the p-moduli of path families (a foundational notion to the approach of Sobolev spaces considered in this book) and probability measures on paths.

The $CD(\kappa, N)$ notion as formulated in the preceding references is not well adapted to the sub-Riemannian setting of Carnot groups. An alternate notion of Bakry–Ledoux-type curvature bounds was extensively developed by Baudoin and Garofalo [25] in the setting of the Heisenberg group and other sub-Riemannian spaces.

References

[1] Adams, D. R. and Hedberg, L. I. 1996. *Function Spaces and Potential Theory*. Grundlehren der Mathematischen Wissenschaften, vol. 314. Berlin: Springer-Verlag.

[2] Adams, R. A. 1975. *Sobolev Spaces*. Pure and Applied Mathematics, vol. 65. New York/London: Academic Press, Harcourt Brace Jovanovich.

[3] Aikawa, H. and Essén, M. 1996. *Potential Theory – Selected Topics*. Lecture Notes in Mathematics, vol. 1633. Berlin: Springer-Verlag.

[4] Aikawa, H. and Ohtsuka, M. 1999. Extremal length of vector measures. *Ann. Acad. Sci. Fenn. Math.*, **24**(1), 61–88.

[5] Alexander, S., Kapovitch, V., and Petrunin, A. *Alexandrov Geometry*. Book in preparation. Draft available at www.math.psu.edu/petrunin/papers/alexandrov-geometry.

[6] Ambrosio, L. 1990. Metric space valued functions of bounded variation. *Ann. Scuola Norm. Sup. Pisa Cl. Sci. (4)*, **17**(3), 439–478.

[7] Ambrosio, L. and Tilli, P. 2000. *Selected Topics on "Analysis in Metric Spaces"*. Appunti dei Corsi Tenuti da Docenti della Scuola, Scuola Normale Superiore, Pisa.

[8] Ambrosio, L. and Tilli, P. 2004. *Topics on Analysis in Metric Spaces*. Oxford Lecture Series in Mathematics and its Applications, vol. 25. Oxford: Oxford University Press.

[9] Ambrosio, L., Gigli, N., and Savaré, G. Bakry–Émery curvature-dimension condition and Riemannian Ricci curvature bounds. Preprint, 2012. Available at lanl.arxiv.org/abs/1209.5786.

[10] Ambrosio, L., Mondino, A., and Savaré, G. On the Bakry–Émery condition, the gradient estimates and the local-to-global property of $RCD(K, N)$ metric measure spaces. Preprint, 2013. Available at lanl.arxiv.org/abs/1309.4664.

[11] Ambrosio, L., Di Marino, S., and Savaré, G. On the duality between p-modulus and probability measures. Preprint, 2013. Available at cvgmt.sns.it/paper/2271/.

[12] Ambrosio, L., Colombo, M., and Di Marino, S. Sobolev spaces in metric measure spaces: reflexivity and lower semicontinuity of slope. Preprint, 2012. Available at http://cvgmt.sns.it/media/doc/paper/2055.

[13] Ambrosio, L., Miranda, Jr., M., and Pallara, D. 2004. Special functions of bounded variation in doubling metric measure spaces, pp. 1–45 in: *Calculus of*

Variations: Topics from the Mathematical Heritage of E. De Giorgi, Quad. Mat., vol. 14. Dept Math., Seconda Univ. Napoli.

[14] Ambrosio, L., Gigli, N., and Savaré, G. 2008. *Gradient Flows in Metric Spaces and in the Space of Probability Measures*. Second edn. Lectures in Mathematics, ETH Zürich. Basel: Birkhäuser Verlag.

[15] Ambrosio, L., Gigli, N., and Savaré, G. 2012. Heat flow and calculus on metric measure spaces with Ricci curvature bounded below – the compact case. *Boll. Unione Mat. Ital. (9)*, **5**(3), 575–629.

[16] Ambrosio, L., Gigli, N., and Savaré, G. 2013. Density of Lipschitz functions and equivalence of weak gradients in metric measure spaces. *Rev. Mat. Iberoam.*, **29**(3), 969–996.

[17] Armitage, D. H. and Gardiner, S. J. 2001. *Classical Potential Theory*. Springer Monographs in Mathematics. London: Springer-Verlag London Ltd.

[18] Assouad, P. 1983. Plongements lipschitziens dans $\mathbf{R}^n$. *Bull. Soc. Math. France*, **111**(4), 429–448.

[19] Auscher, P., Coulhon, Th., and Grigor'yan, A. (eds.). 2003. *Heat Kernels and Analysis on Manifolds, Graphs, and Metric Spaces*. Contemporary Mathematics, vol. 338. Providence, RI: American Mathematical Society.

[20] Balogh, Z. M. and Buckley, S. M. 2005. Sphericalization and flattening. *Conform. Geom. Dynam.*, **9**, 76–101.

[21] Banach, S. 1955. *Théorie des Óperations Linéaires*. New York: Chelsea.

[22] Barlow, M. T., Bass, R. F., Kumagai, T., and Teplyaev, A. 2010. Uniqueness of Brownian motion on Sierpiński carpets. *J. Eur. Math. Soc.*, **12**(3), 655–701.

[23] Barvinok, A. 2002. *A Course in Convexity*. Graduate Studies in Mathematics, vol. 54. Providence, RI: American Mathematical Society.

[24] Bates, D. and Speight, G. 2013. Differentiability, porosity and doubling in metric measure spaces. *Proc. Amer. Math. Soc.*, **141**(3), 971–985.

[25] Baudoin, F. and Garofalo, N. 2011. Perelman's entropy and doubling property on Riemannian manifolds. *J. Geom. Anal.*, **21**(4), 1119–1131.

[26] Benyamini, Y. and Lindenstrauss, J. 2000. *Geometric Nonlinear Functional Analysis, Volume I*. Colloquium Publications, vol. 48. American Mathematical Society.

[27] Beurling, A. 1989. *The Collected Works of Arne Beurling. Vol. 1. Contemporary Mathematicians*. Boston, MA: Birkhäuser.

[28] Beurling, A. and Deny, J. 1959. Dirichlet spaces. *Proc. Nat. Acad. Sci. USA*, **45**, 208–215.

[29] Billingsley, P. 1999. *Convergence of Probability Measures*. Second edn. Wiley Series in Probability and Statistics. New York: John Wiley & Sons.

[30] Bishop, C. and Hakobyan, H. Frequency of dimension distortion under quasisymmetric mappings. Preprint, 2013.

[31] Björn, A. and Björn, J. 2011. *Nonlinear Potential Theory on Metric Spaces*. EMS Tracts in Mathematics, vol. 17. Zürich: European Mathematical Society.

[32] Björn, A., Björn, J., and Shanmugalingam, N. 2003a. The Dirichlet problem for p-harmonic functions on metric spaces. *J. Reine Angew. Math.*, **556**, 173–203.

[33] Björn, A., Björn, J., and Shanmugalingam, N. 2003b. The Perron method for p-harmonic functions in metric spaces. *J. Differential Equations*, **195**(2), 398–429.

[34] Björn, A., Björn, J., and Shanmugalingam, N. 2008. Quasicontinuity of Newton–Sobolev functions and density of Lipschitz functions on metric spaces. *Houston J. Math.*, **34**(4), 1197–1211.

[35] Björn, J. 2006. Wiener criterion for Cheeger p-harmonic functions on metric spaces, pp. 103–115 in: *Adv. Stud. Pure Math.*, vol. 44. Tokyo: Mathematical Society of Japan.

[36] Björn, J., MacManus, P., and Shanmugalingam, N. 2001. Fat sets and pointwise boundary estimates for p-harmonic functions in metric spaces. *J. Anal. Math.*, **85**, 339–369.

[37] Bobkov, S. G. and Houdré, Ch. 1997. Some connections between isoperimetric and Sobolev-type inequalities. *Mem. Amer. Math. Soc.*, **129**(616).

[38] Bogachev, V. I. 1998. *Gaussian Measures*. Mathematical Surveys and Monographs, vol. 62. Providence, RI: American Mathematical Society.

[39] Bonk, M. 2006. Quasiconformal geometry of fractals, pp. 1349–1373 in: *Proc. Int. Congress of Mathematicians. Vol. II.* Zürich: European Mathematical Society.

[40] Bonk, M. and Kleiner, B. 2002. Quasisymmetric parametrizations of two-dimensional metric spheres. *Invent. Math.*, **150**(1), 127–183.

[41] Bonk, M. and Kleiner, B. 2005. Conformal dimension and Gromov hyperbolic groups with 2-sphere boundary. *Geom. Topol.*, **9**, 219–246.

[42] Bourdon, M. and Kleiner, B. 2013. Combinatorial modulus, the combinatorial Loewner property, and Coxeter groups. *Groups Geom. Dyn.*, **7**(1), 39–107.

[43] Bourdon, M. and Pajot, H. 1999. Poincaré inequalities and quasiconformal structure on the boundaries of some hyperbolic buildings. *Proc. Amer. Math. Soc.*, **127**(8), 2315–2324.

[44] Brelot, M. 1971. *On Topologies and Boundaries in Potential Theory*. Lecture Notes in Mathematics, vol. 175. Berlin: Springer-Verlag. Enlarged edition of a course of lectures delivered in 1966.

[45] Bridson, M. R. and Haefliger, A. 1999. *Metric Spaces of Non-Positive Curvature*. Grundlehren der Mathematischen Wissenschaften, vol. 319. Berlin: Springer-Verlag.

[46] Bruckner, A., Bruckner, J., and Thomson, B. 1997. *Real Analysis*. New Jersey: Prentice-Hall.

[47] Buckley, S. M. 1999. Is the maximal function of a Lipschitz function continuous? *Ann. Acad. Sci. Fenn. Math.*, **24**(2), 519–528.

[48] Buckley, S. M., Herron, D. A., and Xie, X. 2008. Metric space inversions, quasihyperbolic distance, and uniform spaces. *Indiana Univ. Math. J.*, **57**(2), 837–890.

[49] Burago, D., Burago, Yu., and Ivanov, S. 2001. *A Course in Metric Geometry*. Graduate Studies in Mathematics, vol. 33. Providence, RI: American Mathematical Society.

[50] Burago, Yu., Gromov, M., and Perel'man, G. 1992. A. D. Aleksandrov spaces with curvatures bounded below. *Uspekhi Mat. Nauk*, **47** 3–51, 222.

[51] Buser, P. 1982. A note on the isoperimetric constant. *Ann. Sci. École Norm. Sup. (4)*, **15**, 213–230.

[52] Capogna, L., Danielli, D., Pauls, S. D., and Tyson, J. T. 2007. *An Introduction to the Heisenberg Group and the Sub-Riemannian Isoperimetric Problem*. Progress in Mathematics, vol. 259. Basel: Birkhäuser Verlag.

[53] Cheeger, J. 1999. Differentiability of Lipschitz functions on metric measure spaces. *Geom. Funct. Anal.*, **9**, 428–517.

[54] Cheeger, J. and Kleiner, B. 2006a. Generalized differential and bi-Lipschitz nonembedding in L^1. *C. R. Math. Acad. Sci. Paris*, **343**(5), 297–301.

[55] Cheeger, J. and Kleiner, B. 2006b. On the differentiability of Lipschitz maps from metric measure spaces to Banach spaces, pp. 129–152 in: *Inspired by S. S. Chern*, Nankai Tracts in Mathematics, vol. 11. Hackensack, NJ: World Science Publishers.

[56] Cheeger, J. and Kleiner, B. 2009. Differentiability of Lipschitz maps from metric measure spaces to Banach spaces with the Radon–Nikodým property. *Geom. Funct. Anal.*, **19**(4), 1017–1028.

[57] Cheeger, J. and Kleiner, B. 2010a. Differentiating maps into L^1, and the geometry of BV functions. *Ann. Math. (2)*, **171**(2), 1347–1385.

[58] Cheeger, J. and Kleiner, B. 2010b. Metric differentiation, monotonicity and maps to L^1. *Invent. Math.*, **182**(2), 335–370.

[59] Cheeger, J. and Kleiner, B. 2013. Realization of metric spaces as inverse limits, and bilipschitz embedding in L_1. *Geom. Funct. Anal.*, **23**(1), 96–133.

[60] Cheeger, J., Kleiner, B., and Naor, A. 2011. Compression bounds for Lipschitz maps from the Heisenberg group to L_1. *Acta Math.*, **207**(2), 291–373.

[61] Clarkson, J. A. 1936. Uniformly convex spaces. *Trans. Amer. Math. Soc.*, **40**, 396–414.

[62] Coifman, R. R. and de Guzmán, M. 1970/1971. Singular integrals and multipliers on homogeneous spaces. *Rev. Univ. Mat. Argentina*, **25**, 137–143.

[63] Coifman, R. R. and Weiss, G. 1971. *Analyse Harmonique Non-commutative sur Certains Espaces Homogènes*. Lecture Notes in Mathematics, vol. 242. Berlin: Springer-Verlag.

[64] Coifman, R. R. and Weiss, G. 1977. Extensions of Hardy spaces and their use in analysis. *Bull. Amer. Math. Soc.*, **83**(4), 569–645.

[65] Colding, T. H. and Minicozzi, W. P., II. 1997. Harmonic functions on manifolds. *Ann. Math. (2)*, **146**, 725–747.

[66] Cucuringu, M. and Strichartz, R. S. 2008. Infinitesimal resistance metrics on Sierpinski gasket type fractals. *Analysis (Munich)*, **28**(3), 319–331.

[67] Danielli, D., Garofalo, N., and Marola, N. 2010. Local behavior of p-harmonic Green's functions in metric spaces. *Potential Anal.*, **32**(4), 343–362.

[68] David, G. and Semmes, S. 1990. Strong A_∞ weights, Sobolev inequalities and quasiconformal mappings, pp. 101–111 in: *Analysis and Partial Differential Equations*. Lecture Notes in Pure and Applied Mathematics, vol. 122. Marcel Dekker.

[69] David, G. and Semmes, S. 1997. *Fractured Fractals and Broken Dreams: Self-Similar Geometry Through Metric and Measure*. Oxford Lecture Series in Mathematics and its Applications, vol. 7. Clarendon Press/Oxford University Press.

[70] Day, M. M. 1941. Reflexive Banach spaces not isomorphic to uniformly convex spaces. *Bull. Amer. Math. Soc.*, **47**, 313–317.

[71] DeJarnette, N. 2013. Self-improving Orlicz–Poincaré inequalities. Ph.D. thesis, University of Illinois at Urbana-Champaign.

[72] Deny, J. 1995. Formes et espaces de Dirichlet. pp. Exp. No. 187, 261–271 in: *Séminaire Bourbaki, Vol. 5*. Paris: Mathematical Society of France.

[73] Deny, J. and Lions, J. L. 1955. Les espaces du type de Beppo Levi. *Ann. Inst. Fourier, Grenoble (5)*, **5**, 305–370.

[74] Diestel, J. 1984. *Sequences and Series in Banach Spaces*. Graduate Texts in Mathematics, vol. 92. New York: Springer-Verlag.

[75] Diestel, J. and Uhl, J. J. 1977. *Vector Measures*. Mathematical Surveys, vol. 15. Providence, R.I.: American Mathematical Society. With a foreword by B. J. Pettis.

[76] Doob, J. L. 1984. *Classical Potential Theory and its Probabilistic Counterpart*. Grundlehren der Mathematischen Wissenschaften, vol. 262. New York: Springer-Verlag.

[77] Dunford, N. and Schwartz, J. T. 1988. *Linear Operators. Part I*. Wiley Classics Library. New York: John Wiley & Sons.

[78] Durand-Cartagena, E. and Shanmugalingam, N. Geometric characterizations of ∞-Poincaré and p-Poincaré inequalities in the metric setting. Preprint, 2014.

[79] Durand-Cartagena, E., Jaramillo, J. A., and Shanmugalingam, N. 2012. The ∞-Poincaré inequality in metric measure spaces. *Michigan Math. J.*, **61**(1), 63–85.

[80] Durand-Cartagena, E., Shanmugalingam, N., and Williams, A. 2012. p-Poincaré inequality versus ∞-Poincaré inequality: some counterexamples. *Math. Z.*, **271**(1–2), 447–467.

[81] Evans, L. C. and Gariepy, R. F. 1992. *Measure Theory and Fine Properties of Functions*. Studies in Advanced Mathematics. Boca Raton, FL: CRC Press.

[82] Fabes, E. B., Kenig, C. E., and Serapioni, R. 1982. The local regularity of solutions to degenerate elliptic equations. *Comm. PDE*, **7**, 77–116.

[83] Federer, H. 1969. *Geometric Measure Theory*. Die Grundlehren der mathematischen Wissenschaften, vol. 153. New York: Springer-Verlag.

[84] Federer, H. and Ziemer, W. P. 1972/73. The Lebesgue set of a function whose distribution derivatives are pth power summable. *Indiana Univ. Math. J.*, **22**, 139–158.

[85] Fitzsimmons, P. J., Hambly, B. M., and Kumagai, T. 1994. Transition density estimates for Brownian motion on affine nested fractals. *Comm. Math. Phys.*, **165**(3), 595–620.

[86] Folland, G. B. 1984. *Real Analysis*. Pure and Applied Mathematics. John Wiley & Sons.

[87] Franchi, B., Hajłasz, P., and Koskela, P. 1999. Definitions of Sobolev classes on metric spaces. *Ann. Inst. Fourier (Grenoble)*, **49**(6), 1903–1924.

[88] Fréchet, M. 1909–10. Les dimensions d'un ensemble abstrait. *Math. Ann.*, **68**, 145–168.

[89] Fuglede, B. 1957. Extremal length and functional completion. *Acta Math.*, **98**, 171–219.

[90] Fukushima, M., Ōshima, Y., and Takeda, M. 1994. *Dirichlet Forms and Symmetric Markov Processes*. Berlin: Walter de Gruyter & Co.

[91] Ghys, E. and de la Harpe, P. 1990. *Sur les Groupes Hyperboliques d'après Mikhael Gromov*. Progress in Mathematics. Boston/Basel/Berlin: Birkhäuser.

[92] Giaquinta, M. and Giusti, E. 1982. On the regularity of the minima of variational integrals. *Acta Math.*, **148**, 31–46.

[93] Giaquinta, M. and Giusti, E. 1984. Quasiminima. *Ann. Inst. H. Poincaré Anal. Non Linéaire*, **1**(2), 79–107.

[94] Gigli, N., Kuwada, K., and Ohta, S.-I. 2013. Heat flow on Alexandrov spaces. *Comm. Pure Appl. Math.*, **66**(3), 307–331.

[95] Giusti, E. 1969. Precisazione delle funzioni di $H^{1,p}$ e singolarità delle soluzioni deboli di sistemi ellittici non lineari. *Boll. Un. Mat. Ital. (4)*, **2**, 71–76.

[96] Gol'dshtein, V. and Troyanov, M. 2001. Axiomatic theory of Sobolev spaces. *Expo. Math.*, **19**(4), 289–336.

[97] Gol'dshtein, V. and Troyanov, M. 2002a. Axiomatic Sobolev spaces on metric spaces. pp. 333–343 in: *Proc. Conf. on Function Spaces, Interpolation Theory and Related Topics (Lund, 2000)*. Berlin: de Gruyter.

[98] Gol'dshtein, V. and Troyanov, M. 2002b. Capacities in metric spaces. *Integral Eq. Operator Theory*, **44**(2), 212–242.

[99] Gong, J. The Lip–lip condition on metric measure spaces. Preprint, 2012. Available at arxiv.org/abs/1208.2869.

[100] Gong, J. Measurable differentiable structures on doubling metric spaces. Preprint, 2012. Available at arxiv.org/abs/1110.4279.

[101] Gong, J. 2012 Rigidity of derivations in the plane and in metric measure spaces. *Illinois J. Math.*, **56**(4), 1109–1147.

[102] Gong, J. 2007. Derivatives and currents on metric (measure) spaces. *Real Anal. Exchange*, 217–223.

[103] Grigor'yan, A. 1995. Heat kernel of a noncompact Riemannian manifold, pp. 239–263 in: *Stochastic Analysis (Ithaca, NY, 1993), Proc. Sympos. Pure Math.*, vol. 57. Providence, RI: American Mathematical Society.

[104] Grigor'yan, A. 1999. Analytic and geometric background of recurrence and non-explosion of Brownian motion on Riemannian manifolds. *Bull. Amer. Math. Soc. (NS)*, **36**(2), 135–249.

[105] Gromov, M. 1987. Hyperbolic groups, pp. 75–265 in: *Essays in Group Theory*. MSRI Publications, Springer-Verlag.

[106] Gromov, M. 1996. Carnot–Carathéodory spaces seen from within, pp. 79–323 in: *Sub-Riemannian Geometry*, Progress in Mathematics, vol. 144. Basel: Birkhäuser.

[107] Gromov, M. 1999. *Metric Structures for Riemannian and Non-Riemannian spaces*. Progress in Mathematics, vol. 152. Boston, MA: Birkhäuser Boston. Based on the 1981 French original, with appendices by M. Katz, P. Pansu, and S. Semmes; translated from the French by Sean Michael Bates.

[108] Hajłasz, P. 1996. Sobolev spaces on an arbitrary metric space. *Potential Anal.*, **5**, 403–415.

[109] Hajłasz, P. 2003. Sobolev spaces on metric-measure spaces, pp. 173–218 in: *Proc. Conf. on Heat Kernels and Analysis on Manifolds, Graphs, and Metric Spaces (Paris, 2002)*. Contemp. Math., vol. 338. Providence, RI: American Mathematical Society.

[110] Hajłasz, P. 2007. Sobolev mappings: Lipschitz density is not a bi-Lipschitz invariant of the target. *Geom. Funct. Anal.*, **17**(2), 435–467.

[111] Hajłasz, P. 2009a. Density of Lipschitz mappings in the class of Sobolev mappings between metric spaces. *Math. Ann.*, **343**(4), 801–823.

[112] Hajłasz, P. 2009b. Sobolev mappings between manifolds and metric spaces, pp. 185–222 in: *Sobolev Spaces in Mathematics. Vol. I.* Int. Math. Ser. (NY), vol. 8. New York: Springer.

[113] Hajłasz, P. and Koskela, P. 1995. Sobolev meets Poincaré. *C. R. Acad. Sci. Paris Sér. I Math.*, **320**, 1211–1215.

[114] Hajłasz, P. and Koskela, P. 2000. Sobolev met Poincaré. *Mem. Amer. Math. Soc.*, **145**(688).

[115] Hakobyan, H. 2010. Conformal dimension: Cantor sets and Fuglede modulus. *Int. Math. Res. Not. IMRN*, 87–111.

[116] Hambly, B. M. and Kumagai, T. 1999. Transition density estimates for diffusion processes on post critically finite self-similar fractals. *Proc. London Math. Soc. (3)*, **78**(2), 431–458.

[117] Hanson, B. and Heinonen, J. 2000. An n-dimensional space that admits a Poincaré inequality but has no manifold points. *Proc. Amer. Math. Soc.*, **128**(11), 3379–3390.

[118] Hebey, E. 1999. *Nonlinear Analysis on Manifolds: Sobolev Spaces and Inequalities.* Courant Lecture Notes in Mathematics, vol. 5. New York: Courant Institute of Mathematical Sciences.

[119] Heinonen, J. 1995. A capacity estimate on Carnot groups. *Bull. Sci. Math.*, **119**, 475–484.

[120] Heinonen, J. 2001. *Lectures on Analysis on Metric Spaces.* New York: Springer-Verlag.

[121] Heinonen, J. 2003. Geometric embeddings of metric spaces. Report, University of Jyväskylä Department of Mathematics and Statistics, vol. 90. University of Jyväskylä.

[122] Heinonen, J. 2007. Nonsmooth calculus. *Bull. Amer. Math. Soc. (NS)*, **44**(2), 163–232.

[123] Heinonen, J. and Koskela, P. 1995. Definitions of quasiconformality. *Invent. Math.*, **120**, 61–79.

[124] Heinonen, J. and Koskela, P. 1996. From local to global in quasiconformal structures. *Proc. Nat. Acad. Sci. USA*, **93**, 554–556.

[125] Heinonen, J. and Koskela, P. 1998. Quasiconformal maps in metric spaces with controlled geometry. *Acta Math.*, **181**, 1–61.

[126] Heinonen, J. and Koskela, P. 1999. A note on Lipschitz functions, upper gradients, and the Poincaré inequality. *New Zealand J. Math.*, **28**, 37–42.

[127] Heinonen, J. and Wu, J.-M. 2010. Quasisymmetric nonparametrization and spaces associated with the Whitehead continuum. *Geom. Topol.*, **14**(2), 773–798.

[128] Heinonen, J., Kilpeläinen, T., and Martio, O. 1993. *Nonlinear Potential Theory of Degenerate Elliptic Equations.* Oxford Science Publications. New York: The Clarendon Press, Oxford University Press.

[129] Heinonen, J., Koskela, P., Shanmugalingam, N., and Tyson, J. T. 2001. Sobolev classes of Banach space-valued functions and quasiconformal mappings. *J. Anal. Math.*, **85**, 87–139.

[130] Herman, P. E., Peirone, R., and Strichartz, R. S. 2004. p-energy and p-harmonic functions on Sierpinski gasket type fractals. *Potential Anal.*, **20**(2), 125–148.

[131] Herron, D. A. Gromov–Hausdorff distance for pointed metric spaces. Preprint, 2011.

[132] Herron, D. A. 2011. Uniform metric spaces, annular quasiconvexity and pointed tangent spaces. *Math. Scand.*, **108**(1), 115–145.

[133] Herron, D. A., Shanmugalingam, N., and Xie, X. 2008. Uniformity from Gromov hyperbolicity. *Illinois J. Math.*, **52**(4), 1065–1109.

[134] Hesse, J. 1975. p-extremal length and p-measurable curve families. *Proc. Amer. Math. Soc.*, **53**(2), 356–360.

[135] Hewitt, E. and Stromberg, K. 1965. *Real and Abstract Analysis. A Modern Treatment of the Theory of Functions of a Real Variable*. New York: Springer-Verlag.

[136] Hildebrandt, T. H. 1953. Integration in abstract spaces. *Bull. Amer. Math. Soc.*, **59**, 111–139.

[137] Holopainen, I. 1990. Nonlinear potential theory and quasiregular mappings on Riemannian manifolds. *Ann. Acad. Sci. Fenn. Ser. A I Math. Dissertationes*, 45.

[138] Jacobs, K. 1978. *Measure and Integral*. New York: Academic Press, Harcourt Brace Jovanovich.

[139] Järvenpää, E., Järvenpää, M., Rogovin, K., Rogovin, S., and Shanmugalingam, N. 2007. Measurability of equivalence classes and MEC$_p$-property in metric spaces. *Rev. Mat. Iberoam.*, **23**(3), 811–830.

[140] Jiang, R. 2012. Lipschitz continuity of solutions of Poisson equations in metric measure spaces. *Potential Anal.*, **37**(3), 281–301.

[141] Jiang, R. and Koskela, P. 2012. Isoperimetric inequality from the Poisson equation via curvature. *Comm. Pure Appl. Math.*, **65**(8), 1145–1168.

[142] John, F. 1948. Extremum problems with inequalities as subsidiary conditions, pp. 187–204 in: *Studies and Essays Presented to R. Courant on his 60th Birthday, January 8, 1948*. New York: Interscience.

[143] Johnson, W. B., Lindenstrauss, J., and Schechtman, G. 1986. Extensions of Lipschitz maps into Banach spaces. *Israel J. Math.*, **54**(2), 129–138.

[144] Kajino, N. 2012a. Heat kernel asymptotics for the measurable Riemannian structure on the Sierpinski gasket. *Potential Anal.*, **36**(1), 67–115.

[145] Kajino, N. 2012b. Time changes of local Dirichlet spaces by energy measures of harmonic functions. *Forum Math.*, **24**(2), 339–363.

[146] Kapovich, I. and Benakli, N. 2002. Boundaries of hyperbolic groups, pp. 39–93 in: *Combinatorial and Geometric Group Theory*, Contemp. Math., vol. 296. Providence, RI: American Mathematical Society.

[147] Kapovitch, V. 2007. Perelman's stability theorem. pp. 103–136 in: *Surveys in Differential Geometry*, vol. 11. Somerville, MA: Int. Press.

[148] Karak, N. Removable sets for Orlicz–Sobolev spaces. Preprint, 2013.

[149] Kechris, A. S. 1995. *Classical Descriptive Set Theory*. Graduate Texts in Mathematics, vol. 156. New York: Springer-Verlag.

[150] Keith, S. 2003. Modulus and the Poincaré inequality on metric measure spaces. *Math. Z.*, **245**(2), 255–292.

[151] Keith, S. 2004a. A differentiable structure for metric measure spaces. *Adv. Math.*, **183**(2), 271–315.

[152] Keith, S. 2004b. Measurable differentiable structures and the Poincaré inequality. *Indiana Univ. Math. J.*, **53**(4), 1127–1150.

[153] Keith, S. and Zhong, X. 2008. The Poincaré inequality is an open ended condition. *Ann. Math. (2)*, **167**(2), 575–599.

[154] Kigami, J. 1994. Effective resistances for harmonic structures on p.c.f. self-similar sets. *Math. Proc. Cambridge Phil. Soc.*, **115**(2), 291–303.

[155] Kigami, J. 2001. *Analysis on Fractals*. Cambridge Tracts in Mathematics, vol. 143. Cambridge: Cambridge University Press.

[156] Kigami, J., Strichartz, R. S., and Walker, K. C. 2001. Constructing a Laplacian on the diamond fractal. *Experimental Math.*, **10**(3), 437–448.

[157] Kinnunen, J. and Latvala, V. 2002. Lebesgue points for Sobolev functions on metric spaces. *Rev. Mat. Iberoam.*, **18**(3), 685–700.

[158] Kinnunen, J. and Shanmugalingam, N. 2001. Regularity of quasi-minimizers on metric spaces. *Manuscripta Math.*, **105**(3), 401–423.

[159] Kinnunen, J. and Shanmugalingam, N. 2006. Polar sets on metric spaces. *Trans. Amer. Math. Soc.*, **358**(1), 11–37.

[160] Kinnunen, J., Korte, R., Shanmugalingam, N., and Tuominen, Heli. 2008. Lebesgue points and capacities via the boxing inequality in metric spaces. *Indiana Univ. Math. J.*, **57**(1), 401–430.

[161] Kinnunen, J., Korte, R., Lorent, A., and Shanmugalingam, N. 2013. Regularity of sets with quasiminimal boundary surfaces in metric spaces. *J. Geom. Anal.*, **23**(4), 1607–1640.

[162] Kleiner, B. 2006. The asymptotic geometry of negatively curved spaces: uniformization, geometrization and rigidity, pp. 743–768 in: *International Congress of Mathematicians, Vol. II*. Zürich: European Mathematical Society.

[163] Kleiner, B. and Mackay, J. 2011. Differentiable structures on metric measure spaces: a primer. Preprint. Available at arxiv.org/abs/1108.1324.

[164] Korevaar, N. J. and Schoen, R. M. 1993. Sobolev spaces and harmonic maps for metric space targets. *Comm. Anal. Geom.*, **1**, 561–659.

[165] Korte, R. 2007. Geometric implications of the Poincaré inequality. *Results Math.*, **50**(1–2), 93–107.

[166] Korte, R., Lahti, P., and Shanmugalingam, N. Semmes family of curves and a characterization of functions of bounded variation in terms of curves. Preprint, 2013. Available at cvgmt.sns.it/paper/2229/.

[167] Koskela, P. 1999. Removable sets for Sobolev spaces. *Ark. Mat.*, **37**(2), 291–304.

[168] Koskela, P. and MacManus, P. 1998. Quasiconformal mappings and Sobolev spaces. *Studia Math.*, **131**, 1–17.

[169] Koskela, P. and Saksman, E. 2008. Pointwise characterizations of Hardy–Sobolev functions. *Math. Res. Lett.*, **15**(4), 727–744.

[170] Koskela, P. and Zhou, Y. 2012. Geometry and analysis of Dirichlet forms. *Adv. Math.*, **231**(5), 2755–2801.

[171] Koskela, P., Rajala, K., and Shanmugalingam, N. 2003. Lipschitz continuity of Cheeger-harmonic functions in metric measure spaces. *J. Funct. Anal.*, **202**(1), 147–173.

[172] Koskela, P., Shanmugalingam, N., and Tyson, J. T. 2004. Dirichlet forms, Poincaré inequalities, and the Sobolev spaces of Korevaar and Schoen. *Potential Anal.*, **21**(3), 241–262.

[173] Koskela, P., Shanmugalingam, N., and Zhou, Y. 2012. L^∞-variational problem associated to Dirichlet forms. *Math. Res. Lett.*, **19**(6), 1263–1275.

[174] Kumagai, T. 1993. Regularity, closedness and spectral dimensions of the Dirichlet forms on PCF. self-similar sets. *J. Math. Kyoto Univ.*, **33**(3), 765–786.

[175] Kuratowski, C. 1935. Quelques problèmes concernant les espaces métriques non-séparables. *Fund. Math.*, **25**, 534–545.

[176] Kuwae, K., Machigashira, Y., and Shioya, T. 2001. Sobolev spaces, Laplacian, and heat kernel on Alexandrov spaces. *Math. Z.*, **238**(2), 269–316.

[177] Laakso, T. J. 2000. Ahlfors Q-regular spaces with arbitrary $Q > 1$ admitting weak Poincaré inequality. *Geom. Funct. Anal.*, **10**(1), 111–123.

[178] Ladyzhenskaya, O. A. and Ural'tseva, N. N. 1984. Estimates of max $|u_x|$ for solutions of quasilinear elliptic and parabolic equations of general type, and some existence theorems. *Zap. Nauchn. Sem. Leningrad. Otdel. Mat. Inst. Steklov. (LOMI)*, **138**, 90–107.

[179] Lang, U. and Schlichenmaier, T. 2005. Nagata dimension, quasisymmetric embeddings, and Lipschitz extensions. *Int. Math. Res. Not.*, 3625–3655.

[180] Lang, U., Pavlović, B., and Schroeder, V. 2000. Extensions of Lipschitz maps into Hadamard spaces. *Geom. Funct. Anal.*, **10**(6), 1527–1553.

[181] Ledoux, M. and Talagrand, M. 1991. *Probability in Banach Spaces*. Ergebnisse der Mathematik und ihrer Grenzgebiete (3), vol. 23. Berlin: Springer-Verlag.

[182] Lee, J. R. and Naor, A. 2005. Extending Lipschitz functions via random metric partitions. *Invent. Math.*, **160**(1), 59–95.

[183] Levi, B. 1906. Sul principio di Dirichlet. *Rend. Circ. Mat. Palermo*, **22**, 293–359.

[184] Li, P. and Tam, L.-F. 1995. Green's functions, harmonic functions, and volume comparison. *J. Differential Geom.*, **41**(2), 277–318.

[185] Li, S. 2014. Coarse differentiation and quantitative nonembeddability for Carnot groups. *J. Funct. Anal.*, **266**(7), 4616–4704.

[186] Li, X. and Shanmugalingam, N. Preservation of bounded geometry under sphericalization and flattening. Preprint 2013. Report No. IML-1314f-16, Institut Mittag-Leffler preprint series. Available at www.mittag-leffler.se/preprints/files/IML-1314f-16.pdf. *Indiana. Math J.*, to appear.

[187] Liu, Y., Lu, G., and Wheeden, R. L. 2002. Some equivalent definitions of high order Sobolev spaces on stratified groups and generalizations to metric spaces. *Math. Ann.*, **323**(1), 157–174.

[188] Lott, J. and Villani, C. 2007. Weak curvature conditions and functional inequalities. *J. Funct. Anal.*, **245**(1), 311–333.

[189] Lott, J. and Villani, C. 2009. Ricci curvature for metric-measure spaces via optimal transport. *Ann. Math. (2)*, **169**(3), 903–991.

[190] Luukkainen, J. and Saksman, E. 1998. Every complete doubling metric space carries a doubling measure. *Proc. Amer. Math. Soc.*, **126**, 531–534.

[191] Mackay, J. M. 2010. Spaces and groups with conformal dimension greater than one. *Duke Math. J.*, **153**(2), 211–227.

[192] Mackay, J. M., Tyson, J. T., and Wildrick, K. 2013. Modulus and Poincaré inequalities on non-self-similar Sierpiński carpets. *Geom. Funct. Anal.*, **23**(3), 985–1034.

[193] Malý, J. and Ziemer, W. P. 1997. *Fine Regularity of Solutions of Elliptic Partial Differential Equations*. Mathematical Surveys and Monographs, vol. 51. Providence, RI: American Mathematical Society.

[194] Malý, L. 2013. Minimal weak upper gradients in Newtonian spaces based on quasi-Banach function lattices. *Ann. Acad. Sci. Fenn. Math.*, **38**(2), 727–745.

[195] Marola, N., Miranda M., Jr., and Shanmugalingam, N. Boundary measures, generalized Gauss–Green formulas and the mean value property in metric measure spaces. Available at cvgmt.sns.it/paper/2129. *Rev. Mat. lberoamericana*, to appear.

[196] Mattila, P. 1973. Integration in a space of measures. *Ann. Acad. Sci. Fenn. Ser. A I*, 1–37.

[197] Mattila, P. 1995. *Geometry of Sets and Measures in Euclidean Spaces*. Cambridge Studies in Advanced Mathematics, vol. 44. Cambridge: Cambridge University Press.

[198] Maz'ya, V. G. 1960. Classes of domains and imbedding theorems for function spaces. *Soviet Math. Dokl.*, **1**, 882–885.

[199] Maz'ya, V. G. 1964. On the theory of the higher-dimensional Schrödinger operator. *Izv. Akad. Nauk SSSR Ser. Mat.*, **28**, 1145–1172.

[200] Maz'ya, V. G. 1970. Classes of sets and measures that are connected with imbedding theorems. Izdat. "Nauka", Moscow, pp. 142–159, 246.

[201] Maz'ya, V. G. 1975. The summability of functions belonging to Sobolev spaces. Izdat. Leningrad. Univ., Leningrad, pp. 66–98.

[202] Maz'ya, V. G. 1985. *Sobolev Spaces*. Berlin: Springer-Verlag. Translated from the Russian by T. O. Shaposhnikova.

[203] McShane, E. J. 1934. Extension of range of functions. *Bull. Amer. Math. Soc.*, **40**, 837–842.

[204] McShane, E. J. 1944. *Integration*. Princeton, NJ: Princeton University Press.

[205] McShane, E. J. 1950. Linear functionals on certain Banach spaces. *Proc. Amer. Math. Soc.*, **1**, 402–408.

[206] Megginson, R. E. 1998. *An Introduction to Banach Space Theory*. Graduate Texts in Mathematics, vol. 183. New York: Springer-Verlag.

[207] Meyers, N. and Serrin, J. 1964. $H = W$. *Proc. Nat. Acad. Sci. USA*, **51**, 1055–1056.

[208] Milman, V. D. and Schechtman, G. 1986. *Asymptotic Theory of Finite-Dimensional Normed Spaces*. Lecture Notes in Mathematics, vol. 1200. Berlin: Springer-Verlag. With an appendix by M. Gromov.

[209] Miranda, M., Jr. 2003. Functions of bounded variation on "good" metric spaces. *J. Math. Pures Appl. (9)*, **82**(8), 975–1004.

[210] Mitchell, J. 1985. On Carnot–Carathéodory metrics. *J. Differential Geom.*, **21**, 35–45.

[211] Montgomery, R. 2002. *A Tour of Subriemannian Geometries, their Geodesics and Applications*. Mathematical Surveys and Monographs, vol. 91. Providence, RI: American Mathematical Society.

[212] Morrey, C. B. 1966. *Multiple Integrals in the Calculus of Variations*. Berlin: Springer-Verlag.

[213] Muckenhoupt, B. 1972. Weighted norm inequalities for the Hardy maximal function. *Trans. Amer. Math. Soc.*, **165**, 207–226.

[214] Munkres, J. R. 1975. *Topology: A First Course*. Englewood Cliffs, NJ: Prentice-Hall.

[215] Nadler, S. B. 1978. *Hyperspaces of Sets*. Monographs and Textbooks in Pure and Applied Mathematics, vol. 49. New York: Marcel Dekker.

[216] Naor, A. 2010. L_1 embeddings of the Heisenberg group and fast estimation of graph isoperimetry, pp. 1549–1575 in: *Proc. Int. Congress of Mathematicians, Vol. III*. New Delhi: Hindustan Book Agency.

[217] Nazarov, F., Treil, S., and Volberg, A. 1998. Weak type estimates and Cotlar inequalities for Calderón–Zygmund operators on nonhomogeneous spaces. *Int. Math. Res. Not.*, 463–487.

[218] Nazarov, F., Treil, S., and Volberg, A. 2003. The Tb-theorem on non-homogeneous spaces. *Acta Math.*, **190**(2), 151–239.

[219] Nikodym, O. M. 1933. Sur une classe de fonctions considérées dans problème de Dirichlet. *Fund. Math.*, **21**, 129–150.

[220] Ohtsuka, M. 2003. *Extremal Length and Precise Functions*. GAKUTO International Series. Mathematical Sciences and Applications, vol. 19. Tokyo: Gakkōtosho. With a preface by Fumi-Yuki Maeda.

[221] Oxtoby, J. C. and Ulam, S. M. 1939. On the existence of a measure invariant under a transformation. *Ann. Math. (2)*, **40**, 560–566.

[222] Pankka, P. and Wu, J.-M. 2014. Geometry and quasisymmetric parametrization of Semmes spaces. *Rev. Mat. Iberoamericana*, **30**(4), 893–960.

[223] Pansu, P. 1989a. Dimension conforme et sphère à l'infini des variétés à courbure négative. *Ann. Acad. Sci. Fenn. Ser. A I Math.*, **14**, 177–212.

[224] Pansu, P. 1989b. Métriques de Carnot–Carathéodory et quasiisométries des espaces symétriques de rang un. *Ann. Math. (2)*, **129**, 1–60.

[225] Perelman, G. 1995. Spaces with curvature bounded below, pp. 517–525 in: *Proc. Int. Congress of Mathematicians, Vols. 1, 2 (Zürich, 1994)*. Basel: Birkhäuser.

[226] Rademacher, H. 1919. Über partielle und totale Differenzierbarkeit von Funktionen mehrerer Variabeln und über die Transformation der Doppelintegrale. *Math. Ann.*, **79**(4), 340–359.

[227] Rajala, K. 2005. Surface families and boundary behavior of quasiregular mappings. *Illinois J. Math.*, **49**(4), 1145–1153.

[228] Rajala, K. and Wenger, S. 2013. An upper gradient approach to weakly differentiable cochains. *J. Math. Pures Appl.*, **100**, 868–906.

[229] Rauch, H. E. 1956. Harmonic and analytic functions of several variables and the maximal theorem of Hardy and Littlewood. *Can. J. Math.*, **8**, 171–183.

[230] Reimann, H. M. 1989. An estimate for pseudoconformal capacities on the sphere. *Ann. Acad. Sci. Fenn. Ser. A I Math.*, **14**, 315–324.

[231] Reshetnyak, Yu. G. 1967. Space Mappings with Bounded Distortion. *Sibirsk. Mat. Z.*, **8**, 629–659.

[232] Reshetnyak, Yu. G. 1989. *Space Mappings with Bounded Distortion*. Translations of Mathematical Monographs, vol. 73. American Mathematical Society. Translated from the Russian by H. H. McFaden.

[233] Reshetnyak, Yu. G. 1997. Sobolev classes of functions with values in a metric space. *Sibirsk. Mat. Zh.*, **38**, 657–675.

[234] Rickman, S. 1993. *Quasiregular Mappings*. Berlin: Springer-Verlag.

[235] Rogers, C. A. 1970. *Hausdorff Measures*. London: Cambridge University Press.

[236] Royden, H. L. 1988. *Real Analysis*. Third edn. Macmillan Publ. Co.

[237] Rudin, W. 1987. *Real and Complex Analysis*. Third edn. New York: McGraw-Hill.

[238] Rudin, W. 1991. *Functional Analysis*. Second edn. New York: McGraw-Hill.

[239] Saksman, E. 1999. Remarks on the nonexistence of doubling measures. *Ann. Acad. Sci. Fenn. Ser. A I Math.*, **24**(1), 155–163.

[240] Saloff-Coste, L. 2002. *Aspects of Sobolev-type Inequalities*. London Mathematical Society Lecture Note Series, vol. 289. Cambridge: Cambridge University Press.

[241] Savaré, G. 2014. Self-improvement of the Bakry–Émery condition and Wasserstein contraction of the heat flow in $RCD(K, \infty)$ metric measure spaces. *Discrete Contin. Dynam. Syst.*, **34**(4), 1641–1661.

[242] Schechter, M. 2002. *Principles of Functional Analysis*. Second edn. Graduate Studies in Mathematics, vol. 36. Providence, RI: American Mathematical Society.

[243] Schwartz, L. 1973. *Radon Measures on Arbitrary Topological Spaces and Cylindrical Measures*. Tata Institute of Fundamental Research Studies in Mathematics, no. 6. London: Published for the Tata Institute of Fundamental Research, Bombay by Oxford University Press.

[244] Semmes, S. 1996a. Finding curves on general spaces through quantitative topology, with applications to Sobolev and Poincaré inequalities. *Selecta Math.*, **2**, 155–295.

[245] Semmes, S. 1996b. Good metric spaces without good parameterizations. *Rev. Mat. Iberoam.*, **12**, 187–275.

[246] Semmes, S. 1996c. On the nonexistence of bi-Lipschitz parameterizations and geometric problems about A_∞-weights. *Rev. Mat. Iberoam.*, **12**, 337–410.

[247] Shanmugalingam, N. 1999. Newtonian spaces: an extension of Sobolev spaces to metric measure spaces. Ph.D. thesis, University of Michigan.

[248] Shanmugalingam, N. 2000. Newtonian spaces: an extension of Sobolev spaces to metric measure spaces. *Rev. Mat. Iberoam.*, **16**(2), 243–279.

[249] Shanmugalingam, N. 2001. Harmonic functions on metric spaces. *Illinois J. Math.*, **45**(3), 1021–1050.

[250] Shanmugalingam, N. 2009. A universality property of Sobolev spaces in metric measure spaces, pp. 345–359 in: *Sobolev Spaces in Mathematics. Vol. I.* International Mathematics Series (NY), vol. 8. New York: Springer.

[251] Shiryaev, A. N. 1996. *Probability*. Second edn. Graduate Texts in Mathematics, vol. 95. New York: Springer-Verlag. Translated from the first (1980) Russian edition by R. P. Boas.

[252] Smith, K. T. 1956. A generalization of an inequality of Hardy and Littlewood. *Can. J. Math.*, **8**, 157–170.

[253] Sobolev, S. L. 1936. On some estimates relating to families of functions having derivatives that are square integrable. *Dokl. Akad. Nauk SSSR*, **1**, 267–270. (In Russian).

[254] Sobolev, S. L. 1938. On a theorem in functional analysis. *Math. Sb.*, **4**, 471–497. (In Russian.)

[255] Sobolev, S. L. 1950. *Some Applications of Functional Analysis in Mathematical Physics*. Izdat. Leningrad. Gos. Univ., Leningrad. (In Russian).

[256] Sobolev, S. L. 1991. *Some Applications of Functional Analysis in Mathematical Physics*. Translations of Mathematical Monographs, vol. 90. Providence, RI: American Mathematical Society. Translated from the third Russian edition by Harold H. McFaden. With comments by V. P. Palamodov.

[257] Stein, E. M. 1993. *Harmonic Analysis: Real-Variable Methods, Orthogonality, and Oscillatory Integrals*. Princeton Mathematical Series. Princeton, NJ: Princeton University Press.

[258] Strichartz, R. S. 1998. Fractals in the large. *Can. J. Math.*, **50**(3), 638–657.

[259] Strichartz, R. S. 1999a. Analysis on fractals. *Not. Amer. Math. Soc.*, **46**(10), 1199–1208.

[260] Strichartz, R. S. 1999b. Some properties of Laplacians on fractals. *J. Funct. Anal.*, **164**(2), 181–208.

[261] Strichartz, R. S. 2001. The Laplacian on the Sierpinski gasket via the method of averages. *Pacific J. Math.*, **201**(1), 241–256.

[262] Strichartz, R. S. 2002. Harmonic mappings of the Sierpinski gasket to the circle. *Proc. Amer. Math. Soc.*, **130**(3), 805–817.

[263] Strichartz, R. S. 2003. Function spaces on fractals. *J. Funct. Anal.*, **198**(1), 43–83.

[264] Sturm, K.-Th. 2005. Generalized Ricci bounds and convergence of metric measure spaces. *C. R. Math. Acad. Sci. Paris*, **340**(3), 235–238.

[265] Sturm, K.-Th. 2006a. A curvature–dimension condition for metric measure spaces. *C. R. Math. Acad. Sci. Paris*, **342**(3), 197–200.

[266] Sturm, K.-Th. 2006b. On the geometry of metric measure spaces. II. *Acta Math.*, **196**(1), 133–177.

[267] Tolsa, X. 1998. Cotlar's inequality without the doubling condition and existence of principal values for the Cauchy integral of measures. *J. Reine Angew. Math.*, **502**, 199–235.

[268] Tolsa, X. 1999. L^2-boundedness of the Cauchy integral operator for continuous measures. *Duke Math. J.*, **98**(2), 269–304.

[269] Tonelli, L. 1926. Sulla quadratura delle superficie. *Atti Reale Accad. Lincei*, **3**, 633–638.

[270] Tuominen, H. 2004. Orlicz–Sobolev spaces on metric measure spaces. *Ann. Acad. Sci. Fenn. Math. Diss.*, **135**, 1–86.

[271] Tyson, J. T. 1998. Quasiconformality and quasisymmetry in metric measure spaces. *Ann. Acad. Sci. Fenn. Ser. A I Math.*, **23**, 525–548.

[272] Tyson, J. T. 1999. Geometric and analytic applications of a generalized definition of the conformal modulus. Ph.D. thesis, University of Michigan.

[273] Väisälä, J. 1971. *Lectures on n-Dimensional Quasiconformal Mappings*. Lecture Notes in Mathematics, vol. 229. Berlin: Springer-Verlag.

[274] Varopoulos, N. 1987. Fonctions harmoniques sur les groupes de Lie. *C. R. Acad. Sci. Paris Sér. I Math.*, **304**, 519–521.

[275] Verdera, J. 2002. The fall of the doubling condition in Calderón–Zygmund theory, pp. 275–292 in: *Proc. 6th Int. Conf. on Harmonic Analysis and Partial Differential Equations (El Escorial, 2000), Publ. Mat.*

[276] Villani, C. 2003. *Topics in Optimal Transportation*. Graduate Studies in Mathematics, vol. 58. Providence, RI: American Mathematical Society.

[277] Vol'berg, A. L. and Konyagin, S. V. 1987. On measures with the doubling condition. *Izv. Akad. Nauk SSSR Ser. Mat.*, **51**, 666–675. English translation: *Math. USSR-Izv.*, **30**, 629–638, 1988.

[278] von Renesse, M.-K. and Sturm, K.-Th. 2005. Transport inequalities, gradient estimates, entropy, and Ricci curvature. *Comm. Pure Appl. Math.*, **58**(7), 923–940.

[279] Vuorinen, M. 1988. *Conformal Geometry and Quasiregular Mappings*. Berlin: Springer-Verlag.

[280] Wallin, H. 1963a. Continuous functions and potential theory. *Ark. Mat.*, **5**, 55–84.

[281] Wallin, H. 1963b. Studies in potential theory. Inaugural dissertation. Acta Universitatis Upsaliensis, Abstracts of Uppsala Dissertations in Science, vol. 26. Almqvist & Wiksells Boktryckeri AB, Uppsala.

[282] Weaver, N. 2000. Lipschitz algebras and derivations. II. Exterior differentiation. *J. Funct. Anal.*, **178**(1), 64–112.

[283] Whitney, H. 1934. Analytic extensions of differentiable functions defined in closed sets. *Trans. Amer. Math. Soc.*, **36**(1), 63–89.

[284] Williams, M. 2012. Geometric and analytic quasiconformality in metric measure spaces. *Proc. Amer. Math. Soc.*, **140**(4), 1251–1266.

[285] Wojtaszczyk, P. 1991. *Banach Spaces for Analysts*. Cambridge: Cambridge University Press.

[286] Yosida, K. 1980. *Functional Analysis*. New York: Springer-Verlag.

[287] Ziemer, W. P. 1967. Extremal length and conformal capacity. *Trans. Amer. Math. Soc.*, **126**, 460–473.

[288] Ziemer, W. P. 1969. Extremal length and p-capacity. *Michigan Math. J.*, **16**, 43–51.

[289] Ziemer, W. P. 1970. Extremal length as a capacity. *Michigan Math. J.*, **17**, 117–128.

[290] Ziemer, W. P. 1989. *Weakly Differentiable Functions*. Graduate Texts in Mathematics, vol. 120. New York: Springer-Verlag.

Notation index

Subject index